Mechanical Engineering Series

Frederick F. Ling
Editor-in-Chief

Jorge Angeles

Dynamic Response of Linear Mechanical Systems

Modeling, Analysis and Simulation

 Springer

Jorge Angeles
Department of Mechanical Engineering
McGill University
Sherbrooke Street West 817
H3A 2K6 Montreal, Québec
Canada
angeles@cim.mcgill.ca

Additional material to this book can be downloaded from http://extras.springer.com

ISSN 0941-5122   e-ISSN 2192-063X
ISBN 978-1-4614-2946-3   ISBN 978-1-4419-1027-1 (eBook)
DOI 10.1007/978-1-4419-1027-1
Springer New York Dordrecht Heidelberg London

Printed on acid-free paper

Springer is part of Springer Science+Business Media (www.springer.com)

To the memory of my son Romain
(1984–2008) who lived the impossible dream

Preface

There are men whose head is full
Of nothing, if not their science;
There are the learned of all trades.
I tell you without pretension:
Rather than learning too much,
Go and learn what really matters.

Hernández, J., 1872, *Martín Fierro,*
Editorial de la Pampa, Buenos Aires (current text
in the original Spanish taken from the 1982 edition
by Bruguera Publishers, Barcelona).[1]

The need to provide instructors and students with a textbook on the classical principles and the modern methods of analysis, modeling and simulation of mechanical systems gave rise to *The Dynamic Response of Linear Mechanical Systems.* I came across this need myself when I was assigned, in the late eighties, the teaching of the undergraduate *Dynamics of Vibrations* course at McGill University's Department of Mechanical Engineering, while one of the instructors was on sabbatical. This was an interesting challenge, as I had never taken an undergraduate vibrations course as such. In fact, I came from the 5-year *Electromechanical Engineering* Program at the National Autonomous University of Mexico (abbreviated UNAM, from its name in Spanish), where the teaching of vibrations was included in the 1-year course on *Applied Mechanics*; this course comprised both kinematics and dynamics of machines. Vibrations being the last topic in the syllabus, the instructor usually rushed through it, the final examination hardly including a question on vibration dynamics. In my senior year the curriculum underwent a radical updating, with courses offered in semesters. This change gave me the opportunity to take a one-semester course on Electromechanical Energy Conversion, which was about my first and only exposure to the discipline of dynamics of systems as an undergraduate.

---

[1] Verses 6923–6928, translated by the author.

My first task as a designated instructor of undergraduate vibrations was to search for the right textbook. I was overwhelmed by the vast bibliography on the subject. To my surprise, with rare exceptions, all textbooks I consulted observed the same pattern, which was originally set by Thomson's classical book, first published in 1948.[2] As I was familiar with system theory, given that this was the minor I took in my graduate studies of Mechanical Engineering (M.Eng., UNAM) and Applied Mechanics (Ph.D., Stanford), I tried to make the connection between these two closely related disciplines. I knew Cannon's book [2], which makes such a connection, this book thus appearing as the right choice. Given the size of the book and its unusually broad scope, however, the students were intimidated, which forced me to look for an alternative, more focused textbook, as I continued teaching that course for several years afterwards. I found other books that somehow integrated system theory with the dynamics of vibration,[3] but I thought that there was more to it that was not as yet in the textbooks. For one thing, I did not like the idea of having to solve a *generalized eigenvalue problem* to find the natural frequencies and the natural modes of a multi-degree-of-freedom system, as the generalized problem does not necessarily lead to real, non-negative eigenvalues. Inman's book had an interesting idea along these lines that I decided to pursue.

I thus undertook an in-depth research on a more intuitive, yet rigorous approach to the teaching of vibration and, for that matter, mechanical-system analysis, that would (a) exploit the mathematical knowledge required from the students, as spelled out in the prerequisites (linear algebra and basic engineering mechanics); (b) connect the theory of vibrations with the more general theory of systems; and (c) resort to well-known graphical techniques of solving engineering problems. This is how I decided to take a departure from traditional approaches to the teaching of vibration dynamics. Features of the book along these lines are listed below:

- Modeling is given due attention—most books appear to take modeling for granted. In doing this, a seven-step procedure is introduced that is aimed at structuring the rather unstructured process of formulating mathematical models of mechanical systems. I always insist in the need for a creative approach to modeling, as no two engineers will always come up with the same model of the same situation. Modeling, while based on sound science, is also an art, that can only be developed by practice.

- A system-theoretic approach is adopted in deriving the time response of the linear mathematical models of the systems under study. In this regard, first what is known in system-theoretic terminology as the *zero-input response*—dubbed the *free response* in vibration analysis—is obtained; next, the *zero-state response* is derived by relying on the impulse response of the system under analysis. The time response of a system to arbitrary excitations is then naturally derived in convolution form upon resorting to a black-box approach that is applicable to all

---

[2]The book has gone through many revisions, e.g., [1], but it keeps its original basic contents.
[3]These were Meirovitch [3] and Inman [4].

linear, time-invariant dynamical systems. The fundamental concept of linearity, which entails superposition and additivity, is stressed throughout the book.

- The modal analysis and the time response of two-degree-of-freedom (two-dof) systems is eased by invoking the standard, symmetric eigenvalue problem. Of the many authors of books on vibrations, only Inman seems to have made a point on the benefits of a symmetric eigenvalue problem, as opposed to its generalized counterpart, in which the matrix in question is not symmetric. My approach requires obtaining the positive-definite square root of the 2×2 mass matrix, which is eased by resorting to (a) facts from linear algebra pertaining to the analytic functions of matrices and (b) a graphical method that relies on the *Mohr circle*, an analysis tool that is learned in elementary courses on solid mechanics. The concept of *frequency matrix*, first found in Inman's book, although not by this name, is introduced here, which helps the student use the Mohr circle in conducting the modal analysis of the systems at hand. This concept was published in a tutorial paper [5].

- The time response of first- and second-order mechanical systems is derived in a *synthetic*, i.e., *constructive* manner. What this means is that I do not follow the classical math-book approach, under which the general response is derived as the sum of a general solution, with undetermined coefficients, and a particular solution. The downside of this approach is that it does not take into account the *causality* of dynamical systems. I exploit causality by deriving the general time response as a sum of the zero-input and the zero-state responses. The former is derived, for first- and second-order systems, by means of an infinite series; this is obtained in turn by successively differentiating the mathematical model at hand, and reducing every higher-order derivative to a multiple of the first derivative of the variable of interest for first-order systems; for second-order systems, every higher-order derivative is reduced to a linear combination of this variable and its first derivative. By evaluating these derivatives at the initial time, numerical values for the coefficients of the series expansion are derived in terms of the initial values of the problem at hand.

- The time response of n-dof systems with a positive-definite stiffness matrix is introduced rather informally, by resorting to the intuitive notion that the response of these systems should be formally analogous to that of single-degree-of-freedom systems. The analogy is achieved by means, again, of the concept of frequency matrix. The time response of n-dof *undamped* systems is then informally derived by replacing the natural frequency of single-dof systems with the frequency matrix of n-dof systems. The underlying informality is then justified by proving that the time response thus obtained is indeed *the integral* of the system of governing ordinary differential equations (ODEs). The foregoing proof is conducted by resorting to a basic concept of system theory: the response thus obtained verifies *both* the ODEs and the initial conditions.

- The time response of n-dof *damped* systems, unfortunately, does not allow for a *straightforward* derivation similar to that applicable to their undamped counterparts. In this light, the time response of these systems is done first by simulation, then by means of the Laplace transform and the concept of *impulse response*.
- Examples and exercises rely on modern computational toolboxes for both numerical and symbolic computations; the powerful capabilities of readily available commercial software for plotting are fully exploited.
- Great care has been taken in producing drawings of mechanical systems, so as to convey the most accurate information graphically. This feature should be appreciated by students and instructors, as inaccurate information in a technical document invariably leads to delays in the completion of a task.
- Emphasis is placed on the logic of computations, and so, wherever needed, procedures that can be implemented with commercial software are included.

While novel techniques are introduced throughout the book, classical approaches are given due attention, as these are needed as a part of the learning process.

A common trend in the literature on the field is to be highlighted: with the aim of bringing computers into the teaching of vibration analysis, many a textbook includes code to calculate the time response of the systems of interest. The problem here is that this code is, more often that not, nothing but the verbatim casting of the time response formulas in computer language, thereby doing away with the actual possibilities offered by computing hardware and software. An alternative approach found in the literature is the numerical integration of the underlying systems of linear ordinary differential equations (ODEs) using a Runge–Kutta algorithm. While there is nothing essentially wrong with this approach, the use of such algorithms is an unnecessary complication. Indeed, Runge–Kutta methods are suitable for the integration of nonlinear ODEs; they do not exploit the linearity of the systems encountered in a first course on dynamics modeling, analysis, and simulation, thereby complicating the issue unnecessarily. We depart from these practices by resorting to the concept of *zero-order hold* and by casting the numerical integration of the underlying mathematical models in the context of discrete-time systems. The outcome is that the problem is reduced to simple operations—additions and multiplications—of arrays of numbers, i.e., vectors and matrices.

The book is accompanied by some Maple^{TM} worksheets that illustrate: (a) the discrete-time response of single-, two-, and three-degree-of-freedom systems; and (b) the use of the Mohr circle in the derivation of the time response of undamped two-dof systems. The worksheets are available at the *Springer Extras* website: http://extras.springer.com/.

A *Solutions Manual* that includes solutions to selected problems accompanies this book; it is made available to instructors.

Before closing, I would like to stress the philosiphy behind this book: knowledge being such a complex, *experiential* phenomenon [6]—it cannot be *downloaded*, contrary to popular belief—it cannot be reduced to a set of ad-hoc rules; it can be transmitted, though, via its underlying *principles*. This is probably what the *gaucho* Martín Fierro had in mind when giving wise advice to his son in the verses quoted above.

For the completion of this manuscript and its supporting materials, many people are to be acknowledged, from undergraduate interns to graduate assistants, postdoctoral fellows and colleagues. In the first versions of the manuscript, as a set of Lecture Notes, Robert Lucyshyn, Meyer Nahon, Abbas Fattah and Eric Martin, then Ph.D. students, made valuable contributions both with rigorous criticism and suggestions to improve the pedagogical value of the material. Svetlana Ostrovskaya played a key role, first as a Ph.D. student and then as a Postdoctoral Fellow, as editor and contributor to the material in general. In the last stages of the editing, Vikram Chopra, an undergraduate intern at the time—now a Ph.D. student—diligently edited the whole manuscript and assembled the Solutions Manual. The version out of which this manuscript was produced is due to the diligent work of Danial Alizadeh, a Ph.D. student in my research group. The professional work behind the figures is credited to Max A. González-Palacios, who set the standards. These individuals, and many others that unavoidably escape my memory, are given due recognition in making this book a reality.

Montreal   Jorge Angeles

References

1. Thomson WT (1993) Theory of vibration with applications. Prentice-Hall, Upper Saddle River
2. Cannon RH (1967) Dynamics of physical systems. McGraw-Hill Book Company, New York
3. Meirovitch L (1986) Elements of vibration analysis. McGraw-Hill Book Company, New York
4. Inman DJ (2007) Engineering vibration. Prentice-Hall, Upper Saddle River
5. Angeles J (1992) On the application of Mohr's circle to two-dof vibration analysis. A tutorial. J Sound Vibr 154(3):556–567
6. Russell B (1992) In: Eames ER (ed) Theory of knowledge. The 1913 manuscript. In collaboration with Blackwell, K., Routledge, London/New York

How to Use this Book

The book is intended for a variety of courses, first and foremost for a course on vibration analysis. The material is thus suitable to both an introductory and an intermediate course, at the junior and senior levels, respectively. A typical introductory course on vibration analysis would cover most of Chaps. 1, 2, 4 and 5, as the material in the last two does not depend on Chap. 3, which can be used to allow for a project on simulation of single-degree-of-freedom systems. At McGill University, I have taught a 13-week course on vibration analysis to junior undergraduates using Chaps. 1 through 5 plus 8, including a quick review of Appendix A. The latter aims to help the student understand Chaps. 2, 3, 5 and beyond.

A senior course on vibration analysis would comprise a review of Chap. 4, followed by Chaps. 6 and 7. This course should be complemented with a simulation term project, similar to the examples in Chap. 7. The instructor could also include vibration analysis of advanced structural elements like plates and shells, which, upon discretization, can be simulated using the material in Chap. 7, even if the students have not been exposed to finite element analysis. Discretization tools here can be simple finite-difference approximations of the partial differential equations involved. If the course includes finite elements, not a part of the book, then this technique can be used to derive the mass and stiffness matrices, that can then be ported into a Matlab program written by the students themselves, to implement the algorithms of Chap. 7 for simulation. This kind of training would give the students more confidence in the use of finite element software, than using this kind of software simply as a black box.

Otherwise, the book can be used to teach a sequence of two junior/senior courses on *linear mechanical analysis*. In this case the instructor can simply follow the order of the chapters, and include in each of the courses a simulation term project, based on the material covered in Chaps. 3 for the junior course, 7 for its senior counterpart.

All chapters include a set of exercises to be used as a complement of the lectures, in tutorials, as homework sets or, in the case of Chaps. 3 and 7, as term projects. Solutions to selected problems for all chapters, except for the two foregoing chapters—term projects are intended to lead to a variety of results, not to a single one—are included. These solutions are provided to instructors upon request.

Material that is not essential for an introductory course on vibration analysis is that pertaining to first-order systems in full detail; however, a cursory look at Sects. 2.2, 2.5.1, 2.6.1 and 2.7.2, is highly advisable to better understand the response of second-order systems.

Finally, a word on notation is in order: consistently throughout the book, vector and matrix arrays are referred to by their variable names in boldface font, lower-case for the former, upper-case for the latter.

Contents

1 The Modeling of Single-dof Mechanical Systems 1
  1.1   Introduction 1
  1.2   Basic Definitions 3
  1.3   The Modeling Process 7
  1.4   The Newton-Euler Equations 8
  1.5   Constitutive Equations of Mechanical Elements 11
      1.5.1   Springs.. 11
      1.5.2   Dashpots.. 17
      1.5.3   Series and Parallel Arrays of Linear Springs 18
      1.5.4   Series and Parallel Arrays of Linear Dashpots 20
  1.6   Planar Motion Analysis.. 21
      1.6.1   Lagrange Equations 25
      1.6.2   Energy Functions... 26
      1.6.3   Kinetic Energy 26
      1.6.4   Potential Energy.. 28
      1.6.5   Power Supplied to a System and Dissipation Function 29
      1.6.6   The Seven Steps of the Modeling Process 34
  1.7   Hysteretic Damping.. 53
  1.8   Coulomb Damping... 54
  1.9   Equilibrium States of Mechanical Systems 59
  1.10 Linearization About Equilibrium States. Stability................... 66
  1.11 Exercises... 76
  References 84

2 Time Response of First- and Second-order Dynamical Systems 85
  2.1   Preamble.. 85
  2.2   The Zero-input Response of First-order LTIS 88
  2.3   The Zero-input Response of Second-order LTIS 91
      2.3.1   Undamped Systems 91
      2.3.2   Damped Systems 97

2.4 The Zero-State Response of LTIS 111
     2.4.1 The Unit Impulse 112
     2.4.2 The Unit Doublet 112
     2.4.3 The Unit Step 113
     2.4.4 The Unit Ramp 114
     2.4.5 The Impulse Response 115
     2.4.6 The Convolution (Duhamel) Integral 126
2.5 Response to Abrupt and Impulsive Inputs 130
     2.5.1 First-order Systems 131
     2.5.2 Second-order Undamped Systems 132
     2.5.3 Second-order Damped Systems 136
     2.5.4 Superposition 140
2.6 The Total Time Response 141
     2.6.1 First-order Systems 141
     2.6.2 Second-order Systems 142
2.7 The Harmonic Response 145
     2.7.1 The Unilateral Harmonic Functions 147
     2.7.2 First-order Systems 149
     2.7.3 Second-order Systems 152
     2.7.4 The Response to Constant and Linear Inputs 159
     2.7.5 The Power Dissipated By a Damped
            Second-order System 161
     2.7.6 The Bode Plots of First- and Second-order Systems 161
     2.7.7 Applications of the Harmonic Response 164
     2.7.8 Further Applications of Superposition 174
     2.7.9 Derivation of $z_b(t)$ 179
2.8 The Periodic Response 181
     2.8.1 Background on Fourier Analysis 182
     2.8.2 The Computation of the Fourier Coefficients 189
     2.8.3 The Periodic Response of First- and
            Second-order LTIS 202
2.9 The Time Response of Systems with Coulomb Friction 207
2.10 Exercises 210
References 231

3 Simulation of Single-dof Systems 233
3.1 Preamble 233
3.2 The Zero-Order Hold (ZOH) 234
3.3 First-Order Systems 235
3.4 Second-Order Systems 239
     3.4.1 Undamped Systems 239
     3.4.2 Damped Systems 245
3.5 Exercises 257
Reference 262

4   Modeling of Multi-dof Mechanical Systems 263
   4.1   Introduction 263
   4.2   The Derivation of the Governing Equations 264
   4.3   Equilibrium States 278
   4.4   Linearization of the Governing Equations
       About Equilibrium States 283
   4.5   Lagrange Equations of Linear Mechanical Systems 288
   4.6   Systems with Rigid Modes 298
   4.7   Exercises... 302
   References 305

5   Vibration Analysis of Two-dof Systems 307
   5.1   Introduction 307
   5.2   The Natural Frequencies and the Natural Modes
       of Two-dof Undamped Systems 308
      5.2.1   Algebraic Properties of the Normal Modes 322
   5.3   The Zero-Input Response of Two-dof Systems 323
      5.3.1   Semidefinite Systems 324
      5.3.2   Systems with a Positive-Definite Frequency Matrix........ 333
      5.3.3   The Beat Phenomenon 344
   5.4   The Classical Modal Method 347
   5.5   The Zero-State Response of Two-dof Systems 353
      5.5.1   Semidefinite Systems 354
      5.5.2   Definite Systems 358
   5.6   The Total Response of Two-dof Systems 363
      5.6.1   The Classical Modal Method Applied to the
         Total Response 364
   5.7   Damped Two-dof Systems... 366
      5.7.1   Total Response of Damped Two-dof Systems 375
   5.8   Exercises... 381
   Reference... 388

6   Vibration Analysis of n-dof Systems 389
   6.1   Introduction 389
   6.2   The Natural Frequencies and the Natural Modes of
       n-dof Undamped Systems 390
      6.2.1   Algebraic Properties of the Normal Modes 405
   6.3   The Zero-input Response of Undamped n-dof Systems............ 406
      6.3.1   The Calculation of the Zero-input Response of
         n-dof Systems Using the Classical Modal Method 409
   6.4   The Zero-state Response of n-dof Systems 412
      6.4.1   The Calculation of the Zero-state Response of
         n-dof Systems Using the Classical Modal Method 414
   6.5   The Total Response of n-dof Undamped Systems.................. 415
   6.6   Analysis of n-dof Damped Systems............................... 415
   6.7   Exercises... 417

7   Simulation of n-dof Systems 419
   7.1   Introduction 419
   7.2   Undamped Systems 420
   7.3   The Discrete-Time Response of Undamped Systems 421
          7.3.1   The Numerical Stability of the Simulation
                    Algorithm of Undamped Systems 426
          7.3.2   On the Choice of the Time Step 428
   7.4   The Discrete-Time Response of Damped Systems 431
          7.4.1   A Straightforward Approach 432
          7.4.2   An Approach Based on the Laplace Transform 435
   7.5   Exercises 452
   References 454

8   Vibration Analysis of Continuous Systems 455
   8.1   Introduction 455
   8.2   Mathematical Modeling 456
          8.2.1   Bars Under Axial Vibration 456
          8.2.2   Bars Under Torsional Vibration 458
          8.2.3   Strings Under Transverse Vibration 460
          8.2.4   Beams Under Flexural Vibration 462
   8.3   Natural Frequencies and Natural Modes 465
          8.3.1   Systems Governed by Second-Order PDE 465
          8.3.2   Systems Governed by Fourth-Order PDEs:
                    Beams Under Flexural Vibration 480
   8.4   The Properties of the Eigenfunctions 484
          8.4.1   Systems Governed by Second-Order PDEs 484
   8.5   Exercises 494
   References 496

A   Matrix Functions 497
   A.1   Introduction 497
   A.2   Preliminary Concepts 497
   A.3   Calculation of Analytic Matrix Functions of a Matrix Argument .. 499
          A.3.1   Special Case: 2×2 Matrices 502
          A.3.2   Examples 505
   A.4   Use of Mohr's Circle to Compute Analytic Matrix Functions 516
          A.4.1   Examples 521
   A.5   Shortcuts for Special Matrices 527
          A.5.1   Example A.5.1 528
          A.5.2   Example A.5.2 529
          A.5.3   Example A.5.3 529
          A.5.4   Example A.5.4 530
   References 530

B   The Laplace Transform 531
   B.1   Introduction 531
       B.1.1   Properties of the Laplace Transform 533
   B.2   Time Response via the Laplace Transform 535
       B.2.1   The Inverse Laplace Transform via
            Partial-Fraction Expansion 538
       B.2.2   The Final- and the Initial-Value Theorems 548
   Reference 550

Index 551

Chapter 1
The Modeling of Single-dof Mechanical Systems

By these seven steps *you have taken with me, you have become my best friend.*
... If you are the lyrics, I am the music. If you are the music I am the lyrics.
If I am the heavenly body, you are the earthly world. ...

A hindi rite of wedding known by its sanskrit name, *saptapadi*, meaning *the seven steps.*[1]

1.1 Introduction

In this chapter, we will cover the items listed below:

- The need of dynamics models in engineering, objectives, scope, and limitations of mechanical analysis in general and of elements of systems (masses, springs, dashpots, energy sources and energy sinks); system; mechanical system; viscous, hysteretic and Coulomb friction.
- Abstraction and idealization of mechanical systems leading to mechanical and iconic models thereof. Physical laws leading to the underlying mathematical models: Newton's and Euler's laws; Newton-Euler governing equations. Constitutive equations of mechanical components. Equivalent spring and equivalent dashpot of parallel and series arrays.
- The Lagrange equations of motion.
- Linear and nonlinear systems; material and geometric nonlinearities; equilibrium configurations of nonlinear systems and linearization about these configurations. An outline of the stability of equilibrium configurations. Natural frequency and damping ratio of linear, stable systems.

The objectives of the book are best summarized in Fig. 1.1, which illustrates the role of modeling within the engineering analysis process. This is done with the aid of an

---

[1]http://varan_bhaath.tripod.com/Pages/Saptapadi.htm.

J. Angeles, *Dynamic Response of Linear Mechanical Systems: Modeling, Analysis and Simulation*, Mechanical Engineering Series, DOI 10.1007/978-1-4419-1027-1_1, © Springer Science+Business Media, LLC 2011

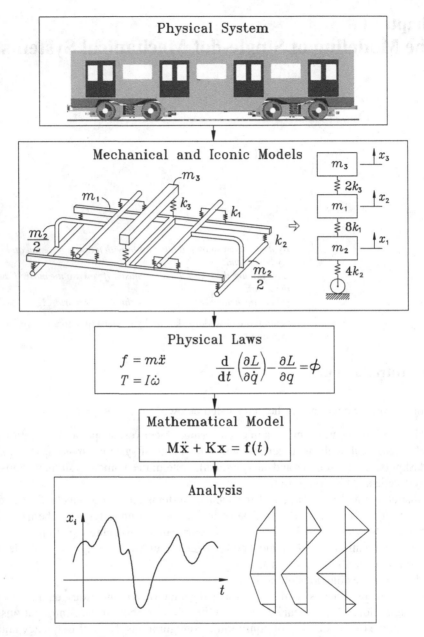

Fig. 1.1 Modeling and analysis of the vertical vibration of subway cars

example pertaining to the analysis of the vertical vibration of the subway car shown in the top box of that figure. Here, we are not interested in the lateral vibrations or the vibrations involving rotations of the car. For this reason, the two halves of the car can be thought of as moving vertically in synchronism, which allows us to study the vibration at hand with one single half-car. We thus 'split' the car into two symmetric parts, the fore and the aft halves. The mass of the car body is then assumed to be $2m_3$, each half m_3 being supported by each of the fore and aft 'bogies' carrying the wheels. Note that each bogie carries four wheels with rubber tires, two on each side, the whole car then being supported by eight wheels. Below the physical system on the top of that figure, we have a box with the *mechanical model* of one of the two bogie-half-car systems; the *iconic model* of the same system is shown at the right of the same box.

After the two foregoing models, which have two different levels of abstraction, come the physical laws used to derive the corresponding mathematical model, namely, the Newton-Euler and the Lagrange equations. The mathematical model, in turn, is one of a three-degree-of-freedom linear mechanical system, and hence, represents a three-dimensional vector ordinary differential equation that is, moreover, linear. As a result of the analysis, we have a plot of one of the *generalized coordinates*, $x_i(t)$, vs. time, as well as a sketch of the three associated normal modes of the system.

The foregoing description pertains to systems with a *finite* degree of freedom. Other class of systems that we will study in this book pertains to *continuous systems*, i.e., systems with an infinite degree of freedom. Paradigms of such systems are bars under longitudinal and torsional vibration, strings, and beams under flexural vibration. More complex structural elements, with more than one spatial dimension, such as plates, shells, and structures with arbitrary geometries, lie beyond the scope of this book.

1.2 Basic Definitions

The objects of our study are *mechanical systems*. Hence, we first have to have an idea of what a *system* at large is. *The Concise Oxford Dictionary* defines a system as a *"complex whole, set of connected things or parts, organized body of material or immaterial things,"* whereas *The Random House College Dictionary* defines the same as an *"assemblage or combination of things or parts forming a complex or unitary whole." Le Petit Robert*, in turn, defines a system as *"Ensemble possédant une structure, constituant un tout organique,"* which can be loosely translated as *"A structured assemblage constituting an organic whole."* In the foregoing definitions we can note that the underlying idea behind the concept of system is that of a set of elements interacting as a whole.

A system *responds* to *excitations*, a.k.a. *inputs*. The *response* of the system, in turn, is often likened with its *output*. However, there is an important difference between response and output: The response refers to the behavior of a system *regardless of the observer*, i.e., regardless of the instruments used to monitor its behavior; the output of the system is the set of variables available to the observer by means of instruments such as encoders to monitor angular displacement, tachometers to monitor angular velocity, or accelerometers to measure point acceleration. Thus, the output of a system may not reveal its whole behavior.

The different kinds of systems are characterized by their different forms of responding to the excitations to which they are subjected.

A *dynamical system* is a special type of system whose response depends not only on the current value, but on the whole past history of the input, dynamical systems thus being said to possess *memory*. Such systems are described mathematically by either partial or ordinary differential equations.[2] Associated with dynamical systems is the notion of *state* of the system. The state of a dynamical system at a certain time t_0 is the information pertaining to the system that completely describes the effect of the whole past input history up to and including time t_0. The output of a dynamical system at some time $t > t_0$ is then uniquely specified if two items are given, namely, (1) the time history of the input between t_0 and t, and (2) the state of the system at t_0. The state of a system at a time t_0 often turns out to be the familiar *initial conditions* of elementary differential equations courses. As we will see later in the book, the notion of state of a system is much deeper and often plays a central role in the analysis of dynamical systems.

In addition to dynamical systems, or systems with memory, we also have *memoryless* systems, i.e., those whose output depends only upon the present value of the input.[3] Such systems can be described mathematically by *algebraic* equations, i.e., equations of the form $f(x,y) = 0$, where x is the input and y is the output. The nature of the response of memoryless systems is then dramatically different from that of dynamical systems. Memoryless systems often occur as subsystems of larger mechanical systems. If we can solve for y explicitly in terms of x in the above equation, then that subsystem can be readily "removed" from the overall system by taking into account the relation between y and x. Transducers, potentiometers, and mechanical transmissions—gears, linkages, cams, etc.—are examples of memoryless systems when the inertia of their elements is negligible.

Obviously, a *mechanical system* is a system composed of mechanical elements. The elements constituting a mechanical system are rigid and deformable solids, as well as compressible and incompressible, inviscid and viscous fluids. These elements can be accurately modeled as continua, thereby leading to what is known as *distributed-parameter models*. The discipline of continuum mechanics is the natural tool to generate such models.

The role of continuum mechanics is to provide a model consisting of a system of partial differential equations (PDEs) describing the behavior of the mechanical system at hand. Such a behavior is determined as the response of the system under given initial conditions, boundary conditions, and applied loads, which calls for an integration in both the time and the space domains of the underlying PDEs. However, the integration of such equations, even if linear, is a highly demanding task that is not suitable for computers, let alone the human brain, even if aided with paper and pencil. As an alternative, engineers have found that, if the response

---

[2]Sometimes, dynamical systems are described by difference equations (case of discrete-time systems), integral equations and even by integro-differential equations.

[3]These systems "forget" the history of their excitation.

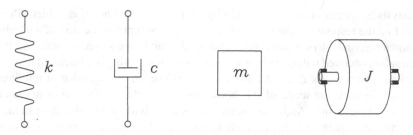

Fig. 1.2 Some lumped-parameter elements of mechanical systems

is approximated in a prescribed form, e.g., as a set of multivariate polynomials with a finite set of time-varying coefficients, then, the response can be found by simply determining how these coefficients evolve in time. These coefficients are determined, in turn, by application of a *variational principle* that leads to a system of ordinary differential equations (ODEs), which are more easily handled with the aid of computers than PDEs. However, integrating a system of ODEs can also be a tremendous task, for these equations often exhibit certain features that make them quite challenging: they are usually many, i.e., various hundreds; they are nonlinear, i.e., unhandleable with general procedures, for nonlinearity often leads to chaotic behavior that is difficult, if not impossible, to predict; often, they are not sufficient to determine the motion of the system at hand and must be complemented with additional algebraic equations, which leads to what is known as *differential-algebraic systems*. The modeling of a continuous system with a system of ODEs is accomplished by resorting to *discretization methods*, such as the Finite Element Method (FEM), which is by far the most broadly accepted, and the Boundary Element Method (BEM). Nowadays, the FEM is the standard approach to solving the most complex engineering and scientific problems, with the BEM gaining some acceptance. We will not be concerned with discretization methods, but will lay the foundations for further study that will ease the grasping of these methods either by self-study or otherwise.

In contrast to distributed-parameter models, we have *lumped-parameter models*. Here, the distributed mass, elasticity, and damping of the various components are divided into separate "lumped" components, the overall system thus being reduced to a system of interconnected particles; rigid bodies; massless, conservative springs; and massless, non-conservative dashpots. Common icons used to represent these elements are illustrated in Fig. 1.2, where we can identify a spring of stiffness k, a dashpot of damping coefficient c, a mass m and a rotor of moment of inertia J about its axis of rotation.

The modeling of lumped-parameter systems leads directly to a set of ordinary differential equations, which are usually analyzed and integrated with the aid of numerical techniques. We will focus on lumped-parameter mechanical systems in this book, but distributed-parameter systems are outlined in the last chapter.

Furthermore, when we discuss the time response of systems in Chap. 2, we will be concerned mostly with *linear* mechanical systems, which are much easier to analyze; moreover, as we will see later in the book, linear systems lend themselves to analysis by means of a powerful technique known as *superposition*, which allows one to find the response of the system to a complicated input as the sum of a number of known responses to simpler inputs. During the modeling stage, we will find that the equations derived will often be nonlinear. Now, integrating nonlinear differential equations is still a task that is far from trivial. However, the behavior of nonlinear systems is beyond the scope of this book, and hence, the integration of nonlinear ODEs will be left aside. Additionally, the detailed behavior of a nonlinear system, to be of any use, requires a highly accurate knowledge of the parameters of the system, but these parameters are most frequently elusive to reliable measurements. On the other hand, in many engineering applications, rather than the global, nonlinear behavior of a system, what is of interest is how the system will respond under 'small' perturbations of the system from a *nominal behavior*, which, often, is simply an equilibrium state. Hence, if we know the behavior of the system under nominal conditions, a *small-perturbation analysis* is all we need. This analysis is undertaken in this book by a simple *linearization* of the nonlinear mathematical model of the system about its equilibrium states.

Thus, although we will likely come across nonlinear systems during *modeling*, we will focus on linear systems during *analysis*. For completeness, the analysis of simple nonlinear systems that occur in the presence of Coulomb friction will be outlined in Chap. 2. Apart from these cases, we will study mechanical systems that lead to systems of linear ordinary differential equations, while outlining techniques to set up the partial differential equations governing distributed-parameter systems. Moreover, the mathematical models that we will derive in this book are naturally cast in the form of second-order ODEs. Occasionally, when some material properties are neglected, first-order ODEs can also occur, as we will show with some examples. In addition, we will be restricting our attention to *time-invariant* or *constant-coefficient* mechanical systems. These are systems whose material properties, e.g., mass, elasticity, and damping, do not vary with time, and whose configuration changes do not affect the constancy of the coefficients appearing in their mathematical model. Such systems lead to constant-coefficient differential equations, whose integrals can be found in closed form.

Mechanical systems usually contain moving elements that induce vibration in the structural components of these systems because of the applied loads. Such loads can take a wide variety of forms. They can be abrupt, impulsive loads, or they can be fast-varying, periodic loads. Moreover, these loads can be the result of external effects acting upon the system, or they can be generated from the moving parts of the system itself. In the preceding cases, vibration is usually an undesirable effect; here, the techniques of vibration analysis aid the engineer in predicting and possibly reducing, or even eliminating the vibratory behavior of the system. In some instances, however, vibratory behavior is a required feature of a system; e.g.,in inertial measurement units (IMU), the techniques of vibration analysis become a useful tool in their design.

This and the ensuing chapter focus on *single-degree-of-freedom systems*. Although these systems may be composed of several elements, their configuration at any time can be described by one single variable, or *generalized coordinate*. The mathematical models of these systems thus consist of one single ODE of the second order, although first-order ODEs are possible in some instances. When the system is described by a set of n independent variables, or independent generalized coordinates, we speak of a mechanical system with n *degrees of freedom* (dof), its mathematical model consisting of n ODEs, usually of the second order. Systems with n dof will be studied in Chaps. 4–7.

1.3   The Modeling Process

The modeling of mechanical systems, or that of systems at large for that matter, is a rather complex process that consists of various steps. Moreover, each of these steps may involve various tasks; sometimes, additionally, the steps are not sequential, but form iterative loops. A procedure to organize these steps is proposed below:

1. Mechanical modeling

    First and foremost, a *simplified version* of the system is proposed that captures the salient features of the system but is more amenable to analysis. This version is known as a *mechanical model* of the system of interest, the process leading to this model being based on *abstraction* and *idealization*.

    Abstraction refers to identifying the relevant mechanical features of the system with regard to the need that motivated the modeling. Such features pertain to:

    (a) The number of bodies of the system[4]
    (b) The nature of these bodies, whether rigid or deformable
    (c) The *constitutive equations* of the deformable bodies of the system, which characterize each body as linearly or nonlinearly elastic, viscoelastic, or elastoplastic and
    (d) The couplings between bodies

    Here, we distinguish between particles and rigid bodies in that the former have position but no orientation. As a consequence, one speaks of angular velocities only when referring to bodies. Particles, by definition, cannot undergo angular displacement, and hence, cannot have angular velocities. In talking of particles or mass points in this book, care should be taken in that the particles we handle are not the same as those of interest to physicists. For example, in studying the orbital motion of a satellite, we may not be interested in its attitude, and hence, can regard the satellite as a mass point rather than a body, even if the satellite has a mass of various hundreds of kilograms, and a size comparable to a truck. However, when studying the maneuvers that this satellite can perform, such as deploying an antenna or a solar panel, we must regard the satellite as a

---

[4]Engineering systems are invariably *multibody systems*, i.e., systems of rigid and deformable bodies. When a rigid body is constrained to move in one single direction, it can be modeled as a particle, but modern software for mechanical analysis regards all elements as bodies, their constraints being included explicitly in the form of algebraic relations in the mathematical model.

multibody system, whose elements are, moreover, deformable. Nevertheless, if the maneuver at hand is performed at a speed that is unlikely to produce inertia forces that will induce structural vibrations, a rigid-body model may be sufficient.

Abstraction comprises what is known as *idealization*, in which various *engineering approximations* are introduced in order to simplify the analysis. For example, a body of a system may be ideally regarded as a continuum, even though it is not, for it is constituted of molecules; the latter, in turn, are constituted of atoms that do not fill the space continuously. Idealization can also pertain to the excitation of the system. For example, if a system is subjected to impact, we may idealize the forces involved as impulses of infinitesimal duration but of infinite amplitude.

The mechanical model of the bogie-half-car system of Fig. 1.1 consists of four rigid bodies that are modeled as particles, for we are interested only in their vertical motions. Moreover, the two symmetric bodies of mass $m_2/2$ are assumed to move as one single body, and hence, can be regarded as a single body of mass m_2. Note that the rotations of the various bodies of this system are irrelevant to the intended analysis in this case. For this reason, the corresponding inertial properties, i.e., moments of inertia, are not needed. Furthermore, these bodies are all coupled by massless, conservative springs.

2. Iconic modeling

   From the mechanical model, we derive next an *iconic model*, i.e., a sketch of the system, consisting of the basic elements that we assumed at the outset, some of which are sketched in Fig. 1.2. Such elements are particles, rigid bodies, massless springs and massless dashpots. Henceforth, we assume that an iconic model of the physical system under modeling is available, our task then being to derive the *mathematical model* of the system at hand. In most engineering science courses, all problems solved in class, tests and assignments assume an iconic model at the outset, which is supposed to represent an actual system. The iconic model of the mechanical model of the system of Fig. 1.1 is shown at the right-hand side of the second block of that figure. This model is composed of three masses coupled to an inertial frame and among themselves by means of springs.

3. Mathematical modeling

   In this book, we resort to both the Newton-Euler and the Lagrange formulations of the mathematical models of the systems under study, while emphasizing the latter. These models are also known as the *governing equations*. In the case of single-dof systems, the mathematical model sought is a single first- or second-order ODE. Moreover, this equation is most likely nonlinear.

1.4   The Newton-Euler Equations

Consider a rigid body together with an inertial reference frame and a body-fixed frame having its origin either (1) at the center of mass C, or (2) at a point O of the body that is *permanently* fixed in the inertial frame, whenever such a point exists. The motion of this body is governed by the Newton-Euler equations, namely,

- If moments are taken with respect to the center of mass C,

$$\mathbf{f} = m\ddot{\mathbf{c}} \tag{1.1a}$$

$$\mathbf{n}_C = \mathbf{I}_C\dot{\omega} + \omega \times \mathbf{I}_C\omega \tag{1.1b}$$

- Now, if moments are taken with respect to a fixed point O, then

$$\mathbf{f} = m\ddot{\mathbf{c}} \tag{1.2a}$$

$$\mathbf{n}_O = \mathbf{I}_O\dot{\omega} + \omega \times \mathbf{I}_O\omega \tag{1.2b}$$

where

m is the mass of the rigid body,
$\ddot{\mathbf{c}}$ is the acceleration of the center of mass of the rigid body with respect to an inertial frame,
$\mathbf{f}$ is the resultant of all external forces acting on the rigid body,
ω is the angular velocity of the body-fixed frame with respect to the inertial frame,
$\mathbf{I}_C$ is the 3×3 inertia matrix of the rigid body with respect to C,
$\mathbf{n}_C$ is the resultant of all external couples and moments of external forces taken about the center of mass C,
$\mathbf{I}_O$ is the 3×3 inertia matrix of the rigid body with respect to O, and
$\mathbf{n}_O$ is the resultant of all external couples and moments of external forces taken about the fixed point O.

Now, we recall Steiner's Theorem[5] below, which relates the inertia matrix $\mathbf{I}_O$ of a rigid body about some point O on the body to the inertia matrix $\mathbf{I}_C$ about the center of mass C namely,

$$\mathbf{I}_O = \mathbf{I}_C + m\left[(\rho^T\rho)\mathbf{1} - \rho\rho^T\right] \tag{1.3}$$

where ρ denotes the vector directed from point O to point C, and $\mathbf{1}$ is the 3×3 identity matrix, i.e.,

$$\mathbf{1} \equiv \begin{bmatrix} 1 & 0 & 0 \\ 0 & 1 & 0 \\ 0 & 0 & 1 \end{bmatrix}$$

Moreover, if x, y, and z denote the components of ρ, then

$$\rho^T\rho = x^2 + y^2 + z^2, \quad \rho\rho^T = \begin{bmatrix} x^2 & xy & xz \\ xy & y^2 & yz \\ xz & yz & z^2 \end{bmatrix}$$

---

[5]a.k.a *parallel-axis theorem*.

Since we will be concerned with systems comprising bodies undergoing planar motion only, we will not need the three-dimensional Newton-Euler equations displayed above. Their planar counterparts take one of the two forms indicated below:

• If moments are taken with respect to the center of mass C,

$$\mathbf{f} = m\ddot{\mathbf{c}} \tag{1.4a}$$

$$n_C = I_C\dot{\omega} \tag{1.4b}$$

• If moments are taken with respect to a fixed point O,

$$\mathbf{f} = m\ddot{\mathbf{c}} \tag{1.5a}$$

$$n_O = I_O\dot{\omega} \tag{1.5b}$$

where variables in boldfaces are now two-dimensional vectors, and

m is the mass of the body,
$\ddot{\mathbf{c}}$ is the acceleration of the center of mass of the rigid body with respect to an inertial frame,
$\mathbf{f}$ is the resultant of all external forces acting on the rigid body,
ω is the scalar angular velocity of the rigid body with respect to an inertial frame,
I_C is the scalar moment of inertia of the rigid body about an axis perpendicular to the plane of motion and passing through the center of mass C,
n_C is the scalar resultant of all external couples and moments of external forces taken about the center of mass C,
I_O is the scalar moment of inertia of the rigid body about an axis perpendicular to the plane of motion and passing through the fixed point O, and
n_O is the scalar resultant of all external couples and moments of external forces taken about the fixed point O.

We note that Eqs. 1.4a and 1.5a are two-dimensional vector equations, while Eqs. 1.4b and 1.5b are scalar. Moreover, the planar version of Steiner's theorem is given below:

$$I_O = I_C + md^2 \tag{1.6}$$

where d is the distance from O to C.

Some important remarks pertaining to the Newton-Euler equations are in order:

• An inertial frame is a coordinate system that remains at rest or translates with a constant velocity with respect to the stars. For most engineering problems, and for all problems we will consider in this book, a frame fixed to the Earth can be regarded as a reasonable approximation to an inertial frame.
• The angular velocity ω of a rigid-body with respect to an inertial frame, for planar motion, can be regarded as the time-rate of change of the angle between any line

Fig. 1.3 Illustration of the
attitude and the angular
velocity of a rigid body with
respect to an inertial frame
for planar motion

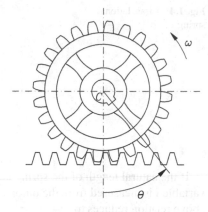

attached to the rigid body and lying in the plane, and a line or a direction, fixed to
the inertial frame. For example, if we consider the rack-and-pinion transmission
of Fig. 1.3, then

$$\omega = \dot{\theta}$$

- The second time-derivative of **c** is taken with respect to an inertial frame, that is,
  c̈ is the *absolute acceleration* of the center of mass C.

The mathematical model of a mechanical system can be derived by a simple
application of the Newton-Euler equations for each body of the system. The
bodies of the systems we will study in this book may be subjected to both elastic
and dissipative forces. We discuss below linear springs and linear dashpots that,
when arrayed in certain layouts, give rise to nonlinear elements, as shown with
examples.

1.5 Constitutive Equations of Mechanical Elements

1.5.1 Springs

A spring is a mechanical element that stores potential energy by virtue of its
elasticity; it is called *linear* if the force F required to stretch or compress it is
directly proportional to its displacement from its natural length l, i.e., with reference
to Fig. 1.4,

$$F = k(s - l) \tag{1.7}$$

where s is the total length of the spring, and k is its *stiffness*, also called the *spring
constant*. Every linear relationship between the force or torque load acting on a
spring and the corresponding translational or angular displacement is known as
Hooke's Law.

Fig. 1.4 A translational
spring

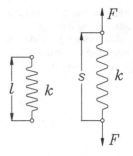

If the natural length of the spring is not mentioned, then we will assume that the variable s is measured from the unloaded configuration of the spring, and hence, the above relation reduces to

$$F = ks \qquad (1.8)$$

It should be noted that the spring of the iconic model of Fig. 1.4 implies a helicoidal type of spring, but the sources of elastic effects of the systems that we will study are more varied than this kind of elements. This icon is rather an abstract representation of the elasticity of the system at hand. As well, notice that linear springs exhibit the same behavior in tension as in compression.

Moreover, elastic elements may be precompressed, or prestretched. For example, a cable is capable of withstanding tensile loads, but not so compressive loads; concrete, in turn, withstands rather high compressive loads, but relatively low tensile loads. However, a crane can be supplied with a cable to support a load. Upon application of the static load, the cable undergoes a prestretching. If the load is supported only by the cable, and held in equilibrium, perturbations can occur—from vibrations transmitted by the floor or from gust winds—that take the load outside of its equilibrium position, the cable then responding with both elongations and compressions from its prestretched state.

A lumped spring is often used to represent the elasticity of continuous structural elements like cables, beams, shafts, rods, plates, and shells. Indeed, when these elements are used to produce stiffness in one single direction, then a simple spring model is sufficient, the spring constant in these cases being derivable from basic elasticity theory or its approximations like beam, plate and shell theories. For example, if a rod or a cable of cross-sectional area A, length l, and modulus of elasticity E is subject to a tension P at its ends, then the cable undergoes an elongation δ under this load, the relation between load and elongation being given by

$$P = \frac{AE}{l}\delta$$

and, hence, the spring stiffness of the element under study takes the form

$$k = \frac{AE}{l} \qquad (1.9)$$

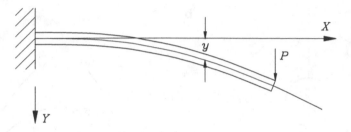

Fig. 1.5 A cantilever beam with a load acting at its free end

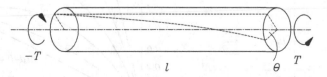

Fig. 1.6 A shaft of circular cross section under torsion

which is a *translational stiffness*, as it is associated with a *translational spring*,[6] its units being N/m.

Shown in Fig. 1.5 is a cantilever beam of length l, modulus of elasticity E, and cross-sectional area moment of inertia I, subject to a vertical load at its free end. The displacement $y(x)$ at any point along the neutral axis is derived from beam theory as

$$y(x) = \frac{P}{EI}\left(\frac{1}{2}lx^2 - \frac{1}{6}x^3\right)$$

and hence, the deflection of the free end, $\delta \equiv y(l)$, is given by

$$\delta = \frac{Pl^3}{3EI}, \quad \text{whence} \quad P = \frac{3EI}{l^3}\delta$$

Therefore, the corresponding spring constant is

$$k = \frac{3EI}{l^3} \tag{1.10}$$

As a further example, consider a shaft of circular cross section, shear modulus G, length l, and torsion constant J, as shown in Fig. 1.6. The relation between the twist angle θ and the torque T applied at its free end, while keeping the other end fixed, is given by

$$T = \frac{GJ}{l}\theta$$

---

[6]*For brevity, translational springs and their corresponding stiffness are referred to as "springs" and "stiffness" when no confusion is possible.*

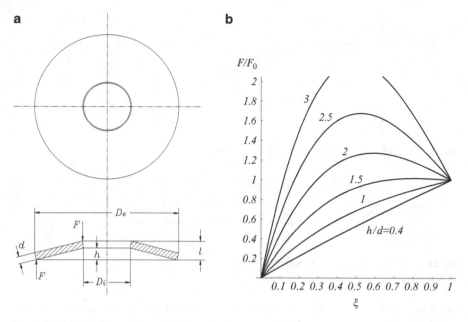

Fig. 1.7   A Belleville spring and its force-displacement relation

and hence, the resulting spring stiffness is

$$k = \frac{GJ}{l} \tag{1.11}$$

which is termed *torsional stiffness*, with units of Nm, the spring at hand being a *torsional spring*.

For the special case in which the shaft is of circular cross section, J is simply the cross-sectional moment of inertia. Thus, if a denotes the radius of the cross section, then

$$J = \frac{1}{2}\pi a^4, \quad \text{and} \quad k = \frac{2G}{l}\pi a^4 \tag{1.12}$$

Yet another example of springs that are not of the helicoidal type is the *Belleville* spring, shown in Fig. 1.7a. This spring is structurally a continuous element fabricated essentially of a washer that has been shaped in such a way that its planar faces have become conical surfaces. Under a perfectly axial load F, the spring exhibits a purely axial displacement x, its force-displacement relation being given by a formula first reported by Almen and László [1], namely,

$$F = F_0 \xi \left[\left(\frac{h}{d}\right)^2 (\xi - 1)(0.5\xi - 1) + 1 \right]$$

where F_0 and ξ are given by

$$F_0 = \frac{4E}{1 - v^2} \frac{hd^3}{KD_e^2}, \quad \xi \equiv \frac{x}{h}$$

while E denotes the modulus of elasticity, v the Poisson ratio, K a dimensionless constant that depends on the ratio D_e/D_i, and all geometric parameters being shown in Fig. 1.7a. Plots of the ratio F/F_0 vs. ξ are shown in Fig. 1.7b for various values of the ratio h/d. Note that F_0 is defined as the value of F when the displacement x that it produces equals h.

Besides linear springs, i.e., those obeying Hooke's Law, we have *nonlinear springs*. The nonlinearity of springs can stem from two sources: their material, in which case one speaks of *material* nonlinearities, or their geometry, in which case one speaks of *geometric* nonlinearities. Note, moreover, that a mechanical system composed of linear springs can exhibit geometric nonlinearities in the presence of large relative displacements at the ends of the springs, as illustrated in Example 1.5.1. Hard springs exhibit a force-displacement curve with a slope whose absolute value increases as the absolute value of the displacement increases; soft springs, correspondingly, exhibit a force-displacement curve with a slope that decreases in absolute value as the absolute value of the displacement increases. The slope at any point of a force-displacement plot represents, in fact, the *local* spring stiffness of the nonlinear spring at hand. Thus, hard springs become stiffer as their deformation increases; soft springs become more compliant as their deformation increases. We shall not discuss material nonlinearities in springs, for these fall beyond the scope of the book, but we will outline the occurrence of geometric nonlinearities in mechanical systems with linear springs.

The three different types of spring force-displacement relations, namely, linear, soft, and hard springs, are illustrated in Fig. 1.8. An example of nonlinear spring made of steel, which is a linearly elastic material under moderate loads, is the Belleville spring of Fig. 1.7, whose nonlinearity is of the geometric type. Note that this spring is linear if the ratio $h/d = 0.4$; larger values of this ratio produce a spring that behaves like that of Fig. 1.9.

Example 1.5.1 (A Nonlinear Spring). Shown in Fig. 1.9a is a mechanical system consisting of a rigid but massless crank of length l and a massless linear spring of stiffness k. The crank and the spring are pinned to the frame of a machine at points a distance l apart. Upon applying a torque τ to the crank, the spring responds with a reaction force that is a function of θ and, together with the reaction force of the pin on the crank, balances τ. Moreover, the spring is unloaded when the distance between the two ends of the spring, s, equals l. Find the relation between τ and θ.

Solution: The free-body diagram of the crank is shown in Fig. 1.9b. The spring force has a magnitude $k\Delta s$, where Δs denotes the change in length of the spring from its natural length, and, since the spring is massless, the force is directed along the axis of the spring. Moreover, the crank is also massless, and hence, the reaction force R at the pin must be equal in magnitude and opposite in direction to the spring

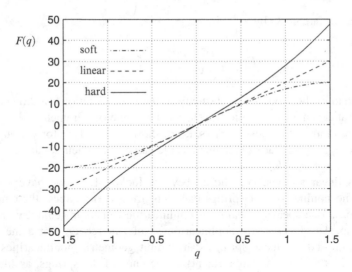

Fig. 1.8 Force-displacement relation of linear and nonlinear springs

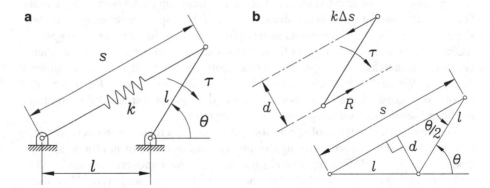

Fig. 1.9 Nonlinearly elastic system

force, as illustrated in the same figure. Now, since the moment of inertia of the crank is neglected, the moments acting on the crank must add up to zero, so that we obtain

$$\tau = (k\Delta s)\,d$$

Furthermore, from Fig. 1.9b, we can express s as

$$s = 2l\cos\left(\frac{\theta}{2}\right)$$

Therefore,

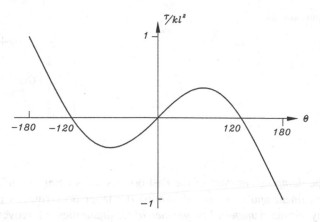

Fig. 1.10 Nonlinear torque-angular displacement relation

$$\Delta s \equiv s - l = l \left[2\cos\left(\frac{\theta}{2}\right) - 1 \right]$$

Furthermore, from the same figure, d can be written as

$$d = l \sin\left(\frac{\theta}{2}\right)$$

Substituting the foregoing expressions for Δs and d in the above expression for τ yields

$$\tau = kl^2 \sin\left(\frac{\theta}{2}\right) \left[2\cos\left(\frac{\theta}{2}\right) - 1 \right]$$

which is apparently a nonlinear relation between τ and θ. The torque-angular displacement relation of the nonlinear spring obtained above is displayed in Fig. 1.10. Note that the spring behaves linearly in the neighborhood of the origin, between $-45°$ and $+45°$, approximately. Beyond this interval, the spring exhibits a soft behavior, up until θ attains values of $\pm70°$, approximately. Beyond this interval, the spring hardens.

1.5.2   Dashpots

A linear dashpot is a *dissipative element*, i.e., one that, rather than storing energy, dissipates it. Dashpots are included in mechanical models in order to take into account the effect of drag forces caused by fluids, such as lubricants, or air. A dashpot is *linear* if the force required to produce a relative velocity $\dot{s}$ between its two ends is proportional to that velocity, i.e., with reference to Fig. 1.11,

$$F = c\dot{s} \tag{1.13}$$

Fig. 1.11 A translational
dashpot

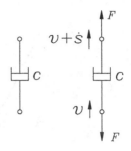

where c is the *damping constant* of the dashpot, its units being Ns/m. Again, we can have both linear and *nonlinear dashpots*, the latter occurring, as in the case of springs, by virtue of *material* or *geometric* nonlinearities. Moreover, we shall leave material nonlinearities in dashpots aside, but will outline the occurrence of nonlinear dashpots by virtue of geometric effects. The discussion of this item will be eased once the concept of dissipation function has been introduced, which is done in Sect. 1.6.5.

The above discussion pertains to *translational dashpots*, the associated coefficient thus being a *translational damping coefficient*. The same concept applies to *rotational dashpots*, the corresponding *rotational damping coefficient* being defined likewise,

$$\tau = c\dot{\theta} \tag{1.14}$$

and hence, has units of Nms.

1.5.3   Series and Parallel Arrays of Linear Springs

Springs can appear in *series* and *parallel* arrays, thereby giving rise to *equivalent springs*. The stiffness of the equivalent spring is different from those of the individual springs of the array, but, obviously, it depends on these. Below we derive expressions for the equivalent stiffness of these two types of arrays. Let us assume that we have two springs of stiffnesses k_1 and k_2, which are laid out in a series array, as shown in Fig. 1.12a[7]

Furthermore, let q_A, q_B and q_C denote the displacements of points A, B and C along the axes of the springs, from a certain reference configuration in which the springs are *unloaded*, i.e., neither stretched nor compressed. What we want is an expression relating the force F applied at points A and C with the relative displacement of A with respect to C, $q_A - q_C$, as

$$F = k_{ser}(q_A - q_C) \tag{1.15}$$

---

[7]FBD is the abbreviation of *free-body diagram*.

Fig. 1.12 (a) Series array of two springs, and (b) the FBDs of the two springs

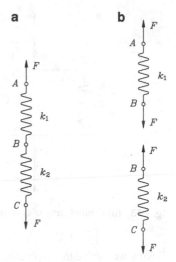

We note that the springs being massless by definition, each is under static equilibrium, and hence, the upper spring exerts a force F upwards on the lower spring, and, by Newton's third law, the lower spring exerts a force F downwards onto the first one, as depicted in Fig. 1.12b.

The force F acting at the ends of each of the two springs can then be correspondingly written in each of two forms, namely,

$$F = k_1(q_A - q_B), \quad F = k_2(q_B - q_C)$$

It is apparent that we have to eliminate q_B from the two above relations, which can be readily done by solving for this variable from each of them, thus obtaining

$$q_B = \frac{k_1 q_A - F}{k_1}, \quad q_B = \frac{F + k_2 q_C}{k_2}$$

Upon equating the above two expressions for q_B, we can obtain a relation between F and $q_A - q_C$. To this end, we solve for F from the equation thus arising, i.e.,

$$F = \frac{k_1 k_2}{k_1 + k_2}(q_A - q_C)$$

whence it is apparent that the series equivalent k_{ser} of the two springs is the coefficient of the relative displacement $q_A - q_C$ in the above expression, namely,

$$k_{ser} = \frac{k_1 k_2}{k_1 + k_2} \tag{1.16}$$

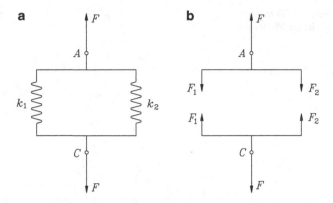

Fig. 1.13 (a) Parallel array of two springs, and (b) FBDs of the two springs

Now we derive the expression for the equivalent stiffness of the parallel array of Fig. 1.13a. We do this by writing the equilibrium equations for each of the supports of the springs, as derived from the free-body diagrams of Fig. 1.13b. In this figure, the force acting at the extremes of the spring of stiffness k_i is denoted by F_i, i.e.,

$$F_1 = k_1(q_A - q_C), \quad F_2 = k_2(q_A - q_C)$$

and, from the static equilibrium of any of the two supports,

$$F = F_1 + F_2 \equiv (k_1 + k_2)(q_A - q_C)$$

from which it is apparent that the equivalent k_{par} of the parallel array is

$$k_{par} = k_1 + k_2 \tag{1.17}$$

The reader is invited to state the implicit assumptions that have allowed us to dispense with the equations of balance of moments in the foregoing analysis. Furthermore, the foregoing discussion applies to translational springs, the same being applicable to their torsional counterparts.

1.5.4 Series and Parallel Arrays of Linear Dashpots

Dashpots can also occur in series and parallel arrays, as shown in Figs. 1.14a and b.

As we did earlier for springs, the equivalent dashpot of the two arrays can be readily determined. It is left to the reader to show that the equivalent dashpot coefficients of the series and parallel arrays, c_{ser} and c_{par}, respectively, take the forms

$$c_{ser} = \frac{c_1 c_2}{c_1 + c_2}; \quad c_{par} = c_1 + c_2 \tag{1.18}$$

Fig. 1.14 (**a**) series and
(**b**) parallel arrays of dashpots

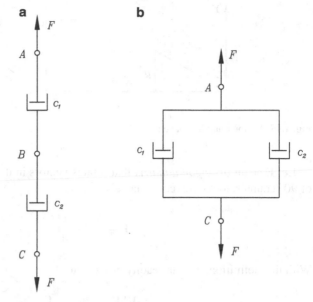

Nonlinear dashpots can occur by virtue of two items, namely, the dissipation mechanism and the geometry. In fact, linear dashpots can give rise to nonlinear ones by virtue of the geometry of the system in which they appear, as we shall show in Example 1.6.2. However, linear dashpots showing geometric nonlinearities give rise to dissipative forces that can be either linear or nonlinear in the generalized coordinate, but *linear* in the generalized speed. Again, the foregoing discussion applies to translational dashpots, the same being applicable to rotational dashpots.

1.6 Planar Motion Analysis

In setting up the equations of planar motion, we will need sometimes to compute expressions such as[8] $\omega \times \mathbf{r}$ or $\mathbf{r} \times \mathbf{f}$, where $\mathbf{r}$ is the position vector of a point of a rigid body in a given coordinate frame. Now, while planar mechanics involves only two-dimensional vectors, the cross product is essentially a three-dimensional vector operation, which prevents us, in principle, from analyzing planar motion with only two-dimensional vectors. What we need is a two-dimensional form of the cross product, as introduced below.

---

[8]Henceforth, vectors are represented with lower-case boldfaces, matrices with upper-case boldfaces, scalars with math italics. Thus, while ω is a vector, ω is a scalar.

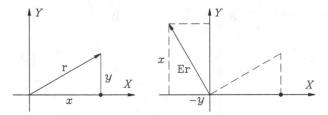

Fig. 1.15 Vector **r** and its image under **E**

Let **E** be an *orthogonal matrix* that rotates vectors in the plane through an angle of 90° counterclockwise (ccw), namely,

$$\mathbf{E} \equiv \begin{bmatrix} 0 & -1 \\ 1 & 0 \end{bmatrix} \tag{1.19a}$$

With this definition, we can readily prove that

$$\mathbf{E}^T \mathbf{E} = \mathbf{E}\mathbf{E}^T = \mathbf{1} \tag{1.19b}$$

in which **1** is now defined as the 2×2 identity matrix. Moreover, note that **E** is *skew-symmetric*, i.e., $\mathbf{E} = -\mathbf{E}^T$, and hence,

$$\mathbf{E}^2 = -\mathbf{1}, \quad \mathbf{E}^{-1} = -\mathbf{E} \tag{1.19c}$$

Also note that, given any vector $\mathbf{r} = [x, y]^T$, its image under **E** is given by

$$\mathbf{Er} = \begin{bmatrix} -y \\ x \end{bmatrix} \tag{1.19d}$$

as illustrated in Fig. 1.15.

Now, let us compute the cross product $\boldsymbol{\omega} \times \mathbf{r}$ for planar motion, where $\boldsymbol{\omega} = \omega\mathbf{k}$, and **k** is a unit vector normal to the plane of motion and pointing towards the viewer. Moreover, if **r** lies in the plane of motion, its z-component vanishes, the cross product thus taking the form

$$\boldsymbol{\omega} \times \mathbf{r} = \det \begin{bmatrix} \mathbf{i} & \mathbf{j} & \mathbf{k} \\ 0 & 0 & \omega \\ x & y & 0 \end{bmatrix} = -\omega y \mathbf{i} + \omega x \mathbf{j}$$

where we have assumed that the unit vectors **i** and **j** are parallel to the X and Y axes, respectively. The two-dimensional form of the foregoing product, then, becomes

$$(\boldsymbol{\omega} \times \mathbf{r})_{2D} = \omega \begin{bmatrix} -y \\ x \end{bmatrix} \equiv \omega\mathbf{Er} \tag{1.20}$$

where we have introduced a subscripted vector product to distinguish it from its 3D counterpart, while recalling Eq. 1.19d.

Likewise, the cross product $\mathbf{r} \times \mathbf{f}$ is a vector perpendicular to the plane of the two-dimensional vectors $\mathbf{r}$ and $\mathbf{f}$, of signed magnitude $\|\mathbf{r}\| \|\mathbf{f}\| \sin(\mathbf{r}, \mathbf{f})$, where $(\mathbf{r}, \mathbf{f})$ denotes the angle between these two vectors, measured from $\mathbf{r}$ to $\mathbf{f}$. Thus, if $\sin(\mathbf{r}, \mathbf{f})$ is positive, the cross-product vector points in the direction of $\mathbf{k}$; otherwise, in the direction of $-\mathbf{k}$. More concretely, let $\mathbf{r}$ be defined as before, $\mathbf{f}$ being defined, in turn, as

$$\mathbf{f} \equiv \begin{bmatrix} f_x \\ f_y \\ 0 \end{bmatrix}$$

Hence,

$$\mathbf{r} \times \mathbf{f} = \det \begin{bmatrix} \mathbf{i} & \mathbf{j} & \mathbf{k} \\ x & y & 0 \\ f_x & f_y & 0 \end{bmatrix} = (xf_y - yf_x)\mathbf{k} \equiv n\mathbf{k}$$

Since we know the direction of $\mathbf{r} \times \mathbf{f}$, i.e., perpendicular to the plane of motion, all we need is the quantity n above, which can be readily recognized as the scalar product[9] of the two-dimensional vectors $\mathbf{Er}$, as given by Eq. 1.19d, and $\mathbf{f}$, i.e.,

$$n = \mathbf{f}^T \mathbf{Er} \equiv (\mathbf{Er})^T \mathbf{f} = -\mathbf{r}^T \mathbf{Ef} \tag{1.21}$$

Therefore, n can be either positive or negative, depending on whether it represents a ccw or a cw (clockwise) moment.

Example 1.6.1 (A Rigid Ring Suspended from a Pin). To illustrate the foregoing concepts, we aim to find the angular velocity of the ring of radius a shown in Fig. 1.16a, when basculating without slipping on a circular pin of radius b. The pin is fixed to a wall, while the ring is subject to the action of gravity, which keeps it in contact with the pin.

Solution: In this problem, θ is the angle made by the line of the ring passing through the contact point Q and the center of the ring C with the vertical. Hence, this line is **not** fixed to the ring; one may be tempted to take $\dot{\theta}$ as the angular velocity ω of the ring, but, as a consequence of the foregoing remark, it turns out that $\omega \neq \dot{\theta}$.

The angular velocity ω can be obtained in many ways, the simplest being based on Willis' formula for epicyclic gear trains.[10] In the absence of knowledge of this formula, one can proceed by a straightforward kinematic analysis. Thus, we realize first that the distance from O to C remains constant, and hence, the ring (R) might as well be coupled to the pin (P) via a rigid arm (A), as sketched in Fig. 1.16b, whose angular velocity is $\dot{\theta}$. Now, C can be regarded either as a point of R or as a point of A. If regarded as a point of R, the velocity $\mathbf{v}_C$ of C can be simply stated as the

---

[9]The scalar product of two vectors $\mathbf{a}$ and $\mathbf{b}$, of the same dimension, also termed the *dot product*, is represented by the two alternative expressions $\mathbf{a} \cdot \mathbf{b}$ and $\mathbf{a}^T \mathbf{b}$.

[10]See a textbook on kinematics of mechanisms.

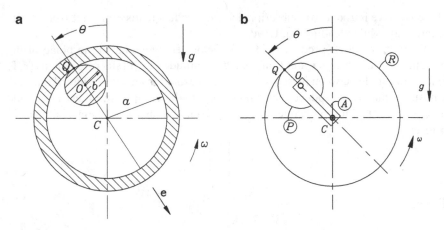

Fig. 1.16 (**a**) A rigid ring suspended from a circular pin, and (**b**) its equivalent model with an additional body, the arm A

cross product of the angular velocity $\omega\mathbf{k}$ by the vector directed from Q to C, $a\mathbf{e}$, because Q is stationary. Thus,

$$\mathbf{v}_C = a\omega\mathbf{Ec}$$

with $\mathbf{e}$ defined as the unit vector parallel to line OC, and $\mathbf{E}$ defined already in Eq. 1.19a. If C is regarded as a point of A, the velocity of C is then obtained as the cross product of the angular velocity of A, namely, $\dot{\theta}\mathbf{k}$, by the vector directed from O to C, i.e.,

$$\mathbf{v}_C = (a-b)\dot{\theta}\mathbf{Ee}$$

Upon equating the coefficients of vector $\mathbf{Ee}$ in the two foregoing expressions, we obtain an equation for ω in the form

$$a\omega = (a-b)\dot{\theta}$$

whence,

$$\omega = \frac{a-b}{a}\dot{\theta}$$

which shows that $\dot{\theta}$ is the angular velocity of the ring only in the very special case in which $b \to 0$, i.e., when the pin becomes a point. Likewise, when $b = a$, the ring is placed tight onto the pin, and cannot move without sliding, thereby obtaining, in that case, that ω vanishes.

The dynamics model of this system finds interesting engineering applications [2]. Indeed, in order to predict wear in journal bearings that are mounted with some play, we need the radius of the bearing, which sometimes is unknown, and difficult to measure accurately with a measuring tape. In this case, the radius of interest can be estimated accurately from the period of the 'small-amplitude' of the array of Fig. 1.16a.

1.6.1 Lagrange Equations

While the use of the Newton-Euler equations is quite straightforward when applied to an isolated rigid body, these equations become too cumbersome when applied to systems composed of many elements. Indeed, the Newton-Euler equations applied to systems invariably require the calculation of *internal forces of constraint* whose sole role is to keep the system together, but develop no power. The occurrence of such forces brings about unnecessary complexities in the models thus derived. For this reason, we recall the Lagrange equations in this section, and note that, within this formulation, internal forces of constraint are eliminated naturally.

The first issue in the Lagrange formulation is how to represent the motion of the system at hand, which brings about the concept of *generalized coordinate*, to be discussed presently. Both the relative layout of the system components and of the whole system with respect to an inertial frame are described by *signed distances* between landmark points in each component, whether rigid body, particle, massless spring or massless dashpot, and *signed angles* between landmark lines in the same components. Of course, the inertial frame must also be supplied with landmark points and lines to determine the position and orientation of each component with respect to it. The foregoing distances and angles are generically termed the *generalized coordinates* of the system. In general, there are no rules to define these coordinates uniquely and hence, the same system can be described by many different sets of generalized coordinates. As well, the number of generalized coordinates describing the configuration of a system is not unique, but a given system always has a minimum number of these below which the configuration is not fully defined. Such a minimum number is the *degree of freedom* (dof) of the system. Single-dof systems are fully described with one single generalized coordinate, henceforth represented as q when the discussion applies to arbitrary systems.

The time-derivatives of the generalized coordinates are the *generalized speeds* of the system. If the number of generalized coordinates equals the degree of freedom of the system, then these are referred to as *independent generalized coordinates*. Correspondingly, the associated generalized speeds are termed the *independent generalized speeds*. Furthermore, the whole set of both independent generalized coordinates and independent generalized speeds constitutes the *state* of the system, the evolution of the configuration of the system, in turn, constituting its *motion*. Notice that the Newton-Euler equations governing the motion of a mechanical system determine the values of the second derivatives of the generalized coordinates, for given values of generalized coordinates and generalized speeds along with given values of the applied forces and moments. Hence, by knowing the state of the system at a given instant, it is possible to predict the state of the system an instant later. Again, the evolution of single-dof systems is fully determined by one single generalized coordinate q and one single generalized speed $\dot{q}$.

The set of independent generalized coordinates and independent generalized speeds is known generically as the *state variables* of the system. Thus, the values of the state variables of a system at a given instant—i.e., the initial

conditions—along with the inputs—applied forces and moments—and the mathematical model derived from the Newton-Euler or the Lagrange equations, *determine*
the motion of the system, which is the reason why we call these systems *deterministic*. The state variables of a single-dof system can then be stored in a
two-dimensional array $\mathbf{x} \equiv [q, \dot{q}]^T$ termed the *state-variable vector* of the system.

1.6.2 Energy Functions

The Lagrange equation for a single-dof system is given below[11]:

$$\frac{d}{dt}\left(\frac{\partial L}{\partial \dot{q}}\right) - \frac{\partial L}{\partial q} = \frac{\partial}{\partial \dot{q}}(\Pi - \Delta) \qquad (1.22)$$

where:

q is the generalized coordinate;
T is the *total kinetic energy* of the system;
V is the *total potential energy* of the system;
L is the *Lagrangian* of the system, given by

$$L = T - V \qquad (1.23)$$

Π is the *power supplied* to the system by motors delivering *controlled forces or
torques*; and
Δ is the *dissipation function* associated with all dissipative forces in the system.

All functions T, V, L, Π, and Δ are generically termed *energy functions*, even
though only the first three have units of energy.

We will discuss below the various terms in the Lagrange equation. While the
Lagrangian formalism can be applied to mechanical systems of particles and both
rigid and flexible bodies, and to other physical systems as well, we will focus
henceforth on mechanical systems composed of rigid bodies undergoing planar
motion. Furthermore, we assume that the system at hand, although having one single
dof, is composed of multiple elements, i.e., r rigid bodies, s massless springs and d
massless dashpots.

1.6.3 Kinetic Energy

The kinetic energy T_i of the ith rigid body can be determined using one of two
expressions: the first is general and applies in all cases; the second is applicable
only in special cases that nevertheless arise frequently, and is simpler than the first

---

[11] The formal derivation of the Lagrange equations lying beyond the scope of the book, the reader is
invited to review this derivation, as pertaining to *systems of particles* in a mechanics book, e.g., [3].

expression. In the first expression, we decompose the motion of the body into two parts, namely, a pure translation of the whole body identical to that of its center of mass, and a pure rotation of the body about its center of mass. This decomposition is known as *König's Theorem*. Thus, we have

Kinetic Energy of the *i*th Rigid Body of a Mechanical System Undergoing Planar Motion

$$T_i = \frac{1}{2} m_i \|\dot{\mathbf{c}}_i\|^2 + \frac{1}{2} I_{Ci} \omega_i^2 \tag{1.24}$$

where:

 m_i is the mass of the *i*th rigid body;
 $\dot{\mathbf{c}}_i$ is the two-dimensional velocity vector of the center of mass of the *i*th rigid body with respect to an inertial frame, $\|\dot{\mathbf{c}}_i\|$ representing its magnitude;
 I_{Ci} is the scalar moment of inertia of the *i*th rigid body with respect to its center of mass; and
 ω_i is the scalar angular velocity of the *i*th rigid body with respect to an inertial frame.

In the alternative expression, we first determine a point O of the rigid body that is *instantaneously* fixed to an inertial frame. In planar motion, this point, called the *instantaneous center*, always exists. If the velocity of each point of the body is referred to O, then the expression for the kinetic energy simplifies as described below:

Kinetic Energy of the *i*th Rigid Body of a Mechanical System Undergoing Planar Motion Based on a Fixed Point O_i of the Body

$$T_i = \frac{1}{2} I_{Oi} \omega_i^2 \tag{1.25}$$

where:

 I_{Oi} is the scalar moment of inertia of the *i*th rigid body with respect to a point O_i of the body that is *instantaneously* fixed to an inertial frame; and
 ω_i is the scalar angular velocity of the *i*th rigid body with respect to an inertial frame.

The kinetic energy of the overall system is the sum of the kinetic energies of all r rigid bodies, i.e.,

$$T = \sum_1^r T_i \tag{1.26}$$

As we will show with examples, the velocity vector of the center of mass of each body and its scalar angular velocity can be always written as *linear functions* of the generalized speed, the associated coefficients being functions of the generalized coordinate. Likewise, the angular velocity of all bodies of the system can always be written as a linear function of the generalized speed. The representation of center-of-mass velocities and body angular velocities requires a careful *kinematic analysis* whose importance cannot be overstated. Under the assumption that the foregoing analysis has been conducted, then, the total kinetic energy of the system under study takes the form

$$T = \frac{1}{2}m(q)\dot{q}^2 + p(q,t)\dot{q} + T_0(q,t) \tag{1.27}$$

where the coefficients $m(q)$ and $p(q,t)$ as well as $T_0(q,t)$ are, in general, functions of the generalized coordinate q; the last two can also be explicit functions of time. However, if the system neither gains nor loses mass—a rocket loses mass as its fuel burns—then m is not an explicit function of time. All systems considered in this book are assumed to be of the first type, i.e., with $m = m(q)$. Note that systems that are externally driven by a source producing a *controlled* motion that does not depend on the generalized speed have a nonzero kinetic energy T_0 even if the generalized speed is set equal to zero. The first term of that expression is *quadratic* in the generalized speed; the second term is linear in this variable; and the third term is independent of the generalized speed, but is a function of the generalized coordinate and, possibly, of time as well. Moreover, this function of q is most frequently nonlinear, while the second and the third terms of the same expression arise in the presence of actuators supplying a *controlled motion* to the system. Furthermore, because of the kinetic energy being essentially positive, the coefficient $m(q)$ of Eq. 1.27 is also positive. It is, hence, called the *generalized mass* of the system. As such, the generalized mass has units of mass or of moment of inertia. In certain cases, when working with nondimensional quantities or with normalized expressions, the generalized mass can even be nondimensional. Other units are possible, depending on the manipulations introduced to derive the governing equation in its final form.

1.6.4 Potential Energy

In setting up the Lagrange equations of the system under study, we need an expression for its potential energy. Here, we assume that we have two possible sources of potential energy: elastic and gravitational. The former appears because of the flexibility of some elements of the system; the latter because of the gravity field. The potential energy due to the gravity field is, in general, a nonlinear function of the generalized coordinate, *but does not depend on the generalized speed*. Thus, we will assume that, in general, the potential energy V_g due to gravity is a nonlinear function of the generalized coordinate q, i.e.,

$$V_g = V_g(q) \tag{1.28}$$

On the other hand, the elastic potential energy is due to springs of all sorts. We will study mainly *linear springs*,[12] i.e., springs requiring a force $k\Delta s$ to stretch them or compress them for an amount Δs, with k denoting the spring stiffness. Here, Δs is the elongation or contraction of the spring from its unloaded configuration. Hence, the potential energy due to the stiffness k_i of the ith spring takes the form

$$V_i = \frac{1}{2}k_i(\Delta s_i)^2 \qquad (1.29a)$$

Δs_i being the elongation or contraction of the ith spring. The foregoing discussion applies to *translational springs*, a similar expression applying to *torsional springs*:

$$V_i = \frac{1}{2}k_{ti}(\Delta \theta_i)^2 \qquad (1.29b)$$

Thus, under the assumption that the system has a total of s springs, the total potential energy is given as

$$V = V_g + \sum_1^s \frac{1}{2}k_i(\Delta q_i)^2 \qquad (1.30)$$

where k_i can be either translational or torsional and q_i is a generalized coordinate.

1.6.5 Power Supplied to a System and Dissipation Function

The energy sources of a mechanical system can be of two kinds, namely, *force—or torque—sources* and *motion sources*, the two being developed by controlled motors. If the controlled item is force or torque, then we speak of a force-controlled source; if motion, then of a motion-controlled source.

A force $\mathbf{f}$ acting at a point of a body that moves with velocity $\mathbf{v}$ develops a power $\mathbf{f} \cdot \mathbf{v}$ onto the body. When the force is applied by *controlled* sources like motors or muscles that impel the body, the power is positive, and we speak of a *driving* or *active force*. On the contrary, when the force is applied by the environment in such a way that it opposes the motion, the power is negative, and we speak of a *load* or a *dissipative force*. However, nothing prevents a motor from acting against the motion, thereby functioning as a brake, i.e., as a load. Therefore, the power supplied by motors can be either positive or negative. In mechanical engineering parlance, when a motor drives the load, one speaks of *forward-driving*; otherwise, of *backward-driving*.

Regarding moments, the same holds, if *velocity* changes to *angular velocity*. In this case, then, the power developed by a moment $\mathbf{n}$ onto a body that rotates with an angular velocity ω is given by $\mathbf{n} \cdot \omega$ or, in the planar case, as $n\omega$. A moment

---

[12]Erroneously, translational springs are sometimes referred to as "linear."

applied about a fixed axis is termed *torque*; it is represented as τ or T, depending on the other variables used, to avoid confusion; the power in this case is given by either $\tau\omega$ or, correspondingly, $T\omega$.

The counterpart of the concept of generalized coordinate is the concept of *generalized force*. Once a set of independent generalized coordinates and speeds has been decided on, the generalized forces are determined *uniquely*. Thus, single-dof systems have a single generalized force associated with them. Now, if we let the total power developed by force-controlled sources be denoted by Π, then, the generalized *active* force ϕ_f associated with the generalized coordinate q is derived as

$$\phi_f = \frac{\partial \Pi}{\partial \dot{q}} \tag{1.31}$$

where $\dot{q}$ is the corresponding generalized velocity. Note that ϕ_f stems only from force-controlled sources.

Systems are not only acted upon by driving forces but also by *dissipative* forces that *intrinsically* oppose the motion of the system. As a rule, dissipative forces are functions of velocity and hence, of the generalized speed of the system. Here, we postulate the existence of a *dissipation function* Δ from which the dissipative force ϕ_d is derived as

$$\phi_d = -\frac{\partial \Delta}{\partial \dot{q}} \tag{1.32}$$

the negative sign taking into account that Δ is *essentially* a positive quantity, while ϕ_d opposes motion. Here, a word of caution is in order: while the dissipation function has units of power, it is not necessarily equal to the power developed by dissipative forces, as we shall see presently. As a matter of fact, power can be either positive or negative, i.e., power is a *sign-indefinite* quantity, while a dissipation function is *positive-definite*.

Dissipation of energy occurs in nature in many forms. The most common mechanisms of energy dissipation are (a) viscous damping, (b) Coulomb or dry-friction damping, (c) hysteretic damping, and (d) flow-induced drag. We will be concerned in the book mainly with the first two forms of energy dissipation, and of these, mostly with the first one. Furthermore, with regard to viscous damping, we will focus on linearly-viscous damping, which can be taken into account by linear dashpots,[13] the counterparts of linear springs. Linear dashpots coupling two elements of a system at points A and B are characterized by *quadratic* dissipation functions. That is, the dissipation function of the ith linear dashpot of a system takes the form

$$\Delta_i = \frac{1}{2}c_i\dot{s}_i^2 \tag{1.33a}$$

---

[13] Again, translational dashpots are sometimes erroneously referred to as "linear."

where $\dot{s}_i$ is the relative velocity of points A and B along the direction of line AB. Note that, by virtue of the form of the dissipation function, this attains the same value whether the dashpot is stretching or contracting at the same rate. The corresponding function for a rotational dashpot is

$$\Delta_i = \frac{1}{2}c_i\dot{\theta}_i^2 \tag{1.33b}$$

Power-dissipation, like energy, is *additive*, and hence, if a mechanical system comprises d dashpots, then the *total* power dissipation of the system is

$$\Delta = \sum_1^d \Delta_i \tag{1.34}$$

Now, upon differentiation of the dissipation function Δ_i of Eq. 1.33a with respect to $\dot{s}_i$, we obtain the generalized force exerted by the dashpot, ϕ_{di}, as

$$\phi_{di} = -c_i\dot{s}_i \tag{1.35}$$

which always opposes the motion, with a similar expression for a rotational dashpot. That is, if the dashpot is stretching, ϕ_{di} acts to compress it; if contracting, then the force acts to stretch it. The constant c_i is known as the *dashpot coefficient* or *damping coefficient*.

Note that the power Π_d dissipated by a linear dashpot of coefficient c is given by

$$\Pi_d = -\phi_d\dot{s} = -c\dot{s}^2 = -2\Delta$$

and hence, equals twice the negative of the dissipation function of the dashpot at hand.

As we shall see with examples, motion-controlled sources induce an active generalized force that we will label $\phi_m = \phi_m(q,\dot{q},t)$, and does not arise from Π, but rather from the Lagrangian of the system. An additional form of generalized force is that stemming from the potential energy $V(q,t)$, that enters in the Lagrangian via this function; these forces, arising from a potential function, are called *conservative forces*, and are denoted by $\phi_p(q,t)$, the subscript reminding us that they stem from a *potential*.

The *total* generalized force ϕ associated with the generalized coordinate q is then the sum of all generalized forces described above, namely,

$$\phi \equiv \phi_p + \phi_m + \phi_f + \phi_d \tag{1.36}$$

Systems free of driving sources, whether of the force- or motion-controlled type, are said to be *autonomous*. Therefore, autonomous systems are characterized by $\phi_m = \phi_f = 0$.

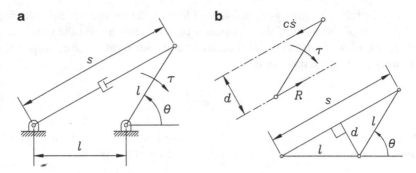

Fig. 1.17 A nonlinear dashpot derived from a linear one

Example 1.6.2 (A Configuration-dependent Dashpot). If the spring of Fig. 1.9 is replaced by a linear dashpot, we have the array of Fig. 1.17. Derive expressions for both the dissipation function of the array and that of the associated dissipative force, while regarding θ as the generalized coordinate.

Solution: Since the dashpot is linear, its dissipation function takes the form

$$\Delta = \frac{1}{2}c\dot{s}^2$$

but we need an expression in terms of θ and $\dot{\theta}$, the generalized coordinate and the generalized speed. This expression is readily derived by noticing that $s = s(\theta)$, and hence,

$$\dot{s} = \frac{ds}{d\theta}\dot{\theta}$$

where $ds/d\theta$ is obtained from the expression for s derived in Example 1.5.1, namely,

$$\frac{ds}{d\theta} = -l\sin\left(\frac{\theta}{2}\right)$$

After simplifications, the foregoing relations lead to

$$\Delta = \frac{1}{4}cl^2(1-\cos\theta)\dot{\theta}^2$$

which is a positive-definite expression that vanishes only when either θ or $\dot{\theta}$ does.

Now the generalized force ϕ_d associated with the dashpot is readily found as

$$\phi_d = -\frac{\partial\Delta}{\partial\dot{\theta}} = -\frac{1}{2}cl^2(1-\cos\theta)\dot{\theta}$$

which is a nonlinear function of θ, although linear in $\dot{\theta}$. For this reason, we say that the associated damping coefficient is *configuration-dependent*.

Example 1.6.3 (The Dissipation Function of an Epicyclic Gear Train). If we fix the internal gear R of Fig. 1.16b to an inertial frame, and let the arm A rotate freely, while carrying the pinion P, that now translates and rotates simultaneously, we obtain an epicyclic gear train. Under the assumption that the joints at O and C are lubricated with a linearly viscous fluid, these joints produce dissipative torques that can be derived from a dissipation function Δ. Find an expression for Δ in terms of the angular velocity $\dot{\theta}$ of the arm A.

Solution: Let the angular velocity of the pinion be denoted by ω_P, which is cw if that of the arm, $\dot{\theta}$, is ccw. Now, in order to set up the dissipation function of this system, it is important to note that this is the sum of two quadratic expressions in *relative* angular velocities, one for each joint. The reason why we have emphasized here the word *relative* is that motions of a mechanical system as a single rigid body do not produce any dissipation. It is the relative motion that matters when it comes to energy dissipation. Therefore, we have

$$\Delta = \frac{1}{2}c_C\dot{\theta}^2 + \frac{1}{2}c_O\omega_{P/A}^2$$

where c_C and c_O denote the damping coefficients at joints C and O, respectively. Moreover,

$$\omega_{P/A} \equiv \omega_P - \omega_A = \omega_P - \dot{\theta}$$

The velocity of point Q can be expressed as

$$v_Q = v_O + v_{Q/O}$$

where v_O and $v_{Q/O}$ are given as

$$v_O = (a-b)\dot{\theta} \quad \text{and} \quad v_{Q/O} = b\,\omega_P$$

Upon setting the velocity of point Q equal to zero, we have

$$v_{Q/O} = -v_O$$

or

$$\omega_P = -\frac{a-b}{b}\dot{\theta}$$

and hence,

$$\omega_{P/A} = -\frac{a}{b}\dot{\theta}$$

the dissipation function thus becoming

$$\Delta = \frac{1}{2}\left[c_C + \left(\frac{a}{b}\right)^2 c_O\right]\dot{\theta}^2$$

which is the expression sought.

1.6.6   The Seven Steps of the Modeling Process

We derive below the general form of the Lagrange equations, Eq. 1.22, as pertaining to single-dof systems. To this end, we use Eq. 1.27 to derive the Lagrangian in the form

$$L = \frac{1}{2}m(q)\dot{q}^2 + p(q,t)\dot{q} + T_0(q,t) - V(q)$$

Hence,

$$\frac{\partial L}{\partial \dot{q}} = m(q)\dot{q} + p(q,t)$$

and

$$\frac{d}{dt}\left(\frac{\partial L}{\partial \dot{q}}\right) = m(q)\ddot{q} + m'(q)\dot{q}^2 + \frac{\partial p}{\partial q}\dot{q} + \frac{\partial p}{\partial t}$$

Likewise,

$$\frac{\partial L}{\partial q} = \frac{1}{2}m'(q)\dot{q}^2 + \frac{\partial p}{\partial q}\dot{q} + \frac{\partial T_0}{\partial q} - \frac{\partial V}{\partial q}$$

Moreover, the right-hand side of Eq. 1.22 is nothing but the sum of the terms due to force-controlled sources and dissipation, i.e,

$$\frac{\partial}{\partial \dot{q}}(\Pi - \Delta) = \phi_f + \phi_d$$

Therefore, the Lagrange equations take the form

$$m(q)\ddot{q} + \underbrace{\frac{1}{2}m'(q)\dot{q}^2}_{h(q,\dot{q})} = \underbrace{-\frac{\partial p}{\partial t} + \frac{\partial T_0}{\partial q}}_{\phi_m} \underbrace{-\frac{\partial V}{\partial q}}_{\phi_p} + \phi_f + \phi_d \qquad (1.37)$$

That is,

$$h(q,\dot{q}) = \frac{1}{2}m'(q)\dot{q}^2 \qquad (1.38a)$$

$$\phi_m(q,\dot{q},t) = -\frac{\partial p}{\partial t} + \frac{\partial T_0}{\partial q} \qquad (1.38b)$$

$$\phi_p(q,t) = -\frac{\partial V}{\partial q} \qquad (1.38c)$$

In summary, then, the Lagrange equation for single-dof systems takes the form

$$m(q)\ddot{q} + h(q,\dot{q}) = \phi(q,\dot{q},t) \qquad (1.39a)$$

where

$$h(q,\dot{q}) = \frac{1}{2}m'(q)\dot{q}^2 \tag{1.39b}$$

$$\phi(q,\dot{q},t) = \phi_p(q,t) + \phi_m(q,\dot{q},t) + \phi_f(q,\dot{q},t) + \phi_d(q,\dot{q},t) \tag{1.39c}$$

i.e., the governing model is a second-order ordinary differential equation in the generalized coordinate q. In this equation, the configuration-dependent coefficient $m(q)$, the generalized mass, was already introduced in Eq. 1.27, while the second term of the left-hand side, $h(q,\dot{q})$, contains inertia forces stemming from Coriolis and centrifugal accelerations. For this reason, this term is sometimes called the term of *Coriolis and centrifugal forces*. The right-hand side, in turn, is the sum of four different generalized-force terms, namely, (1) $\phi_p(q,t)$, a generalized force stemming from the potential energy of the Lagrangian; (2) $\phi_m(q,\dot{q},t)$, a generalized force stemming from motion-controlled sources and contributed by the Lagrangian as well; (3) $\phi_f(q,\dot{q},t)$, a generalized force stemming from force-controlled sources, and contributed by the power Π; and (4) $\phi_d(q,\dot{q},t)$, a generalized force of dissipative forces, stemming from the dissipation function Δ. The integration of Eq. 1.39a with prescribed initial values $q(0) = q_0$ and $\dot{q}(0) = \dot{q}_0$, and given generalized force $\phi(q,\dot{q},t)$ can be a challenging task. We will not be concerned with this task in this book. What we will do is study the *time response* of the foregoing system under 'small' perturbations from its equilibrium state. The techniques to derive the associated time response will be studied in Chap. 2. To illustrate the modeling process we resort to an example, the process comprising a sequence of *seven steps*.

Example 1.6.4 (A Locomotive Wheel Array). Derive the Lagrange equation of the system shown in Fig. 1.18. This system consists of two identical wheels of mass m and radius a that can be modeled as uniform disks. Furthermore, the two wheels are coupled by a slender, uniform, rigid bar of mass M and length l, pinned to the wheels at points a distance b from the wheel centers. We can safely assume that the wheels roll without slipping on the horizontal rail and that the only nonnegligible dissipative effects arise from the lubricant at the bar-wheel pins. These pins produce dissipative moments proportional to the *relative angular velocity* of the bar with respect to each wheel.

Solution: Under the no-slip assumption, it is clear that a single generalized coordinate, such as θ, suffices to describe the configuration of the entire system at any instant. The system thus has a dof $= 1$. We now introduce the *seven-step* procedure to derive the mathematical model sought:

1. **Kinematics.** This is the most important part of the process, because the remainder depends on its correctness. Moreover, this step is also the most challenging, for it requires sound knowledge of basic mechanics and geometry. What is important here is to realize that the simplest approach is always the easiest to implement and the most reliable. In general, we aim here to find expressions of center-of-mass (c.o.m) velocities and angular velocities that are linear in the

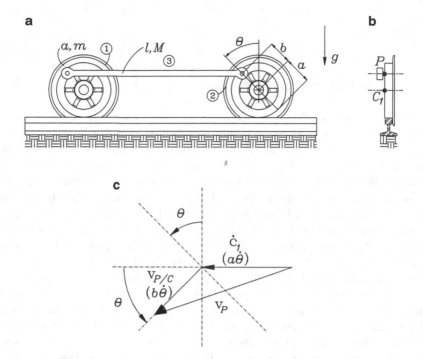

Fig. 1.18 A mechanical system with three moving bodies and one single dof

generalized speed $\dot{\theta}$. A previous inspection of the quantities required will reveal, in many instances, that c.o.m velocities themselves are not needed, but rather the squared magnitudes of these velocities. Since the latter are scalar, their derivation is far simpler than that of the former. Upon numbering the bodies of the system as shown in Fig. 1.18a, we realize that the motions of bodies 1 and 2 are identical, that of body 3 being a pure translation. Moreover, all points of body 3 describe circles of radius b.

Now, if we denote by P the center of the pin connecting the wheel 1 with the coupler bar, we have

$$\mathbf{v}_P = \dot{\mathbf{c}}_1 + \mathbf{v}_{P/C}$$

where $\mathbf{v}_{P/C}$ denotes the relative velocity of P with respect to C_1, the center of mass and centroid of wheel 1, a relation that is shown in Fig. 1.18c. In this figure, we show in parentheses the signed magnitude of each velocity term of the right-hand side of the foregoing equation. Note that, from this velocity triangle we could determine $\mathbf{v}_P$ as such, but, as a matter of fact, we do not need it; what we need is rather the squared magnitude of this velocity vector. Moreover, the coupler bar undergoing a pure translation, its center of mass C_3 has a velocity identical to $\mathbf{v}_P$, and hence,

$$\dot{\mathbf{c}}_3 = \dot{\mathbf{c}}_1 + \mathbf{v}_{P/C}$$

In fact, from the expression for T_i of Eq. 1.24, all that we need to find T_3 is $\|\dot{\mathbf{c}}_3\|^2$, which can be done with much less effort than finding $\dot{\mathbf{c}}_3$ itself. This is done from the velocity triangle of Fig. 1.18c and the 'cosine law', namely,

$$\|\dot{\mathbf{c}}_3\|^2 = a^2\dot{\theta}^2 + b^2\dot{\theta}^2 + 2ab\dot{\theta}^2\cos\theta$$

and, clearly, $\omega_3 = 0$.

2. **Kinetic energy**. Here, our goal is to derive an expression for the kinetic energy that is of the form of Eq. 1.27. To this end, we use the forms (1.24 and 1.25) and so, I_i denotes the moment of inertia of the ith body about its center of mass C_i. Moreover,

$$I_1 = I_2 = \frac{1}{2}ma^2, \quad I_3 = \frac{1}{12}Ml^2$$

As a matter of fact, I_3 will not be needed because body 3 undergoes a pure translation. We have written it down only for completeness, and for future reference, because this is a useful expression. Thus, from Eqs. 1.24 and 1.25, we obtain, respectively,

$$T_1 = T_2 = \frac{1}{2}ma^2\dot{\theta}^2 + \frac{1}{2}\left(\frac{1}{2}ma^2\right)\dot{\theta}^2 = \frac{1}{2}m\left(\frac{3a^2}{2}\right)\dot{\theta}^2$$

$$T_3 = \frac{1}{2}M(a^2 + b^2 + 2ab\cos\theta)\dot{\theta}^2$$

whence,

$$T = \left[\frac{3}{2}ma^2 + \frac{1}{2}M(a^2 + 2ab\cos\theta + b^2)\right]\dot{\theta}^2$$

which is, indeed, a quadratic expression in $\dot{\theta}$. Notice that this expression, in this particular example, contains neither a linear term in the generalized speed nor an independent term.

3. **Potential energy**. First, we note that the sole source of potential energy is gravity. Moreover, the centers of mass of the two wheels remain at the same level, and hence, it is convenient to use this level as a reference. Therefore,

$$V_g = Mgb\cos\theta$$

4. **Lagrangian**. This is now readily derived as

$$L = T - V = \left[\frac{3}{2}ma^2 + \frac{1}{2}M(a^2 + 2ab\cos\theta + b^2)\right]\dot{\theta}^2 - Mgb\cos\theta$$

5. **Power supplied**. Apparently, the system is not subjected to any force-controlled source, and hence, $\Pi = 0$.

6. **Power dissipation**. What is important here is to find the angular velocity of the bar with respect to each of the wheels; this velocity will determine the dissipation function occurring by virtue of the lubricant at the pins. Since the bar moves under pure translation, its relative angular velocity with respect to each of the wheels is $-\dot{\theta}$, and hence,

$$\Delta = 2 \times \frac{1}{2} c \dot{\theta}^2$$

where the factor of 2 is included to account for the two pins.

7. **Lagrange equation**. Now, all we need is to evaluate the partial derivatives involved in Eq. 1.22, namely,

$$\frac{\partial L}{\partial \dot{\theta}} = \left[3ma^2 + M(a^2 + 2ab\cos\theta + b^2) \right] \dot{\theta}$$

Hence,

$$\frac{d}{dt}\left(\frac{\partial L}{\partial \dot{\theta}} \right) = \left[3ma^2 + M(a^2 + 2ab\cos\theta + b^2) \right] \ddot{\theta} - 2Mab(\sin\theta)\dot{\theta}^2$$

and

$$\frac{\partial L}{\partial \theta} = -Mab(\sin\theta)\dot{\theta}^2 + Mgb\sin\theta$$

Furthermore,

$$\frac{\partial \Pi}{\partial \dot{\theta}} = 0, \quad \frac{\partial \Delta}{\partial \dot{\theta}} = 2c\dot{\theta}$$

the equation sought thus being

$$\left[3ma^2 + M(a^2 + 2ab\cos\theta + b^2) \right] \ddot{\theta} - Mab(\sin\theta)\dot{\theta}^2 = Mgb\sin\theta - 2c\dot{\theta}$$

which is termed the governing equation of the system at hand.

In the equation derived above, we can readily identify the functions $m(\theta)$ and $h(\theta, \dot{\theta})$ of Eq. 1.39a. Thus, the generalized mass of the system is

$$m(\theta) \equiv 3ma^2 + M(a^2 + 2ab\cos\theta + b^2)$$

Here, note that $m(\theta)$ has units of moment of inertia. Moreover, the generalized mass is, in this case, configuration-dependent, for it is apparently a function of angle θ. Likewise, the term of Coriolis and centrifugal forces is readily identified as

$$h(\theta, \dot{\theta}) = -Mab(\sin\theta)\dot{\theta}^2$$

which is quadratic in $\dot{\theta}$ and, hence, arises from centrifugal forces. Finally, the right-hand side contains one term of gravity forces, $Mgb\sin\theta$, and one that is dissipative, $-2c\dot{\theta}$. No active force occurs in this example because the system is

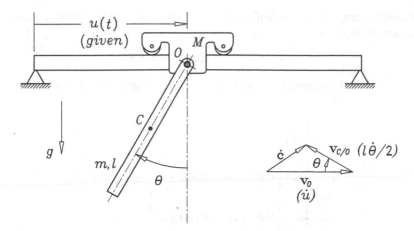

Fig. 1.19 Overhead crane driven by a motion-controlled source

neither force- nor motion-driven by controlled sources. The only driving force here is gravity, and this is taken into account in the Lagrangian. We thus have

$$\phi_p(\theta) = Mgb\sin\theta, \quad \phi_m = \phi_f = 0, \quad \phi_d(\dot{\theta}) = -2c\dot{\theta}$$

Example 1.6.5 (An Overhead Crane). Now we want to derive the Lagrange equation of the overhead crane of Fig. 1.19 that consists of a cart of mass M that is driven with a controlled motion $u(t)$. A slender rod of length l and mass m is pinned to the cart at point O by means of roller bearings producing a resistive torque that can be assumed to be equivalent to that of a linear dashpot of coefficient c.

Solution: We proceed as in the foregoing example, in seven steps, namely,

1. **Kinematics.** Since the motion of the cart is controlled, $u(t)$ is not a generalized coordinate, but rather a *control variable*, also known as an *input* to the system, the only generalized coordinate thus being θ. Moreover, the cart undergoes pure translation, and so, its kinematics is rather trivial. The rod, however, undergoes both translation and rotation. Let $\dot{c}$ denote the velocity of the c.o.m of the rod, and ω its scalar angular velocity. We would like to have both $\dot{c}$ and ω as linear functions of $\dot{\theta}$, but, just as in Example 1.6.4, we do not actually need $\dot{c}$ itself, but rather its magnitude-squared, which turns out to be much simpler to derive, as we will show presently. With regard to $\dot{c}$, then, at this stage we only set up the velocity triangle of Fig. 1.19, from which we will derive later the desired magnitude-squared. In that triangle, $\mathbf{v}_O$ and $\mathbf{v}_{C/O}$ denote, respectively, the velocity of point O and the relative velocity of C with respect to O. Hence,

$$\dot{c} = \mathbf{v}_O + \mathbf{v}_{C/O}, \quad \omega = \dot{\theta}$$

2. **Kinetic energy**. Here, we will need $\|\dot{\mathbf{c}}\|^2$, which is readily obtained from the velocity triangle as

$$\|\dot{\mathbf{c}}\|^2 = \frac{l^2}{4}\dot{\theta}^2 - l(\cos\theta)\dot{u}\dot{\theta} + \dot{u}^2$$

Now, let T_c and T_r denote the kinetic energies of the cart and the rod, respectively, i.e.,

$$T_c = \frac{1}{2}M\dot{u}^2$$

$$T_r = \frac{1}{2}m\left[\frac{1}{4}l^2\dot{\theta}^2 - l(\cos\theta)\dot{u}\dot{\theta} + \dot{u}^2\right] + \frac{1}{2}\frac{1}{12}ml^2\dot{\theta}^2$$

$$= \frac{1}{2}m\left[\frac{1}{3}l^2\dot{\theta}^2 - l(\cos\theta)\dot{u}\dot{\theta} + \dot{u}^2\right]$$

and hence,

$$T = \frac{1}{2}m\left[\frac{1}{3}l^2\dot{\theta}^2 - l(\cos\theta)\dot{u}\dot{\theta}\right] + \frac{1}{2}(M+m)\dot{u}^2$$

which contains one quadratic term in the generalized speed, one linear term in the same variable, and one independent term. Compared with Eq. 1.27, the foregoing expression yields

$$p(\theta,t) = -\frac{1}{2}ml(\cos\theta)\dot{u}(t), \quad T_0(t) = \frac{1}{2}(M+m)\dot{u}^2$$

where the physical interpretation of $p(\theta,t)$ becomes apparent: it is the angular momentum of a particle of mass m, moving with a velocity $\dot{u}(t)$, with respect to a point lying a distance $(1/2)l\cos\theta$ from the particle. Note that the latter is the difference of level between C and O. Likewise, T_0 represents the kinetic energy of a particle of mass $M+m$ moving with a velocity $\dot{u}(t)$.

3. **Potential energy**. This is only gravitational and pertains to the rod, the potential energy of the cart remaining constant, and, therefore, can be assumed to be zero. Hence, if we use the level of the pin as a reference,

$$V_g = -mg\frac{l}{2}\cos\theta$$

4. **Lagrangian**. This is simply

$$L = \frac{1}{2}m\left[\frac{1}{3}l^2\dot{\theta}^2 - l(\cos\theta)\dot{u}\dot{\theta}\right] + \frac{1}{2}(M+m)\dot{u}^2 + mg\frac{l}{2}\cos\theta$$

5. **Power supplied**. Again, we have no driving force, and hence, $\Pi = 0$.
6. **Power dissipation**. Here, the only sink of energy occurs in the pin, and hence,

$$\Delta = \frac{1}{2}c\dot{\theta}^2$$

7. **Lagrange equations**. Now it is a simple matter to calculate the partial derivatives of the foregoing functions:

$$\frac{\partial L}{\partial \dot{\theta}} = \frac{1}{3}ml^2\dot{\theta} - \frac{1}{2}ml(\cos\theta)\dot{u}$$

$$\frac{d}{dt}\left(\frac{\partial L}{\partial \dot{\theta}}\right) = \frac{1}{3}ml^2\ddot{\theta} + \frac{1}{2}ml(\sin\theta)\dot{u}\dot{\theta} - \frac{1}{2}ml(\cos\theta)\ddot{u}$$

$$\frac{\partial L}{\partial \theta} = \frac{1}{2}ml(\sin\theta)\dot{u}\dot{\theta} - \frac{1}{2}mgl\sin\theta$$

In order to complete the Lagrange equations, we need only the generalized force. Since the system is driven under a controlled motion, the active component of the generalized force stems from the Lagrangian, and will be made apparent when we set up the governing equations, the dissipative component being linear in $\dot{\theta}$, i.e.,

$$\phi_d = -c\dot{\theta}$$

Therefore, the governing equation becomes

$$\frac{1}{3}ml^2\ddot{\theta} - \frac{1}{2}ml(\cos\theta)\ddot{u} + \frac{1}{2}mgl\sin\theta = -c\dot{\theta}$$

that can be rearranged in the form

$$\frac{1}{3}ml^2\ddot{\theta} = -\frac{1}{2}mgl\sin\theta + \frac{1}{2}ml(\cos\theta)\ddot{u} - c\dot{\theta}$$

Now we can readily identify the generalized mass as

$$m(\theta) = \frac{1}{3}ml^2$$

which, again, has units of moment of inertia. Moreover, this quantity turns out to be, in this case, constant and is identical to the moment of inertia of the rod about point O. Likewise,

$$h(\theta, \dot{\theta}) = 0$$

and so, the system at hand contains neither Coriolis nor centrifugal forces. Finally, the right-hand side is composed of three terms: (1) a gravity term $\phi_p(\theta)$ that is solely a function of θ, but not of $\dot{\theta}$; (2) a generalized active force $\phi_m(\theta, t)$ that is provided by a motion-controlled source, i.e., the motor driving the cart with a controlled displacement $u(t)$; and (3) a dissipative term. All these terms are displayed below:

$$\phi_p(\theta) = -\frac{1}{2}mgl\sin\theta, \quad \phi_m(\theta, t) = \frac{1}{2}ml(\cos\theta)\ddot{u}, \quad \phi_f = 0, \quad \phi_d(\dot{\theta}) = -c\dot{\theta}$$

Apparently, the motion of the slider induces an active force $(1/2)ml(\cos\theta)\ddot{u}$ onto the system. This force, however, is not derived from a power function, because the system at hand is not force-driven, but rather motion-driven. As well, note that the gravity term now bears a negative sign, whereas in Example 1.6.4, a positive sign. The reader is invited to comment on the reason for this difference in signs.

Example 1.6.6 (A Force-driven Overhead Crane). Describe how the mathematical model of Example 1.6.5 changes if the pin is substituted by a motor that supplies a controlled torque $\tau(t)$, in order to control the orientation of the rod for purposes of manipulation tasks. Here, we assume that this torque is accompanied by a dissipative torque linear in $\dot{\theta}$, as in Example 1.6.5.

Solution: The motor-supplied torque now produces a power $\Pi = \tau(t)\dot{\theta}(t)$ onto the system, the generalized force now containing an *active* controlled-force component $\phi_f(\theta,\dot{\theta},t)$, namely,

$$\phi_f(\theta,\dot{\theta},t) = \frac{\partial \Pi}{\partial \dot{\theta}} = \tau(t)$$

all other terms remaining the same. Therefore, the governing equation becomes now

$$\frac{1}{3}ml^2\ddot{\theta} = -\frac{1}{2}mgl\sin\theta + \tau(t) + \frac{1}{2}ml(\cos\theta)\ddot{u} - c\dot{\theta}$$

Note that this system has two inputs, $u(t)$ and $\tau(t)$, but, apparently, one single dof. Nevertheless, the system can still be driven without conflict between the two inputs because $u(t)$ actually drives the second degree of freedom of the system, that of the cart translation. If the motion $\theta(t)$, instead of $\tau(t)$, were the second input, then the system would lose its sole degree of freedom, i.e., it would no longer be dynamical.

In particular, for *linear systems*, which will be the focus of the book, h vanishes, because it contains higher-order terms. Moreover, ϕ_p becomes the product of a constant times q, this constant taking positive, negative, or even zero values. If it is positive, then it represents the stiffness of the system. Note that this stiffness is, in general, a combination of the stiffnesses of the individual springs of the system and includes gravity effects as well.

Example 1.6.7 (An Eccentric Circular Plate). Shown in Fig. 1.20a is an eccentric circular plate whose c.o.m C is located a distance d from the center of its circular bore of radius a, as depicted in that figure. Moreover, the plate rolls on the pin of radius b without slipping. Under the above conditions, derive the Lagrange equation that governs the oscillations of the plate, with ψ as generalized coordinate.

Solution: Again, we proceed in seven steps to derive the Lagrange equation sought.

1. **Kinematics.** Consider the position of the plate shown with a dashed circle in Fig. 1.20b, in which the point of contact Q_0 of the plate with the pin is aligned with the center of the bore, O_0 and the center of mass of the plate, C_0. The three

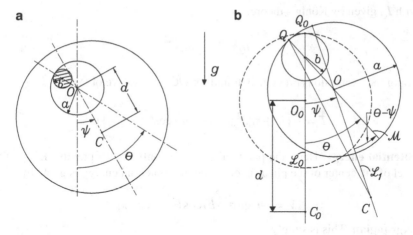

Fig. 1.20 An eccentric plate suspended from a pin

foregoing points thus define a vertical line $\mathscr{L}_0$ of the plate. Moreover, considering a general position of the plate, as the one shown with a solid line in the same figure, line $\mathscr{M}$ is defined as passing through the instantaneous contact point Q and the centroid of the bore O, and is, consequently, not fixed to the plate. Furthermore, $\mathscr{L}_1$ denotes line $\mathscr{L}_0$ in the general position. The angles ψ and θ are defined as the angles made by lines $\mathscr{L}_1$ and $\mathscr{M}$, respectively, with the vertical, as shown in Fig. 1.20b. The angular velocity of the plate is identical to that of the ring of Example 1.6.1 and corresponds to ψ. So, we have

$$\omega = \dot{\psi} = \frac{a-b}{a}\dot{\theta}$$

At the dashed position of the plate, we have $\psi = \theta = 0$. Therefore, after integration of the above relation, we obtain

$$\psi = \frac{a-b}{a}\theta \quad \text{or} \quad \theta = \frac{a}{a-b}\psi$$

To obtain the kinetic energy expression, the distance $\overline{QC}$ is required and is now obtained. With the aid of triangle QOC of Fig. 1.20b, this distance is readily obtained, using the cosine law, as

$$\overline{QC}^2 = a^2 + d^2 + 2ad\cos(\theta - \psi)$$

2. **Kinetic energy.** Here, we use Eq. 1.25 to obtain the kinetic energy of the plate based on the contact point Q, that is instantaneously fixed to the pin and hence, to an inertial frame. Thus

$$T = \frac{1}{2}I_Q\omega^2$$

with I_Q given by König's theorem as

$$I_Q = I_C + m\overline{QC}^2$$

Using the expressions previously found for $\overline{QC}^2$ and ω, we obtain

$$T = \frac{1}{2}\{I_C + m[a^2 + d^2 + 2ad\cos(\theta - \psi)]\}\dot{\psi}^2$$

3. **Potential energy**. Here, the potential energy is only due to gravity. Taking the level of the center of the pin as a reference, the potential energy is given by

$$V = -mg[(a - b)\cos\theta + d\cos\psi]$$

4. **Lagrangian**. This is simply

$$L = T - V = \frac{1}{2}\{I_C + m[a^2 + d^2 + 2ad\cos(\theta - \psi)]\}\dot{\psi}^2$$
$$+ mg[(a - b)\cos\theta + d\cos\psi]$$

5. **Power supplied**. Again, we have no driving force, and hence, $\Pi = 0$.
6. **Power dissipation**. Apparently, the system is not subjected to any *working* friction force,[14] and hence, $\Delta = 0$.
7. **Lagrange equations**. Now, all we need is to evaluate the partial derivatives involved in Eq. 1.22, namely,

$$\frac{\partial L}{\partial \dot{\psi}} = \{I_C + m[a^2 + d^2 + 2ad\cos(\theta - \psi)]\}\dot{\psi}$$

$$\frac{d}{dt}\left(\frac{\partial L}{\partial \dot{\psi}}\right) = \{I_C + m[a^2 + d^2 + 2ad\cos(\theta - \psi)]\}\ddot{\psi}$$
$$-2mad\left(\frac{d\theta}{d\psi} - 1\right)\sin(\theta - \psi)\dot{\psi}^2$$

$$\frac{\partial L}{\partial \psi} = -mad\left(\frac{d\theta}{d\psi} - 1\right)\sin(\theta - \psi)\dot{\psi}^2 - mg(a - b)\frac{d\theta}{d\psi}\sin\theta - mgd\sin\psi$$

Furthermore,

$$\frac{\partial \Pi}{\partial \dot{\psi}} = \frac{\partial \Delta}{\partial \dot{\psi}} = 0$$

---

[14] Actually, friction force between pin and plate is present, but it develops no work because the point of application Q of this force is stationary. The sole role of the friction force here is to prevent sliding.

the equation sought thus being

$$\{I_C + m\left[a^2 + d^2 + 2ad\cos(\theta - \psi)\right]\}\ddot{\psi} - mad\left(\frac{d\theta}{d\psi} - 1\right)\sin(\theta - \psi)\dot{\psi}^2$$

$$+ mg(a - b)\frac{d\theta}{d\psi}\sin\theta + mgd\sin\psi = 0$$

Now, from the relation between θ and ψ,

$$\frac{d\theta}{d\psi} = \frac{a}{a - b}, \quad \frac{d\theta}{d\psi} - 1 = \frac{b}{a - b}$$

whence the Lagrange equation can be rearranged as

$$\{I_C + m[a^2 + d^2 + 2ad\cos(\theta - \psi)]\}\ddot{\psi}$$

$$- mad\frac{b}{a - b}\sin(\theta - \psi)\dot{\psi}^2 = -mg(a\sin\theta + d\sin\psi)$$

The generalized mass can be readily identified as

$$m(\psi) = I_C + m[a^2 + d^2 + 2ad\cos(\theta - \psi)]$$

which has units of moment of inertia and is configuration-dependent.
    Likewise, we can readily identify the term of Coriolis and centrifugal forces as

$$h(\psi, \dot{\psi}) = -mad\frac{b}{a - b}\sin(\theta - \psi)\dot{\psi}^2$$

thus obtaining, again, an expression quadratic in the generalized speed $\dot{\psi}$ for this term. Finally, the right-hand side is composed of a sole term that is due to gravity forces; we thus have

$$\phi_p(\psi) = -mg(a\sin\theta + d\sin\psi)$$

while

$$\phi_m = \phi_f = \phi_d = 0$$

Example 1.6.8 (A Simplified Model of an Actuator). Shown in Fig. 1.21 is the iconic model of an actuator used to rotate a load—e.g., a control surface in an aircraft, a robot link, a door, or a valve—of mass m, represented by link AB of length a, about a point A. The driving mechanism consists of three elements in parallel, namely, a linear spring of stiffness k, a linear dashpot of coefficient c, and a hydraulic cylinder exerting a controlled force $F(t)$, to lower or raise the load. Under the assumption that the spring is unloaded when $s = a$, and that the link is a slender rod, find the

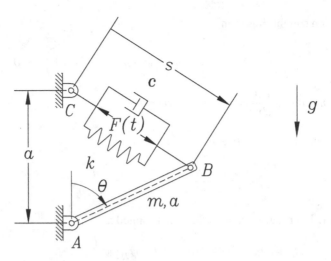

Fig. 1.21 The iconic model of an actuator

mathematical model of the foregoing system in terms of θ, which is to be used as the generalized coordinate in this example.

Solution: We proceed as in the previous cases, i.e., following the usual seven-step procedure:

1. **Kinematics**. This step is quite simple, for the only element capable of storing kinetic energy is the load, which undergoes a pure rotation about point A, and hence, its kinetic energy can be derived in terms of its angular velocity ω only; the angular velocity is given simply by

$$\omega = \dot{\theta}$$

where we have adopted the positive direction of θ, and hence, of ω, clockwise. Besides the foregoing item, we will need a relation between the length of the spring, s, and the generalized variable, θ. This is readily found from the geometry of Fig. 1.21, namely,

$$s = 2a\sin\left(\frac{\theta}{2}\right)$$

2. **Kinetic energy**. Here, we use Eq. 1.25 to obtain the kinetic energy of the load based on the center A of the pivot fixed to the inertial frame. Thus,

$$T = \frac{1}{2}I_A\omega^2 = \frac{1}{2}I_A\dot{\theta}^2$$

with I_A given by

$$I_A = \frac{1}{3}ma^2$$

Thus,

$$T = \frac{1}{6}ma^2\dot{\theta}^2$$

3. **Potential energy**. We now have both gravitational and elastic forms of potential energy. If we take the horizontal position of AB as the reference level to measure the potential energy due to gravity, then

$$V = mg\frac{a}{2}\cos\theta + \frac{1}{2}k(s-a)^2$$

or, in terms of the generalized coordinate alone,

$$V = mg\frac{a}{2}\cos\theta + \frac{1}{2}k\left[2a\sin\left(\frac{\theta}{2}\right) - a\right]^2$$

4. **Lagrangian**. This is simply

$$L = T - V = \frac{1}{6}ma^2\dot{\theta} - mg\frac{a}{2}\cos\theta - \frac{1}{2}k\left[2a\sin\left(\frac{\theta}{2}\right) - a\right]^2$$

5. **Power supplied**. The system is driven under a controlled force $F(t)$ that is applied at a speed $\dot{s}$, and hence, the power supplied to the system is

$$\Pi = F(t)\dot{s}$$

In order to express the foregoing power in terms of the generalized speed $\dot{s}$, we differentiate the relation between s and θ with respect to time:

$$\dot{s} = a\dot{\theta}\cos\left(\frac{\theta}{2}\right)$$

thereby obtaining the desired expression:

$$\Pi = F(t)a\dot{\theta}\cos\left(\frac{\theta}{2}\right)$$

6. **Power dissipation**. Power is dissipated only by the dashpot of the hydraulic cylinder, and hence,

$$\Delta = \frac{1}{2}c\dot{s}^2 = \frac{1}{2}ca^2\dot{\theta}^2\cos^2\left(\frac{\theta}{2}\right)$$

7. **Lagrange equations**. We first evaluate the partial derivatives of the Lagrangian:

$$\frac{\partial L}{\partial \dot\theta} = \frac{1}{3}ma^2\dot\theta, \quad \Rightarrow \quad \frac{d}{dt}\left(\frac{\partial L}{\partial \dot\theta}\right) = \frac{1}{3}ma^2\ddot\theta$$

$$\frac{\partial L}{\partial \theta} = \frac{a}{2}mg\sin\theta - ka^2\left[2\sin\left(\frac{\theta}{2}\right) - 1\right]\cos\left(\frac{\theta}{2}\right)$$

$$= amg\sin\left(\frac{\theta}{2}\right)\cos\left(\frac{\theta}{2}\right) - ka^2\left[2\sin\left(\frac{\theta}{2}\right) - 1\right]\cos\left(\frac{\theta}{2}\right)$$

where we have written all partial derivatives in terms of angle $\theta/2$. Furthermore,

$$\frac{\partial \Pi}{\partial \dot\theta} = F(t)a\cos\left(\frac{\theta}{2}\right), \quad \frac{\partial \Delta}{\partial \dot\psi} = ca^2\dot\theta\cos^2\left(\frac{\theta}{2}\right)$$

the equation sought thus being

$$\frac{1}{3}ma^2\ddot\theta = -a\left\{ka\left[2\sin\left(\frac{\theta}{2}\right) - 1\right] - mg\sin\left(\frac{\theta}{2}\right)\right\}\cos\left(\frac{\theta}{2}\right)$$

$$+F(t)a\cos\left(\frac{\theta}{2}\right) - ca\dot\theta\cos\left(\frac{\theta}{2}\right)$$

The generalized mass can be readily identified from the above model as

$$m(\theta) = \frac{1}{3}ma^2$$

which is nothing but the moment of inertia of the bar AB about the pivot center A. Likewise, we can readily identify the term of Coriolis and centrifugal forces as zero, which is understandable because of the simple motion undergone by the bar. Note that the centrifugal inertia force of the bar is directed toward point A, i.e., this force is applied at a point of zero velocity, and hence, does no work on the system, for which reason it does not appear in the mathematical model. Also note that the generalized-force terms are identified below as:

$$\phi_p(\theta) = -a\left[(2ka - mg)\sin\left(\frac{\theta}{2}\right) - ka\right]\cos\left(\frac{\theta}{2}\right)$$

while

$$\phi_d(\theta) = -ca\dot\theta\cos\left(\frac{\theta}{2}\right), \quad \phi_f(\theta) = F(t)a\cos\left(\frac{\theta}{2}\right), \quad \phi_m = 0$$

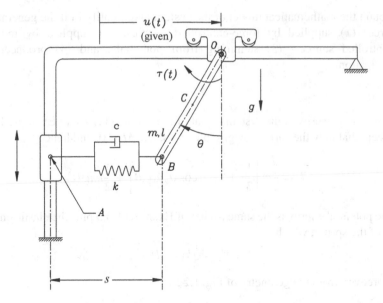

Fig. 1.22 A simple model of an aircraft control surface and its actuation mechanism

for the system is not driven by any motion-controlled source of energy. Note that the mechanical interpretation of the generalized force is the moment of the actuator force $F(t)$ about point A.

Finally, the equation of motion can be rearranged as

$$\ddot{\theta} = -\frac{3}{ma}\left[(2ka - mg)\sin\left(\frac{\theta}{2}\right) - ka\right]\cos\left(\frac{\theta}{2}\right)$$
$$+\frac{3F(t)}{ma}\cos\left(\frac{\theta}{2}\right) - \frac{3c}{ma}\dot{\theta}\cos\left(\frac{\theta}{2}\right)$$

Example 1.6.9 (Motion-driven Control Surface). Shown in Fig. 1.22 is a highly simplified model of the actuator mechanism of an aircraft control surface—e.g., ailerons, rudder, etc. In the model, a massless slider is positioned by a stepper motor at a displacement $u(t)$. The inertia of the actuator-aileron system is lumped in the rigid, slender, uniform bar of length l and mass m, while all the stiffness and damping is lumped in a parallel spring-dashpot array whose left end A is pinned to a second massless slider that can slide without friction on a vertical guideway. Moreover, it is known that the spring is unloaded when $u = l$ and $\theta = 0$. A second motor, mounted on the first slider, exerts a torque $\tau(t)$ on the bar.

(a) Derive the Lagrangian of the system.
(b) Give expressions for the power Π supplied to the system and for the dissipation function Δ.

(c) Obtain the mathematical model of the system and identify in it the generalized
   forces (1) supplied by force-controlled sources; (2) supplied by motion-
   controlled sources; (3) stemming from potentials; and (4) produced by
   dissipation.

Solution:

(a) The kinetic energy of the system is that of the overhead crane of Example 1.6.5,
   except that now the cart has negligible mass, i.e., $M = 0$, and hence,

$$T = \frac{1}{2}m\left[\frac{1}{3}l^2\dot{\theta}^2 - l(\cos\theta)\dot{u}(t)\dot{\theta}\right] + \frac{1}{2}m\dot{u}^2(t)$$

The potential energy is the same as that of Example 1.6.5 plus the elastic energy
V_e of the spring, which is

$$V_e = \frac{1}{2}k(s - l)^2$$

Moreover, from the geometry of Fig. 1.22,

$$s = u(t) - l\sin\theta$$

whence,

$$V = -\frac{1}{2}mgl\cos\theta + \frac{1}{2}k[u(t) - l\sin\theta - l]^2$$

and

$$L = \frac{1}{2}m\left[\frac{1}{3}l^2\dot{\theta}^2 - l(\cos\theta)\dot{u}(t)\dot{\theta}\right] + \frac{1}{2}m\dot{u}^2(t) + mg\frac{l}{2}\cos\theta - \frac{1}{2}k[u(t) - l\sin\theta - l]^2$$

(b)

$$\Pi = \tau(t)\dot{\theta}, \quad \Delta = \frac{1}{2}c\dot{s}^2$$

and $\dot{s}$ is computed by differentiation of the foregoing expression for s:

$$\dot{s} = \dot{u}(t) - l\dot{\theta}\cos\theta$$

Therefore,

$$\Delta = \frac{1}{2}c[\dot{u}(t) - l\dot{\theta}\cos\theta]^2$$

(c)

$$\frac{\partial L}{\partial\dot{\theta}} = \frac{1}{3}ml^2\dot{\theta} - \frac{1}{2}ml(\cos\theta)\dot{u}(t)$$

$$\frac{d}{dt}\left(\frac{\partial L}{\partial\dot{\theta}}\right) = \frac{1}{3}ml^2\ddot{\theta} + \frac{1}{2}ml(\sin\theta)\dot{u}(t)\dot{\theta} - \frac{1}{2}ml(\cos\theta)\ddot{u}(t)$$

$$\frac{\partial L}{\partial\theta} = \frac{1}{2}ml(\sin\theta)\dot{u}(t)\dot{\theta} - \frac{1}{2}mgl\sin\theta - k[u(t) - l\sin\theta - l](-l\cos\theta)$$

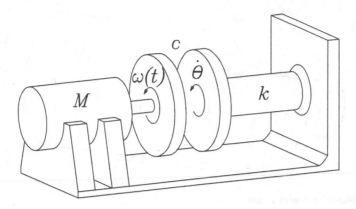

Fig. 1.23 A fluid clutch

$$\frac{\partial L}{\partial \theta} = \frac{1}{2}ml(\sin\theta)\dot{u}(t)\dot{\theta} - \frac{1}{2}mgl\sin\theta + kl[u(t) - l\sin\theta - l]\cos\theta$$

$$\frac{\partial \Pi}{\partial \dot{\theta}} = \tau(t)$$

$$\frac{\partial \Delta}{\partial \dot{\theta}} = c[\dot{u}(t) - l\dot{\theta}\cos\theta](-l\cos) = cl[l\dot{\theta}\cos\theta - u(t)]\cos\theta$$

Thus, the mathematical model is derived as

$$\frac{1}{3}ml^2\ddot{\theta} + \frac{1}{2}ml(\sin\theta)\dot{u}(t)\dot{\theta} - \frac{1}{2}ml(\cos\theta)\ddot{u}(t) - \frac{1}{2}ml(\sin\theta)\dot{u}(t)\dot{\theta} + \frac{1}{2}mgl\sin\theta$$

$$-kl[u(t) - l\sin\theta - l]\cos\theta = \tau(t) - cl[l\dot{\theta}\cos\theta - u(t)]\cos\theta$$

or,

$$\frac{1}{3}ml^2\ddot{\theta} = \overbrace{\frac{1}{2}ml(\cos\theta)\ddot{u}(t) + cl\dot{u}(t)\cos\theta + klu(t)\cos\theta}^{(ii)}$$

$$\underbrace{-\frac{1}{2}mgl\sin\theta - kl^2(1+\sin\theta)\cos\theta}_{(iii)} + \overbrace{\tau(t)}^{(i)} - \overbrace{cl^2\dot{\theta}\cos^2\theta}^{(iv)}$$

Example 1.6.10 (A First-order Model—A Hydraulic Clutch). Given in Fig. 1.23 is the iconic model of a hydraulic clutch undergoing tests. In this setting, a motor drives the left disk of the clutch at a controlled rate $\omega(t)$, while the right disk is coupled to a linearly elastic shaft of torsional stiffness k and negligible inertia. Moreover, the right end of the shaft is clamped to a fixed wall. Derive the mathematical model of the device.

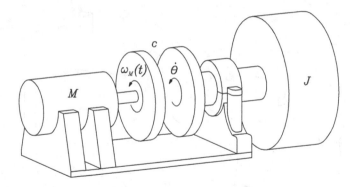

Fig. 1.24 Fluid clutch with attached rotor

Solution: Similar to Example 1.6.5, this system is motion-driven, and hence, $\omega(t)$ is an input to the system, rather than a generalized speed. Moreover, from the conditions of the problem, it is apparent that the system cannot store kinetic energy, its only energy-storing capability being in the form of elastic potential energy. The fluid in the clutch, on the other hand, dissipates energy and is, hence, equivalent to a dashpot that, to keep the model simple, is assumed to be linear. Note also that there is no power input to the system stemming from a controlled-force source. Thus,

$$T = 0, \quad V = \frac{1}{2}k\theta^2, \quad \Pi = 0, \quad \Delta = \frac{1}{2}c(\omega - \dot{\theta})^2$$

With these energy functions already derived, we proceed now to obtain the Lagrange equation of the system, which leads to a mathematical model in the form

$$\dot{\theta} = -\frac{k}{c}\theta + \omega(t)$$

Note that this is a first-order ODE. Similar to second-order systems, we can identify, in this case, a 'generalized force' stemming from a potential-energy function on the right-hand side, and one stemming from a controlled-motion source. However, on the left-hand side we do not have a generalized mass, because no generalized acceleration occurs in this equation, which is the result of the inability of the system to store kinetic energy.

Example 1.6.11 (One More First-order Model). We now unclamp the right disk of the clutch of Fig. 1.23 and attach it rigidly to a rotor of moment of inertia J, thereby ending up with the system of Fig. 1.24. Furthermore, we assume in this case that the shaft is rigid, and derive the mathematical model governing the motion of the rotor in terms of the generalized coordinate θ.

Solution: In this case, inertia is no longer neglected, but the energy-storing capability of the shaft is. Thus, the kinetic energy is nonzero in this case, but the potential energy vanishes. Otherwise, the conditions remain as in Example 1.6.10. The energy functions are, thus,

$$T = \frac{1}{2}J\dot{\theta}^2, \quad V = 0, \quad \Pi = 0, \quad \Delta = \frac{1}{2}c(\dot{\theta} - \omega)^2$$

and hence, the Lagrange equation of the system leads to

$$J\ddot{\theta} = c\omega(t) - c\dot{\theta}$$

where the generalized mass is the moment of inertia of the rotor. Moreover, in the right-hand side we can readily identify a generalized force stemming from a controlled-motion source, namely, the first term, the second term being a dissipative generalized force. Now, if we let $\dot{\theta} = \omega_R$ and divide both sides of the foregoing equation by J, we obtain

$$\dot{\omega}_R + \frac{c}{J}\omega_R = \frac{c}{J}\omega(t)$$

which is, again, a first-order ODE but, this time, in a generalized speed, rather than in a generalized coordinate.

1.7 Hysteretic Damping

Hysteretic damping occurs in solids by virtue of internal friction, which arises, in turn, when the solid undergoes deformations that vary with time. This kind of damping can be accounted for by the introduction of an equivalent dashpot coefficient whose numerical value is obtained via experiments. In these experiments, the structural element whose hysteresis is to be accounted for by means of an equivalent dashpot is subjected to harmonic loads that induce correspondingly harmonic deformations. This experiment can be performed on a machine commonly used for fatigue tests, various kinds of which are commercially available. Tests performed on structural elements have shown that, when the element undergoes a series of cycles of harmonic deformations, the energy dissipated by virtue of internal friction is proportional to the square of the amplitude A of the harmonic deformation, i.e., if we denote the dissipation of energy per cycle by E_c, then we have

$$E_c = \alpha A^2 \tag{1.40a}$$

where α is a factor with units of stiffness, i.e., N/m, if A has units of meter; this factor is determined experimentally.

Now, in order to find the equivalent dashpot coefficient c_{eq}, what we do is calculate the energy dissipated per cycle when the linear dashpot is subjected to a harmonic motion of the same amplitude and the same frequency. We shall show in

Chap. 2. that the energy dissipated per cycle by a mass-spring-dashpot system, with a dashpot coefficient c_{eq}, is

$$E_c = c_{eq}\pi\omega A^2 \tag{1.40b}$$

in which ω is the frequency of the excitation signal. Upon equating the two expressions for E_c in Eqs. 1.40a and b, we can readily solve for c_{eq}, namely,

$$c_{eq} = \frac{\alpha}{\pi\omega} \tag{1.41}$$

1.8 Coulomb Damping

Coulomb damping is more difficult to model than viscous damping. In fact, the mechanism by which power is dissipated when two solids move one with respect to the other in direct contact, i.e., without an intermediate layer of lubricant, is extremely complex. Simple models have been proposed to account for Coulomb friction, the simplest one assuming that the friction force is of a *saturation* type and always opposes either relative motion or the trend towards it. We assume that we have two bodies A and B undergoing planar motion, in contact at a common plane, and their velocities, denoted by scalars v_A and v_B, are parallel to the plane of motion and to the contact plane, as shown in Fig. 1.25. Note that the vector representation of these velocities is not needed because their directions are assumed constant. The relative velocity of A with respect to B, when the two bodies move in pure translation, is $v_R \equiv v_A - v_B$. The Coulomb friction force exerted by body B onto body A acts in the direction opposite to the relative velocity and can then be approximated as $f_{BA} = -\mu N \text{sgn}(v_R)$, where μ is the coefficient of *dynamic* friction, N is the normal force exerted by one body onto the other, and $\text{sgn}(\cdot)$ represents the *signum* function, defined as $+1$ or -1, depending on whether the argument is positive or negative. If the argument is zero, the signum function is so far undefined, but this does not bother us for, in this case, the friction force can take on any value between $-\mu N$ and $+\mu N$. For concreteness, let us assume that we have a body with the form of a block, resting on a horizontal surface, in direct contact with the surface; besides,

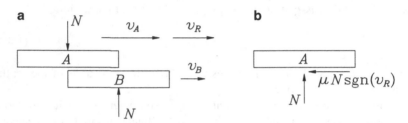

Fig. 1.25 Two bodies in relative motion under direct contact

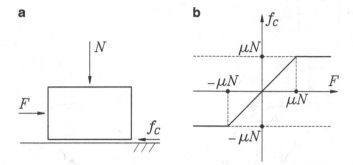

Fig. 1.26 Relation between applied force and Coulomb friction force

a vertical force N presses the block against the surface. If we now apply a force F varying from 0 to any value, whether positive or negative, as indicated in Fig. 1.26a, then we observe that the block remains at rest until a saturation level is reached. Beyond this level, the block starts sliding in the direction of the force. Moreover, the force balancing the block when at rest, denoted by f_C, then, accommodates to the value of F, i.e., $f_C = F$, until saturation is reached. Once the threshold value at saturation has been reached, experience tells that f_C remains virtually constant. In reality, f_C shows a decrease as the relative velocity of the block with respect to the surface increases. For purposes of our study, however, we will assume that this force remains constant. We therefore assume that the relation between F and f_C is of the *saturation* type shown in Fig. 1.26b, i.e.,

$$f_C = \mu N \mathrm{sat}\left(\frac{F}{\mu N}\right) \tag{1.42}$$

where $\mathrm{sat}(x)$ is the *saturation function*, defined as

$$\mathrm{sat}(x) \equiv \begin{cases} x, & \text{for} \quad |x| \le 1; \\ \mathrm{sgn}(x), & \text{for} \quad |x| \ge 1. \end{cases} \tag{1.43}$$

with x being a dimensionless, unbounded real variable.

The power dissipated by the Coulomb friction force, Π_C, with the definitions of Fig. 1.25, is then readily calculated as

$$\Pi_C = -f_{BA} v_R = -\mu N |v_R| \tag{1.44}$$

Note that, if v_R is identical to the generalized speed $\dot{q}$, and the power Π_C is defined as the associated Coulomb dissipation function, then the friction force f_{BA} can be calculated as the partial derivative of Π_C with respect to $\dot{q}$, for we have, for any real number x,

$$\frac{\partial |x|}{\partial x} = \mathrm{sgn}(x) \tag{1.45}$$

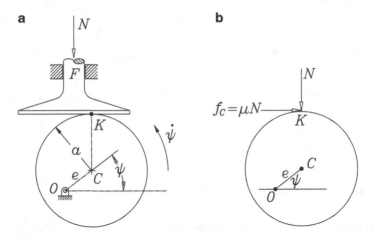

Fig. 1.27 Cam-follower mechanism with Coulomb friction

In this case the dissipation function is just the negative of the dissipated power, i.e.,

$$\Delta = \mu N |v_R|$$

Example 1.8.1 (Coulomb Friction Cum Geometric Nonlinearity). Illustrated in Fig. 1.27a is a cam plate with a flat face follower. The cam is an eccentric circular plate with radius a and center C that rotates at an angular speed $\dot{\psi}$ about point O a distance e from C. In the absence of lubricant, a Coulomb friction force, proportional to the normal force N, is developed at the contact point, the friction coefficient being μ. Moreover, angle ψ denotes the orientation of a line fixed to the cam with respect to a line fixed to the frame. Derive an expression for the generalized friction force thus developed, while using angle ψ as the generalized coordinate.

Solution: If the cam plate rotates ccw, the friction force f_C is directed to the right, as shown in Fig. 1.27b; otherwise, f_C is directed to the left. In any event, the magnitude of the force is μN; hence, the friction force is

$$f_C = -\mu N \mathrm{sgn}(v_R)$$

where v_R is the relative velocity of the cam plate with respect to the follower F at the contact point K in the direction of f_C. As such, this quantity can be positive or negative. We will designate v_R as positive when it produces a friction force onto the cam that is directed to the right, i.e., v_R is positive when it points to the left.

We now determine v_R. To this end, let $\mathbf{v}_{KC}$ and $\mathbf{v}_{KF}$ denote the vector velocities of the contact point K as belonging to the cam and the follower, respectively. Moreover, the relative-velocity vector $\mathbf{v}_R$ of point KC of the cam with respect to point KF of the follower is given by

$$\mathbf{v}_R \equiv \mathbf{v}_{KC} - \mathbf{v}_{KF}$$

v_R being the signed magnitude of $\mathbf{v}_R$, which is positive when $\mathbf{v}_R$ points to the left.

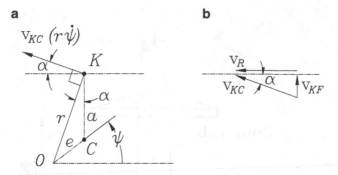

Fig. 1.28 Velocity relations of the cam-follower mechanism

From Fig. 1.28a, it is apparent that the magnitude of $\mathbf{v}_{KC}$ is $r\dot\psi$, with r defined as the distance from O to K. This distance can be determined from the same figure, of course, by the 'cosine law', but, as we will show presently, we will not need it. Moreover, we assume that there is neither separation nor penetration between cam plate and follower, the relative velocity $\mathbf{v}_R$ then having a zero component in the direction of the common normal. The velocity triangle thus resulting is shown in Fig. 1.28b. From this triangle, then,

$$v_R = r(\cos\alpha)\dot\psi$$

where $r\cos\alpha$ is readily found from Fig. 1.28a as

$$r\cos\alpha = a + e\sin\psi$$

Hence,

$$v_R = (a + e\sin\psi)\dot\psi$$

Therefore,

$$f_C = -\mu N \mathrm{sgn}\left[(a + e\sin\psi)\dot\psi\right] = -\mu N \mathrm{sgn}(\dot\psi)$$

where the factor $(a + e\sin\psi)$ has been deleted from the signum function because it is positive-definite, i.e., it is positive for any value of ψ, which is apparent from Fig. 1.27.

Now, the power dissipated by f_C is simply $\Pi_C = f_C v_R$ and, if we recall the expressions of f_C and v_R, then,

$$\Pi_C = -\mu N(a + e\sin\psi)[\mathrm{sgn}(\dot\psi)]\dot\psi$$

However, for any real number x,

$$[\mathrm{sgn}(x)]x = |x|$$

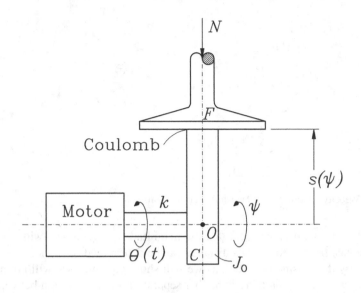

Fig. 1.29 Cam-follower mechanism with Coulomb friction

and so,

$$\Pi_C = -\mu N(a + e\sin\psi)|\dot{\psi}|$$

Moreover, if we recall relation (1.45), then

$$\frac{\partial \Pi_C}{\partial \dot{\psi}} = -\mu N(a + e\sin\psi)\mathrm{sgn}(\dot{\psi})$$

This means that, in this case, the power dissipated is simply the negative of the dissipation function Δ. Thus, if we denote by ϕ_C the generalized Coulomb force associated with the generalized coordinate ψ, we have

$$\phi_C = -\frac{\partial \Delta}{\partial \dot{\psi}} \equiv \frac{\partial \Pi_C}{\partial \dot{\psi}} = -\mu N(a + e\sin\psi)\mathrm{sgn}(\dot{\psi})$$

Example 1.8.2 (The Governing Equations of a Motor-cam Transmission). Now, assume that the cam mechanism of Fig. 1.27a is driven by a motor that delivers a controlled motion $\theta(t)$ at the left end of its elastic shaft of stiffness k, as shown in Fig. 1.29, the cam plate having a moment of inertia with respect to its axis of rotation J_O. Under the assumption that a *constant* load N acts on the follower, which has a negligible mass, and that contact between cam and follower is direct, i.e., without any intermediate lubricant, derive the mathematical model of the shaft-cam-follower system.

Solution: While the kinematics of this system is quite simple, we have to start by establishing any relations of this kind that we may need. As a matter of fact, all we need here is $s(\psi)$, which was derived in Example 1.8.1. For quick reference, we reproduce this relation below, accompanied by an expression for $\dot{s}$ in terms of $\dot{\psi}$:

$$s(\psi) = a + e\sin\psi, \quad \dot{s} = e\dot{\psi}\cos\psi$$

It is now a trivial matter to derive the Lagrangian of the system, namely,

$$L \equiv T - V = \frac{1}{2}J_O\dot{\psi}^2 - \frac{1}{2}k(\psi - \theta)^2$$

Moreover, the power supplied to the system from force-controlled sources is simply

$$\Pi = -N\dot{s} = -Ne\dot{\psi}\cos\psi$$

where the minus sign arises from the opposite directions of $\dot{s}$—which is upwards—and N—which is downwards. The dissipation function, in turn, was derived in Example 1.8.1, and is here repeated for quick reference:

$$\Delta = \mu N(a + e\sin\psi)|\dot{\psi}|$$

After performing all differentiations involved, the governing equation turns out to be

$$J_O\ddot{\psi} = k\psi + k\theta - Ne\cos\psi - \mu N(a + e\sin\psi)\mathrm{sgn}(\dot{\psi})$$

In this model, we can readily identify J_O as the generalized mass of the system, while $h = 0$. Furthermore, $\phi_p(\psi) = -k\psi$; $\phi_m(t) = k\theta(t)$; $\phi_f(\psi, \dot{\psi}) = -Ne\cos\psi$; and $\phi_d(\psi, \dot{\psi}) = -\mu N(a + e\sin\psi)\mathrm{sgn}(\dot{\psi})$. It should be pointed out that the sign of the ϕ_f term is alternating; this term becomes positive whenever $90° \leq \psi \leq 270°$. As such, this term acts alternately as a source and as a sink of energy. Therefore, the same term cannot stem from a dissipation function, which is invariably a sink of energy; it is accounted for in the power supplied to the system, and hence, when this power is negative, the term becomes one of power *extracted* from the system, and the load is said to be *backward-driven*; otherwise, the load is said to be *forward-driven*.

1.9 Equilibrium States of Mechanical Systems

Henceforth, a mechanical system will be said to be in an equilibrium state if both $\dot{q} = 0$ and $\ddot{q} = 0$. This condition implies that, at equilibrium, q attains a constant value q_E.

Now, given the form of $h(q,\dot{q})$, as displayed in Eq. 1.38a, it is apparent that, at equilibrium, h vanishes. Since the first term of Eq. 1.39a also vanishes at equilibrium, we derive the equilibrium equation in the form

$$\phi(q_E,0) = 0 \qquad (1.46)$$

Usually, q_E is not known, its value being found from the roots of the above equation.

Moreover, Eq. 1.46 is nonlinear, and hence, it may admit a solution or none. If it does admit a solution, usually this is not unique, which means that, in general, multiple equilibrium states are possible. Additionally, nonlinear equations admit closed-form solutions only occasionally. In general, *iterative* numerical methods are needed to find the equilibrium states, but, in many engineering applications, rough estimates of these roots suffice, which can, moreover, be obtained by inspection of a plot.

Example 1.9.1 (Equilibrium Analysis of the Overhead Crane). Find the equilibrium configuration(s) of the system introduced in Example 1.6.5 under the condition that the cart is driven with a constant acceleration $\ddot{u} = a$.

Solution: We first set $\ddot{u} = a$, $\ddot{\theta} = 0$ and $\dot{\theta} = 0$ in the governing equation derived in that example, thereby obtaining the equilibrium equation

$$-\frac{1}{2}mgl\sin\theta_E + \frac{1}{2}ml(\cos\theta_E)a = 0$$

which can be written as

$$\tan\theta_E = \frac{a}{g}$$

and hence,

$$\theta_E = \tan^{-1}\left(\frac{a}{g}\right)$$

which is a double-valued relation. The system, therefore, admits two equilibrium configurations, one with the rod above and one with the rod below the pin. These configurations will be referred to as the *rod-up* and *rod-down* configurations, the two θ values differing by $180°$. The foregoing relation among the constant horizontal acceleration a, the vertical gravity acceleration g and the equilibrium angle θ_E is best illustrated in Fig. 1.30, where the force triangle, composed of the pin force f_P, the weight of the rod mg, and the inertia force $-ma$, is sketched. Note that the same force triangle holds for the two equilibrium configurations.

Moreover, from Fig. 1.30, it is apparent that

$$\cos\theta_E = \frac{g}{\sqrt{a^2 + g^2}}, \quad \sin\theta_E = \frac{a}{\sqrt{a^2 + g^2}}$$

thereby completing the equilibrium analysis of the crane.

Fig. 1.30 Force triangle of
rod of overhead crane in
equilibrium

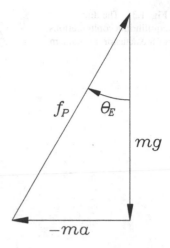

Example 1.9.2 (Equilibrium States of the Actuator Mechanism).   Determine *all* the
equilibrium states of the actuator mechanism introduced in Example 1.6.8. To this
end, assume that

$$\frac{mg}{ka} = 2 - \sqrt{2} \quad \text{and} \quad F(t) = 0$$

Solution: Upon setting $\theta = \theta_E$, $\dot{\theta} = 0$, and $\ddot{\theta} = 0$ in the mathematical model derived
in Example 1.6.8, the equilibrium equation is obtained as

$$\left[(2ka - mg) \sin\left(\frac{\theta_E}{2}\right) - ka \right] \cos\left(\frac{\theta_E}{2}\right) = 0$$

If we now introduce the given relation between the gravitational and the elastic
forces into the above equation, this simplifies to

$$\left[\sqrt{2} \sin\left(\frac{\theta_E}{2}\right) - 1 \right] \cos\left(\frac{\theta_E}{2}\right) = 0$$

or, even further,

$$\left[\sin\left(\frac{\theta_E}{2}\right) - \frac{\sqrt{2}}{2} \right] \cos\left(\frac{\theta_E}{2}\right) = 0$$

which vanishes under any of the conditions given below:

$$\sin\frac{\theta_E}{2} = \frac{\sqrt{2}}{2} \quad \text{or} \quad \cos\frac{\theta_E}{2} = 0$$

Fig. 1.31 The three
equilibrium configurations
of the actuator mechanism

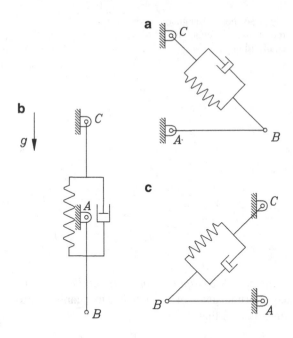

That is, equilibrium is reached whenever θ_E attains any of the three values given below:

$$\theta_E = 90°, \quad 180°, \quad 270°$$

The foregoing values yield the equilibrium configurations of Fig. 1.31.

Example 1.9.3 (Equilibrium Analysis of the Eccentric Plate). Here, we want to determine the equilibrium configuration(s) of the eccentric plate introduced in Example 1.6.7.

Solution: The equilibrium equation is obtained by setting $\theta = \theta_E$, $\psi = \psi_E$, $\dot{\psi} = 0$, and $\ddot{\psi} = 0$ in the governing equation, which leaves us only with

$$\phi_p(\psi_E) \equiv -mg[a\sin\theta_E + d\sin\psi_E] = 0$$

or

$$a\sin\theta_E + d\sin\psi_E = 0$$

The geometric interpretation of the above equation is straightforward if we look at Fig. 1.20. What this equation states is that the c.o.m of the plate must lie on the vertical of the center of the pin, a rather plausible result. This condition is thus fulfilled by

$$\psi_E = 0, \pi$$

If $\psi_E = 0$, then $\theta_E = 0$ from the geometric relation between θ and ψ, and the equilibrium condition is satisfied; however, if $\psi_E = \pi$, then $\theta_E = [a/(a-b)]\pi$ from

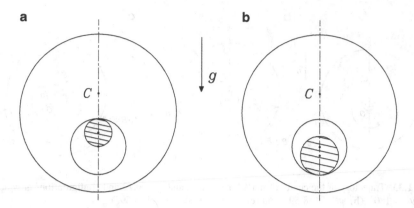

Fig. 1.32 The equilibrium configurations of the eccentric plate for $\psi_E = \pi$, with (a) $b = a/2$ and (b) $b = 2a/3$

the geometric relation between the angles mentioned above, but the equilibrium condition is not necessarily satisfied in this case. Indeed, upon substitution of the last two values into the equilibrium equation, we obtain

$$a \sin \left(\frac{a}{a-b} \pi \right) = 0$$

which is verified if and only if the argument of the sine function is an integer multiple of π, i.e., if

$$\frac{a}{a-b} = N$$

where N is an integer. Hence, if $\psi_E = \pi$ is an equilibrium position at all, we must have

$$b = \frac{N-1}{N} a$$

Therefore, unless b is given the above value for some integer N, an equilibrium position with the center of mass C above the center of the pin is not possible. This type of equilibrium configurations, for $N = 2$ and $N = 3$, or $b = a/2$ and $b = 2a/3$, respectively, is shown in Figs. 1.32a and b.

Apparently, the configuration of Fig. 1.32b is physically impossible because, although the center of mass lies on the vertical of the center of the pin, the pin cannot exert a pull on the plate.

Moreover, if we substitute $\theta_E = 2\psi_E$, for a value of $b = a/2$ in the equilibrium equation, and expand $\sin(2\psi_E)$, we obtain, after rearranging of terms,

$$(2a \cos \psi_E + d) \sin \psi_E = 0$$

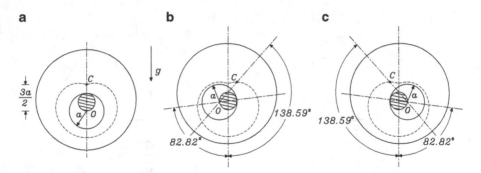

Fig. 1.33 Trajectory of the center of mass for $b = a/2$ and $d = 3a/2$ and configurations at which (**a**) $\psi_E = 180°$; (**b**) $\psi_E = 138.59°$; and (**c**) $\psi_E = 221.41°$ (or $-138.59°$)

which is satisfied not only for $\psi_E = 0$, but also for

$$\psi_E = \arccos\left(-\frac{d}{2a}\right), \quad \text{if} \quad d \leq 2a$$

An equilibrium position other than $\psi_E = 0$ or π is obviously outside of the question, although it is apparently possible in light of the above equilibrium equation. The apparent paradox can be resolved if we trace the path of the center of mass, as shown in Fig. 1.33a, for $b = a/2$ and $d = (3/2)a$. It is apparent that the two configurations stemming from $\cos \psi_E = -3/4$, namely, $\psi_{E1} = 138.59°$ and $\psi_{E2} = 221.41°$, are symmetric with respect to the vertical. Moreover, $\theta_{E1} = 277.18°$ or $-82.82°$, and $\theta_{E2} = 442.82°$ or $82.82°$. Therefore, the two above equilibrium configurations are symmetric with respect to the vertical, which is plausible, for one equilibrium configuration should be the mirror-image of the other with respect to the vertical of the center of the pin. From the path of C it is apparent that the two configurations satisfying the equilibrium equation at values other than $\psi_E = 0$ or π correspond to maxima of the potential energy of the plate, but they are not equilibrium configurations, for the center of mass of the plate in these configurations does not respect the equilibrium condition of lying on the vertical of the center of the pin. Shown in Figs. 1.33b and c are the configurations at which the potential energy of the system at hand attains local maxima. Note that the configuration of the plate for $\psi_E = \pi$ in this example is one of a minimum potential energy and is an equilibrium configuration.

Example 1.9.4 (A System with a Time-varying Equilibrium State). Derive the equilibrium states of the rack-and-pinion transmission of Fig. 1.3.

Solution: Let us assume that the moment of inertia of the pinion about its center of mass is I, and that its mass is m, its radius being a. Thus, the velocity of its c.o.m is $a\dot{\theta}$, and hence, the kinetic energy becomes

$$T = \frac{1}{2}ma^2\dot{\theta}^2 + \frac{1}{2}I\dot{\theta}^2 = \frac{1}{2}(I + ma^2)\dot{\theta}^2$$

Moreover, no changes in the potential energy are apparent, since the center of mass of the pinion remains at the same level, and hence, we can set $V = 0$, which leads to $L = T$ in this case. Now the Lagrange equation of the system is, simply,

$$(I + ma^2)\ddot{\theta} = 0$$

and hence, the equilibrium equation, obtained when we set $\ddot{\theta} = 0$, becomes an identity, namely, $0 = 0$. This means that the system is in equilibrium under any constant angular velocity of the pinion, which is sometimes referred to as an *indifferent* equilibrium configuration.

Example 1.9.5 (One More System with a Time-varying Equilibrium State). We study here the epicyclic gear train of Fig. 1.16b, which, we assume, has all its axes vertical, and hence, gravity does not come into the picture. Under these conditions, and with the outer gear fixed to an inertial base, we will determine the equilibrium states of this system. Furthermore, we assume that the arm A is a uniform, slender rod of length $a - b$ and mass m, while the planet P is a uniform disk of mass M and radius b.

Solution: First, we need the Lagrange equations of the system, which requires, in the first place, the kinetic energy of all its moving parts. In Example 1.6.3, we found the angular velocity ω_P of the planet P, which turned out to be

$$\omega_P = -\frac{a - b}{b}\dot{\theta}$$

The kinetic energy is given by

$$T = \frac{1}{2}\frac{1}{3}m(a - b)^2\omega_A^2 + \frac{1}{2}M(a - b)^2\dot{\theta}^2 + \frac{1}{2}\frac{1}{2}Mb^2\omega_P^2$$

Upon substitution of the above expression for ω_P in the foregoing formula and the angular velocity $\dot{\theta}$ of the arm, and performing some simplifications, we have

$$T = \frac{1}{12}(2m + 9M)(a - b)^2\dot{\theta}^2$$

Moreover, no source of potential energy is apparently present in the system, and hence, we have

$$V = 0$$

the Lagrange equation of the system thus reducing to

$$\frac{1}{6}(2m + 9M)(a - b)^2\ddot{\theta} = 0$$

Therefore, the equilibrium equation reduces, again, to the identity $0 = 0$, the system thus admitting equilibrium states wherever $\dot{\theta}_E = $ const., not necessarily zero.

What we can see as a common feature in the last two examples is that their Lagrangians are independent of the generalized coordinate. Furthermore, both are conservative, autonomous systems, i.e., neither subject to dissipation nor to driving forces or motions. As a matter of fact, any conservative, autonomous, single-dof system whose Lagrangian is independent of the generalized coordinate admits time-varying equilibrium states.

1.10   Linearization About Equilibrium States. Stability

Linear systems do not exist in real life. They are abstractions that scientists and engineers have created in order to derive models amenable to analysis. In fact, nonlinear systems, in general, exhibit an extremely complex and, many a time, unexpected behavior. On the contrary, linear systems, particularly those with constant coefficients, also known as *linear time-invariant* or *stationary* systems, which are the focus of this book, exhibit a simple pattern of behavior, and hence, are easier to analyze than their nonlinear counterparts. Most sources of linear systems are nonlinear systems that are *linearized about their equilibrium states*.

If a system is in an equilibrium state and is perturbed *slightly*, then, the system may respond in one of three possible ways:

- The system returns *eventually* to its equilibrium state
- The system never returns to its equilibrium state, from which it wanders farther and farther
- The system neither returns to its equilibrium state nor escapes from it; rather, the system *oscillates* about the equilibrium state

In the first case, the equilibrium state is said to be *stable* or, more precisely, *asymptotically stable*; in the second case, the equilibrium state is *unstable* or *asymptotically unstable*. The third case is a borderline case between the two foregoing cases. This case, then, leads to what is known as a *marginally stable* equilibrium state.

In studying the stability of a system, we can gain insight into this issue by resorting to energy arguments, as we do below. The energy of the system under study may (1) be dissipated into unrecoverable heat, as predicted by the *Second Law of Thermodynamics*, (2) grow due to a source, or even (3) remain constant. In the first case, we have a *stable* equilibrium state; in the second case, we have an *unstable* equilibrium state, while, in the third case, we have a *marginally stable* equilibrium state. Note that, in the third case, the energy remains constant. Thus, the third case is *conservative*, for it preserves its energy at a constant value.

We analyze below a single-dof system governed by the equation

$$m(q)\ddot{q} + h(q,\dot{q}) = \phi(q,\dot{q},t) \qquad (1.47)$$

Thus, we denote the value of q at its equilibrium state by q_E and linearize the system about this state, as described below. Moreover, at equilibrium, $\dot{q} = 0$.

First, we perturb *slightly* the equilibrium state, which we do by adding an amount δq to q_E, this perturbation thus inducing a perturbation $\delta \dot{q}$ of $\dot{q}$, and $\delta \ddot{q}$ of $\ddot{q}$. We then have

$$q \equiv q_E + \delta q, \quad \dot{q} \equiv \delta \dot{q}, \quad \ddot{q} \equiv \delta \ddot{q} \tag{1.48}$$

where δq and its time-derivatives are all functions of time.

It is noteworthy that the nonlinear functions h and ϕ will have to be evaluated at their perturbed values. A direct evaluation of these functions is possible only in special cases, which nevertheless occur frequently in applications. These cases are those in which the said functions are either trigonometric, hyperbolic or polynomial, for which explicit formulas for the evaluation of sums appearing in their arguments are readily available. Otherwise, one can resort to a *series expansion*. In any instance, we recall the assumption that the perturbation is small, and hence, the functions involved are readily calculated via their first-order approximations. Thus, if a series expansion is introduced, we have, up to first-order terms[15]

$$m(q_E + \delta q) \approx m_E + m'(q_E)\delta q \tag{1.49a}$$

$$h(q_E + \delta q, \delta \dot{q}) \approx h(q_E, 0) + \left.\frac{\partial h}{\partial q}\right|_E \delta q + \left.\frac{\partial h}{\partial \dot{q}}\right|_E \delta \dot{q} \tag{1.49b}$$

$$\phi(q_E + \delta q, \delta \dot{q}, t) \approx \phi(q_E, 0, t) + \left.\frac{\partial \phi}{\partial q}\right|_E \delta q + \left.\frac{\partial \phi}{\partial \dot{q}}\right|_E \delta \dot{q} + \delta \phi(t) \tag{1.49c}$$

where the subscripted vertical bar indicates that the quantity to its left is evaluated at equilibrium, i.e., at the state $\mathbf{x}_E = [q_E, 0]^T$, m_E and ϕ_E denoting functions m and ϕ evaluated at the equilibrium state; moreover, we have taken into account that h vanishes at equilibrium, by virtue of Eq. 1.38a. Further, ϕ also vanishes at equilibrium, by virtue of the equilibrium Eq. 1.46, and so,

$$m_E \equiv m(q_E) \tag{1.50}$$

The same applies to the partial derivatives of m, h and ϕ with respect to q and $\dot{q}$ with the symbol $|_E$.

An *equilibrium configuration* is understood throughout the book as a configuration of the system governed by $q = q_E$, at which we have assumed that $\dot{q} = 0$.

In evaluating the partial derivatives of h with respect to q and $\dot{q}$ at equilibrium, we have, from Eq. 1.38a,

$$\left.\frac{\partial h}{\partial q}\right|_E = \left.\frac{1}{2}m''(q)\dot{q}^2\right|_E = 0, \quad \left.\frac{\partial h}{\partial \dot{q}}\right|_E = m'(q)\dot{q}|_E = 0$$

---

[15]The expressions (1.49a–c) are said to be "first-order" because all variations δq and $\delta \dot{q}$ appear *linearly* therein.

Therefore,

$$h(q_E + \delta q, \delta \dot{q}) = 0 \tag{1.51}$$

Upon substituting Eqs. 1.49a and c, along with Eq. 1.51, into Eq. 1.47, we have

$$\left(m_E + m'(q_E)\delta q\right)\delta\ddot{q} = \left.\frac{\partial\phi}{\partial q}\right|_E \delta q + \left.\frac{\partial\phi}{\partial\dot{q}}\right|_E \delta\dot{q} + \delta\phi(t) \tag{1.52}$$

Under the *small-perturbation* assumption, the quadratic term involving the product $\delta q \delta \ddot{q}$ in the left-hand side of the above equation is *too small* with respect to the linear terms, and hence, is neglected. The perturbed equation of motion thus reduces to

$$m_E\delta\ddot{q} + c_E\delta\dot{q} + k_E\delta q = \delta\phi(t) \tag{1.53a}$$

with the definitions below:

$$c_E \equiv -\left.\frac{\partial\phi}{\partial\dot{q}}\right|_E, \quad k_E \equiv -\left.\frac{\partial\phi}{\partial q}\right|_E \tag{1.53b}$$

From the foregoing discussion, m_E is a mass and is, hence, positive. However, nothing in the above derivations prevents the other two coefficients from taking any real values, including negative ones. If they are both positive, then they represent the equivalent damping and the equivalent stiffness of a dashpot and a spring, respectively; otherwise, they behave like active elements that introduce energy into the system, which thus gives rise to instabilities. We do not elaborate on this issue here, but rather limit ourselves to asymptotically stable and marginally stable systems.

The system of Eq. 1.53a is asymptotically stable if both k_E and c_E are positive. If $k_E = 0$, then we can still have asymptotic stability, provided that $c_E > 0$. However, if $c_E = 0$ and $k_E > 0$, then all we have is marginal stability. All other cases are asymptotically unstable. In summary, we have the conditions below:

Given the linearized equation of a one-dof system,

$$m_E\delta\ddot{q} + c_E\delta\dot{q} + k_E\delta q = \delta\phi(t) \tag{1.54}$$

where $m_E > 0$, the system

- is asymptotically stable if and only if c_E is positive and k_E is non-negative;
- is marginally stable if and only if $c_E = 0$ and $k_E > 0$;
- is asymptotically unstable otherwise.

We assume now that none of the coefficients of Eq. 1.54 is negative. Moreover, since m_E is positive, we can divide all terms of that equation by m_E, thereby obtaining

$$\delta\ddot{q} + 2\zeta\omega_n\delta\dot{q} + \omega_n^2\delta q = \delta f(t) \tag{1.55}$$

Fig. 1.34 A linear
mass-spring-dashpot system

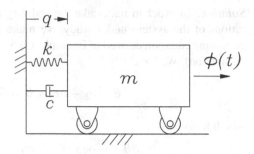

which is the *normal form* of the second-order ODE governing the motion of the
linear system under study, with ω_n and ζ, the *natural frequency* and the *damping
ratio* of the system, respectively, defined as

$$\omega_n \equiv \sqrt{\frac{k_E}{m_E}}, \qquad \zeta \equiv \frac{c_E}{2m_E\omega_n} = \frac{c_E}{2\sqrt{k_E m_E}} \qquad (1.56a)$$

and

$$\delta f(t) \equiv \frac{\delta\phi(t)}{m_E} \qquad (1.56b)$$

For brevity, we shall write Eqs. 1.53a and 1.54 in a simpler form, i.e., by dropping
the subscript E and the δ symbol, whenever the equilibrium configuration is either
self-understood or immaterial, namely,

$$m\ddot{q} + c\dot{q} + kq = \phi(t) \qquad (1.57)$$

or, in normal form, as

$$\ddot{q} + 2\zeta\omega_n\dot{q} + \omega_n^2 q = f(t) \qquad (1.58a)$$

with ω_n and ζ defined now as

$$\omega_n \equiv \sqrt{\frac{k}{m}}, \qquad \zeta \equiv \frac{c}{2m\omega_n} \qquad (1.58b)$$

and, obviously, $f(t)$ defined as

$$f(t) \equiv \frac{\phi(t)}{m} \qquad (1.58c)$$

The mathematical model appearing in Eq. 1.57 corresponds to the iconic model
of Fig. 1.34.

Example 1.10.1 (Stability Analysis of the Overhead Crane). Determine the stability
of the equilibrium configurations of the overhead crane introduced in Example 1.6.5.

Solution: In order to undertake the stability analysis of the equilibrium configurations of the system under study, we make the substitutions given below in the governing equation derived in Example 1.6.5, with $\ddot{u} = a = \text{const}$ and $c = 0$ (neglect dissipation). We have

$$\theta = \theta_E + \delta\theta, \quad \dot{\theta} = \delta\dot{\theta}, \quad \ddot{\theta} = \delta\ddot{\theta}$$

which leads to

$$\delta\ddot{\theta} + \frac{3g}{2l}\sin(\theta_E + \delta\theta) = \frac{3a}{2l}\cos(\theta_E + \delta\theta)$$

We now invoke the assumption that $\delta\theta$ is "small", and hence, $\cos\delta\theta \to 1$ and $\sin\delta\theta \to \delta\theta$, the linearized model thus taking the form

$$\delta\ddot{\theta} + \frac{3}{2l}\left(g\cos\theta_E + a\sin\theta_E\right)\delta\theta = 0$$

Finally, we recall the expressions derived for $\cos\theta_E$ and $\sin\theta_E$ in Example 1.9.1 above, thereby obtaining

$$\delta\ddot{\theta} \pm \frac{3}{2l}\sqrt{a^2 + g^2}\,\delta\theta = 0$$

We thus have that the equilibrium configuration at which angle θ_E lies between $0°$ and $90°$, i.e., the rod-down configuration, is marginally stable. However, the equilibrium configuration for which θ_E lies between $180°$ and $270°$, i.e., the rod-up configuration, is unstable. Note that the natural frequency of the marginally stable configuration is readily derived from the above linearized equation as

$$\omega_n = \sqrt{\frac{3\sqrt{a^2 + g^2}}{2l}}$$

which is reminiscent of the natural frequency of a simple pendulum of length l, $\sqrt{g/l}$.

Example 1.10.2 (Stability Analysis of the Actuator Mechanism). Decide whether each of the equilibrium configurations of Fig. 1.31, of the actuator mechanism, found in Example 1.9.2 is stable or unstable, when the system is unactuated—i.e., when $F(t) = 0$. For the stable cases, whether stability is asymptotic or marginal, and find, in each case, the equivalent natural frequency and, if applicable, the damping ratio.

Solution: What we have to do is linearize the governing equation about each equilibrium configuration, with $F(t) = 0$. To do this, we substitute the values below into the Lagrange equation of the system at hand:

$$\theta = \theta_E + \delta\theta, \quad \dot{\theta} = \delta\dot{\theta} + \delta\theta, \quad \ddot{\theta} = \delta\ddot{\theta}, \quad F(t) = 0$$

The equation of motion thus becomes

$$\delta\ddot{\theta} = -3\sqrt{2}\frac{k}{m}\left[\sin\left(\frac{\theta_E + \delta\theta}{2}\right) - \frac{\sqrt{2}}{2}\right]\cos\left(\frac{\theta_E + \delta\theta}{2}\right) - \frac{3c}{ma}\delta\dot{\theta}\cos\left(\frac{\theta_E + \delta\theta}{2}\right)$$

We analyze below each of the three equilibrium configurations found in Example 1.9.2:

(a) $\theta_E = \pi/2$: In this case,

$$\sin\left(\frac{\theta_E + \delta\theta}{2}\right) = \sin\left(\frac{\pi}{4} + \frac{\delta\theta}{2}\right) = \frac{\sqrt{2}}{2}\left(1 + \frac{\delta\theta}{2}\right)$$

$$\cos\left(\frac{\theta_E + \delta\theta}{2}\right) = \cos\left(\frac{\pi}{4} + \frac{\delta\theta}{2}\right) = \frac{\sqrt{2}}{2}\left(1 - \frac{\delta\theta}{2}\right)$$

the linearized equation about the equilibrium configuration considered here thus becoming, after simplifications,

$$\delta\ddot{\theta} = -\frac{3\sqrt{2}}{2}\frac{k}{m}\frac{\delta\theta}{2}\left(1 - \frac{\delta\theta}{2}\right) - \frac{3\sqrt{2}}{2}\frac{c}{ma}\delta\dot{\theta}\left(1 - \frac{\delta\theta}{2}\right)$$

Upon dropping the quadratic terms of the above expression, and rearranging the expression thus resulting in normal form, we obtain

$$\delta\ddot{\theta} + \frac{3\sqrt{2}}{2}\frac{c}{ma}\delta\dot{\theta} + \frac{3\sqrt{2}}{2}\frac{k}{m}\frac{\delta\theta}{2} = 0$$

Now it is apparent that all coefficients of the above equation are positive, and hence, the system equilibrium state is asymptotically stable. The natural frequency and the damping ratio associated with the linearized system are thus

$$\omega_n = \sqrt{\frac{3\sqrt{2}}{2}\frac{k}{m}}, \quad \zeta = \frac{3c}{4a}\sqrt{\frac{2\sqrt{2}}{3km}}$$

which are reminiscent of the natural frequency and the damping ratio of a mass-spring-dashpot system, $\sqrt{k/m}$ and $c/(2\sqrt{km})$, respectively.

(b) $\theta_E = \pi$: now we have

$$\sin\left(\frac{\theta_E + \delta\theta}{2}\right) = \sin\left(\frac{\pi}{2} + \frac{\delta}{2}\right) = 1, \quad \cos\left(\frac{\pi}{2} + \frac{\delta}{2}\right) = -\frac{\delta\theta}{2}$$

Upon substitution of the foregoing values into the Lagrange equation, we obtain, after simplification,

$$\delta\ddot{\theta} = -3\sqrt{2}\frac{k}{m}\left(1-\frac{\delta\theta}{2}\right)\left(-\frac{\delta\theta}{2}\right) - \frac{3c}{ma}\delta\dot{\theta}\left(-\frac{\delta\theta}{2}\right)$$

Next, we delete the quadratic terms from the above equation, and rewrite it in normal form, thus obtaining

$$\delta\ddot{\theta} - 3\frac{k}{m}\frac{2\sqrt{2}-2}{2}\frac{\delta\theta}{2} = 0$$

whence it is apparent that the coefficient of $\delta\theta$ is negative, the equilibrium configuration at hand thus being unstable.

(c) $\theta_E = 3\pi/2$: this configuration is the mirror-image of the first one, $\theta_E = \pi/2$, and hence, is bound to have the same properties as that one. Without further analysis,[16] we can conclude that this configuration is asymptotically stable, with its natural frequency and damping ratio identical to those found for its mirror-image in the first case analyzed above.

Example 1.10.3 (Stability Analysis of the Eccentric Plate). Analyze the equilibrium configurations found in Example 1.9.3 for the eccentric plate, find its natural frequency and, if applicable, the damping ratio of the stable configuration(s).

Solution: In Example 1.9.3 we found that the system at hand admits general and special equilibrium configurations. The general equilibrium configuration is $\psi_E = \theta_E = 0$. We thus set

$$\psi = \delta\psi, \quad \theta = \delta\theta = \frac{a}{a-b}\delta\psi, \quad \dot{\psi} = \delta\dot{\psi}, \quad \ddot{\psi} = \delta\ddot{\psi}$$

in the Lagrange equation derived in Example 1.6.7:

$$\{I_C + m[a^2 + d^2 + 2ad\cos(\delta\theta - \delta\psi)]\}\delta\ddot{\psi} - mad\frac{b}{a-b}\sin(\delta\theta - \delta\psi)\delta\dot{\psi}^2$$
$$= -mg(a\sin\delta\theta + d\sin\delta\psi)$$

---

[16]One equilibrium configuration can be obtained from the other by looking at the latter with the aid of a mirror. Mirror-imaging, of course, shouldn't affect the intrinsic properties of the system.

where

$$\cos(\delta\theta - \delta\psi) = \cos\delta\theta\cos\delta\psi + \sin\delta\theta\sin\delta\psi \approx 1$$

$$\sin(\delta\theta - \delta\psi) \approx \delta\theta - \delta\psi = \frac{b}{a-b}\delta\psi$$

$$\sin\delta\theta \approx \delta\theta = \frac{a}{a-b}\delta\psi$$

Therefore, upon dropping the higher-order terms from the perturbed governing equation and performing the foregoing substitutions, we have

$$[I_C + m(a+d)^2]\delta\ddot{\psi} + mg\left(\frac{a^2}{a-b}+d\right)\delta\psi = 0$$

and, since $a > b$, the two above coefficients are positive, with the coefficient of $\delta\psi$ vanishing, the equilibrium configuration thus being marginally stable, its natural frequency being

$$\omega_n = \sqrt{\frac{mg[a^2 + (a-b)d]}{(a-b)[I_C + m(a+d)^2]}}$$

Furthermore, we analyze below the special equilibrium configurations. For concreteness, let us analyze the configuration of Fig. 1.32a, for $b = a/2$, which thus yields the special equilibrium configuration

$$\psi_E = \pi, \quad \theta_E = 2\psi_E = 2\pi \quad \text{or} \quad 0$$

upon linearization about this configuration, we have

$$\psi = \pi + \delta\psi, \quad \theta = \delta\theta = 2\delta\psi, \quad \theta - \psi = \delta\psi - \pi, \quad \dot{\psi} = \delta\dot{\psi}, \quad \ddot{\psi} = \delta\ddot{\psi}$$

and hence, the perturbed governing equation takes the form

$$\{I_C + m[a^2 + d^2 + 2ad\cos(\delta\psi - \pi)]\}\delta\ddot{\psi} - mad\sin(\delta\psi - \pi)\delta\dot{\psi}^2$$
$$+ mg[a\sin 2\delta\psi + d\sin(\pi + \delta\psi)] = 0$$

or

$$[I_C + m(a-d)^2]\delta\ddot{\psi} + mg(2a - d)\delta\psi = 0$$

whence it is apparent that the equilibrium configuration under study is stable as long as $d < 2a$. Let $d_C \equiv 2a$ be the *critical* value of d, beyond which the equilibrium configuration becomes unstable. Here we have a case of an equilibrium configuration with the center of mass above the support, which is nevertheless stable. In order to gain insight into the nature of the equilibrium configuration at

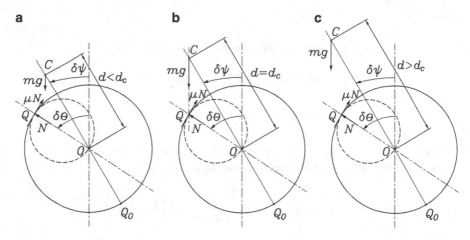

Fig. 1.35 The equilibrium configuration $\psi_E = \pi$, $\theta_E = 0$, for (**a**) $d < d_C$, (**b**) $d = d_C$, and (**c**) $d > d_C$

hand, for the three cases (a) $d < d_C$, (b) $d = d_C$, and (c) $d > d_C$, we display a slightly perturbed[17] configuration from equilibrium for each of these cases in Fig. 1.35a–c.

It is apparent from Fig. 1.35 that the moment of the normal force N acting on the eccentric plate with respect to O is zero. Now, when $d < d_C$, the moments of the two remaining forces, the friction force μN and the weight mg, are opposite to each other, with μN having a larger moment arm, thereby allowing for stability. On the contrary, in the case $d > d_C$, the moment arm of μN is smaller than that of the weight mg, which thus leads to instability.

Finally, if $d < d_C$, the natural frequency of the plate for small-amplitude oscillations around the configurations $\psi_E = 0$ and $\psi_E = \pi$ are

$$\omega_n = \sqrt{\frac{mg(2a+d)}{I_C + m(a+d)^2}}, \quad \text{for} \quad \psi_E = 0$$

and

$$\omega_n = \sqrt{\frac{mg(2a-d)}{I_C + m(a-d)^2}}, \quad \text{for} \quad \psi_E = \pi$$

Example 1.10.4 (A Mass-spring-dashpot System in a Gravity Field). Illustrated in Fig. 1.36 is a mass-spring-dashpot system suspended from a rigid ceiling. For this system, discuss the difference in the models resulting when the displacement of the mass is measured (a) from the configuration where the spring is unloaded and (b) from the static equilibrium configuration.

---

[17]For ease of visualization, $\delta\theta$ and $\delta\psi$ are exaggerated in the figure, but they are both assumed to be "small".

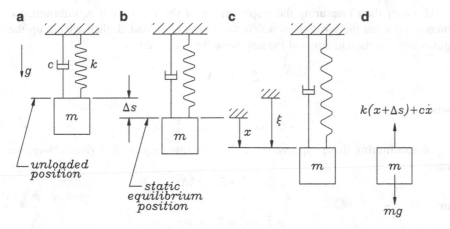

Fig. 1.36 Mass-spring-dashpot system in a gravity field

Solution: In the foregoing examples it became apparent that, when linearizing a system at an equilibrium state, constant terms drop from the linearized equation by virtue of the equilibrium equation. As a matter of fact, the same happens with linear systems acted upon by constant forces, e.g., when subjected to a gravity field.

In Fig. 1.36a the system is displayed with the spring unloaded, while the same system is shown in its *static equilibrium position* in Fig. 1.36b, this position being that at which the spring force balances the weight of the mass, and hence,

$$k\Delta s = mg$$

Apparently, Δs is the corresponding deflection of the spring. Further, we define the origin of the generalized coordinate x to be at the position of static equilibrium, as illustrated in Fig. 1.36c. Now we aim to derive the equation of motion of the system by assuming that the mass undergoes a positive displacement, as shown in Fig. 1.36c, which thus leads to the free-body diagram of Fig. 1.36d. We see that the only forces acting on the mass are the weight mg of the block and the upwards spring force $k(x+\Delta s)$, where $x+\Delta s$ is the extension of the spring from its natural length. Application of Newton's equation in the vertical direction now gives

$$m\ddot{x} = mg - k(x+\Delta s) - c\dot{x}$$
$$= mg - k\Delta s - kx - c\dot{x}$$

Thus, the equilibrium equation derived above reduces to

$$m\ddot{x} + c\dot{x} + kx = 0$$

If, rather than measuring the displacement of the mass from equilibrium, we measure it from the position in which the spring is unloaded, then we set up the governing equation in terms of the new variable ξ, defined as

$$\xi \equiv x + \Delta s$$

whence

$$x = \xi - \Delta s, \quad \dot{\xi} = \dot{x}, \quad \ddot{\xi} = \ddot{x}$$

Upon substituting these expressions in the governing equation derived above, we have

$$m\ddot{\xi} + c\dot{\xi} + k\xi = k\Delta s$$

or

$$m\ddot{\xi} + c\dot{\xi} + k\xi = mg$$

Note that a simple change of variable makes the difference between a *homogeneous* ODE, i.e., one with zero in the right-hand side, and an ODE with a constant input. Homogeneous ODEs are associated with *autonomous* systems.

1.11   Exercises

1.1. A robotic link is modeled as a slender rigid bar of length l and uniformly distributed mass m, as shown in Fig. 1.37. The link is actuated by an electric motor via a gear train with gear and pinion of N_g and N_p teeth, respectively. Moreover, the gear is coupled to the link by means of a massless, rigid shaft. Establish the mathematical model of the system for a given torque $\tau(t)$ applied by the motor, in terms of the angle of rotation ψ of the pinion, if the mounting of the shaft on its bearings produces a linearly viscous torque of coefficient c.

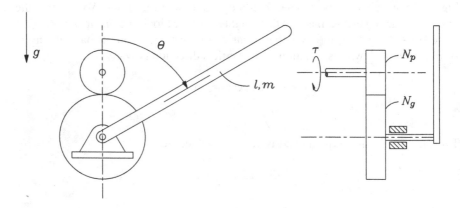

Fig. 1.37 A simple model of a robotic link and its mechanical transmission

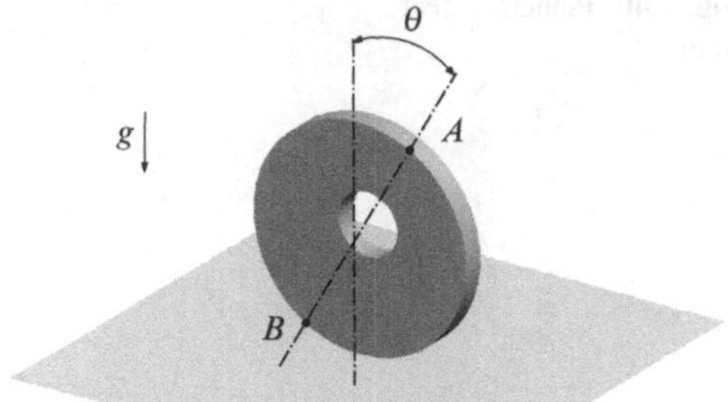

Fig. 1.38 A disk with an eccentric bore on a rough surface

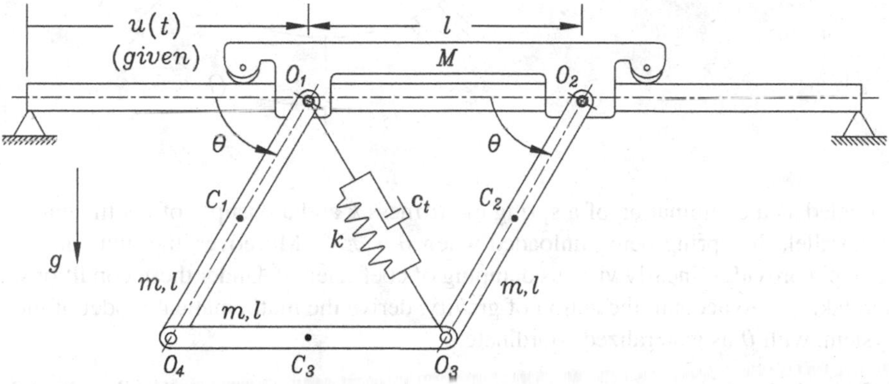

Fig. 1.39 An overhead transport mechanism

1.2. With reference to the system of Fig. 1.37, assume now that the pinion is driven with a controlled angle $\psi(t)$, and that the shaft is linearly elastic, all other conditions remaining as in Problem 1.1. Derive the mathematical model of the system in terms of the generalized coordinate θ

1.3. Shown in Fig. 1.38 is a uniform disk of mass m and radius a with an eccentric bore of radius $b < a$, the bore center being a distance e from the center of the disk. While trying to place the disk in its stable equilibrium state by hand, an operator incurs an error that we can safely assume to be *small*. Moreover, we can also assume that the disk and the table surfaces are rough enough, so as to prevent any sliding. Under these conditions, find the natural frequency of the ensuing motion in terms of the given parameters.

1.4. Shown in Fig. 1.39 is the iconic model of an overhead mechanism used for material handling. This mechanism is composed of a bogie that is driven under a controlled displacement $u(t)$, and carries three pin-joined identical links of mass m and length l, which can be modeled as slender rigid bars of uniformly distributed mass. The mechanism is "stiffened" by means of a transverse element that can be

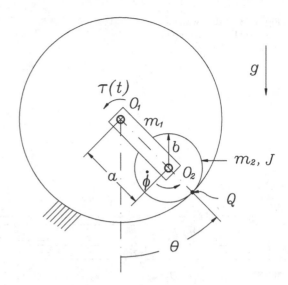

Fig. 1.40 Planetary gear train

modeled as a combination of a spring of stiffness k and a dashpot of coefficient c_t in parallel, the spring being unloaded when $\theta = \pi/2$. Moreover, the lubricant of each pin provides linearly viscous damping of coefficient c. Under these conditions, and taking into account the action of gravity, derive the mathematical model of the system, with θ as generalized coordinate.

1.5. With the transverse element removed and a particular constant acceleration of the bogie, show that the mathematical model of the system of Fig. 1.39 takes the form

$$\ddot{\theta} + \frac{12}{5}\frac{c}{ml^2}\dot{\theta} + \frac{2}{5}\frac{g}{l}(4\sin\theta - 3\cos\theta) = 0$$

where damping arises from the lubricant in the four joints. Now, using the above model,

(a) Find the equilibrium configuration(s) of the system
(b) Find the constant acceleration of the bogie
(c) Identify the stable equilibrium configuration(s), and determine the corresponding natural frequency and damping coefficient for "small" deviations from equilibrium

1.6. Shown in Fig. 1.40 is the iconic model of a planetary gear train. The planet is modeled as a rigid disk of radius b, mass m_2 and moment of inertia $J = m_2 b^2/2$ with respect to its center of mass O_2. Moreover, the planet gear is pinned to the carrier of mass m_1 at O_2, while the carrier is pinned in turn to the mechanism frame at O_1 and driven by a motor that supplies a torque $\tau(t)$. Furthermore, the lubricant at the joints O_1 and O_2 is assumed to provide linear damping of coefficients c_1 and c_2,

Fig. 1.41 Simplified model
of an automobile suspension

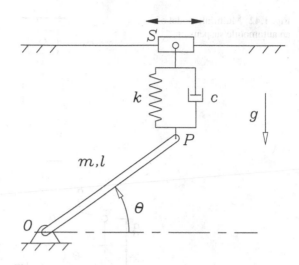

respectively. Finally, the planet rolls without slipping on the internal gear of radius
$a + b$ that is fixed to an inertial frame, with an angular velocity $\dot{\phi}$.

(a) Derive the mathematical model of the system.
(b) Now, in the absence of driving torque, find all equilibrium states.
(c) Decide whether each equilibrium state is stable, unstable or marginally stable.
(d) For the stable or marginally stable state(s), find the corresponding natural
    frequency and the damping ratio under *small* oscillations about the equilibrium
    state.

1.7. Shown in Fig. 1.41 is a highly simplified iconic model of an automobile
suspension. It consists of a slender, homogeneous, rigid rod of length l and
mass m pinned to an inertial frame at O by means of a frictionless hinge.
Moreover, the end P of the rod is supported by a spring-dashpot array whose
upper end is free to move horizontally and without friction by virtue of a massless
slider S.

Under the assumption that the system is in equilibrium when $\theta = 0$, and that,
in this state, the spring is stretched by an amount $l/2$, show that this configuration
is stable. Then, find the natural frequency and the damping ratio of the system for
small-amplitude oscillations around the same equilibrium configuration.   □

For the systems described in Exercises 1.8–1.12:

(a) Derive the corresponding mathematical model in terms of the generalized
    coordinate indicated either in the text or in the accompanying figure
(b) Derive the linearized equation for 'small-amplitude' oscillations about the
    stable equilibrium configuration(s)
(c) Find expressions for the natural frequency and damping ratio associated with
    the linearized equation

Fig. 1.42 Multilink model of
an automobile suspension

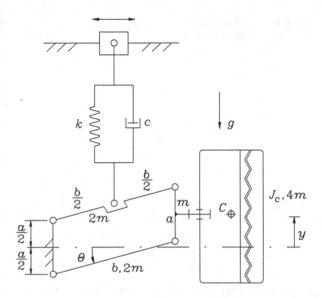

1.8. Multilink model of an automobile suspension with a slider that translates without friction along a horizontal guideway, as shown in Fig. 1.42. Assume that the system overall damping is lumped in the dashpot of coefficient c, and that the system is in equilibrium when the parallel bars of length b are in the horizontal position, in which case $y = 0$. Use y as the generalized coordinate. What is the value of y in terms of the given physical parameters when the spring is unloaded?

1.9. Repeat Problem 1.8 with θ as generalized coordinate.

1.10. The positioning mechanism of a machine tool, consisting of a disk of mass m with center of mass at point C a distance e from O, and moment of inertia J about O, is shown in Fig. 1.43. Moreover, the disk is coupled to the machine frame via a viscoelastic element that is lumped as a spring in parallel with a dashpot. The disk rolls without slipping on a horizontal surface. Use x as generalized coordinate.

1.11. Loading system consisting of a vehicle with mass M and prescribed motion $x(t)$, together with a mass m rolling on wheels that provide linearly viscous damping c_w (Ns/m), and pulled by an inextensible cable of negligible mass, as shown in Fig. 1.44. The spring of stiffness k is intended to account for the elasticity of the physical cable. Assume that the dissipation provided by the lubricant at the pulley of radius r, which has negligible inertia, is also linear, of coefficient c_p (Nms). Lump all viscous damping in one single dissipation function. Derive the mathematical model of the system under the assumption that $u = 0$ when $x = r$, and that l_E is the length of the spring when the system is in equilibrium.

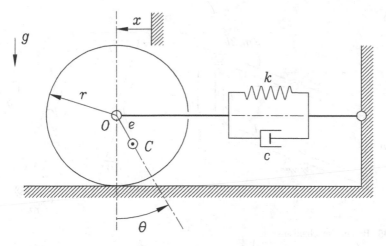

Fig. 1.43 Iconic model of the positioning mechanism of a machine tool

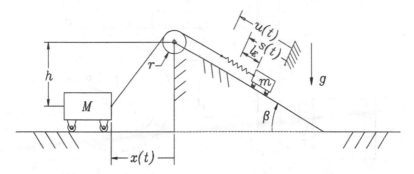

Fig. 1.44 A loading system

1.12. The balance mechanisms of Figs. 1.45a and b, consisting of uniform, slender bars of masses and lengths indicated in the figures. Assume linearly viscous damping of coefficient c in every joint.

1.13. An eccentric circular cam of radius a having moment of inertia J_O about point O rotates about point O a distance e from the center C, as indicated in Fig. 1.46. The follower is of mass m, while the cam, which can be safely modeled as a circular disk of uniform density, is driven by a motor via a rigid shaft. Under the assumptions that (1) the motor supplies a controlled torque $\tau(t)$ to the shaft and that (2) the interface between cam and follower is lubricated, so that Coulomb friction is neglected, but the lubricant provides a linearly viscous force, derive

(a) An expression for the mass of the cam, m_C, in terms of the data, and
(b) The equation of motion of the system, with θ as generalized coordinate.

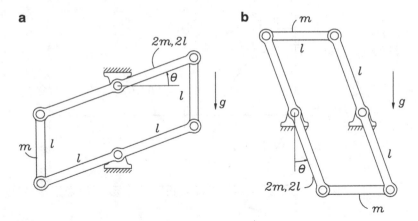

Fig. 1.45 Balance mechanisms

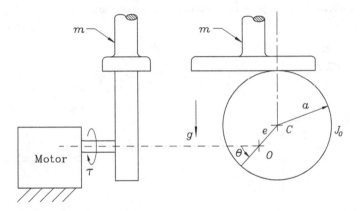

Fig. 1.46 Eccentric cam mechanism

1.14. Now, replace the rigid shaft of the system of Fig. 1.20 by an elastic shaft of torsional stiffness k and drive the shaft with a controlled angle $\psi(t)$. If all other conditions remain the same, derive the underlying mathematical model.

1.15. With reference to Fig. 1.47, which represents the iconic model of a machine-tool positioning mechanism, assume that the disk, of mass m and moment of inertia J about its c.o.m, is statically balanced, so that its centroid O coincides with its center of mass. Moreover, the disk represents a pinion that rolls over a horizontal rack without slipping, while the slider at the right end of the linear spring-dashpot array is free to move along a vertical, smooth guideway. If it is known that the spring is unloaded when $\theta = 0$,

(a) Set up the mathematical model of the system with θ as generalized coordinate; then,

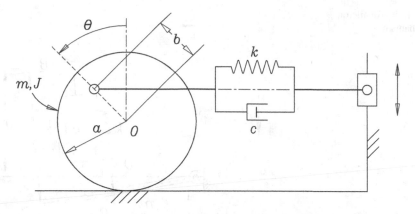

Fig. 1.47 Iconic model of a machine-tool positioning mechanism

Fig. 1.48 A deploying
mechanism

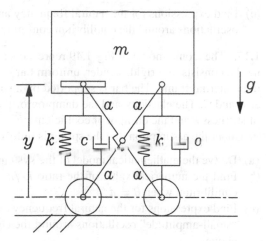

(b) Show that a possible equilibrium configuration of the system is $\theta = 0$. Are there other equilibrium configurations? *Hint: the line $y = -mx$, for $m > 0$ intersects the curve $y = sin(x)$ at other point than $x = 0$ if and only if $m \le 1$.*

(c) Decide whether the equilibrium configuration $\theta = 0$ is stable; if it is, find the damping ratio and the natural frequency of the system for small-amplitude motions.

1.16. The iconic model of Fig. 1.48 represents a deploying mechanism used in aerospace applications. Under the assumption that the spring is unloaded when $y = a$,

(a) Derive the mathematical model of the system
(b) Find the value of y at equilibrium and

Fig. 1.49 An aileron
mechanism

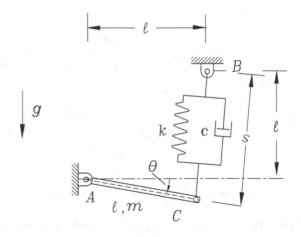

(c) Find expressions for the natural frequency and damping ratio of the model, for oscillations around the equilibrium configuration

1.17. The iconic model of Fig. 1.49 represents an aircraft aileron mechanism. The model consists of a rigid, slender, uniform bar of length ℓ and mass m, pivoted at A to an inertial frame. The bar is supported by a spring-dashpot parallel array pinned at B and C. The elasticity and the damping of the system are lumped in the spring of stiffness k and the dashpot of coefficient c.

Under the assumption that the spring is unloaded when the bar is horizontal,

(a) Derive the mathematical model of the system
(b) Find the numerical value of the ratio $k\ell/mg$ required for the system to be in equilibrium when $\theta = \pi/4$
(c) Find expressions for the natural frequency and damping ratio of the model, for "small-amplitude" oscillations around the equilibrium configuration described above
(d) Find the value of c/m in terms of g/l so that the "small-amplitude" model will be *critically-damped*, i.e., so that $\zeta = \sqrt{2}/2$

References

1. Almen J, László A (1936) The uniform-section disk spring. Trans ASME 58:305–314
2. Di Benedetto A and Pennestrì E (1993) Introduzione alla cinematica dei meccanismi. Moti infinitesimi. Casa Editrice Ambrosiana, Milan
3. Meriam JL, Kraige LG (1992) Engineering mechanics. Dynamics, vol 2, 5th edn. Wiley, New York

Chapter 2
Time Response of First- and Second-order Dynamical Systems

Science seems to have uncovered a set of laws that,
within the limits set by the uncertainty principle,
tell us how the universe will develop with time,
if we know its state *at any one time.*

Hawking, S.W., 1988, *A Brief History of Time,*
Bantam Books, Toronto-New York-London-Sydney-Auckland.

2.1 Preamble

How physical systems, e.g., aircraft, *respond* to a given input, such as a gust wind, under given initial conditions, like cruising altitude and cruising speed, is known as the *time response* of the system. Here, we have two major items that come into the picture when determining the time response, namely, the mathematical model of the system and the *history* of the input. The mathematical model, as studied in Chap. 1, is given as an ODE in the generalized coordinate of the system. Moreover, this equation is usually nonlinear, and hence, rather cumbersome to handle with the purpose of predicting how the system will respond under given initial conditions and a given input. However, if we first find the equilibrium states of the system, e.g., the altitude, the aircraft angle of attack, and cruising speed in our example above, and then linearize the model about this equilibrium state, then we can readily obtain the information sought, as described here.

We will thus start by assuming that the system model at hand has been linearized about an equilibrium state. That is, we will be concerned *mainly* with the time response of linear, time-invariant (LTI) dynamical systems, also termed *linear time-invariant systems* (LTIS), when subjected to arbitrary initial conditions and inputs. This class of systems is also known as *stationary systems* and *systems with constant coefficients*, and so, we will use these terms interchangeably. We will discuss in this chapter only first- and second-order systems, i.e., systems that are described by first- and second-order ODEs, respectively, more general systems being discussed in later

J. Angeles, *Dynamic Response of Linear Mechanical Systems: Modeling, Analysis and Simulation*, Mechanical Engineering Series, DOI 10.1007/978-1-4419-1027-1_2,
© Springer Science+Business Media, LLC 2011

chapters. However, notice that we have emphasized above that we will be concerned mainly with LTIS, thereby indicating other kinds of system models that do not fall into the same category. Such models arise, as we saw in Chap. 1, in the presence of Coulomb friction forces. In this case, the models at hand are not linear but, by adopting a simple model of the friction forces, as we did in Chap. 1, the models thus derived turn out to be *piecewise linear*, and hence, lend themselves to an analysis with the tools of LTI dynamical systems.

Applications of this analysis are numerous, e.g., vibration isolation; prediction of vibration transmitted by a moving base; design of instruments; vehicle dynamics; etc. We will outline applications in these fields.

We consider first systems to which no input is applied, these systems being subject only to initial conditions. The response of such systems, in the realm of system theory, is known as the *zero-input response*. In the language of mechanical vibration, the same goes by the name of *free response*. After this, we will study the response of the same systems to an arbitrary input, with zero initial conditions, which is called, in the realm of system theory again, the *zero-state response*. In some instances, this response is called the *forced response*. Here, note that the terminology of system theory is more accurate, for it presupposes zero initial conditions; the forced response presupposes only a nonzero-input, but says nothing about the initial conditions. Hence, we will adopt the terminology of system theory for the sake of language accuracy. The response of these systems to arbitrary initial conditions and arbitrary input is then obtained by *superposition*, namely, as the sum of their zero-input and zero-state responses.

Since we will study LTIS intensively, we need first a definition of these. In Chap. 1 we introduced a definition of linear systems from the point of view of the structure of their mathematical models, namely, *systems that give rise to governing equations where the generalized coordinate, the generalized speed and the time rate of change of the latter appear linearly, i.e., to the first power, multiplied by constants, that are termed the stiffness, the damping coefficient and the mass of the system, respectively.*

From the viewpoint of their time response, the systems of interest to our study in this chapter have the properties listed below: let the time response of a system to an input $f(t)$, with zero initial conditions, be denoted by $x_{f(t)}(t)$, which will represent the time history of the generalized coordinate $x(t)$, for $t \geq 0$. Thus, the time response of linear time-invariant dynamical systems

1. is *linearly homogeneous*, i.e., for any real number α,

$$x_{\alpha f(t)}(t) = \alpha x_{f(t)}(t)$$

   a property that is called *linear homogeneity*, or *homogeneity*, for brevity;
2. is *additive*, i.e., for a second input $g(t)$,

$$x_{f(t)+g(t)}(t) = x_{f(t)}(t) + x_{g(t)}(t)$$

a property that is called *additivity*;

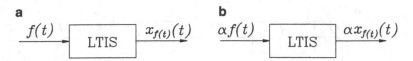

Fig. 2.1 Homogeneity of a LTIS

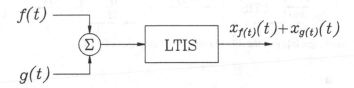

Fig. 2.2 Additivity of a LTIS

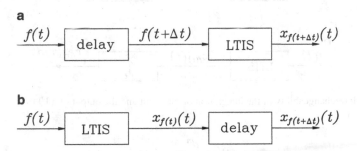

Fig. 2.3 Time invariance of a LTIS

3. is *time invariant*, i.e., for any time-interval Δt,

$$x_{f(t+\Delta t)}(t) = x_{f(t)}(t+\Delta t) \tag{2.1}$$

a property that is called *time invariance*.

The foregoing properties are best illustrated with the aid of a *black-box* representation of a LTIS, as in Fig. 2.1a. In this figure, the system at hand is represented as a block with an input $f(t)$ and a response, also known as *output*, $x_{f(t)}(t)$. Homogeneity is illustrated in Fig. 2.1b, while additivity and time invariance in Figs. 2.2 and 2.3, respectively. In fact, time invariance is best illustrated with the aid of an *artifact* called a *time delay*.

Moreover, let

$$f^{(n)}(t) \equiv \frac{d^n}{dt^n} f(t), \quad F(t) \equiv \underbrace{\int_0^t \cdots \int_0^t f(\tau) d\tau}_{n \ \text{times}}$$

Then, as a consequence of the additivity property, we have two further properties, namely,

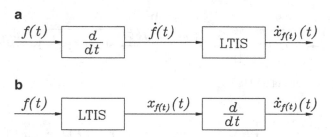

Fig. 2.4 Interchangeability of the differentiation and the input of a LTIS

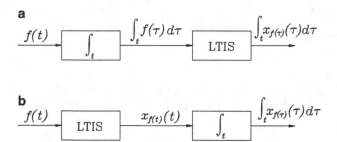

Fig. 2.5 Interchangeability of the integration of the input and the output of a LTIS

$$x_{f^{(n)}(t)}(t) = \frac{d^n}{dt^n} x_{f(t)}(t) \tag{2.2}$$

$$x_{F(t)}(t) = \underbrace{\int_0^t \cdots \int_0^t x_{f(\tau)}(\tau) d\tau}_{n \quad \text{times}} \tag{2.3}$$

We will find that the foregoing properties allow us to obtain the time response of this class of systems in a systematic way, sometimes with substantial time savings and, quite important, in a safer manner, less error prone. The two foregoing properties are illustrated in Figs. 2.4 and 2.5, for one single step of differentiation and integration, respectively.

2.2   The Zero-input Response of First-order LTIS[1]

Consider the first-order, linear, stationary dynamical system described below:

$$\dot{x} = -ax, \quad t > 0, \quad x(0) = x_0 \tag{2.4}$$

---

[1]In a course focusing on vibrations, this section can be skipped. However, its reading is strongly recommended because it helps gain insight into the response of second-order damped systems, studied in Sect. 2.3.2

where a and x_0 are given constants. We want to determine $x(t)$ *explicitly*; some would say *analytically*, but the latter is vague and very often misused. What we mean here by *explicitly*, as opposed to *numerically*, is also termed *symbolically*, for we are after an algebraic expression of the function $x(t)$. To this end, we assume that this function is *analytic*, i.e., that it has a series expansion of the form:

$$x(t) = x(0) + \dot{x}(0)t + \frac{1}{2}\ddot{x}(0)t^2 + \ldots + \frac{x^{(k)}(0)}{k!}t^k + \ldots \tag{2.5}$$

In order to evaluate the coefficients of the above series, we differentiate both sides of the differential equation of (2.4) infinitely many times with respect to time, which yields

$$\dot{x} = -ax$$

$$\ddot{x} = -a\dot{x} = a^2x$$

$$x^{(3)} = a^2\dot{x} = -a^3x$$

$$\vdots$$

$$x^{(k)} = (-1)^k a^k x, \quad \text{etc.} \tag{2.6}$$

Upon evaluation of the foregoing expressions at $t = 0$, we obtain

$$\dot{x}(0) = -ax_0$$

$$\ddot{x}(0) = a^2x_0$$

$$x^{(3)}(0) = -a^3x_0$$

$$\vdots$$

$$x^{(k)}(0) = (-1)^k a^k x_0, \quad \text{etc.} \tag{2.7}$$

If we substitute the expressions appearing in Eq. 2.7 into Eq. 2.5, the desired expression is readily derived, namely,

$$x(t) = \left[1 - at + \frac{1}{2}a^2t^2 - \ldots + (-1)^k\frac{a^k t^k}{k!} + \ldots\right]x_0$$

The series appearing inside the brackets in the above equation is readily identified as the exponential of $-at$, i.e.,

$$x(t) = e^{-at}x_0, \quad t \geq 0 \tag{2.8}$$

which is the time response sought.

Needless to say, the time response of Eq. 2.8 is *proportional* to the initial value x_0. Moreover, the time-varying part of the above expression is apparently dependent

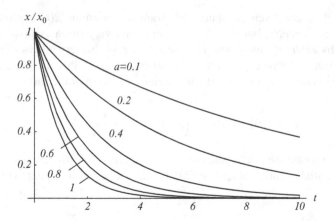

Fig. 2.6 Zero-input response of a first-order system for various values of its time constant

Fig. 2.7 Tugboat towing a barge: (**a**) physical system; and (**b**) iconic model

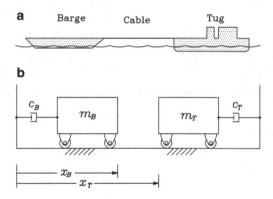

upon one single parameter, namely, coefficient a, which determines the nature of the response. Note that this coefficient has units of frequency, its reciprocal, labelled τ, having units of time. In fact, when $a > 0$, τ being positive as well, is termed the *time constant* of the system.

Furthermore, from Eq. 2.8, it is apparent that (1) if $a > 0$, the time response decays with time and does so the faster the greater a is, i.e., the time response of a system of this kind fades away faster for smaller time constants τ; (2) if $a < 0$, the time response grows unbounded, i.e., we are in the presence of an unstable system. We show in Fig. 2.6 various plots of the response of LTI first-order dynamical systems for various values of positive a, with $x(t)$ divided by x_0.

Example 2.2.1 (Collision Detection). Figure 2.7a shows a tugboat towing a barge at a uniform speed v_0. Due to an engine failure, the tugboat undergoes a shock that breaks the towing cable. Upon the assumption that the boat and the barge will continue on the same course after failure, an insurance company wants to know under which conditions on the relevant physical parameters involved an eventual collision will not occur.

Solution: Let m_T and m_B denote the mass of the tugboat and the barge, respectively. Moreover, assume that wind and water drag forces are known to be proportional to traveling speeds, the proportionality factors depending on water and wind conditions as well as on hull shape and hull materials. For the prevailing wind and water conditions, and hull properties, the drag constants are denoted by c_T and c_B for the tugboat and barge, respectively. After the failure, each vessel behaves as a first-order system and can be modeled as a mass-dashpot system, as shown in Fig. 2.7b, namely,

$$m_B \dot{v}_B + c_B v_B = 0, \quad m_T \dot{v}_T + c_T v_T = 0, \quad v_B(0) = v_T(0) = v_0$$

or

$$\dot{v}_B + \frac{1}{\tau_B} v_B = 0, \quad \dot{v}_T + \frac{1}{\tau_T} v_T = 0, \quad v_B(0) = v_T(0) = v_0 \tag{2.9}$$

The time constants of the two foregoing systems, τ_B and τ_T, are thus,

$$\tau_B = \frac{m_B}{c_B}, \quad \tau_T = \frac{m_T}{c_T}$$

Obviously, to avoid collisions, the barge speed should decay faster than the tugboat speed. Hence, the company should grant insurance only if the time constant of the barge is smaller than that of the tugboat. The condition sought is, then,

$$\tau_B < \tau_T \quad \text{or} \quad \frac{m_B}{c_B} < \frac{m_T}{c_T}$$

Note that tugboats are usually less massive than barges but, fortunately for insurance companies, barges are usually not as streamlined as tugboats and, hence, present a larger drag coefficient than tugboats, thereby making unlikely a collision under the circumstances described above.

2.3 The Zero-input Response of Second-order LTIS

In this section we are concerned with LTI dynamical systems described by a second-order ODE. We shall thus distinguish among *undamped*, *underdamped*, *critically damped* and *overdamped* systems. For brevity, we focus on the first two types, the last two being left as exercises.

2.3.1 Undamped Systems

The model associated with these systems is displayed below:

$$\ddot{x} = -\omega_n^2 x, \quad t > 0, \quad x(0) = x_0, \quad \dot{x}(0) = v_0 \tag{2.10}$$

which is the equation governing the motion of the *harmonic oscillator*. Again, in order to derive the time response of this system explicitly, we differentiate both sides of the differential equation of (2.10) infinitely many times, which yields

$$x^{(3)} = -\omega_n^2 \dot{x}$$

$$x^{(4)} = -\omega_n^2 \ddot{x} = \omega_n^4 x$$

$$x^{(5)} = \omega_n^4 \dot{x}$$

$$x^{(6)} = \omega_n^4 \ddot{x} = -\omega_n^6 x$$

$$\vdots$$

$$x^{(2k)} = (-1)^k \omega_n^{2k} x$$

$$x^{(2k+1)} = (-1)^k \omega_n^{2k} \dot{x}, \quad \text{etc.} \tag{2.11}$$

Upon evaluation of the foregoing derivatives at $t = 0$, and substitution of these values into the series expansion of $x(t)$, we derive the expression given below:

$$x(t) = \left[1 - \frac{\omega_n^2 t^2}{2!} + \frac{\omega_n^4 t^4}{4!} - \ldots + (-1)^k \frac{\omega_n^{2k} t^{2k}}{2k!} + \ldots \right] x_0$$

$$+ \frac{1}{\omega_n} \left[\omega_n t - \frac{\omega_n^3 t^3}{3!} + \frac{\omega_n^5 t^5}{5!} - \ldots + (-1)^k \frac{\omega_n^{2k+1} t^{2k+1}}{(2k+1)!} + \ldots \right] v_0 \tag{2.12}$$

The terms in brackets multiplying x_0 and v_0 in Eq. 2.12 are readily recognized to be the series expansions of $\cos \omega_n t$ and $\sin \omega_n t$, respectively. Hence, the expression sought for $x(t)$ is readily derived as

$$x(t) = (\cos \omega_n t) x_0 + \frac{1}{\omega_n} (\sin \omega_n t) v_0, \quad t \geq 0 \tag{2.13}$$

Apparently, the response of undamped second-order systems is a linear combination of the two initial conditions x_0 and v_0, the associated coefficients being time-varying. Moreover, the first coefficient is simply the cosine function with a frequency equal to the natural frequency of the system; the second coefficient is the sine function with the same frequency, but multiplied by the reciprocal of the natural frequency. Note that, regardless of the nature of the generalized coordinate $x(t)$, whether a translational or an angular displacement, the two terms of the expression of Eq. 2.13 are dimensionally homogeneous. Indeed, the first coefficient is dimensionless and multiplies a constant with units of displacement; the second coefficient has units of time but multiplies a constant with units of velocity. As a consequence, the participation of the velocity initial condition on the response becomes more relevant as the ratio v_0/ω_n grows with respect to the initial condition x_0. Shown in Fig. 2.8 are plots of time responses for various values of $v_0/(\omega_n x_0)$.

Fig. 2.8 Zero-input response of undamped second-order system for various values of $v_0/\omega_n x_0$

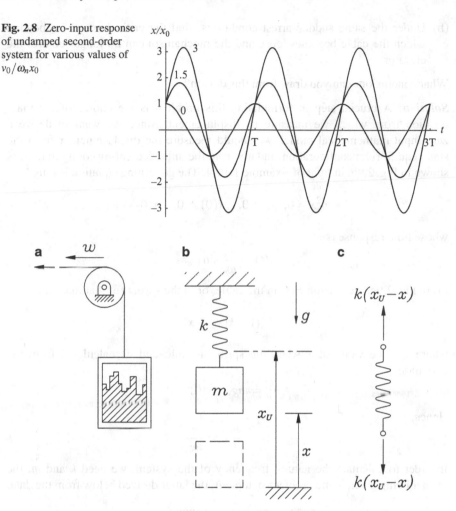

Fig. 2.9 The lifting mechanism of an elevator: (**a**) the mechanical model; (**b**) its iconic model; and (**c**) the FBD of the cable

Example 2.3.1 (Elevator Design). Shown in Fig. 2.9a is a crude iconic model of the lifting mechanism of an elevator. Under the assumption that the damping in the mechanism is negligible, one can model the mechanism-load system as the mass-spring layout shown in Fig. 2.9b. The cable is made of steel fibers that give it a stiffness of $1{,}000\,kN/m$, the weight of the empty elevator is $4{,}905\,N$ and the maximum allowable load is 12 passengers, under the assumption that the average passenger weighs $735.75\,N$.

(a) Determine the speed w_1 that produces a tension in the cable that is five times the static load in the presence of a sudden arrest.

(b) Under the same sudden-arrest conditions, find the critical speed w_2 beyond which the cable becomes loose and the mechanism can no longer control the elevator.

What conclusions can you draw from this design?

Solution: A crucial step in formulating this problem is the choice of the mass position from which we measure its displacement. Since we want to derive a zero-input mathematical model, we should measure the displacement x from the static equilibrium configuration, and not from the unloaded-spring configuration, as shown in Fig. 2.9b, in light of Example 1.10.4. The governing equation is thus

$$\ddot{x} + \omega_n^2 x = 0, \quad t \geq 0, \quad x(0) = 0, \quad \dot{x}(0) = w_1$$

whose time response is

$$x(t) = \frac{w_1}{\omega_n} \sin \omega_n t$$

From Fig. 2.9c, the tension $F(t)$ in the cable, or in the spring for that matter, is

$$F(t) = k(x_U - x)$$

where x_U, the value of x when the spring is unloaded, is calculated from the condition

$$F(0) = mg, \quad x(0) = 0$$

Hence,

$$x_U = \frac{m}{k} g = \frac{g}{\omega_n^2}$$

In order to calculate the natural frequency of the system, we need k and m, the former being given in the problem statement, the latter derived below from the data:

$$m = \frac{5000 + 750 \times 12}{9.81} = \frac{14000}{9.81} = 1427 \text{ Kg}$$

the natural frequency of the system thus being

$$\omega_n = \sqrt{\frac{k}{m}} = \sqrt{\frac{1 \times 10^6}{1427}} = 26.47 \text{ s}^{-1} = 4.213 \text{ Hz}$$

(a) Upon substitution of x and x_U in the expression for $F(t)$, we obtain

$$F(t) = k \left(\frac{g}{\omega_n^2} - \frac{w_1}{\omega_n} \sin \omega_n t \right)$$

which apparently attains a maximum when $\sin \omega_n t$ attains a minimum, i.e., at an instant t_1 at which $\sin \omega_n t_1 = -1$. Therefore,

$$F_{max} = k \left(\frac{g}{\omega_n^2} + \frac{w_1}{\omega_n} \right)$$

By setting $F_{max} = 5mg$, we obtain

$$k \left(\frac{g}{\omega_n^2} + \frac{w_1}{\omega_n} \right) = 5mg$$

Therefore,

$$w_1 = \frac{4g}{\omega_n} = 1.4824 \text{ m/s}$$

a speed with which passengers are lifted at the rate of about one-half storey per second.[2]

(b) The above expression for $F(t)$ obviously attains a minimum when $\sin \omega_n t$ attains a maximum, i.e., at an instant t_2 at which $\sin \omega_n t_2 = 1$. Therefore,

$$F_{min} = k \left(\frac{g}{\omega_n^2} - \frac{w}{\omega_n} \right)$$

Thus, the cable becomes loose when $F_{min} = 0$, i.e., when

$$\frac{g}{\omega_n^2} = \frac{w}{\omega_n}$$

Hence,

$$w = \frac{g}{\omega_n} = 0.3706 \text{ m/s}$$

which is a speed at which passengers are lifted at the rate of about one storey every 4 s, i.e., unacceptably low. From the above result, it is obvious that the design is poor. Now, if we observe the two expressions for w, it is apparent that this value is inversely proportional to the quantity $\sqrt{k/m}$. Now, the mass is fixed, because the load is specified by the client; what we can do is choose a more compliant cable. For example, if we want to increase the value of w obtained in item (b) by a factor of 25, which would give $w = 9.265$ m/s as a speed that would make the cable loose upon a sudden arrest, then the cable should be specified with a stiffness of 200 kN/m.

Alternatively, we can solve this problem using an energy approach. Indeed, the total energy E_0 of the system can be calculated from the conditions at $t = 0$, while using the position of the mass at this instant as the reference level for the potential

---

[2]A typical average lifting speed is about 1 storey (≈ 3 m) per second.

energy due to gravity. Thus, the potential energy due to gravity at $t = 0$ vanishes, but the spring is stretched by a length x_U, and hence,

$$E_0 = \frac{1}{2}mw^2 + \frac{1}{2}kx_U^2 = \frac{1}{2}m\left(w^2 + \frac{g^2}{\omega_n^2}\right)$$

From the expression for $F(t)$ derived above, it is apparent that the maximum value of the tension in the cable is attained at the minimum value of x at the bottom position of the mass, x_b, at which $\dot{x} = 0$. The tension F_b at $x = x_b$ is, thus,

$$F_b = k(x_U - x_b) = 5mg$$

Hence,

$$x_b = x_U - 5\frac{mg}{k} = -4\frac{mg}{k} \equiv -4\frac{g}{\omega_n^2}$$

Therefore, the energy E_b at the bottom position is given by

$$E_b = \frac{1}{2}k(x_U - x_b)^2 + mgx_b = \frac{25}{2}\frac{m^2g^2}{k} - 4\frac{m^2g^2}{k}$$

i.e.,

$$E_b = \frac{17}{2}\frac{m^2}{k}g^2$$

Upon equating E_b with E_0, we obtain

$$\frac{17}{2}\frac{m^2}{k}g^2 = \frac{1}{2}m\left(w^2 + \frac{g^2}{\omega_n^2}\right)$$

Hence,

$$w_1 = 4\frac{g}{\omega_n}$$

which is exactly the same result obtained in item (a). Now, to solve item (b), we note that the cable becomes loose if $x = x_U$ at the top position, in which $\dot{x} = 0$. The energy E_{top} at this position is

$$E_{\text{top}} = mgx_U = mg\left(\frac{m}{k}\right)g$$

Thus, if we equate E_{top} with E_0, we obtain

$$\left(\frac{m^2}{k}\right)g^2 = \frac{1}{2}m\left(w^2 + \frac{g^2}{\omega_n^2}\right)$$

whence

$$w_2 = \frac{g}{\omega_n}$$

a result identical to that obtained in (b).

2.3.2 Damped Systems

Now we consider a damped system, namely,

$$\ddot{x} + 2\zeta\omega_n\dot{x} + \omega_n^2 x = 0, \quad t > 0, \quad x(0) = x_0, \quad \dot{x}(0) = v_0 \qquad (2.14)$$

with both ζ and ω_n non-negative. In trying to find an expression for $x(t)$ by following the procedure introduced in the first two cases, we would readily find that this does not apply as directly. The reason is that now we would have to factor two infinite series, a task that is extremely cumbersome. However, we can circumvent this problem by writing the given system in *state-variable* form, i.e., by transforming it into a system of two first-order, linear, constant-coefficient ordinary differential equations. This is readily done by letting

$$z_1 \equiv x, \quad z_2 \equiv \dot{x} \qquad (2.15)$$

whence the two-dimensional *state-variable* vector $\mathbf{z}(t)$ is defined as

$$\mathbf{z}(t) \equiv \begin{bmatrix} z_1(t) \\ z_2(t) \end{bmatrix} = \begin{bmatrix} x \\ \dot{x} \end{bmatrix}, \quad \mathbf{z}(0) \equiv \begin{bmatrix} x_0 \\ v_0 \end{bmatrix} \qquad (2.16)$$

Thus, the system appearing in Eq. 2.14 can be readily expressed as

$$\dot{z}_1 = z_2 \qquad (2.17a)$$
$$\dot{z}_2 = -\omega_n^2 z_1 - 2\zeta\omega_n z_2 \qquad (2.17b)$$
$$z_1(0) = x_0, \quad z_2(0) = v_0 \qquad (2.17c)$$

or, in vector form,

$$\dot{\mathbf{z}} = \mathbf{A}\mathbf{z}, \quad t > 0, \quad \mathbf{z}(0) = \mathbf{z}_0 \qquad (2.18a)$$

with matrix $\mathbf{A}$ defined as

$$\mathbf{A} \equiv \begin{bmatrix} 0 & 1 \\ -\omega_n^2 & -2\zeta\omega_n \end{bmatrix} \qquad (2.18b)$$

Now, the solution $\mathbf{z}(t)$ can be expressed by means of its series expansion, namely,

$$\mathbf{z}(t) = \mathbf{z}(0) + \dot{\mathbf{z}}(0)t + \frac{1}{2!}\ddot{\mathbf{z}}(0)t^2 + \ldots + \frac{1}{k!}\mathbf{z}^{(k)}(0)t^k + \ldots \qquad (2.19)$$

On the other hand, the time derivatives of $\mathbf{z}$ appearing in Eq. 2.19 can be readily derived by successively differentiating both sides of Eq. 2.18a, namely,

$$\dot{\mathbf{z}} = \mathbf{Az}$$

$$\ddot{\mathbf{z}} = \mathbf{A}\dot{\mathbf{z}} \equiv \mathbf{A}^2\mathbf{z}$$

$$\vdots$$

$$\mathbf{z}^{(k)} = \mathbf{A}^k\mathbf{z}, \quad \text{etc.}$$

Upon evaluation of the foregoing derivatives at $t = 0$ and substitution of these values into Eq. 2.19, a series expansion for $\mathbf{z}(t)$ is obtained, namely,

$$\mathbf{z}(t) = \left(\mathbf{1} + \mathbf{A}t + \frac{1}{2!}\mathbf{A}^2t^2 + \ldots + \frac{1}{k!}\mathbf{A}^kt^k + \ldots \right)\mathbf{z}_0 \qquad (2.20)$$

where $\mathbf{1}$ denotes the 2×2 identity matrix.

The series inside the parentheses multiplying $\mathbf{z}_0$ is readily identified as the exponential[3] of $\mathbf{A}t$, and hence,

$$\mathbf{z}(t) = e^{\mathbf{A}t}\mathbf{z}_0, \quad t \geq 0 \qquad (2.21)$$

which is the expression sought.

The problem of computing the time response of the given system has thus been reduced to computing the exponential of $\mathbf{A}t$. This computation can be done in many ways, 19 of which were discussed by Moler and Van Loan [1], but there are many more. Conceptually, the simplest way is via the Cayley-Hamilton Theorem, as discussed in Appendix A. The exponential of $\mathbf{A}t$, as computed therein, is given below. We distinguish here three cases, namely,

1. Underdamped case: $\zeta < 1$. Here we have

$$e^{\mathbf{A}t} = \frac{e^{-\zeta\omega_n t}}{\sqrt{1-\zeta^2}} \begin{bmatrix} \sqrt{1-\zeta^2}\cos\omega_d t + \zeta\sin\omega_d t & (\sin\omega_d t)/\omega_n \\ -\omega_n\sin\omega_d t & \sqrt{1-\zeta^2}\cos\omega_d t - \zeta\sin\omega_d t \end{bmatrix}$$
$$(2.22a)$$

---

[3]The exponential of a matrix is formally identical to that of a real argument: both have the same expansion.

Fig. 2.10 Relation among ω_n, ω_d and ζ

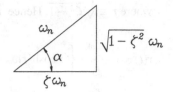

where ω_d, termed the *damped natural frequency*, is defined as

$$\omega_d \equiv \omega_n \sqrt{1 - \zeta^2} \qquad (2.22b)$$

and hence, the time response of underdamped systems is given by the two components of $\mathbf{z}(t)$, $x(t)$ and $\dot{x}(t)$:

$$x(t) = \frac{e^{-\zeta \omega_n t}}{\sqrt{1 - \zeta^2}} \left(\sqrt{1 - \zeta^2} \cos \omega_d t + \zeta \sin \omega_d t \right) x_0$$

$$+ \frac{e^{-\zeta \omega_n t}}{\omega_d} (\sin \omega_d t) v_0 \qquad (2.23a)$$

$$\dot{x}(t) = -\frac{\omega_n e^{-\zeta \omega_n t}}{\sqrt{1 - \zeta^2}} (\sin \omega_d t) x_0 + \frac{e^{-\zeta \omega_n t}}{\sqrt{1 - \zeta^2}} \left(\sqrt{1 - \zeta^2} \cos \omega_d t - \zeta \sin \omega_d t \right) v_0$$

$$(2.23b)$$

The relation among ω_n, ω_d and ζ is best illustrated in Fig. 2.10, whence it is apparent that $\omega_d < \omega_n$.

2. Critically damped case: $\zeta = 1$. Now we have

$$e^{\mathbf{A}t} = e^{-\omega_n t} \begin{bmatrix} 1 + \omega_n t & t \\ -\omega_n^2 t & 1 - \omega_n t \end{bmatrix} \qquad (2.24a)$$

and hence, the time response of critically damped systems is

$$x(t) = e^{-\omega_n t} [(1 + \omega_n t) x_0 + v_0 t] \qquad (2.24b)$$

$$\dot{x}(t) = e^{-\omega_n t} [-\omega_n^2 t x_0 + (1 - \omega_n t) v_0] \qquad (2.24c)$$

3. Overdamped case: $\zeta > 1$. In this case,

$$e^{\mathbf{A}t} = \frac{e^{-\zeta \omega_n t}}{r} \begin{bmatrix} r \cosh(r \omega_n t) + \zeta \sinh(r \omega_n t) & \frac{1}{\omega_n} \sinh(r \omega_n t) \\ -\omega_n \sinh(r \omega_n t) & r \cosh(r \omega_n t) - \zeta \sinh(r \omega_n t) \end{bmatrix}$$

$$(2.25a)$$

where $r \equiv \sqrt{\zeta^2 - 1}$. Hence, the time response of overdamped systems is

$$x(t) = \frac{e^{-\zeta \omega_n t}}{r} \left\{ [r\cosh(r\omega_n t) + \zeta \sinh(r\omega_n t)]x_0 + \frac{1}{\omega_n} \sinh(r\omega_n t)v_0 \right\} \qquad (2.25b)$$

$$\dot{x}(t) = \frac{e^{-\zeta \omega_n t}}{r} \left\{ -\omega_n \sinh(r\omega_n t)x_0 + \frac{e^{-\zeta \omega_n t}}{r} [r\cosh(r\omega_n t) - \zeta \sinh(r\omega_n t)]v_0 \right\}$$

$$\qquad (2.25c)$$

In all three above cases it is apparent that the response is a linear combination of the two initial conditions x_0 and v_0, such as in the undamped case. Contrary to the undamped case, in the case of damped systems we have one more parameter that comes into the picture, namely, the damping ratio ζ, which brings about another parameter, the damped frequency ω_d of underdamped systems. Shown in Figs. 2.11a, c and e are plots of time responses of underdamped, critically damped and overdamped systems, respectively, for various values of damping ratio, with $x_0 = 1$ and $v_0 = 0$. Figures 2.11b, d and f show the same responses for initial conditions $x_0 = 0$ and $v_0 = 1$. In all cases, x_0 has units of displacement, whether translational or angular, while v_0 has units of the corresponding velocity.

2.3.2.1   Identification of Damping from the Time Response

Knowing the time response of an underdamped system allows us to determine its damping ratio, as we will show presently. Let us take the time response of Fig. 2.11b for $x_0 = 0$ and $v_0 = 1$, and measure the displacements x_k and x_{k+1} at two different instants t_k and t_{k+1}, respectively, separated by a full period T, i.e., $t_{k+1} = t_k + T$, where $T \equiv 2\pi/\omega_d$. We then have, from expression (2.23a),

$$\frac{x_k}{x_{k+1}} = \frac{e^{-\zeta \omega_n t_k} \sin \omega_d t_k}{e^{-\zeta \omega_n t_{k+1}} \sin \omega_d t_{k+1}}$$

However, by virtue of the relationship between t_k and t_{k+1},

$$\sin \omega_d t_{k+1} = \sin(\omega_d t_k + 2\pi) \equiv \sin \omega_d t_k$$

and hence, the foregoing ratio becomes

$$\frac{x_k}{x_{k+1}} = \frac{e^{-\zeta \omega_n t_k}}{e^{-\zeta \omega_n t_{k+1}}} \equiv e^{\zeta \omega_n T}$$

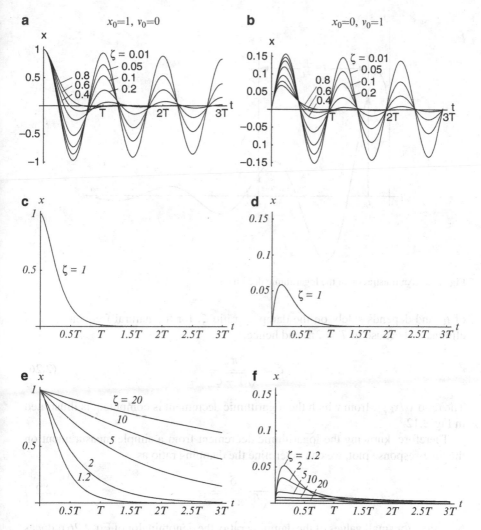

Fig. 2.11 Zero-input response of various second-order systems for various initial conditions

Now, if we take the logarithm of both sides of the foregoing equation, we obtain an interesting relation, namely,

$$\ln\left(\frac{x_k}{x_{k+1}}\right) = \zeta\omega_n T \equiv \frac{2\pi\zeta\omega_n}{\omega_d} \equiv 2\pi\frac{\zeta}{\sqrt{1-\zeta^2}}$$

The left-hand side of the latter expression is known as the *logarithmic decrement* of the motion and is denoted by δ. Note that it is a constant, regardless of the value

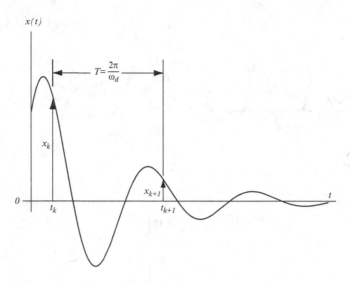

Fig. 2.12   An illustration of the logarithmic decrement

of t_k, and depends solely on the damping ratio ζ, for the natural frequency ω_n is eliminated because $\omega_d T = 2\pi$, and hence,

$$\delta = \frac{2\pi\zeta}{\sqrt{1-\zeta^2}} \tag{2.26}$$

The ratio x_k/x_{k+1}, from which the logarithmic decrement is computed, is illustrated in Fig. 2.12.

Therefore, knowing the logarithmic decrement from a simple measurement on the time-response plot, we can determine the damping ratio as

$$\zeta = \frac{\delta}{\sqrt{(2\pi)^2 + \delta^2}} \tag{2.27}$$

Note that, for small values of the damping ratio, the denominator of Eq. 2.26 reduces to unity, and hence, the logarithmic decrement and the damping ratio obey a linear relation, namely,

$$\text{For} \quad \zeta \ll 1, \quad \zeta \approx \frac{\delta}{2\pi} \tag{2.28}$$

Finally, note that the above results do not depend on the initial conditions; we should thus be able to derive them with another set of initial conditions. Moreover, if the displacement is measured N cycles apart, it is not difficult to show that the logarithmic decrement can then be expressed as

$$\delta = \frac{1}{N} \ln\left(\frac{x_k}{x_{k+N}}\right) \tag{2.29}$$

The above derivation is left as an **exercise**.

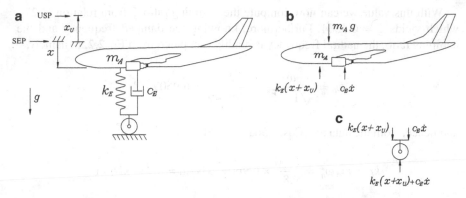

Fig. 2.13   A landing gear: (**a**) its iconic model (USP: unloaded spring position; SEP: static equilibrium position); (**b**) the FBD of the aircraft; and (**c**) the FBD of the landing gear

Example 2.3.2 (Design of a Landing Gear).   If we assume that, upon landing, all wheels of an aircraft touch the landing strip simultaneously, and that the aircraft body is rigid, the system consisting of aircraft body and landing gear can be modeled as a mass-spring-dashpot system with a mass m_A accounting for the mass of the aircraft body plus the payload, an equivalent stiffness k_E and an equivalent damping coefficient c_E, as shown in Fig. 2.13a.[4] For an aircraft weighing 1,000 kN, find the equivalent stiffness and the equivalent damping coefficient that, upon landing, will produce oscillations of the body

(a) of 1 Hz and with an amplitude of its second cycle of 5% that of the first cycle.
(b) Moreover, determine the deflection of the landing gear when the oscillations have settled and the system is in equilibrium.
(c) As well, determine the peak force exerted on the ground upon landing, if the vertical approach velocity is of 10 m/s (this value is about 2.5 times that of a rough landing, but we want to design for extreme conditions).

Solution: To begin with, we set up the mathematical model of the system at hand, with the generalized coordinate x measured from the static equilibrium position, the value of x when the spring is unstretched being denoted by x_U, as shown in Fig. 2.13a. The mathematical model thus becomes

$$m_A\ddot{x} + c_E\dot{x} + k_Ex = 0, \quad x(0) = x_0, \quad \dot{x}(0) = v_0$$

Now, from the second design requirement, we can readily determine the logarithmic decrement δ by application of relation (2.26), namely,

$$\delta = \ln\left(\frac{100}{5}\right) = 2.9957$$

---

[4]For this simple model, the wheels are assumed to be massless, rigid disks.

With this value we can now compute the damping ratio ζ from relation (2.27), which yields $\zeta = 0.4304$. Furthermore, knowing the damped frequency and the damping ratio, the natural frequency is readily computed from Eq. 2.22b, namely,

$$\omega_n = \frac{\omega_d}{\sqrt{1-\zeta^2}} = \frac{2\pi}{0.9026} = 6.9609 \text{ s}^{-1}$$

and hence, from the data and Eqs. 1.56a or 1.58b,

$$k_E = m_A \omega_n^2 = \frac{1000}{9.81} \times 6.9609^2 \text{ kN/m} = 4\,939 \text{ kN/m}$$

and

$$c_E = 2\zeta m_A \omega_n = 2 \times 0.4304 \times \frac{1000}{9.81} \times 6.9609 = 610.8 \text{ kNs/m}$$

thereby determining the design requirements for the overall landing-gear system.[5]

Further, the deflection of the landing-gear system at its equilibrium state is equal to that of the mass-spring-dashpot with which it is modeled. This deflection, denoted x_U, can be readily computed from the relation

$$k_E x_U = m_A g$$

and hence,

$$x_U = \frac{m_A g}{k_E} = \frac{1000}{4\,939} \text{ m} = 0.2024 \text{ m}$$

Now we proceed to determine the peak force transmitted by the landing aircraft to the ground. To this end, we note that the force transmitted by the spring-dashpot array to the mass is equal to that transmitted to the ground, as illustrated in Figs. 2.13b, c.

Thus, if we let $f_T(t)$ denote the transmitted force, we have

$$f_T(t) = c_E \dot{x}(t) + k_E(x + x_U) = c_E \dot{x} + k_E x + m_A g$$

The maximum force, occurring at a time t_M, as yet to be determined, is found upon zeroing the derivative of $f_T(t)$ with respect to time. This derivative is calculated below:

$$\dot{f}_T(t) = c_E \ddot{x} + k_E \dot{x}(t)$$

---

[5]Aircraft landing gears, like the suspension of terrestrial vehicles are designed with a natural frequency of around 1 Hz.

Upon solving for $\ddot{x}$ from the mathematical model and inserting the result into the foregoing equation, we have, after simplification,

$$\dot{f}_T(t) = m_A \left[\left(\frac{k_E}{m_A} - \frac{c_E^2}{m_A^2} \right) \dot{x} - \frac{c_E}{m_A} \omega_n^2 x \right]$$

Now, if we recall definitions (1.58b), the zeroing of the foregoing expression then leads to

$$(1 - 4\zeta^2)\dot{x} - 2\zeta\omega_n x = 0$$

which is satisfied at time $t = t_M$.

Next, we resort to the time response obtained above, i.e., Eqs. 2.23a, b, and substitute the expressions for $x(t)$ and $\dot{x}(t)$ of those relations into the foregoing equations, which yields, after simplifications,

$$D \sin \omega_d t_M = -\sqrt{1 - \zeta^2} N \cos \omega_d t_M$$

with D and N defined as

$$D \equiv (1 - 2\zeta^2)\omega_n x_0 + \zeta(3 - 4\zeta^2)v_0$$
$$N \equiv 2\zeta\omega_n x_0 - (1 - 4\zeta^2)v_0$$

and hence, t_M is determined as

$$t_M = \frac{1}{\omega_d} \tan^{-1}\left(-\sqrt{1 - \zeta^2}\frac{N}{D} \right)$$

and the maximum transmitted force, f_M, is found as

$$f_M \equiv f_T(t_M) = c_E \dot{x}(t_M) + k_E x(t_M) + m_A g$$

Upon substitution of the relation between x and $\dot{x}$ at $t = t_M$ found above, into the foregoing expression, we obtain

$$f_M = m_A \left(\frac{4\zeta^2}{1 - 4\zeta^2} + 1 \right) \omega_n^2 x(t_M) + m_A g$$

which readily simplifies to

$$f_M = \frac{m_A \omega_n^2}{1 - 4\zeta^2} x(t_M) + m_A g$$

This expression is valid for $1 - 4\zeta^2 \neq 0$. The case $1 - 4\zeta^2 = 0$ is left as an **exercise**.

All we need now is $x(t_M)$, which we evaluate below. We first note that

$$x(t_M) = \frac{e^{-\zeta \omega_n t_M}}{\sqrt{1-\zeta^2}} \left[\left(\sqrt{1-\zeta^2} \cos \omega_d t_M + \zeta \sin \omega_d t_M \right) x_0 + \frac{\sin \omega_d t_M}{\omega_n} v_0 \right]$$

Next, we substitute the relation between $\cos \omega_d t_M$ and $\sin \omega_d t_M$ obtained above, which yields

$$x(t_M) = \frac{e^{-\zeta \omega_n t_M}}{\sqrt{1-\zeta^2}} \left[\sqrt{1-\zeta^2}(\cos \omega_d t_M) x_0 - \left(\zeta x_0 + \frac{v_0}{\omega_n} \right) \cos \omega_d t_M \right]$$

Upon simplification,

$$x(t_M) = e^{-\zeta \omega_n t_M} \frac{x_0 D - (\zeta x_0 + v_0/\omega_n)N}{D} \cos \omega_d t_M$$

where $\cos \omega_d t_M$ is derived from the value of $\tan \omega_d t_M$ obtained above, i.e.,

$$\cos \omega_d t_M = \frac{D}{\sqrt{(1-\zeta^2)N^2 + D^2}}, \quad 0 \le \omega_d t_M \le \pi$$

Therefore.

$$x(t_M) = e^{-\zeta \omega_n t_M} \frac{x_0 D - (\zeta x_0 + v_0/\omega_n)N}{\sqrt{(1-\zeta^2)N^2 + D^2}}$$

Thus, all we need now to compute $x(t_M)$ is the initial conditions x_0 and v_0. The value of x_0 is the length of the spring upon touching down; at this instant, the spring is stretched a small amount due to the weight of the landing gear, which is not known. However, if we realize that the latter is negligible when compared to the weight of the aircraft body, we can then safely assume that the spring is unloaded upon touching down, and hence,

$$x_0 = -x_U = -0.2024 \text{ m}$$

Further, v_0 is the vertical component of the aircraft velocity upon touching down, i.e.,

$$v_0 = 10 \text{ m/s}$$

which thus yields $x(t_M) = 0.2956$ m, and hence, the peak force is given by

$$f_M = \frac{m_A \omega_n^2}{1 - 4\zeta^2} x(t_M) + m_A g = \frac{1000 \times 6.9609^2}{g(1 - 4 \times 0.4304^2)} \times (0.2956) + 1000 = 6{,}636 \text{ kN}$$

Note that the maximum force exerted by the landing gear on the ground is about seven times the weight of the aircraft, which indicates that dynamical effects are

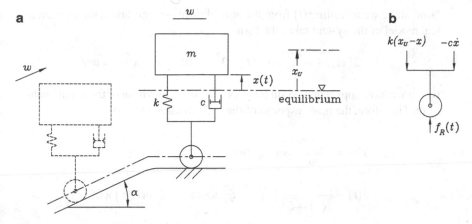

Fig. 2.14 The iconic model of a terrestrial vehicle: (**a**) upon climbing up a slope; and (**b**) the free-body diagram of its suspension

of the utmost importance both for the mechanical engineer designing landing gears and the fuselage structure, and for the civil engineer designing the landing strip.

Example 2.3.3 (Suspension Analysis). A rather crude iconic model of a terrestrial vehicle is shown in Fig. 2.14a, upon overcoming a slope of angle α, assumed to be "small," at a uniform speed w. (a) Determine the force transmitted to the ground *just after* the vehicle has overcome the slope and finds itself on level ground, in terms of the given parameters; (b) for a slope of 2%, $\zeta = \sqrt{2}/2$, and $\omega_n = 1$ Hz, find the speed w that will cause the vehicle "to fly" upon reaching the top of the slope; and (c) for the given numerical values, find the maximum and the minimum values of the force transmitted to the ground and the instants at which these extrema occur, during the first cycle of oscillations.

Solution:

(a) A free-body diagram of the lower part of the suspension is shown in Fig. 2.14b, from which the force $f_T(t)$ transmitted to the ground can be expressed as the negative of the reaction force $f_R(t)$ on the vehicle. Moreover, such as in Example 2.3.2, x_U denotes the value of x when the spring is unloaded, and hence,

$$x_U = \frac{m}{k}g = \frac{g}{\omega_n^2}$$

Therefore,

$$f_R(t) = -f_T(t) = -(kx + c\dot{x}) + mg \equiv -m\left(\omega_n^2 x + 2\zeta\omega_n\dot{x} - g\right)$$

For convenience, we work with $f_R(t)$, which is positive upwards, and hence, this function is non-negative.

Now, since we measure $x(t)$ from the equilibrium configuration, the mathematical model of the system takes the form

$$\ddot{x} + 2\zeta\omega_n\dot{x} + \omega_n^2 x = 0, \quad t \geq 0, \quad x(0) = 0, \quad \dot{x}(0) = w\alpha$$

where we have approximated $\sin\alpha$ by its argument, by virtue of the small[6] value of α. Therefore, the time response of the system takes the form

$$x(t) = \frac{e^{-\zeta\omega_n t}}{\omega_d}(\sin\omega_d t)w\alpha$$

$$\dot{x}(t) = \frac{e^{-\zeta\omega_n t}}{\sqrt{1-\zeta^2}}\left(\sqrt{1-\zeta^2}\cos\omega_d t - \zeta\sin\omega_d t\right)w\alpha$$

Now, the force transmitted to the ground just after the vehicle has overcome the slope is

$$f_R(0) = -m\left[\omega_n^2 x(0) + 2\zeta\omega_n\dot{x}(0) - g\right] = -m(2\zeta\omega_n w\alpha - g)$$

(b) For the vehicle body "to fly," $f_R(0)$ must vanish, and hence, we must have

$$w = \frac{g}{2\zeta\omega_n\alpha} = \frac{9.81}{2(\sqrt{2}/2)\times 2\pi \times 0.02} = 55 \text{ m/s}$$

or 198.7 km/h, which is a rather unlikely speed in North-American roads, and hence, the design, defined by the values of ζ and ω_n, is suitable.

(c) In order to determine the maximum—or minimum for that matter—of the force transmitted to the ground, we must set $\dot{f}_R(t) = 0$, i.e.,

$$\dot{f}_R(t) = -m(\omega_n^2\dot{x} + 2\zeta\omega_n\ddot{x}) = 0$$

By taking into account the mathematical model of the system at hand into the above equation, we obtain

$$\dot{f}_R(t) = -m\left[\omega_n^2\dot{x} + 2\zeta\omega_n\left(-2\zeta\omega_n\dot{x} - \omega_n^2 x\right)\right] = 0$$

or

$$(1 - 4\zeta^2)\dot{x} = 2\zeta\omega_n x$$

---

[6]Roads are typically designed with slopes below 6%, although exceptionally roads with slopes of 10% and even 20% exist. The validity of the approximation $\sin x \approx x$ is limited by $\pi/30$ or $6°$, while a slope of 6% entails an angle $\alpha = 3.4°$.

and, since $1 - 4\zeta^2 = 1 - 4(1/2) = -1 \neq 0$,

$$\dot{x} = \frac{2\zeta\omega_n x}{1 - 4\zeta^2}$$

If we now substitute the expressions for $x(t)$ and $\dot{x}(t)$ obtained above, into the foregoing expressions, we have

$$\sqrt{1 - \zeta^2} \cos \omega_d t_0 - \zeta \sin \omega_d t_0 = \frac{2\zeta\omega_n}{1 - 4\zeta^2} \frac{\sin \omega_d t_0}{\omega_n}$$

where t_0 is the instant at which a stationary value of $f_R(t)$ occurs. After simplifications,

$$\tan \omega_d t_0 = \frac{\sqrt{1 - \zeta^2}(1 - 4\zeta^2)}{(3 - 4\zeta^2)\zeta}$$

For the given numerical value of ζ, $\sqrt{2}/2$, we obtain

$$\tan \omega_d t_0 = -1$$

and hence, stationary values, i.e., maxima and minima, of $f_R(t)$ occur at the values given below:

$$\omega_d t_0 = \frac{3\pi}{4}, \quad \frac{7\pi}{4}, \quad \frac{11\pi}{4}, \dots, \text{etc.}$$

Now, in order to determine the instants corresponding to the foregoing angular values, we need ω_d, which is readily calculated as

$$\omega_d = \sqrt{1 - \zeta^2}\,\omega_n = \frac{\sqrt{2}}{2} 2\pi = \sqrt{2}\pi$$

Thus, $f_R(t)$ attains extremum values—local maxima and minima—at a sequence of values t_i given by

$$t_i = \frac{\sqrt{2}}{2}\left(\frac{3}{4} + i\right), \quad i = 1, 2, \dots$$

or

$$t_1 = 0.5303 \text{ s}, \quad t_2 = 1.2374 \text{ s}, \quad t_3 = 1.9445 \text{ s}, \dots, \quad \text{etc.}$$

Now, in order to determine whether a stationary value of $f_R(t)$ is a maximum or a minimum, or even a saddle point, we calculate $\ddot{f}_R(t)$ and investigate its sign. The general expression for the derivative of interest is

$$\ddot{f}_R(t) = -m\left[\left(\omega_n^2 - 4\zeta^2\omega_n^2\right)\ddot{x} - 2\zeta\omega_n^3\dot{x}\right]$$

Upon introducing the mathematical model into the above derivative, we obtain

$$\dot{f}_R(t) = m\omega_n^3 \left[4\zeta(1-2\zeta^2)\dot{x}+(1-4\zeta^2)\omega_n x\right]$$

Now, at a stationary value, i.e., at $t = t_i$, for $i = 1, 2, \ldots$,

$$(1-4\zeta^2)\dot{x}(t_i) = 2\zeta\omega_n x(t_i)$$

and hence,

$$\ddot{f}_R(t_i) = m\frac{\omega_n^4}{1-4\zeta^2}x(t_i) = -19.739x(t_i)$$

whence,

$$\text{sgn}[\ddot{f}_R(t_i)] = -\text{sgn}[x(t_i)]$$

Now we evaluate $x(t_1)$:

$$x(t_1) = x\left(\frac{3\pi}{4\omega_d}\right) = \frac{e^{-\zeta\omega_n t_1}}{\omega_d}\sin\left(\frac{3\pi}{4}\right)w\alpha = \frac{\sqrt{2}}{2}\frac{e^{-\zeta\omega_n t_1}}{\omega_d}w\alpha > 0$$

the sign of $\ddot{f}_R(t_1)$ being therefore negative, which means that $f_R(t_1)$ is a maximum, namely,

$$f_R(t_1) = -m\left[\omega_n^2 x(t_1) + 2\zeta\omega_n \dot{x}(t_1) - g\right]$$

and, if we take into account the relation between x and $\dot{x}$ at stationary values of $f_R(t)$, we obtain

$$f_R(t_1) = -m\left[\frac{\omega_n^2}{1-4\zeta^2}x(t_1) - g\right]$$

where $x(t_1)$ is given by

$$x(t_1) = \frac{e^{-\zeta\omega_n t_1}}{\omega_d}(\sin\omega_d t_1)w\alpha = \frac{e^{-\zeta\omega_d t_1/\sqrt{1-\zeta^2}}}{\omega_d}(\sin\omega_d t_1)w\alpha$$

whence,

$$x(t_1) = 3.0169 \times 10^{-4}w$$

Therefore,

$$f_R(t_1) = -m\left(\frac{4\pi^2}{1-2} \times 3.0169 \times 10^{-4}w - g\right) = m(0.1191w + g)$$

Now we investigate the sign of $x(t_2)$:

$$x(t_2) = \frac{e^{-\zeta \omega_n t_2}}{\omega_d} \sin\left(\frac{7\pi}{4}\right) w\alpha = -\frac{\sqrt{2}}{2} \frac{e^{-\zeta \omega_n t_2}}{\omega_d} w\alpha < 0$$

Therefore, $f_R(t_2)$ attains a minimum value given by

$$f_R(t_2) = -m\left[\omega_n^2 x(t_2) + 2\zeta \omega_n \dot{x}(t_2) - g\right]$$

Again, upon considering the relation between x and $\dot{x}$ at a stationary value of $f_R(t)$, we obtain

$$f_R(t_2) = m\left[\frac{\omega_n^2 x(t_2)}{4\zeta^2 - 1} + g\right]$$

with $x(t_2)$ readily calculated as

$$x(t_2) = -1.3037 \times 10^{-5} w$$

and hence,

$$f_R(t_2) = \max\{(-5.1468 \times 10^{-4} w + g), 0\}$$

where a provision has been taken in order to avoid negative values of the reaction force, which are physically inadmissible.

2.4   The Zero-State Response of LTIS

The response of a system to the sole action of external excitations, i.e., under the assumption that the system is at rest prior to the application of the excitations, is termed the *zero-state response*. This term indicates that the state of the system—position and velocity in mechanical systems—is equal to zero. In this section, we derive the zero-state response of the systems under study for an arbitrary excitation by resorting to the property of superposition of linear systems. To this end, we decompose the arbitrary input as a continuous train of amplitude-modulated impulses. Hence, we start by studying the response of the systems at hand to a *unit impulse*. It will become apparent that this form of excitation is the simplest one. The unit impulse, like its derivative and its integral, the doublet and the unit step, respectively, belong to a class of functions termed *discontinuous*. These functions pose interesting challenges to the analyst [2], that need not be discussed in an introductory book. These functions will be handled with due care in the balance of the book.

Fig. 2.15 Unit impulse
applied at $t = 0$

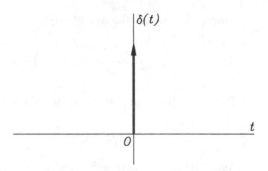

2.4.1 The Unit Impulse

The unit impulse is defined as a function of time that is zero everywhere, except
for an infinitesimally small neighborhood around the origin, in which the function
attains unbounded values. However, its time integral from $-\infty$ to $+\infty$ is exactly $+1$.
Thus, if we denote by 0^- and 0^+ the instants *just before* and *just after* 0, respectively,
the unit-impulse function, represented as $\delta(t)$, is then defined as:

$$\delta(t) \begin{cases} = 0 & \text{for } t \le 0^-; \\ \to \infty & \text{for } 0^- < t < 0^+; \\ = 0 & \text{for } t \ge 0^+ \end{cases} \tag{2.30}$$

and

$$\int_{-\infty}^{+\infty} \delta(t)dt = 1 \tag{2.31}$$

the unity on the right-hand side of the last equation being dimensionless. As a
consequence, the unit-impulse function has units of s^{-1}, i.e., of frequency. This
function is also called the *delta function* or the *Dirac function*, and is represented as
a vertical arrow at the origin, as in Fig. 2.15, of unit length in the scale adopted.

Note that, if the unit-impulse function is multiplied by a constant A to obtain
$A\delta(t)$, it follows from Eq. 2.31 that its time integral from $-\infty$ to $+\infty$ is equal to A.
We then say that $A\delta(t)$ is an *impulse function* of magnitude A, the "magnitude" being, in fact, the area under the impulse. Such a non-unit impulse is thus represented
as an arrow of height A in the scale adopted. The height of the arrow, then, denotes
the time integral of the associated impulse function on the whole real axis.

2.4.2 The Unit Doublet

The time derivative of a unit impulse, $\dot{\delta}(t)$, is called the *doublet* function. It is
formally defined as

$$\dot{\delta}(t) = \frac{d}{dt}\delta(t) \tag{2.32}$$

Fig. 2.16 Doublet applied at
$t = 0$

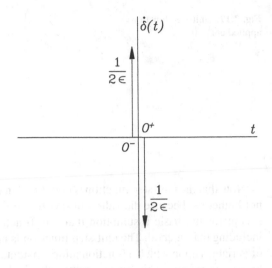

Using the definition of the time-derivative, Eq. 2.32 can be written as

$$\dot{\delta}(t) = \lim_{\varepsilon \to 0} \frac{\delta(t+\varepsilon) - \delta(t-\varepsilon)}{2\varepsilon} = \lim_{\varepsilon \to 0} \left[\frac{1}{2\varepsilon} \delta(t+\varepsilon) - \frac{1}{2\varepsilon} \delta(t-\varepsilon) \right]$$

with ε being defined as

$$2\varepsilon \equiv 0^+ - 0^-$$

Therefore, the physical interpretation of a doublet is the limiting case of two impulses of ∞ amplitude, one applied at $t = 0^-$ and the other at $t = 0^+$, the latter being the negative of the former. The doublet is sketched in Fig. 2.16.

The units of the doublet are, of course, s^{-2}, i.e., frequency-squared. As will be shown later, the doublet can be used to describe inputs that cause the position of the mass of a mass-spring-dashpot system to change instantaneously without undergoing any change in its velocity.

The *triplet*, the *quadruplet*, etc. are functions defined likewise. In general, the $(n-1)$st derivative of the unit impulse, denoted by $\delta^{(n-1)}(t)$, is termed the *n-tuplet* function. Of interest to us are mainly the impulse and the doublet functions.

2.4.3   The Unit Step

Further, the *unit-step* function, or *Heaviside function*, is defined below:

$$u(t) \equiv \begin{cases} 0, & \text{for } t \leq 0^-, \\ +1, & \text{for } t \geq 0^+ \end{cases} \tag{2.33}$$

This function is sketched in Fig. 2.17.

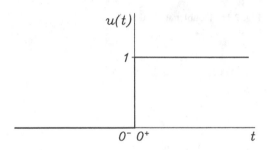

Fig. 2.17 Unit-step function applied at $t = 0$

Note that the unit-step function is undefined in the interval $0^- < t < 0^+$. This does not bother us, because the values of this function in that interval are never needed, except for the basic assumption that this function remains bounded everywhere, including that interval. The unit-step function is needed to represent *abrupt* changes of variables upon which a function jumps instantaneously from one value to another by a *finite* amount. This corresponds to physical situations such as a constant, finite force applied suddenly onto a mass, the sudden closing of a switch in a circuit driven by a battery, the sudden exposure of a body at a given temperature to a constant, finite temperature, different from that of the body, and so on.

The relationship between the unit-step and the unit-impulse functions is obviously

$$u(t) = \int_{-\infty}^{t} \delta(\theta)d\theta \qquad (2.34)$$

where θ is a dummy variable of integration. Hence, the unit-step function is dimensionless. Furthermore,

$$\delta(t) = \frac{du(t)}{dt} \qquad (2.35)$$

2.4.4   The Unit Ramp

One more function of interest is the *unit-ramp function*, $r(t)$, defined as

$$r(t) \equiv \begin{cases} 0, & \text{for} \quad t \leq 0^- \\ t, & \text{for} \quad t \geq 0^+. \end{cases} \qquad (2.36)$$

The unit-ramp function is sketched in Fig. 2.18.

Note that $r(t)$ can be interpreted as the integral of $u(t)$, i.e.,

$$r(t) \equiv \int_{-\infty}^{t} u(\theta)d\theta \qquad (2.37)$$

Fig. 2.18 Unit-ramp
function applied at $t = 0$

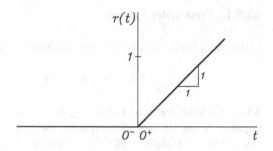

with θ, again, being a dummy variable of integration. Moreover, we have the relations:

$$u(t) = \frac{dr(t)}{dt}, \quad \delta(t) = \frac{d^2 r(t)}{dt^2} \tag{2.38}$$

Obviously, the unit-ramp function has units of time, i.e., second.

We will now derive explicit expressions for the impulse responses of first- and second-order LTI dynamical systems.

2.4.5   The Impulse Response

In this subsection we consider the response of linear time-invariant dynamical systems to a unit impulse, the response at hand being termed the *impulse response*—of the system at hand, of course. The importance of the impulse response cannot be overstated. For example, knowing the impulse response of a LTIS we can obtain, by superposition, the response of the same system to *any* input, provided that the initial conditions are zero in all cases. On the other hand, the unit impulse models quite effectively inputs that have a very short duration but a very large amplitude and hence, their effect on the behavior of the system under study is not negligible. Examples of such inputs appear whenever collisions occur. For example, when hitting a ball with a tennis racket, the ball is acted upon by an infinitely large force during an infinitesimally small interval of time. As a consequence, the ball undergoes a finite change in its velocity that implies a finite change in its momentum. This finite change can be explained as the result of a finite impulse acting on the ball.

The impulse responses of first- and second-order LTI systems are derived in this order. After this, we show how, by exploiting the properties of *linearity* and *time-invariance* of the systems at hand, the impulse response of such systems can be used to determine the response of the same systems to *any* type of input by means of the *convolution integral*. The responses of first- and second-order LTI systems for any arbitrary input will then be derived.

2.4.5.1 First-order Systems

In this case, the associated differential equation is

$$\dot{x} = -ax + \delta(t), \quad t > 0^-, \quad x(0^-) = 0 \tag{2.39}$$

where we have assumed that the system is at rest prior to the application of the impulse. A simple way of computing the response of the foregoing system is by calculating its state $x(t)$ at $t = 0^+$ and rewriting Eq. 2.39 for $t \geq 0^+$, when the impulse function is zero, thereby reducing the response of the system to the zero-input response, which was already found. In order to find $x(0^+)$, Eq. 2.39 is rewritten in the form:

$$\dot{x} + ax = \delta(t), \quad t > 0^-, \quad x(0^-) = 0$$

The presence of a delta function on the right-hand side of the previous equation requires a delta function on the left-hand side. Now, let us look for this function there. It can be either in x or in $\dot{x}$. If it were in x, then $\dot{x}$ would be proportional to a doublet, which should be balanced by a doublet on the right-hand side. Since no such doublet is present there, this possibility is discarded. Then, the impulsive function must be in the $\dot{x}$ term, which implies that x is a multiple of the unit-step function, and hence, at the origin, x undergoes a finite jump. This means that, in the neighborhood of $t = 0$, x remains bounded. Now we have enough information on all terms appearing in Eq. 2.39 to perform their integral from $t = 0^-$ to $t = 0^+$, which is done below:

$$\int_{0^-}^{0^+} \dot{x} dt = -a \int_{0^-}^{0^+} x dt + \int_{0^-}^{0^+} \delta(t) dt \tag{2.40}$$

The integral of the left-hand side is readily identified as $x(0^+)$; the first integral appearing on the right-hand side vanishes, for $x(t)$, not containing any impulsive component, is finite in the neighborhood of $t = 0$. Finally, the last integral of that equation is simply $+1$ by definition, and hence, one has

$$x(0^+) = 1$$

Now, Eq. 2.39 can be rewritten as:

$$\dot{x} = -ax, \quad t \geq 0^+, \quad x(0^+) = 1 \tag{2.41}$$

with zero-input term and nonzero initial conditions. The response of the system appearing in Eq. 2.41 was already found in Sect. 2.2. If we now just replace 0 of Eq. 2.8 with 0^+, and recall the initial condition of the system at hand, we readily derive

$$x(t) = \begin{cases} 0, & \text{for } t \leq 0^-; \\ e^{-at}, & \text{for } t \geq 0^+ \end{cases} \tag{2.42a}$$

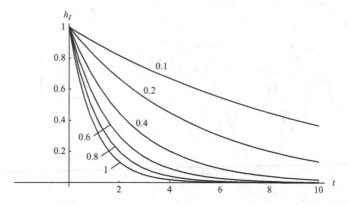

Fig. 2.19   Impulse response of a first-order system for various values of a

The response given in Eq. 2.42a is the impulse response of a first-order system; we shall denote this response by $h_I(t)$, and thus, in compact form,

$$h_I(t) \equiv e^{-at} u(t) \tag{2.42b}$$

which is sketched in Fig. 2.19, for various values of parameter a. Notice that, in the figure, $h_I(t) = 0$, for $t < 0^-$, an important feature of the response at hand that escapes to methods relying on math-book solutions to ODEs.

Note that the integral of the differential equation (2.39) given in Eq. 2.42a should verify (1) the differential equation and (2) the initial conditions. Below we show that it indeed verifies both. In fact, upon differentiation of $x(t)$ as given by Eq. 2.42b, we obtain

$$\dot{h}_I(t) = -ae^{-at} u(t) + e^{-at} \delta(t)$$

where relation (2.35) has been recalled. Now, since we are considering the response of the system for $t \geq 0^+$ only, at which $\delta(t) = 0$ and $u(t) = 1$, the foregoing expression readily reduces to

$$\dot{h}_I(t) = -ae^{-at} = -ah_I(t), \quad t \geq 0^+$$

thereby showing that the solution found verifies indeed the differential equation. Moreover,

$$h_I(0^+) = \underbrace{e^{-a0^+}}_{1} \underbrace{u(0^+)}_{1} = 1$$

and hence, h_I also verifies the initial conditions, the proposed response, Eq. 2.42b, thus being a valid solution to the *initial-value problem* (2.41), and hence to the original problem (2.39).

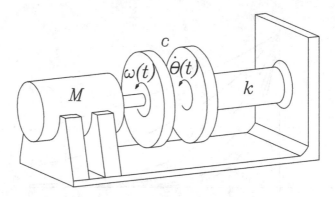

Fig. 2.20   Fluid clutch

Example 2.4.1 (Fluid-clutch Tests). Consider the fluid clutch of Example 1.6.10. The system is reproduced in Fig. 2.20 for quick reference. If the left-end disk is displaced suddenly through an angle ϕ_0, determine the ensuing motion of the right-end plate, $\theta(t)$. If we let $\phi(t)$ denote the angular displacement of the left-end plate, then $\omega(t) = \dot{\phi}(t)$.

Solution: The mathematical model of this system can be readily shown to be

$$\dot{\theta} + \frac{1}{\tau}\theta = \omega(t), \quad \theta(0^-) = 0, \quad \omega(t) = \dot{\phi}(t), \quad \phi(t) = \phi_0 u(t), \quad t > 0^-$$

where the time constant τ is defined as $\tau \equiv c/k$ and the input angle $\phi(t)$ has been modeled as a unit-step function. Hence,

$$\omega = \dot{\phi}(t) = \phi_0 \delta(t)$$

and the model now takes on the form

$$\dot{\theta} + \frac{1}{\tau}\theta = \phi_0 \delta(t), \quad \theta(0) = 0, \quad t > 0$$

In the next step, we transform the above model into zero-input form by expressing it as a homogeneous ODE with non-zero initial condition. To this end, we integrate both sides of the equation between 0^- and 0^+, while taking into account that $\theta(t)$ is not impulsive. We realize this by resorting to the arguments introduced above. From this integration we obtain

$$\theta(0^+) - \theta(0^-) = \phi_0$$

and, by virtue of the original initial condition, $\theta(0^-) = 0$ and hence, $\theta(0^+) = \phi_0$, the homogeneous ODE thus taking the form

$$\dot{\theta} + \frac{1}{\tau}\theta = 0, \quad \theta(0^+) = \phi_0, \quad t > 0^+$$

Therefore, the time response sought is

$$\theta(t) = \phi_0 e^{-t/\tau}, \quad t > 0^+$$

2.4.5.2 Second-order Undamped Systems

In this case, the governing equation is

$$\ddot{x} + \omega_n^2 x = \delta(t), \quad x(0^-) = 0, \quad \dot{x}(0^-) = 0, \quad t > 0^- \qquad (2.42c)$$

As in the case of first-order systems, the impulse on the right-hand side must be balanced with an impulse on the left-hand side. Moreover, this impulse cannot lie in the x term. If it did, then $\dot{x}$ would necessarily contain a doublet and $\ddot{x}$ a *triplet*, which should be balanced in the right-hand side. Since neither triplet nor doublet appear in that side, x is not impulsive, and hence, is bounded at the origin. The only impulsive term of the LHS[7] of Eq. 2.42c is, then, the first term. Next, we integrate both sides of the same equation between 0^- and 0^+, thereby obtaining

$$\int_{0^-}^{0^+} \ddot{x} \, dt + \omega_n^2 \int_{0^-}^{0^+} x \, dt = \int_{0^-}^{0^+} \delta(t) \, dt \qquad (2.42d)$$

The first integral of the left-hand side of Eq. 2.42d yields, as in the first-order case,

$$\int_{0^-}^{0^+} \ddot{x} \, dt = \dot{x}(0^+) - \dot{x}(0^-) = \dot{x}(0^+) \qquad (2.42e)$$

the second integral vanishing because its integrand is finite at the origin and the integration is performed over an infinitesimally small time-interval around the origin. Finally, the integral of the right-hand side yields 1. Thus, Eq. 2.42d reduces to

$$\dot{x}(0^+) = 1$$

Moreover, since $\dot{x}(t)$ is bounded at the origin, $x(t)$ does not undergo any finite jump at the origin, and hence,

$$x(0^+) = x(0^-) = 0$$

Thus, Eq. 2.42c can be rewritten as

$$\ddot{x} + \omega_n^2 x = 0, \quad x(0^+) = 0, \quad \dot{x}(0^+) = 1, \quad t \geq 0^+$$

---

[7] Abbreviation of *left-hand side*.

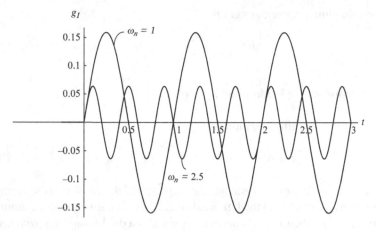

Fig. 2.21 Impulse response of a second-order undamped system

Therefore, the time response of the system under study can be obtained from the general expression, Eq. 2.13, as

$$x(t) = \begin{cases} 0, & \text{for} \quad t \leq 0^-; \\ (\sin \omega_n t)/\omega_n, & \text{for} \quad t \geq 0^+ \end{cases}$$

The impulse response of the given undamped second-order system is represented henceforth by $g_I(t)$, i.e., in compact form,

$$g_I(t) \equiv \frac{1}{\omega_n}(\sin \omega_n t)u(t) \tag{2.42f}$$

which is plotted in Fig. 2.21, for $\omega_n = 1$ s^{-1} and $\omega_n = 2.5$ s^{-1}. Again, notice that the foregoing time response vanishes for values of $t \leq 0^-$.

2.4.5.3 Second-order Damped Systems

In this case, the governing equation is

$$\ddot{x} + 2\zeta\omega_n\dot{x} + \omega_n^2 x = \delta(t), \quad x(0^-) = 0, \quad \dot{x}(0^-) = 0, \quad t > 0^- \tag{2.42g}$$

As in the previous case, the impulse on the right-hand side must be balanced with an impulse on the left-hand side. Moreover, this impulse can neither lie in the x nor in the $\dot{x}$ term. If it lied, say, in the latter, then $\ddot{x}$ would necessarily contain a doublet, which should be balanced in the right-hand side. Since no doublet appears in that side, $\dot{x}$ and, consequently, x do not contain impulsive functions, and are, hence, bounded at the origin. Next, we integrate both sides of Eq. 2.42g between 0^- and 0^+, thereby obtaining

$$\int_{0^-}^{0^+} \ddot{x}dt + 2\zeta\omega_n \int_{0^-}^{0^+} \dot{x}dt + \omega_n^2 \int_{0^-}^{0^+} xdt = \int_{0^-}^{0^+} \delta(t)dt \qquad (2.42h)$$

The first integral of the left-hand side of Eq. 2.42h is readily evaluated, namely,

$$\int_{0^-}^{0^+} \ddot{x}dt = \dot{x}(0^+) - \dot{x}(0^-) = \dot{x}(0^+)$$

where the given initial conditions of Eq. 2.42g were taken into account.

Moreover, the second and the third integrals of the left-hand side of Eq. 2.42h vanish because their integrands are finite at the origin and the integration is performed over an infinitesimally small time-interval around the origin. Finally, the integral of the right-hand side yields 1, Eq. 2.42h thus reducing to

$$\dot{x}(0^+) = 1$$

Furthermore, since $\dot{x}(t)$ is bounded at the origin, $x(t)$ does not undergo any finite jump at the origin, and hence,

$$x(0^+) = x(0^-) = 0$$

Thus, Eq. 2.42g can be rewritten as

$$\ddot{x} + 2\zeta\omega_n\dot{x} + \omega_n^2 x = 0, \quad x(0^+) = 0, \quad \dot{x}(0^+) = 1, \quad t \geq 0^+$$

Therefore, the time response of the system under study can be obtained from the general expressions derived in Sect. 2.3.2. For *underdamped systems*, for example, we have, from Eq. 2.23a,

$$x(t) = \begin{cases} 0, & \text{for } t \leq 0^-; \\ (e^{-\zeta\omega_n t}\sin\omega_d t)/\omega_d, & \text{for } t \geq 0^+ \end{cases}$$

By analogy with the previous cases, the response of the underdamped second-order system to a unit impulse is termed the *impulse response* of that system, and is denoted by $j_I(t)$, i.e., in compact form,

$$j_I(t) \equiv \frac{e^{-\zeta\omega_n t}}{\omega_d}(\sin\omega_d t)u(t) \qquad (2.42i)$$

The impulse response derived above is represented graphically in Fig. 2.22 for various values of ζ, with ω_n fixed. Needless to say, $j_I(t) = 0$, for $t \leq 0^-$.

The impulse responses $k_I(t)$ and $l_I(t)$ of critically damped and overdamped systems, respectively, are derived likewise, the details of their derivation being left

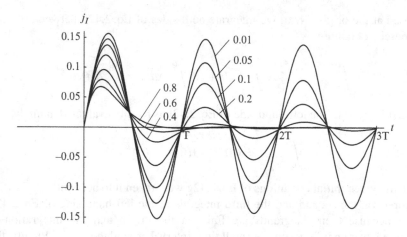

Fig. 2.22 Impulse response of a second-order underdamped system for various values of ζ

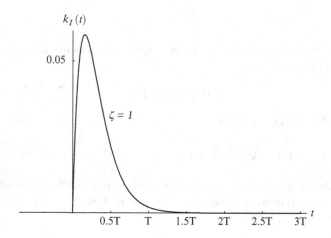

Fig. 2.23 Impulse response of a second-order critically damped system

as an exercise to the reader. These responses are displayed in Figs. 2.23 and 2.24, respectively, the latter for various values of ζ, expressions for these being included in Eqs. 2.91 and 2.92a. It goes without saying that $k_I(t) = 0$ and $l_I(t) = 0$, for $t \le 0^-$.

Example 2.4.2 (Damping Identification from the Impulse Response). A test pad[8] is modeled as shown in Fig. 2.25. In this model, the steel pad of mass m_1 is mounted on relatively soft springs. The light damping of the springs, as yet unknown but assumed to be linear, is represented in the model by the dashpot. In order to determine the physical parameters of the pad-suspension system, it has been

---

[8]Taken, with some modifications, from [3].

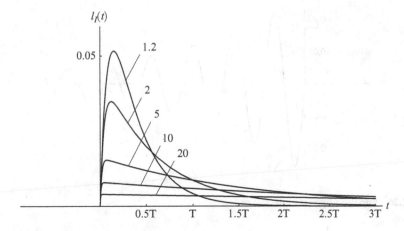

Fig. 2.24 Impulse response of a second-order overdamped system

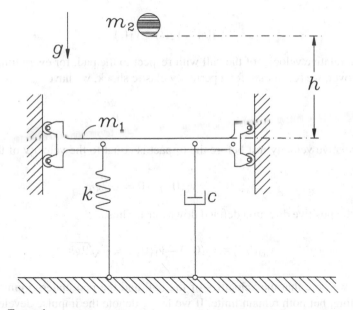

Fig. 2.25 Test pad

proposed to record its impulse response. To this end, a steel ball of mass m_2 is dropped onto the center of the pad from a height h, and then caught by an observer on the first bounce. The ensuing motion of the pad is then recorded, as appearing in Fig. 2.26. Under the assumption that the collision of the ball with the pad is perfectly elastic, (a) estimate, from the plot of this figure, the natural frequency and the damping ratio of the system, and (b) calculate the maximum excursion of the pad in its first oscillation. It is known that the ratio m_2/m_1 is 0.01 and $h = 1$ m.

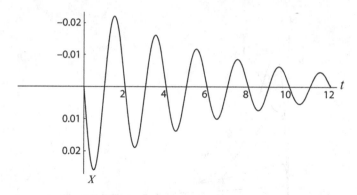

Fig. 2.26 Impulse response of the test pad

Solution:

(a) Let

$$v_{\mathrm{rel}}(t) \equiv v_2(t) - v_1(t)$$

be the relative velocity of the ball with respect to the pad, for every time $t \geq 0$. Moreover, we recall that, for a perfectly elastic shock, we have

$$v_{\mathrm{rel}}(0^+) = -v_{\mathrm{rel}}(0^-)$$

The relative velocity just before the impact is equal to the velocity of the ball, i.e.,

$$v_{\mathrm{rel}}(0^-) = v_2(0^-) - 0 = \sqrt{2gh}$$

with the positive direction defined downward. Thus,

$$v_{\mathrm{rel}}(0^+) \equiv v_2(0^+) - v_1(0^+) = -\sqrt{2gh} \tag{2.43}$$

Now, we note that the velocities of the ball and the pad undergo jumps upon colliding, but both remain finite. If we let j_d denote the impulse developed by the dashpot and j_s denote that developed by the spring, then

$$j_d = \int_{0^-}^{0^+} cv_1(t)\,dt = 0 \quad \text{because } v_1 \text{ is finite} \tag{2.44}$$

$$j_s = \int_{0^-}^{0^+} kx_1(t)\,dt = 0 \quad \text{because } x_1 \text{ is finite} \tag{2.45}$$

so that the total impulse is zero. Thus, the total change in linear momentum, Δp, is zero as well, which means

$$\Delta p = m_1 v_1(0^+) + m_2 v_2(0^+) - m_2 v_2(0^-) = 0$$

from which we obtain, upon dividing all terms of the above equation by m_1,

$$v_1(0^+) + \alpha v_2(0^+) = \alpha \sqrt{2gh} \tag{2.46}$$

where $\alpha \equiv m_2/m_1 = 0.01$. Now, solving Eqs. 2.43 and 2.46 for $v_1(0^+)$ and $v_2(0^+)$ gives

$$v_1(0^+) = \frac{2\alpha}{1+\alpha} \sqrt{2gh}, \qquad v_2(0^+) = -\frac{1-\alpha}{1+\alpha} \sqrt{2gh}$$

Noting that $x_1(0^+) = 0$, the time response of Eq. 2.23a takes the form

$$x_1(t) = \frac{e^{-\zeta \omega_n t}}{\omega_d} (\sin \omega_d t) v_1(0^+)$$

Moreover,

$$v_1(0^+) = \frac{0.02}{1.01} \sqrt{2 \times 9.81 \times 1.0} = 0.08771 \text{ m/s}$$

From the plot of Fig. 2.26, we can readily find that the damped frequency ω_d of the system is 0.5 Hz. Now, the damping ratio ζ of the system can be found via the logarithmic decrement δ, which can be estimated from the same plot. Here, it is apparent that the system is, in fact, lightly damped, and so, the decrement of the amplitude of the oscillations from one cycle to the next one is very small. Measuring it from the plot, then, can lead to substantial error. In order to cope with this uncertainty, we measure the decrement through various cycles. From the figure, it is apparent that we have information of up to six cycles, and so, if we measure the first and the sixth negative peaks, we obtain $x_1 = 0.022$ m and $x_6 = 0.005$ m, which, when plugged into Eq. 2.29, with $k = 1$ and $N = 5$, yield

$$\delta = \frac{1}{5} \ln \left(\frac{x_1}{x_6} \right) = \frac{1}{5} \ln(4.4) = 0.296$$

For light damping, we have

$$\zeta \approx \frac{\delta}{2\pi} = 0.04711$$

Now, the natural frequency ω_n is simply obtained as

$$\omega_n = \frac{\omega_d}{\sqrt{1-\zeta^2}} = \frac{2\pi/T}{\sqrt{1-\zeta^2}} = \frac{2\pi/2}{\sqrt{1-0.04711^2}} = 3.1451 \text{ rad/s}$$

where, from inspection of the plot of Fig. 2.26, $T = 2$ s, thereby completing the identification of the parameters involved.

(b) Now, in order to calculate the maximum excursion of the pad at the first oscillation, all we need is calculate $x_1(t)$ at the instant t_M at which $\dot{x}_1$ first vanishes. From the above expression for $x_1(t)$, we can readily derive

$$\dot{x}_1(t) = \frac{e^{-\zeta\omega_n t}}{\sqrt{1-\zeta^2}} \left(\sqrt{1-\zeta^2}\cos\omega_d t - \zeta\sin\omega_d t \right) v_1(0^+)$$

and hence, t_1 can be found as the instant at which the term inside the parentheses of the above expression vanishes, i.e., from

$$\sqrt{1-\zeta^2}\cos\omega_d t_1 - \zeta\sin\omega_d t_1 = 0$$

whence,

$$\tan\omega_d t_1 = \frac{\sqrt{1-\zeta^2}}{\zeta}$$

Upon substituting the foregoing value of ζ, 0.04711, into the above expression, we obtain

$$\tan\omega_d t_1 = 21.1843 \quad \Rightarrow \quad \omega_d t_1 = 87.2974° = 1.5236 \text{ rad}$$

and hence, $\sin\omega_d t_1 = 0.9989$, which yields

$$\omega_n t_1 = \frac{\omega_d t_1}{\sqrt{1-\zeta^2}} = \frac{1.5236}{0.9989} = 1.5253 \text{ rad}$$

Therefore,

$$x_1(t_1) = x_{max} = \frac{e^{-0.04711\times1.5253}}{2\pi\times0.5} \times 0.9989 \times 0.08771 = 0.02595$$

with an accuracy that could not have been obtained from the plot, thereby completing the solution.

2.4.6   The Convolution (Duhamel) Integral

The *unit impulse* was shown above to be useful in the modeling of impulsive inputs of extremely short duration, but the usefulness of this concept goes far beyond that.

Specifically, we can use the impulse response, along with the properties of *linearity* and *time invariance* of LTI systems, to determine the response of these systems to *any* input *under zero initial conditions*. The time response thus obtained is what is known as the *zero-state response* of the system under analysis, also known as the forced response, as pointed out already in the Preamble.

As a matter of fact, if we have the impulse response of a LTIS, then we do not even have to know the system in detail in order to determine its zero-state response to any input. To illustrate this statement, we take a black-box approach; we shall thus see that the zero-state response of the system under study can be obtained as the *convolution*, to be defined presently, of the impulse response of the system with the input.

We shall resort below to a fundamental identity, that we shall prove in detail. Prior to this, we need the relation

$$f(t)\delta(t) \equiv f(0)\delta(t) \tag{2.47a}$$

which follows because the delta function is zero everywhere, except at the origin, where it attains an infinite value. Note, moreover, that, if the impulse is applied at $t = \tau > 0$, rather than at $t = 0$, then, Eq. 2.47 takes the form

$$f(t)\delta(t - \tau) \equiv f(\tau)\delta(t - \tau) \equiv f_\tau\delta(t - \tau) \tag{2.47b}$$

where f_τ is a constant, for fixed τ. Thus,

$$\int_0^t f(\tau)\delta(t - \tau)d\tau \equiv \int_0^t f(t)\delta(t - \tau)d\tau \equiv f(t)\int_0^t \delta(t - \tau)d\tau$$

the last identity following because $f(t)$ is independent from the integration variable, τ. Now we evaluate the integral appearing in the rightmost-hand side of the above equation. To do this, we let $\theta \equiv t - \tau$ and regard t as fixed, which thus leads to $d\theta = -d\tau$, and hence,

$$\int_0^t \delta(t - \tau)d\tau = -\int_t^0 \delta(\theta)d\theta \equiv \int_0^t \delta(\theta)d\theta$$

Furthermore, we recall relation (2.34) and the definition of $\delta(t)$, Eq. 2.30, the last integral thus yielding

$$\int_0^t \delta(\theta)d\theta \equiv \int_{-\infty}^t \delta(\theta)d\theta \equiv u(t)$$

But we are interested in positive values of t, for we are studying the response of the system to an excitation applied at $t = 0$ and observing its behavior thereafter. Thus, the unit-step function becomes 1, thereby proving that

$$f(t) \equiv \int_0^t f(\tau)\delta(t - \tau)d\tau \tag{2.48}$$

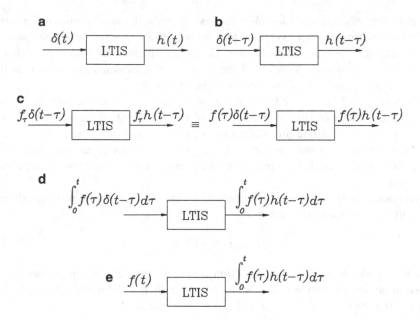

Fig. 2.27 (**a**) The impulse response of a LTIS; (**b**) the same response to a delayed impulse; (**c**) the response of the system to a modulated, delayed impulse; (**d**) the response of the system to a continuous train of modulated, delayed impulses; (**e**) the zero-state response as the convolution of the input with the impulse response of the system

which is the identity sought. i.e., $f(t)$ *can be regarded as a continuous train of impulses applied at instants* τ, *of amplitude* $f_\tau \equiv f(\tau)$, *for* $0 \leq \tau \leq t$.

Now, in order to obtain the response $x(t)$ of the system to an arbitrary input $f(t)$, we regard the latter as the integral appearing in the right-hand side of Eq. 2.48. Moreover, we denote the impulse response of the same system, which can be *any* system, as long as it is linear and time-invariant, by $h(t)$, as indicated with the block diagram of Fig. 2.27a. Now, by time-invariance, the response of the system to a delayed input $\delta(t - \tau)$ is a delayed impulse response $h(t - \tau)$, as shown in Fig. 2.27b. If, now, the delayed impulse is modulated with an amplitude f_τ, the corresponding response is, by homogeneity, $f_\tau h(t - \tau)$, as illustrated in Fig. 2.27c, with f_τ defined as in Eq. 2.47b. Furthermore, by additivity, the response of the system to a *continuous* train of delayed, modulated impulses, is, in turn, a *continuous* train of delayed, modulated impulse responses, both trains being represented as integrals between $\tau = 0$ and $\tau = t$, because of the continuity of the train, as illustrated in Fig. 2.27d. Finally, by virtue of the identity shown in Eq. 2.48, the train of inputs of Fig. 2.27d is readily identified as $f(t)$, thereby obtaining the input–output relation of Fig. 2.27e, namely,

$$x(t) = \int_0^t f(\tau)h(t - \tau)d\tau \qquad (2.49)$$

The integral appearing in Eq. 2.49 is called the *convolution*, or the *Duhammel integral*, of the two functions, $f(t)$ and $h(t)$. More generally, given any two functions $\phi(t)$ and $\psi(t)$, their convolution, represented as $\phi(t) * \psi(t)$, is defined as

$$\phi(t) * \psi(t) \equiv \int_0^t \phi(\tau)\psi(t-\tau)d\tau \qquad (2.50)$$

Important properties of the convolution are given below:

1. Commutativity:

$$\phi(t) * \psi(t) = \psi(t) * \phi(t) \qquad (2.51a)$$

2. Distributivity:

$$\phi(t) * [g_1(t) + g_2(t)] = \phi(t) * g_1(t) + \phi(t) * g_2(t) \qquad (2.51b)$$

3. Linear homogeneity:

$$\phi(t) * [\alpha\psi(t)] = \alpha\phi(t) * \psi(t) \qquad (2.51c)$$

Note that commutativity implies the identity shown below, whose **proof** is left to the reader:

$$\int_0^t \phi(\tau)\psi(t-\tau)d\tau \equiv \int_0^t \phi(t-\tau)\psi(\tau)d\tau \qquad (2.52)$$

distributivity and homogeneity following from the properties of the Riemann integral.

From the above equation, we can rewrite Eq. 2.49 alternatively as

$$x(t) = \int_0^t f(t-\tau)h(\tau)d\tau \qquad (2.53)$$

In summary, we have seen that the response of a linear time-invariant system to an arbitrary input $f(t)$ is equal to the convolution of the input with the impulse response of the system. Thus, the response of a LTIS to *any input* can be determined from its impulse response. We can then say that the behavior of a LTI system is characterized completely by its impulse response.

Given the convolution theorem for a general LTIS, the responses of first- and second-order systems to an arbitrary input can be readily derived, as shown below.

2.4.6.1   First-order Systems

Inserting the expression for the impulse response $h_I(t)$ given in Eq. 2.42b into Eq. 2.53, we obtain

$$x(t) = \int_0^t f(t-\tau)e^{-a\tau}d\tau \qquad (2.54)$$

2.4.6.2 Second-order Undamped Systems

Likewise, if we recall the expression for the impulse response $g_I(t)$ given in Eq. 2.42f, and substitute it into Eq. 2.53, we obtain

$$x(t) = \int_0^t f(t - \tau) \frac{1}{\omega_n} (\sin \omega_n \tau) \, d\tau \tag{2.55}$$

2.4.6.3 Second-order Damped Systems

We consider first the convolution of underdamped systems. To derive this convolution, we recall the expression for the pertinent impulse response $j_I(t)$, as given in Eq. 2.42i, and substitute it into Eq. 2.53, thus obtaining

$$x(t) = \int_0^t f(t - \tau) \frac{e^{-\zeta \omega_n \tau}}{\omega_d} (\sin \omega_d \tau) \, d\tau \tag{2.56}$$

the convolution for critically damped systems being derived likewise, i.e.,

$$x(t) = \int_0^t f(t - \tau) \tau e^{-\omega_n \tau} d\tau \tag{2.57}$$

Finally, the convolution for overdamped systems takes the form

$$x(t) = \int_0^t f(t - \tau) \frac{e^{-\zeta \omega_n \tau}}{\sqrt{1 - \zeta^2}} \sinh \left(\sqrt{1 - \zeta^2} \omega_n \tau \right) d\tau \tag{2.58}$$

Note that the foregoing results could have been obtained alternatively by inserting the appropriate impulse response function in Eq. 2.49.

2.5 Response to Abrupt and Impulsive Inputs

The step function models applications of *abrupt* inputs, such as the sudden closing of a circuit or the sudden arrest of a body undergoing a perfectly plastic collision with a stationary body of infinite mass, e.g., a speeding car hitting a wall.[9] The impulse function, on the other hand, models forces of a very short duration that nevertheless affect the state of the system due to the finite amount of energy transferred to or from the system in infinitesimal amounts of time. While we have already studied the response of first- and second-order LTI dynamical systems to

---

[9]Note that an abrupt change in the velocity implies an impulse in the acceleration.

unit impulses, the response to step inputs is yet to be studied. In addition, other situations occurring in real life, such as the sudden relocation of a heavy body, without a noticeable change in its velocity, such as a table driven by an indexing mechanism—a Geneva mechanism, for example—have not yet been studied. In this section, we shall determine the responses of the system of interest to various types of impulsive and abrupt inputs. These inputs can be used to model certain types of physical phenomena, as discussed earlier. The responses can be found using three different approaches:

- By using the convolution theorem
- By direct integration of the differential equation, with the appropriate input function inserted
- By making use of the properties illustrated in Figs. 2.4 and 2.5

We determine below the responses of first- and second-order systems to unit-doublet, unit-step, and unit-ramp inputs using either or both of the last two methods. The use of the convolution will be discussed in Sect. 2.7.

2.5.1   First-order Systems

2.5.1.1   Step Response

If the system under study is acted upon by a unit-step input, then Eq. 2.39 becomes

$$\dot{x} = -ax + u(t), \quad t \geq 0^-, \quad x(0^-) = 0 \tag{2.59}$$

By linearity, the response of the system under study to the unit step is the integral of the response of the same system to the unit impulse, since the former is the integral of the latter. We next denote the response to the unit step by $h_S(t)$, i.e.,

$$h_S(t) = \int_0^t h_I(\theta)d\theta \tag{2.60a}$$

where θ is a dummy variable of integration. Thus,

$$h_S(t) \equiv \begin{cases} 0, & t \leq 0^-; \\ (1 - e^{-at})/a, & t \geq 0^+ \end{cases} \tag{2.60b}$$

or, in a more compact form,

$$h_S(t) \equiv \frac{1 - e^{-at}}{a}u(t) \tag{2.60c}$$

which can be called the *step response* of the first-order system under study. This response is plotted in Fig. 2.28 for various values of positive a. Note that systems

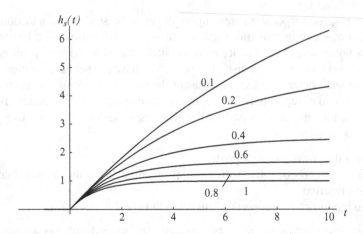

Fig. 2.28 Unit-step response of a first-order system for various positive values of a

with a small time constant are capable of following an input faster than those with a large time constant. That is, systems with a large time constant are slower than systems with a small one.

2.5.1.2 Ramp Response

The mathematical model now takes the form

$$\dot{x} = -ax + r(t), \quad t > 0^-, \quad x(0^-) = 0 \tag{2.61}$$

In Exercise 2.30 the reader is asked to derive the *ramp response* of the system at hand, denoted by $h_R(t)$, as shown below:

$$h_R(t) \equiv \frac{1}{a}\left[r(t) - \frac{1}{a}(1 - e^{-at})u(t) \right] \tag{2.62}$$

2.5.2 Second-order Undamped Systems

2.5.2.1 Doublet Response

In this case, the governing equation is

$$\ddot{x} + \omega_n^2 x = \dot{\delta}(t), \quad x(0^-) = 0, \quad \dot{x}(0^-) = 0, \quad t > 0^- \tag{2.63}$$

The doublet appearing in the right-hand side of Eq. 2.63 must be balanced with a corresponding term in the left-hand side, which cannot appear in the x-term. Indeed, if it did, then the $\dddot{x}$ term would comprise a quadruplet, which does not appear in the right-hand side. Thus, the doublet appears only in the $\ddot{x}$ term, and hence, the x term comprises a step function and is thus finite in a neighborhood around the origin. Now, both sides of Eq. 2.63 are integrated between $t = 0^-$ and $t > 0^-$, thus obtaining

$$\dot{x} + \omega_n^2 \int_{0^-}^t x\, d\tau = \delta(t), \quad x(0^-) = 0, \quad t > 0^- \qquad (2.64a)$$

where the initial condition on $\dot{x}(t)$ is no longer needed, for the highest-order derivative appearing in the foregoing equation is $\dot{x}$. Next, the two sides of Eq. 2.64a are integrated between 0^- and 0^+, thereby obtaining

$$\int_{0^-}^{0^+} \dot{x}\, dt + \omega_n^2 \int_{0^-}^{0^+} \int_{0^-}^t x\, d\tau\, dt = \int_{0^-}^{0^+} \delta(t)\, dt \qquad (2.64b)$$

The first integral of the left-hand side of Eq. 2.64b is readily recognized as $x(0^+) - x(0^-)$, the second integral vanishing because x is bounded at the origin, as proven above, and hence, its integral is bounded in a neighborhood around the origin as well. Furthermore, the integral of the right-hand side reduces to 1, and hence, Eq. 2.64b takes the form

$$x(0^+) - x(0^-) = 1$$

Moreover, if we take into account the initial conditions of Eq. 2.63, then

$$x(0^+) = 1$$

The above discussion shows that a unit doublet input indeed produces an instantaneous change in position of unit magnitude. A doublet of magnitude $A\dot{\delta}(t)$ would produce an instantaneous position change of magnitude A, without affecting its velocity. Furthermore, since $x(t)$ is a multiple of the unit-step function, $\dot{x}(t)$ is a multiple of the unit-impulse function, and hence, $\dot{x}(0^+) = \dot{x}(0^-) = 0$, i.e., it attains infinite values inside the interval $0^- < t < 0^+$, but is zero elsewhere. Then, Eq. 2.63 reduces to the form:

$$\ddot{x} + \omega_n^2 x = 0, \quad x(0^+) = 1, \quad \dot{x}(0^+) = 0, \quad t \geq 0^+$$

whose time response was already found in Eq. 2.13 for a more general case. Taking the foregoing equation into account, we can write

$$x(t) = \begin{cases} 0, & \text{for } t \leq 0^-; \\ \cos \omega_n t, & \text{for } t \geq 0^+ \end{cases} \qquad (2.65a)$$

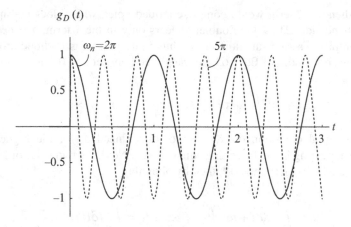

Fig. 2.29 Doublet response of an undamped second-order system

By analogy with Eq. 2.60c, the response of Eq. 2.65a can be called the *doublet response* of the given second-order system, and is denoted by $g_D(t)$, i.e., in compact form,

$$g_D(t) \equiv (\cos \omega_n t)\, u(t) \tag{2.65b}$$

its plot being included in Fig. 2.29, for two different values of ω_n.

Alternatively, since the unit doublet is the first derivative of the unit impulse, by Eq. 2.2 we can obtain the doublet response by differentiating the impulse response, namely,

$$g_D(t) = \frac{d}{dt} g_I(t)$$

i.e.,

$$g_D(t) = \frac{d}{dt}\left(\frac{1}{\omega_n}(\sin \omega_n t)\, u(t)\right) = (\cos \omega_n t)\, u(t) + \frac{1}{\omega_n}(\sin \omega_n t)\, \delta(t) \tag{2.66}$$

Recalling the definitions of $\delta(t)$ and $u(t)$, we see that the last term in the rightmost-hand side is zero for $t \geq 0^+$ and that $g_D(t) = 0$ for $t \leq 0^-$. Thus, $g_D(t)$, as obtained in Eq. 2.66, reduces to Eq. 2.65b.

A mass-spring system is acted upon by a doublet when its mass is given a sudden relocation, without affecting its velocity. For example, when a crank turning at a very high speed performs an indexing operation on a Geneva[10] wheel, the latter undergoes a finite rotation in no time at all. This situation can be modeled as the system appearing in Eq. 2.63, when the elasticity of the shaft attached to the crank is taken into account.

---

[10]See Fig. 3.2.

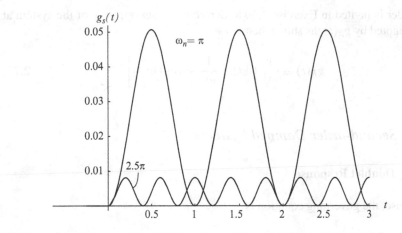

Fig. 2.30 Unit-step response of a second-order undamped system

2.5.2.2 Step Response

If now the system is acted upon by a unit step, we have

$$\ddot{x} + \omega_n^2 x = u(t), \quad t > 0^-, \quad x(0^-) = 0, \quad \dot{x}(0^-) = 0 \qquad (2.67)$$

Again, by linearity, we can express the response of the system under study as the integral of the impulse response of the same system, i.e.,

$$x(t) = \int_0^t g_I(\theta)d\theta = \frac{1 - \cos \omega_n t}{\omega_n^2} u(t) \qquad (2.68)$$

Expression (2.68) can be called the *step response* of the undamped second-order system under study. We will represent this response as $g_S(t)$, i.e.,

$$g_S(t) \equiv \frac{1 - \cos \omega_n t}{\omega_n^2} u(t) \qquad (2.69)$$

This response is plotted in Fig. 2.30.

2.5.2.3 Ramp Response

We now have

$$\ddot{x} + \omega_n^2 x = r(t), \quad t > 0^-, \quad x(0^-) = 0, \quad \dot{x}(0^-) = 0 \qquad (2.70)$$

The reader is invited in Exercise 2.30 to derive the *ramp response* of the system at hand, denoted by $g_R(t)$, as shown below:

$$g_R(t) \equiv \frac{1}{\omega_n^2}\left[r(t) - \frac{1}{\omega_n}(\sin\omega_n t)u(t)\right] \tag{2.71}$$

2.5.3 Second-order Damped Systems

2.5.3.1 Doublet Response

In this case, the governing equation is

$$\ddot{x} + 2\zeta\omega_n\dot{x} + \omega_n^2 x = \dot{\delta}(t), \quad x(0^-) = 0, \quad \dot{x}(0^-) = 0, \quad t > 0^- \tag{2.72}$$

The doublet appearing in the right-hand side of Eq. 2.72 must be balanced with a corresponding term in the left-hand side, which can neither appear in the second nor in the third term of that side. Indeed, if it appeared in the $\dot{x}$ term, then the $\ddot{x}$ term would comprise a triplet, which does not appear in the right-hand side. Likewise, if the doublet appeared in the x term, then the $\ddot{x}$ term would have a quadruplet, which should be balanced in the right-hand side. Thus, the doublet appears only in the $\ddot{x}$ term, and hence, the $\dot{x}$ term comprises an impulse and the x term, a step function. Now, both sides of Eq. 2.72 are integrated between $t = 0^-$ and $t > 0^-$, thus obtaining

$$\dot{x} + 2\zeta\omega_n x + \omega_n^2\int_{0^-}^{t} x\,d\tau = \delta(t), \quad x(0^-) = 0, \quad t > 0^- \tag{2.73}$$

where, again the initial condition on $\dot{x}(t)$ is not needed, for the highest-order derivative appearing in the foregoing equation is $\dot{x}$.

Further, the two sides of Eq. 2.73 are integrated between 0^- and 0^+, which yields

$$\int_{0^-}^{0^+}\dot{x}\,dt + 2\zeta\omega_n\int_{0^-}^{0^+} x\,dt + \omega_n^2\int_{0^-}^{0^+}\int_{0^-}^{t} x\,d\tau = \int_{0^-}^{0^+}\delta(t)\,dt \tag{2.74}$$

The first integral of the left-hand side of Eq. 2.74 is readily recognized as $x(0^+) - x(0^-)$, the second and the third integrals of the same side vanishing because x and its integral were found to be bounded at the origin. The integral of the right-hand side reduces to 1, and hence,

$$x(0^+) - x(0^-) = 1$$

which, in light of the initial conditions of Eq. 2.72, leads to

$$x(0^+) = 1$$

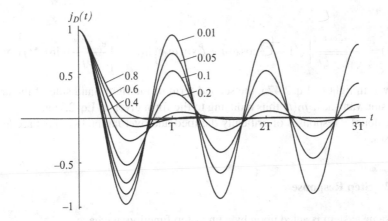

Fig. 2.31 Doublet response of an underdamped second-order system for various values of ω_n

Since $x(t)$ is a multiple of the unit-step function, $\dot{x}(t)$ is a multiple of the unit-impulse function, and hence, $\dot{x}(0^+) = \dot{x}(0^-) = 0$, i.e., it attains infinite values inside the interval $0^- < t < 0^+$, but is zero everywhere else. In summary, Eq. 2.72 reduces to a homogeneous equation with nonzero initial conditions, namely,

$$\ddot{x} + 2\zeta\omega_n\dot{x} + \omega_n^2 x = 0, \quad x(0^+) = 1, \quad \dot{x}(0^+) = 0, \quad t \geq 0^+ \qquad (2.75)$$

whose time response was already found in Eq. 2.23a for a more general case, and hence, for *underdamped* systems,

$$x(t) = \begin{cases} 0, & \text{for } t \leq 0^-; \\ \dfrac{e^{-\zeta\omega_n t}}{\sqrt{1-\zeta^2}}\left(\zeta\sin\omega_d t + \sqrt{1-\zeta^2}\cos\omega_d t\right), & \text{for } t \geq 0^+ \end{cases}$$

The response of a damped, second-order system to a unit doublet can be called the *doublet response* of this system, and is, henceforth, represented by $j_D(t)$, i.e., in compact form,

$$j_D(t) \equiv \frac{e^{-\zeta\omega_n t}}{\sqrt{1-\zeta^2}}\left(\zeta\sin\omega_d t + \sqrt{1-\zeta^2}\cos\omega_d t\right)u(t) \qquad (2.76a)$$

The doublet response of the system under study is plotted in Fig. 2.31, for various values of ζ between 0 and 1.

Alternatively, as done with the undamped second-order system, we can also obtain the doublet response by taking the first derivative of the impulse response, namely,

$$j_D(t) = \frac{d}{dt} j_I(t) = \frac{d}{dt}\left[\frac{e^{-\zeta\omega_n t}}{\omega_d}(\sin\omega_d t)u(t)\right]$$

i.e.,

$$j_D(t) = \frac{e^{-\zeta\omega_n t}}{\sqrt{1-\zeta^2}}\left[\sqrt{1-\zeta^2}\cos\omega_d t + \zeta\sin\omega_d t\right] u(t) + \left[\frac{e^{-\zeta\omega_n t}}{\omega_d}(\sin\omega_d t)\right]\delta(t)$$

However, in light of Eq. 2.47a, the second term of the right-hand side of the above expression vanishes, $j_D(t)$ thus reducing to the form given in Eq. 2.76a.

The doublet responses of critically damped and overdamped systems are left as exercises.

2.5.3.2 Step Response

If the same system is acted upon by a unit-step function, we have

$$\ddot{x} + 2\zeta\omega_n\dot{x} + \omega_n^2 x = u(t), \quad t > 0^-, \quad x(0^-) = 0, \quad \dot{x}(0^-) = 0 \qquad (2.77)$$

whose response can be found, again by superposition, as the integral of the response of the same system to a unit impulse. Thus, for *underdamped* systems we have

$$x(t) = \int_0^t j_I(\theta)d\theta = \int_0^t \frac{e^{-\zeta\omega_n\theta}}{\omega_d}(\sin\omega_d\theta)d\theta \qquad (2.78)$$

The integral in Eq. 2.78 can be readily evaluated if $\sin\omega_d\theta$ is expressed as

$$\sin\omega_d\theta \equiv \frac{e^{j\omega_d\theta} - e^{-j\omega_d\theta}}{2j}, \quad j = \sqrt{-1}$$

thereby obtaining

$$x(t) = \frac{1}{2j\omega_d}\int_0^t \left[e^{-(\zeta\omega_n - j\omega_d)\theta} - e^{-(\zeta\omega_n + j\omega_d)\theta}\right]d\theta$$

and hence

$$x(t) = \begin{cases} 0, & t \leq 0^-; \\ \left[1 - e^{-\zeta\omega_n t}(\cos\omega_d t + \frac{\zeta\omega_n}{\omega_d}\sin\omega_d t)\right]/\omega_n^2, & t \geq 0^+ \end{cases}$$

Again, we call the response displayed above the *step response* of the second-order underdamped system, and denote it by $j_S(t)$, i.e., in compact form,

$$j_S(t) \equiv \frac{1}{\omega_n^2}\left[1 - e^{-\zeta\omega_n t}\left(\cos\omega_d t + \frac{\zeta}{\sqrt{1-\zeta^2}}\sin\omega_d t\right)\right]u(t) \qquad (2.79)$$

which is plotted in Fig. 2.32, for fixed ω_n and various values of ζ.

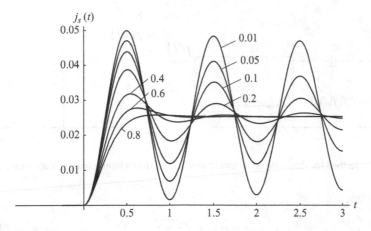

Fig. 2.32 Step response of a second-order underdamped system for various values of ζ

The step responses of critically damped and overdamped systems are left, again, as **exercises**.

2.5.3.3 Ramp Response

If now the system is acted upon by a ramp function, we have

$$\ddot{x} + 2\zeta\omega_n\dot{x} + \omega_n^2 x = r(t), \quad t > 0^-, \quad x(0^-) = 0, \quad \dot{x}(0^-) = 0 \tag{2.80}$$

As the reader can verify, the response of a second-order underdamped system to a ramp, denoted by $j_R(t)$, is given by

$$j_R(t) \equiv \frac{1}{\omega_n^2}\left[r(t) - \frac{2\zeta}{\omega_n}u(t)\right] + \frac{e^{-\zeta\omega_n t}}{\omega_n^3}\left(2\zeta\cos\omega_d t + \frac{2\zeta^2 - 1}{\sqrt{1 - \zeta^2}}\sin\omega_d t\right)u(t) \tag{2.81}$$

Likewise, the response of a second-order, critically damped system to a ramp, denoted by $k_R(t)$, is given by

$$k_R(t) \equiv \frac{1}{\omega_n^2}\left(1 - e^{-\omega_n t}\right)r(t) - \frac{2}{\omega_n^3}\left(1 - e^{-\omega_n t}\right)u(t) \tag{2.82}$$

Finally, the ramp response of an overdamped second-order system, denoted by $l_R(t)$, is given by

$$l_R(t) \equiv \frac{1}{\omega_n^2}\left(r(t) - \frac{2\zeta}{\omega_n}u(t)\right) + \frac{e^{-\zeta\omega_n t}}{\omega_n^3}\left[2\zeta\cosh(\rho\omega_n t) + \frac{2\zeta^2 - 1}{\rho}\sinh(\rho\omega_n t)\right]u(t) \tag{2.83a}$$

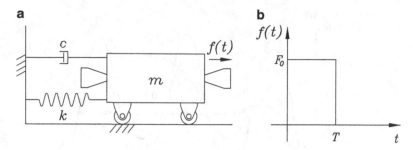

Fig. 2.33 Testbed for thrusters (**a**) its iconic model, and (**b**) its input under a firing thruster

where

$$\rho \equiv \sqrt{\zeta^2 - 1} \tag{2.83b}$$

An example of input that can be represented as a superposition of inputs studied here is the *pulse*, introduced below: A *pulse* $p(t)$ is a signal, i.e., a function of time, that is characterized by its *short duration*. Formally,

$$p(t) = \begin{cases} 0, & t < 0^- \\ \Pi(t), & 0^+ \le t \le T^- \\ 0, & t > T^+ \end{cases} \tag{2.84}$$

where provisions here have been made to allow for possible discontinuities in the function $\Pi(t)$. As well, T is a "small" time-interval with respect to the observation time.

2.5.4 Superposition

Sometimes, a LTIS is acted upon by an input that is abrupt, impulsive, or a combination of both, but is none of the above functions. Nevertheless, this input can be decomposed into a linear combination of doublets, impulses, steps, and ramps. By homogeneity, additivity, and time-invariance, the response can then be readily obtained by *superposition* as the corresponding linear combination of the individual responses to the inputs studied in this section.

Example 2.5.1 (Response of a Flexible System Under Thruster-firing). The iconic model of a mechanical system used to test on-off thrusters for space applications is shown in Fig. 2.33a. Such a thruster fires for a given finite time-interval T, thus providing a constant force F_0 to the mass, during that period, as shown in Fig. 2.33b, the force thus being a pulse with $\Pi(t) = F_0$. Find the time response of the underdamped system to this pulse.

Solution: The pulse of Fig. 2.33b can be represented as the sum of two step functions of the same amplitude F_0 and opposite signs, the positive step being applied at $t = 0$, the negative at $t = T$, i.e.,

$$f(t) = F_0 u(t) - F_0 u(t - T)$$

the mathematical model of the system being

$$m\ddot{x} + c\dot{x} + kx = f(t)$$

As the system is underdamped, its step response is $j_S(t)$, as given by Eq. 2.79. We can, therefore, write the time response sought in the form

$$x(t) = \frac{F_0}{m} j_S(t) - \frac{F_0}{m} j_S(t - T)$$

Since we have already found an expression for $j_S(t)$, the foregoing expression would suffice for our purposes. For completeness, we expand this response as

$$x(t) = \frac{F_0}{m\omega_n^2} \left\{ \left[1 - e^{-\zeta \omega_n t} \left(\cos \omega_d t + \frac{\zeta}{\sqrt{1 - \zeta^2}} \sin \omega_d t \right) \right] u(t) \right.$$
$$\left. - \left[1 - e^{-\zeta \omega_n (t - T)} \left(\cos \omega_d (t - T) + \frac{\zeta}{\sqrt{1 - \zeta^2}} \sin \omega_d (t - T) \right) \right] u(t - T) \right\}$$

This total time response is sketched in Fig. 2.34 as the sum of two step responses.

2.6   The Total Time Response

The *total time response* of dynamical systems is the time response under non-zero input and non-zero initial conditions. By superposition, the total time response of linear dynamical systems and, in particular, of LTI dynamical systems, is simply the sum of the zero-input and the zero-state responses. We summarize below these results for each of the systems under study.

2.6.1   *First-order Systems*

Here the system of interest is modeled by

$$\dot{x} = -ax + f(t), \quad t > 0^-, \quad x(0^-) = x_0 \tag{2.85}$$

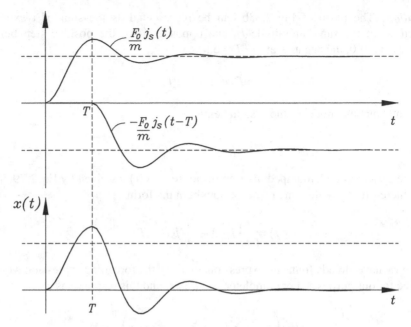

Fig. 2.34 Time response of the testbed to a thruster pulse

If we recall the zero-input response of Eq. 2.8 and the zero-state response given by the convolution of Eq. 2.54, the required time response is

$$x(t) = e^{-at}x_0 + \int_0^t f(t-\tau)e^{-a\tau}d\tau \tag{2.86}$$

2.6.2 Second-order Systems

As in previous sections, here we distinguish among the usual cases, as described below.

2.6.2.1 Undamped Systems

For second-order undamped systems, we have

$$\ddot{x} + \omega_n^2 x = f(t), \quad x(0^-) = x_0, \quad \dot{x}(0^-) = v_0, \quad t > 0^- \tag{2.87}$$

The time response of this system is then the sum of the zero-input response of Eq. 2.13 and the zero-state response given by the corresponding convolution, as appearing in Eq. 2.55, namely,

$$x(t) = (\cos \omega_n t)x_0 + \frac{1}{\omega_n}(\sin \omega_n t)v_0 + \int_0^t f(t-\tau)\frac{1}{\omega_n}(\sin \omega_n \tau)\,d\tau \qquad (2.88)$$

2.6.2.2   Damped Systems

A second-order damped system acted upon by a non-zero input and non-zero initial conditions is represented below:

$$\ddot{x} + 2\zeta \omega_n \dot{x} + \omega_n^2 x = f(t), \quad x(0^-) = x_0, \quad \dot{x}(0^-) = v_0, \quad t > 0^- \qquad (2.89)$$

For underdamped systems, the time response is derived as the sum of the zero-input response of Eq. 2.23a and the zero-state response given by the convolution of Eq. 2.56, namely,

$$x(t) = \frac{e^{-\zeta \omega_n t}}{\sqrt{1-\zeta^2}}\left(\sqrt{1-\zeta^2}\cos \omega_d t + \zeta \sin \omega_d t\right)x_0 + \frac{e^{-\zeta \omega_n t}}{\omega_d}(\sin \omega_d t)v_0$$

$$+ \int_0^t f(t-\tau)\frac{e^{-\zeta \omega_n \tau}}{\omega_d}(\sin \omega_d \tau)\,d\tau \qquad (2.90)$$

Correspondingly, the total responses of critically damped and overdamped systems are shown below:

$$x(t) = e^{-\omega_n t}\left[(1+\omega_n t)x_0 + tv_0\right] + \int_0^t f(t-\tau)\tau e^{-\omega_n \tau}\,d\tau \qquad (2.91)$$

and

$$x(t) = \frac{e^{-\zeta \omega_n t}}{r}\left[[r\cosh(r\omega_n t) + \zeta \sinh(r\omega_n t)]x_0 + \frac{1}{\omega_n}\sinh(r\omega_n t)v_0\right]$$

$$+ \int_0^t f(t-\tau)\frac{e^{-\zeta \omega_n \tau}}{r\omega_n}\sinh(r\omega_n \tau)\,d\tau \qquad (2.92a)$$

$$r \equiv \sqrt{\zeta^2 - 1} \qquad (2.92b)$$

In simulation studies it is useful to have the foregoing total responses for all three cases of damped systems in a single expression, which can be done with the aid of the concept of state variable introduced in Sect. 2.3.2. The zero-input response of damped systems in state-variable form is given in Eq. 2.21. The zero-state response in state-variable form can be derived from the convolution, but we have three different formulas for this, namely, for each of the three associated cases. A generally applicable convolution expression in terms of state variables is derived below. To this end, we first cast the governing equation, Eq. 2.89, in state-variable form, namely,

$$\dot{\mathbf{z}} = \mathbf{A}\mathbf{z} + \mathbf{b}f(t), \quad t > 0, \quad \mathbf{z}(0) = \mathbf{z}_0 \tag{2.93}$$

with $\mathbf{z}$ and $\mathbf{A}$ defined in Eqs. 2.15, 2.16, 2.18a, b, while $\mathbf{b}$ as defined as

$$\mathbf{b} \equiv \begin{bmatrix} 0 \\ 1 \end{bmatrix} \tag{2.94}$$

Thus, the impulse response of the system in state-variable form is defined as the response of the system under $\mathbf{z}_0 = \mathbf{0}$ and $f(t) = \delta(t)$. We can determine, then, the impulse response of the system at hand in exactly the same manner as we did for the scalar first-order system in Sect. 2.4.5. That is, we transform Eq. 2.93 with zero initial condition and acted upon by a unit impulse into a system with zero input and non-zero initial condition at time 0^+. We do this by integration of both sides of the aforementioned equation, with $f(t) = \delta(t)$, between $t = 0^-$ and $t = 0^+$, which yields

$$\mathbf{z}(0^+) - \mathbf{z}(0^-) = \int_{0^-}^{0^+} \mathbf{A}\mathbf{z}(t)dt + \int_{0^-}^{0^+} \mathbf{b}\delta(t)dt \tag{2.95}$$

In the foregoing equation, the left-hand side reduces to $\mathbf{z}(0^+)$, while the first integral of the right-hand side vanishes because $\mathbf{z}(t)$ remains bounded at the origin. Moreover, the second integral reduces to $\mathbf{b}$ by virtue of the definition of the impulse function, and hence,

$$\mathbf{z}(0^+) = \mathbf{b}$$

Thus, the system under study takes the form

$$\dot{\mathbf{z}} = \mathbf{A}\mathbf{z}, \quad \mathbf{z}(0^+) = \mathbf{b}, \quad t > 0^+ \tag{2.96}$$

whose time response, henceforth represented by $\mathbf{z}_I(t)$, can be readily derived from Eq. 2.21, namely,

$$\mathbf{z}_I(t) = e^{\mathbf{A}t}\mathbf{b}u(t) \tag{2.97}$$

which is the impulse response of the system under study. Now, obtaining the zero-state response for an arbitrary input is straightforward if we *convolve* the foregoing impulse response with a given input $f(t)$, thus deriving

$$\mathbf{z}(t) = \int_0^t e^{\mathbf{A}\tau}\mathbf{b}f(t - \tau)d\tau \tag{2.98}$$

and hence, the total response of the system takes the form

$$\mathbf{z}(t) = e^{\mathbf{A}t}\mathbf{z}_0 + \int_0^t e^{\mathbf{A}\tau}\mathbf{b}f(t - \tau)d\tau \tag{2.99a}$$

or, by virtue of the commutativity of the convolution,

$$\mathbf{z}(t) = e^{\mathbf{A}t}\mathbf{z}_0 + \int_0^t e^{\mathbf{A}(t-\tau)}\mathbf{b}f(\tau)d\tau \tag{2.99b}$$

The two above expressions for the total response of damped systems are valid, in fact, for arbitrary n-degree-of-freedom (n-dof) systems, in which the state-variable vector $\mathbf{z}$ becomes $2n$-dimensional, while vector $\mathbf{b}$ $2n$-dimensional, and matrix $\mathbf{A}$ of $2n \times 2n$.

2.7 The Harmonic Response

Besides impulsive and abrupt inputs, dynamical systems can be subjected to persistent, time-varying inputs that behave in an unpredictable manner. This is the case in aircraft subjected to gust winds and turbulence or terrestrial vehicles traveling on imperfect roads. While these systems are subjected to time-varying inputs that are difficult, if not impossible, to describe in the form of an *explicit* function $f(t)$, these inputs can be regarded as the summation of infinitely-many inputs of a much simpler structure. In fact, as the French mathematician Joseph Fourier (1768–1830) showed in his famous work *Théorie analytique de la chaleur*, published in 1812, any periodic function can be decomposed into an infinite sum of sines and cosines of multiples of a fundamental frequency. Such an expansion of a periodic function is known as a *Fourier series*. By extension, the same decomposition can be applied to arbitrary functions using a continuous distribution of frequencies that can be determined using what is known as the *Fourier transform*. Thus, given any input, not necessarily periodic, it is always possible to find its frequency content by a study that is known as *spectral analysis*. For linear systems, such a decomposition, whether in a continuum of frequencies or in a discrete, although infinite set of multiples of a fundamental frequency, is extremely useful because it allows the analyst, by superposition, to find the response of the system to that arbitrary input. Indeed, by means of superposition, the said response can be found as an infinite sum, i.e., a series of responses to harmonic (sinusoidal and cosinusoidal) inputs. These facts allow us to understand the importance of the response of dynamical systems to simple harmonic inputs. We will term, henceforth, such a response the *harmonic response* of the system at hand, which also goes by the name of *frequency response*.

Below we study the response of mechanical first- and second-order systems to harmonic inputs and then, using a Fourier expansion, we derive the response of these systems to any periodic function. While the Fourier transform of arbitrary, possibly random functions allows the study of the response of dynamical systems to unpredictable or *stochastic* inputs, we will skip in this book the Fourier transform and focus on the Fourier series.

The philosophy behind the analysis that follows is that we are interested in the behavior of systems *after a very long time* has elapsed since an initial time, i.e., since the system started being perturbed or excited with an input of the nature mentioned above. Hence, initial conditions become immaterial and can be set arbitrarily. Now, when we speak of *long periods*, we must realize that we are speaking of a time span that is large with respect to the *natural time scale* of the system under study. In fact, every single-dof system studied in Chap. 1 has one distinct time scale, namely, the *time constant* for first-order systems or the *natural period*, i.e., the inverse of the natural frequency, for second-order systems. Moreover, all physical systems always contain a certain amount of damping, that takes care of any nonzero initial conditions whose effect on the system response becomes negligible after a certain finite time. Thus, we begin by deriving the zero-state response of the systems at hand to harmonic excitations. For example, when we study the vibrations in the fuselage of aircraft under turbulence occurring at cruising conditions, the initial conditions become irrelevant. We might as well consider, then, that the system, in this case the fuselage, started its motion at time $-\infty$.

We will assume a harmonic input, i.e., the forcing term in the governing equations is assumed to have the form

$$f(t) = A\cos \omega t + B\sin \omega t \tag{2.100}$$

We can study the response of the systems at hand to the input of Eq. 2.100 by superposition, i.e., by finding first the response to an input of the form $\cos \omega t$, then that to one of the form $\sin \omega t$ and then express the total input as a suitable linear combination of the two foregoing responses. Of course, by linearity, the coefficients of that linear combination are the corresponding coefficients A and B of Eq. 2.100. Coefficients A and B are termed the *amplitude* of the cosine and the sine signals, respectively.

Below we study the time response of first- and second-order systems and show that this response can be decomposed into a *transient part* and a *steady-state part*. The transient part decays with time and, after a finite time T has elapsed, it becomes negligible. What remains is a harmonic response that constitutes the *steady-state response*. In particular, for a harmonic excitation of the form

$$f(t) = A\cos \omega t \tag{2.101}$$

the steady-state response, represented by $x_{CS}(t)$, takes the form

$$x_{CS}(t) = M\cos(\omega t + \phi) \tag{2.102}$$

where M and ϕ are the *magnitude* and *phase* of the response under study. Thus, the harmonic response is fully characterized by these two parameters, as shown in Fig. 2.35. It will become apparent that both the magnitude and the phase are functions of: (a) the amplitude and the frequency of the input and (b) the parameters—τ for first-order, ζ and ω_n for second-order systems. Below we derive

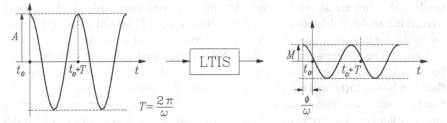

Fig. 2.35 Harmonic response of a linear, time-invariant dynamical system

expressions for the magnitude and phase of the harmonic response of the systems of interest and plot what is known as their *frequency response plots*. These plots are also known as the *Bode plots* of the system at hand.

Prior to our study, we recall some useful trigonometric identities, namely,

$$C_C \cos \omega t + C_S \sin \omega t = K \cos(\omega t + \alpha_C) \qquad (2.103a)$$

where

$$K = \sqrt{C_C^2 + C_S^2}, \qquad \alpha_C = -\arctan\left(\frac{C_S}{C_C}\right) \qquad (2.103b)$$

and

$$C_C \cos \omega t + C_S \sin \omega t = K \sin(\omega t + \alpha_S) \qquad (2.104a)$$

where

$$\alpha_S = \arctan\left(\frac{C_C}{C_S}\right) \qquad (2.104b)$$

with K as defined in Eq. 2.103b.

Note that if the harmonic functions of the left-hand side of Eqs. 2.103a and 2.104a involve themselve a phase angle β, then the phase angle of the harmonic functions in the right-hand side changes correspondingly, i.e.,

$$C_C \cos(\omega t + \beta) + C_S \sin(\omega t + \beta) = K \cos(\omega t + \alpha_C') \quad \text{or} \quad K \sin(\omega t + \alpha_S') \quad (2.105a)$$

where the coefficient K is the same as that of Eq. 2.103b, but the phase angles are now

$$\alpha_C' = \alpha_C + \beta, \qquad \alpha_S' = \alpha_S + \beta \qquad (2.105b)$$

2.7.1 The Unilateral Harmonic Functions

It will be useful to derive a relationship between the zero-state response of a system to inputs of the forms $(\cos \omega t)u(t)$ and $(\sin \omega t)u(t)$. These two functions are zero up until $t = 0^-$; they are identical to the harmonic functions afterwards, and henceforth termed the *unilateral harmonic functions*. Note that, if observed at instants $t \gg 0$, the unilateral harmonic functions become the usual harmonic functions.

Let $x_C(t)$ be the zero-state response (ZSR) of any linear system to an input of the form $(\cos \omega t)u(t)$; likewise, let $x_S(t)$ be the ZSR of the same system to an input of the form $(\sin \omega t)u(t)$; and let $x_I(t)$ be the impulse response of the same system. Note that here we do not distinguish among undamped, underdamped, critically damped, and overdamped systems; neither do we distinguish between first- and second-order systems. In the process, we will need the time-derivatives of the unilateral harmonic functions, which we readily derive:

$$\frac{d}{dt}[(\cos \omega t)u(t)] = -\omega(\sin \omega t)u(t) + (\cos \omega t)\delta(t)$$

But, from Eq. 2.47a,

$$(\cos \omega t)\delta(t) = (\cos(0))\delta(t) \equiv \delta(t)$$

and so,

$$\frac{d}{dt}[(\cos \omega t)u(t)] = -\omega(\sin \omega t)u(t) + \delta(t) \tag{2.106}$$

or, solving for $(\sin \omega t)u(t)$,

$$(\sin \omega t)u(t) = -\frac{1}{\omega}[(\cos \omega t)u(t)] + \frac{1}{\omega}\delta(t) \tag{2.107}$$

Thus, the response of the system to the unilateral sine function can be derived as a linear combination of the response to the unilateral cosine function and the impulse response, by merely mimicking the foregoing relation among inputs in terms of the corresponding relation among responses. This we can do by exploiting the properties of LTIS, namely,

$$x_S(t) = -\frac{1}{\omega}\frac{d}{dt}x_C(t) + \frac{1}{\omega}x_I(t) \tag{2.108}$$

Likewise,

$$\frac{d}{dt}[(\sin \omega t)u(t)] = \omega(\cos \omega t)u(t) + (\sin \omega t)\delta(t)$$

Again, from Eq. 2.47,

$$(\sin \omega t)\delta(t) \equiv (\sin(0))\delta(t) \equiv 0$$

and so,

$$\frac{d}{dt}[(\sin \omega t)u(t)] = \omega(\cos \omega t)u(t) \tag{2.109}$$

whence,

$$(\cos \omega t)u(t) = \frac{1}{\omega}\frac{d}{dt}[(\sin \omega t)u(t)] \qquad (2.110)$$

Thus, we can find the response of a LTIS to the unilateral sine function in terms of its response to the unilateral cosine function by simply mimicking the foregoing expression, in exactly the same way as done previously, i.e.,

$$x_C(t) = \frac{1}{\omega}\frac{d}{dt}x_S(t) \qquad (2.111)$$

The presence of the impulse response in Eq. 2.108 is to be highlighted. This term appears in that equation by virtue of the corresponding impulsive term in Eq. 2.106.

2.7.2 First-order Systems

The equation of motion of a first-order system is recalled below:

$$\dot{x} + ax = f(t) \qquad (2.112)$$

In the subsequent discussion, we shall assume that the parameter a characterizing the system is positive. Moreover, the zero-state response of a first-order system to the $(\cos \omega t)u(t)$ input, denoted, as above, by $x_C(t)$, will be derived. Now, since the unit step function is dimensionless, the input function in Eq. 2.112 is dimensionless, which means that its left-hand side is also dimensionless, and hence, $x(t)$ **is expressed in units of time**. This should not bother the reader, for actual physical excitations always carry physical units that are accounted for by the units of the *amplitudes A* and *B* of Eq. 2.100.

The response sought can be expressed in terms of the convolution, namely,

$$x_C(t) = \int_0^t \cos \omega(t - \tau)h_I(\tau)d\tau \qquad (2.113a)$$

where $h_I(t)$ is the impulse response of the system under study, i.e.,

$$h_I(t) = e^{-at}u(t) \qquad (2.113b)$$

which is recalled to be dimensionless.

Now, the integral of Eq. 2.113a can be obtained in many ways. If a straightforward approach is used, we would substitute the cosine function of the convolution by its exponential representation, namely,

$$\cos \omega t \equiv \frac{e^{j\omega t} + e^{-j\omega t}}{2} \qquad (2.114)$$

with j defined as the imaginary unity, i.e., as $j \equiv \sqrt{-1}$. Alternatively, one can simply search in a table of integrals or resort to *symbolic computations*, using scientific software—e.g., *Maple*, *Mathematica* or *Macsyma*. The result is, in any case, as indicated below:

$$x_C(t) = \frac{1}{a^2 + \omega^2} \left(-a e^{-at} + a \cos \omega t + \omega \sin \omega t \right) u(t) \qquad (2.115)$$

from which we can readily identify the transient and the steady-state components of the time response. Indeed, since we assumed at the outset that $a > 0$, the exponential term in the above expression decays with time, and hence, the transient part of the response, $x_{CT}(t)$, is readily identified, namely,

$$x_{CT}(t) = \frac{-a}{a^2 + \omega^2} e^{-at} u(t) \qquad (2.116)$$

while the steady-state component, $x_{CS}(t)$, is given by

$$x_{CS}(t) = \frac{1}{a^2 + \omega^2} \left(a \cos \omega t + \omega \sin \omega t \right) u(t)$$

Since we are interested in this section in studying the behavior of the systems at hand after a long time has elapsed, we can assume that the system started being excited in the distant past. For the time frame of observation in which we are interested, we can write $u(t) = 1$, and the above equation can then be rewritten in the form

$$x_{CS}(t) = \left(\frac{a}{a^2 + \omega^2} \right) \cos \omega t + \left(\frac{\omega}{a^2 + \omega^2} \right) \sin \omega t \qquad (2.117)$$

Using Eqs. 2.103a, b, we may rewrite this expression in the form

$$x_{CS}(t) = M \cos(\omega t + \phi) \qquad (2.118a)$$

where

$$M = \frac{1}{\sqrt{a^2 + \omega^2}} \qquad \phi = -\arctan\left(\frac{\omega}{a} \right) \qquad (2.118b)$$

For completeness, we derive now the time response of the same system to an input of the form $(\sin \omega t) u(t)$ using Eq. 2.108. First, we calculate

$$\frac{d}{dt} (x_C(t)) = \frac{1}{a^2 + \omega^2} \left(a^2 e^{-at} - a \omega \sin \omega t + \omega^2 \cos \omega t \right) u(t)$$

$$+ \frac{1}{a^2 + \omega^2} \left(-a e^{-at} + a \cos \omega t + \omega^2 \sin \omega t \right) \delta(t) \qquad (2.119)$$

which, by virtue of Eq. 2.47, simplifies to

$$\frac{d}{dt}(x_C(t)) = \frac{1}{a^2 + \omega^2}\left(a^2 e^{-at} - a\omega \sin \omega t + \omega^2 \cos \omega t\right) u(t)$$

Moreover, we recall that

$$x_I(t) = e^{-at} u(t)$$

Thus, from Eq. 2.108, we have

$$x_S(t) = -\frac{1}{\omega}\frac{1}{a^2 + \omega^2}\left(a^2 e^{-at} - a\omega \sin \omega t + \omega^2 \cos \omega t\right) u(t) + \frac{1}{\omega}e^{-at}u(t)$$

$$= -\frac{1}{a^2 + \omega^2}\left[-\omega e^{-at} - a\sin \omega t + \omega \cos \omega t\right] u(t) \tag{2.120}$$

The transient component of the response $x_S(t)$, denoted by $x_{ST}(t)$, is readily singled out from Eq. 2.120, namely,

$$x_{ST}(t) = \frac{\omega}{a^2 + \omega^2}e^{-at}u(t) \tag{2.121}$$

Likewise, the steady-state component of the same response, denoted by $x_{SS}(t)$, is identified from Eq. 2.120 as

$$x_{SS}(t) = \frac{1}{a^2 + \omega^2}\left(a\sin \omega t - \omega \cos \omega t\right) u(t)$$

After a long time t has elapsed, the unit-step function is identical to unity, and hence, the above expression becomes

$$x_{SS}(t) = \frac{1}{a^2 + \omega^2}\left(a\sin \omega t - \omega \cos \omega t\right) \tag{2.122}$$

Correspondingly, the magnitude and the phase of the $x_{SS}(t)$ response can be readily identified from the above expression for this signal, thereby obtaining

$$x_{SS}(t) = \left(\frac{a}{a^2 + \omega^2}\right)\sin \omega t - \left(\frac{\omega}{a^2 + \omega^2}\right)\cos \omega t \tag{2.123}$$

Using Eq. 2.104a, we may write this equation as

$$x_{SS}(t) = M \sin(\omega t + \phi) \tag{2.124a}$$

where

$$M = \frac{1}{\sqrt{a^2 + \omega^2}} \quad \phi = -\arctan\left(\frac{\omega}{a}\right) \tag{2.124b}$$

which are essentially the same as those derived for the cosine input, as they should.

2.7.3 Second-order Systems

We attempt first to find the harmonic response of undamped systems using the approach introduced for first-order systems.

2.7.3.1 The Response to the Unilateral Cosine Function

For a unilateral cosine input, we have

$$\ddot{x} + \omega_n^2 x = (\cos \omega t) u(t) \tag{2.125}$$

In the above equation, the right-hand side is, again, dimensionless, and so is its left-hand side, which implies that $x(t)$ **is represented in units of time-squared**.

Further, we recall the impulse response of second-order undamped systems, namely,

$$g_I(t) = \frac{1}{\omega_n}(\sin \omega_n t) u(t)$$

which is recalled to bear units of time.

Upon calculation of the convolution of the $(\cos \omega t) u(t)$ input with the foregoing impulse response, we obtain

$$x(t) = \frac{1}{\omega_n^2 - \omega^2}(\cos \omega t - \cos \omega_n t) u(t) \tag{2.126}$$

from which we cannot identify a decaying term, as we did for first-order systems, and our attempt to follow the previous approach stops here.

An alternative approach to finding the harmonic response of undamped second-order systems is now introduced. We first derive the harmonic response of underdamped systems, the response associated with undamped systems being obtained as the limiting case, when $\zeta \to 0$, of the harmonic response thus found.

Let us then consider the second-order damped system excited by a unilateral cosine function:

$$\ddot{x} + 2\zeta \omega_n \dot{x} + \omega_n^2 x = (\cos \omega t) u(t) \tag{2.127}$$

Moreover, the impulse response of the above system for the underdamped case is

$$j_I(t) = \frac{e^{-\zeta \omega_n t}}{\omega_d} (\sin \omega_d t) u(t) \qquad (2.128)$$

In order to obtain the response of the system of Eq. 2.127, we resort to the convolution of the $(\cos \omega t) u(t)$ input with the foregoing impulse response. This convolution is calculated in exactly the same manner as done for first-order systems, thereby obtaining the desired expression $x_C(t)$ for the given input, namely,

$$x_C(t) = \frac{N(t)}{D} u(t) \qquad (2.129a)$$

with $N(t)$ and D defined as

$$N(t) \equiv \left[\frac{-\zeta(\omega^2 + \omega_n^2)}{\sqrt{1 - \zeta^2}} \sin \omega_d t + (\omega^2 - \omega_n^2) \cos \omega_d t \right] e^{-\zeta \omega_n t}$$
$$- \left[(\omega^2 - \omega_n^2) \cos \omega t - 2\zeta \omega \omega_n \sin \omega t \right] \qquad (2.129b)$$
$$D \equiv \omega^4 - 2\omega^2 \omega_n^2 + \omega_n^4 + 4\zeta^2 \omega^2 \omega_n^2 \equiv (\omega^2 - \omega_n^2)^2 + 4\zeta^2 \omega^2 \omega_n^2 \qquad (2.129c)$$

We thus have

$$x_L(t) = \frac{1/\omega_n^2}{(1 - r_f^2)^2 + 4\zeta^2 r_f^2} \left\{ e^{-\zeta \omega_n t} \left[\frac{-\zeta(1 + r_f^2)}{\sqrt{1 - \zeta^2}} \sin \omega_d t - (1 - r_f^2) \cos \omega_d t \right] \right.$$
$$\left. + \left[2\zeta r_f \sin \omega t + (1 - r_f^2) \cos \omega t \right] \right\} u(t) \qquad (2.130a)$$

with the *frequency ratio* r_f defined as

$$r_f \equiv \frac{\omega}{\omega_n} \qquad (2.130b)$$

Since $\zeta > 0$ and $\omega_n > 0$, we can see that the transient term is that comprising the exponential in Eq. 2.130a, namely

$$x_{CT}(t) = \frac{e^{-\zeta \omega_n t}}{\omega_n^2 \left[(1 - r_f^2)^2 + (2\zeta r_f)^2 \right]} \left[\frac{-\zeta(1 + r_f^2)}{\sqrt{1 - \zeta^2}} \sin \omega_d t - (1 - r_f^2) \cos \omega_d t \right] u(t)$$

The steady-state component x_{CS} of the response given in Eq. 2.130a can now be recognized as

$$x_{CS}(t) = \frac{1}{\omega_n^2 \left[(1 - r_f^2)^2 + (2\zeta r_f)^2 \right]} \left[2\zeta r_f \sin \omega t + (1 - r_f^2) \cos \omega t \right] u(t)$$

Again, we are interested only in the response after a long time has elapsed since the system was first excited, and so, the unit-step function can be replaced by unity in the above expression, thus obtaining

$$x_{CS}(t) = \frac{2\zeta r_f}{\omega_n^2\left[(1-r_f^2)^2+(2\zeta r_f)^2\right]}\sin\omega t + \frac{1-r_f^2}{\omega_n^2\left[(1-r_f^2)^2+(2\zeta r_f)^2\right]}\cos\omega t$$

(2.131)

Using Eq. 2.103a we may rewrite this expression in the form

$$x_{CS} = M\cos(\omega t + \phi)$$

(2.132a)

where

$$M = \frac{1/\omega_n^2}{\sqrt{(1-r_f^2)^2+(2\zeta r_f)^2}}\qquad \phi = -\arctan\left(\frac{2\zeta r_f}{1-r_f^2}\right)$$

(2.132b)

Now the total response of an undamped second-order system to a unilateral cosine input $(\cos\omega t)u(t)$ can be derived by setting $\zeta = 0$ in Eq. 2.130a, thereby obtaining

$$x_C(t) = \frac{1/\omega_n^2}{1-r_f^2}(\cos\omega t - \cos\omega_n t)\,u(t)$$

(2.133)

which is identical to the expression derived directly from the convolution theorem and displayed in Eq. 2.126.

The corresponding harmonic response of undamped systems can now be derived by setting $\zeta = 0$ in Eq. 2.131, which yields the associated steady-state response as

$$x_{CS}(t) = \frac{1/\omega_n^2}{1-r_f^2}\cos\omega t$$

(2.134)

From the above expression, it is clear that the magnitude of the response is the absolute value of the coefficient of $\cos\omega t$. For the phase, we must consider two cases. First, when $r_f < 1$, the denominator of the said coefficient is positive and, hence, the coefficient is also positive. Thus, the response to the input $(\cos\omega t)u(t)$ is exactly the same signal, except for a difference in magnitude; the response is therefore in phase with the input, i.e., $\phi = 0$. On the other hand, when $r_f > 1$, the denominator and, hence, the whole coefficient of $\cos\omega t$, are negative. Thus, the response of the system, apart from a difference in amplitude, is opposite in sign to the input, which corresponds to $\phi = \pm 180°$. However, the positive sign does not make physical sense, since it implies that the response "leads" the input, i.e., the system responds to the input before it is applied. The negative sign implies a lag in phase, which is consistent with the actual behavior of dynamical systems. The response of the undamped system to the given input is, then,

$$x_{CS}(t) = M \cos(\omega t + \phi) \tag{2.135a}$$

where

$$M = \frac{1/\omega_n^2}{|1 - r_f^2|}, \qquad \phi = \begin{cases} 0° & \text{if } r_f < 1; \\ -180° & \text{if } r_f > 1. \end{cases} \tag{2.135b}$$

The foregoing expressions for the amplitude and phase of the harmonic response of second-order systems were derived from the steady-state response of underdamped systems. However, a close look at those derivations reveals that nothing prevents us from applying them to critically damped and overdamped systems as well. In fact, $M \cos(\omega t + \phi)$ is a particular solution of the second-order damped system excited by the input $\cos(\omega t)$.

As a matter of fact, the steady-state responses derived above are nothing but *particular solutions* of the associated ODEs, as the reader can readily verify. Upon substituting the foregoing expressions for the harmonic response of second-order systems into the corresponding mathematical model, it will become apparent that those expressions verify indeed the model, if with particular initial conditions. In the realm of the harmonic response, we are interested in the behavior of the system at arbitrarily large values of t, and so, the initial conditions in this case become irrelevant.

2.7.3.2 Resonance

One special situation occurs when $r_f = 1$, or $\omega = \omega_n$. From Eq. 2.135a, we see that M attains unbounded values when this occurs. Thus, when a harmonic input is applied to a second-order undamped system at a frequency exactly equal to its natural frequency, the magnitude of its response is infinite. This phenomenon is called *resonance* and does not occur in first-order systems. Of course, such a situation could never occur in a real-life system since, first, damping is always present to some extent in actual systems, and, more importantly, the assumption of the system being linear would no longer apply, as large amplitudes in the response would occur, thus making the underlying model no longer valid. Still, the concept is of extreme practical importance since, in actual systems, a very large and violent response is produced whenever an exciting force is near the natural frequency of the system. Thus, great care must go into the design of mechanical systems that are required to remain as unperturbed as possible in the presence of harmonic disturbances. What is needed in these cases is, apparently, a design that will ensure that the system at hand is not subjected to excitations bearing frequencies near its natural frequency at any time during their normal operating conditions.

2.7.3.3 The Response to the Unilateral Sine Function

For completeness, and for further reference, we derive below the response of underdamped systems to a unilateral sine input, $(\sin \omega t)u(t)$; the desired response is denoted here by $x_S(t)$. We do this by resorting to the same idea introduced earlier for first-order systems, namely, by recalling Eq. 2.108. To this end, we must first differentiate $x_C(t)$ as given by Eqs. 2.129a, b, i.e.,

$$\frac{d}{dt}(x_C(t)) = \frac{\dot{N}(t)}{D}u(t) + \frac{N(t)}{D}\delta(t)$$

where $N(t)$ and D are given in Eqs. 2.129b, d; it will become more convenient to write the latter in the form

$$D = \omega_n^4\left[(1 - r_f^2)^2 + 4\zeta^2 r_f^2\right] \tag{2.136}$$

The impulsive term in the above expression for the time-derivative of $x_C(t)$ reduces to

$$\frac{N(t)}{D}\delta(t) = \frac{N(0)}{D}\delta(t)$$

but $N(0)$ vanishes, and hence,

$$\frac{d}{dt}(x_C(t)) = \frac{\dot{N}(t)}{D}u(t) \tag{2.137}$$

The time derivative $\dot{N}(t)$ is readily found to be, with the aid of computer algebra,

$$\dot{N}(t) = -\zeta\omega_n e^{-\zeta\omega_n t}\left[-\frac{\zeta(\omega^2 + \omega_n^2)}{\sqrt{1 - \zeta^2}}\sin\omega_d t + (\omega^2 - \omega_n^2)\cos\omega_d t\right]$$

$$+ e^{-\zeta\omega_n t}\left[-\frac{\zeta\omega_d(\omega^2 + \omega_n^2)}{\sqrt{1 - \zeta^2}}\cos\omega_d t - \omega_d(\omega^2 - \omega_n^2)\sin\omega_d t\right]$$

$$+ \left[2\zeta\omega^2\omega_n\cos\omega t + \omega(\omega^2 - \omega_n^2)\sin\omega t\right]$$

or, after rearrangement of terms,

$$\dot{N}(t) = e^{-\zeta\omega_n t}\left\{\left[\frac{\zeta^2\omega_n(\omega^2 + \omega_n^2)}{\sqrt{1 - \zeta^2}} - \sqrt{1 - \zeta^2}\omega_n(\omega^2 - \omega_n^2)\right]\sin\omega_d t\right.$$

$$\left. - \zeta\omega_n(\omega^2 - \omega_n^2 + \omega^2 + \omega_n^2)\cos\omega_d t\right\}$$

$$+ A\omega_n^3\left[2\zeta r_f^2\cos\omega t + r_f(r_f^2 - 1)\sin\omega t\right]$$

Upon further rearranging of terms, the above expression reduces to

$$\dot{N}(t) = \omega_n^3 e^{-\zeta \omega_n t} \left(\frac{1 - r_f^2 + 2\zeta^2 r_f^2}{\sqrt{1 - \zeta^2}} \sin \omega_d t - 2\zeta r_f^2 \cos \omega_d t \right)$$

$$+ \omega_n^3 \left[2\zeta r_f^2 \cos \omega t + r_f(r_f^2 - 1) \sin \omega t \right] \quad (2.138)$$

We can now substitute Eq. 2.137 and the corresponding impulse response, $j_I(t)$, into Eq. 2.108. For quick reference, we recall below the impulse response of underdamped systems:

$$j_I(t) = \frac{e^{-\zeta \omega_n t}}{\omega_d} (\sin \omega_d t)$$

where $\omega_d = \omega_n \sqrt{1 - \zeta^2}$. After substituting the previous results into Eq. 2.108, and performing some algebraic manipulations, we obtain

$$x_S(t) = -\frac{1/\omega_n^2}{(1 - r_f^2)^2 + (2\zeta r_f)^2} H(t) u(t) \quad (2.139a)$$

where

$$H(t) \equiv e^{-\zeta \omega_n t} r_f \left[\frac{1 - r_f^2 - 2\zeta^2}{\sqrt{1 - \zeta^2}} \sin \omega_d t - 2\zeta \cos \omega_d t \right]$$

$$+ 2\zeta r_f \cos \omega t + (r_f^2 - 1) \sin \omega t \quad (2.139b)$$

Now, the transient response $x_{ST}(t)$ of the system at hand to the unilateral sine input is readily identified from the above expression, namely,

$$x_{ST}(t) = \frac{-(1/\omega_n^2)}{(1 - r_f^2)^2 + (2\zeta r_f)^2} e^{-\zeta \omega_n t} r_f \left[\frac{1 - r_f^2 - 2\zeta^2}{\sqrt{1 - \zeta^2}} \sin \omega_d t - 2\zeta \cos \omega_d t \right] u(t)$$

$$(2.140)$$

while the corresponding steady-state response is

$$x_{SS}(t) = \frac{-(1/\omega_n^2)}{(1 - r_f^2)^2 + (2\zeta r_f)^2} \left[2\zeta r_f \cos \omega t - (1 - r_f^2) \sin \omega t \right] u(t)$$

Again, we are interested, in the steady state, at times t very large, at which $u(t)$ can be replaced by unity, the above response then becoming

$$x_{SS}(t) = \frac{-(1/\omega_n^2)}{(1 - r_f^2)^2 + (2\zeta r_f)^2} \left[2\zeta r_f \cos \omega t - (1 - r_f^2) \sin \omega t \right] \quad (2.141)$$

and hence, $x_{SS}(t)$ has the form

$$x_{SS}(t) = M \sin(\omega t + \phi) \tag{2.142a}$$

where

$$M = \frac{1/\omega_n^2}{\sqrt{(1 - r_f^2)^2 + (2\zeta r_f)^2}}, \quad \phi = -\arctan\left(\frac{2\zeta r_f}{1 - r_f^2}\right) \tag{2.142b}$$

The total response of an undamped system to a unilateral sine input can be derived by setting $\zeta = 0$ in the expression for $x_S(t)$ given by Eq. 2.139 to obtain

$$x_S(t) = \frac{1/\omega_n^2}{1 - r_f^2}\left(\sin \omega t - r_f \sin \omega_n t\right) u(t) \tag{2.143}$$

Likewise, the steady-state response of an undamped system to a sinusoidal input is now derived by setting $\zeta = 0$ in the expression derived for $x_{SS}(t)$, thereby obtaining

$$x_{SS}(t) = M_S \sin(\omega t + \phi) \tag{2.144a}$$

where

$$M_S = \frac{(1/\omega_n^2)}{|1 - r_f^2|}, \quad \phi = \begin{cases} 0° & \text{if } r_f < 1; \\ -180° & \text{if } r_f > 1 \end{cases} \tag{2.144b}$$

It is instructive to analyze the total response of an undamped system when excited by a harmonic input with a frequency identical to that of the system, i.e., when $\omega = \omega_n$ or, equivalently, when $r_f = 1$. For example, if we set $\omega = \omega_n$ in Eq. 2.133 or Eq. 2.143, we obtain an indeterminacy, i.e., $0/0$. In order to resolve this indeterminacy, we apply L'Hospital's rule to Eq. 2.133, that we rewrite in the form

$$x_C(t) = \frac{1/\omega_n^2}{1 - r_f^2}(\cos \omega_n r_f t - \cos \omega_n t) \equiv F(r_f) \equiv \frac{N(r_f)}{D(r_f)}$$

and we regard this expression as a function of r_f only. By L'Hospital's rule, then,

$$F(1) = \lim_{r_f \to 1} \frac{N'(r_f)}{D'(r_f)} = \left.\frac{-(1/\omega_n^2)\omega_n t \sin \omega_n r_f t}{-2r_f}\right|_{r_f=1}$$

which yields

$$F(1) = \frac{1}{2\omega_n} t \sin \omega_n t$$

The function $2\omega_n F(t)$ is plotted in Fig. 2.36 for $\omega_n = 1\,\text{Hz} = 2\pi\,\text{rad/s}$, which shows a dramatic case of instability. Note that the amplitude of the above response grows linearly with time.

Fig. 2.36 Resonant response of an undamped second-order system

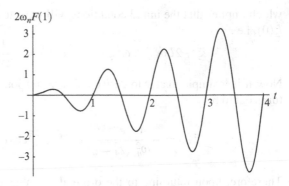

2.7.4 The Response to Constant and Linear Inputs

We consider a load that has been suspending from a crane by means of a cable during a time long enough as to have allowed all transient components of its time response to have faded away. If the crane is subjected to the sole weight of the load, we have an example of a mass-spring system acted upon by a constant input, the weight of the load. Likewise, a terrestrial vehicle, like that of Example 2.3.3, that has been climbing up a slope for a time long enough as to have allowed for the decay of the transient components of its time response, is an instance of a mass-spring-dashpot system acted upon by a linear input. In this subsection we study the time response of these systems, which is quite straightforward to obtain.

For the sake of brevity, we skip first-order and undamped second-order systems here, and focus on second-order underdamped systems. Moreover, in all cases considered here we assume zero initial conditions, with the purpose of being able to apply superposition. We thus have

$$\ddot{x} + 2\zeta\omega_n\dot{x} + \omega_n^2 x = A, \quad x(0) = 0, \quad \dot{x}(0) = 0 \tag{2.145}$$

where A is a constant, and initial conditions are given at $t = 0$, without mentioning explicitly whether at 0^+ or at 0^-, a difference that now becomes irrelevant because the excitation is smooth, and hence, does not contain any impulsive or abrupt components. The simplest way of obtaining the time response of the system at hand is by rewriting the foregoing equation in the form

$$\ddot{x} + 2\zeta\omega_n\dot{x} + \omega_n^2 \left(x - \frac{1}{\omega_n^2}A \right) = 0, \quad x(0) = 0, \quad \dot{x}(0) = 0$$

It is thus apparent that we can cast the foregoing system in a zero-input form with a simple change of variable, namely,

$$\xi = x - \frac{1}{\omega_n^2}A, \quad \text{or} \quad x = \xi + \frac{1}{\omega_n^2}A \tag{2.146}$$

which implies that the initial conditions will have to be given in terms of $\xi(0)$ and $\dot{\xi}(0)$, i.e.,

$$\ddot{\xi} + 2\zeta\omega_n\dot{\xi} + \omega_n^2\xi = 0, \quad \xi(0) = -\frac{1}{\omega_n^2}A, \quad \dot{\xi}(0) = 0$$

Now it is a simple matter to derive the time response of the above system, which turns out to be

$$\xi(t) = -\frac{A}{\omega_n^2}\frac{e^{-\zeta\omega_n t}}{\sqrt{1-\zeta^2}}(\sqrt{1-\zeta^2}\cos\omega_d t + \zeta\sin\omega_d t)$$

Therefore, upon returning to the original variable $x(t)$, the foregoing expression leads to the time response of an underdamped second-order system to a constant input, namely,

$$x_{\text{const}}(t) = \frac{A}{\omega_n^2}\left[1 - \frac{e^{-\zeta\omega_n t}}{\sqrt{1-\zeta^2}}(\sqrt{1-\zeta^2}\cos\omega_d t + \zeta\sin\omega_d t)\right] \qquad (2.147)$$

which the reader is invited to compare with the corresponding step response.

Now we derive the time response of the same system to an input that is linear with time, under zero initial conditions, i.e.,

$$\ddot{x} + 2\zeta\omega_n\dot{x} + \omega_n^2 x = At, \quad x(0) = 0, \quad \dot{x}(0) = 0 \qquad (2.148)$$

Let us call $x_{\text{lin}}(t)$ the time response of the system under study to a linear input, with zero initial conditions. Since the linear input is the integral of the constant input, by linearity we have

$$x_{\text{lin}}(t) = \int_0^t x_{\text{const}}(\theta)d\theta$$

Upon performing the above integration with the aid of computer algebra, we have

$$x_{\text{lin}}(t) = \frac{A}{\omega_n^3}\left\{\omega_n t + e^{-\zeta\omega_n t}\left[2\zeta\cos\omega_d t + \frac{(2\zeta^2 - 1)}{\sqrt{1-\zeta^2}}\sin\omega_d t\right] - 2\zeta\right\} \qquad (2.149)$$

whence it is apparent that the initial condition $x_{\text{lin}}(0) = 0$ is satisfied. The time-derivative of the foregoing expression need not be computed, for it is identical to $x_{\text{const}}(t)$, by definition. Since this function verifies the initial condition $x_{\text{const}}(0) = 0$, we have that the expression for $x_{\text{lin}}(t)$ indeed satisfies the given zero initial conditions, thereby completing the computation of the time responses of second-order damped systems to constant and linear inputs.

2.7.5 The Power Dissipated By a Damped Second-order System

Damping can be determined experimentally if the energy dissipated by a damped second-order system and the motion undergone by the system are both known. In particular, when the system is excited by a harmonic force, we know that the system undergoes harmonic motion as well. Below we derive the relation that allows us to determine the damping coefficient c when we know the amount of energy E_C dissipated by a damped second-order system per cycle as well as its amplitude M and the frequency ω.

The system at hand is assumed to undergo a motion of the form

$$x(t) = M\cos(\omega t + \phi) \tag{2.150}$$

and hence,

$$\dot{x}(t) = -\omega M \sin(\omega t + \phi) \tag{2.151}$$

Thus, from Chap. 1, the power Π_d dissipated by the dashpot takes the form

$$\Pi_d(t) = c\dot{x}^2(t) = c\,\omega^2 M^2 \sin^2(\omega t + \phi) \tag{2.152}$$

which can be further expressed as

$$\Pi_d(t) = \frac{1}{2}c\,\omega^2 M^2 \left[1 - \cos 2(\omega t + \phi)\right] \tag{2.153}$$

Now, E_C is obtained simply by integration of the power $\Pi_d(t)$ dissipated throughout a complete cycle, i.e.,

$$E_C = \int_0^{2\pi/\omega} \Pi_d(t)dt = \frac{1}{2}c\,\omega M^2 \int_0^{2\pi} \left[1 - \cos 2(\omega t + \phi)\right] \omega dt \tag{2.154}$$

the integral thus reducing to 2π, and hence,

$$E_C = \pi c\,\omega M^2$$

Therefore, if all parameters E_C, ω and M are known, the damping coefficient can be readily found as

$$c = \frac{E_C}{\pi\omega M^2} \tag{2.155}$$

2.7.6 The Bode Plots of First- and Second-order Systems

In the preceding section we derived expressions for the magnitude of the response of first- and second-order systems to a harmonic input. This magnitude has a unit associated with it, which is the same as that of the variable x. In engineering practice, it is often advantageous to work with dimensionless quantities. Thus, we will define a dimensionless quantity μ called the *magnification factor*. For first-order systems, M has units of a^{-1}, and hence, Ma or, equivalently, M/τ, is dimensionless and plays the role of the magnification factor μ, i.e.,

$$\mu_f \equiv Ma \equiv \frac{M}{\tau} \tag{2.156}$$

Likewise, for second-order systems, M has units of time squared, and hence, we can render it dimensionless by multiplying it by a quantity with units of frequency-squared. The obvious candidate is ω_n^2, i.e., the square of the natural frequency of the system at hand. Thus, for second-order systems, the magnification factor is defined as

$$\mu_s = M\omega_n^2 \tag{2.157}$$

With the preceding definitions, we can readily derive the magnification factors for first- and second-order systems, using the expressions derived for M previously. These are listed below, along with the corresponding expression for the phase angle. As before, for second-order systems, only the undamped and underdamped cases are considered, the critically damped and overdamped cases being left as exercises.

For first-order systems, we have

$$\mu_f = \frac{1}{\sqrt{1 + \left(\frac{\omega}{a}\right)^2}}, \qquad \phi_f = -\arctan\left(\frac{\omega}{a}\right) \tag{2.158}$$

while, for second-order undamped systems,

$$\mu_{su} = \frac{1}{|1 - r_f^2|}, \qquad \phi_{su} = \begin{cases} 0° & \text{if } r_f < 1; \\ -180° & \text{if } r_f > 1 \end{cases} \tag{2.159}$$

Finally, for second-order underdamped, critically damped and overdamped systems,

$$\mu_{sd} = \frac{1}{\sqrt{(1 - r_f^2)^2 + (2\zeta r_f)^2}}, \qquad \phi_{sd} = -\arctan\left(\frac{2\zeta r_f}{1 - r_f^2}\right) \tag{2.160}$$

Plots of μ and ϕ versus $\log \omega$ are known as *Bode plots* of the system and provide all the information of the harmonic response of the system under study. However, the $\log(\cdot)$, like the exponential $e^{(\cdot)}$, of an argument with physical units is meaningless, which calls for a *normalization* of ω in the Bode plots, as explained below. Figures 2.37 and 2.38 show the Bode plots of first- and second-order systems. Note that, for first-order systems, we have plotted μ and ϕ versus ω/a, the latter being a dimensionless quantity similar to the frequency ratio of second-order systems.

Fig. 2.37 Magnification factor and phase plots for first-order system

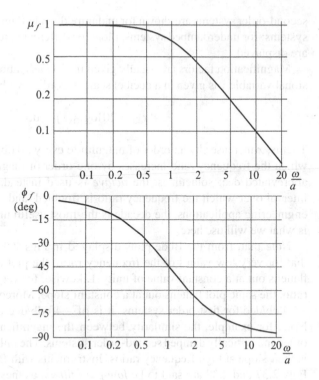

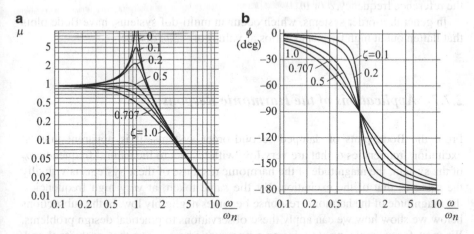

Fig. 2.38 Bode plots of second-order system for different values of damping ratio ζ: (a) magnification factor; and (b) phase

Similarly, for second-order systems, μ and ϕ are plotted versus the frequency ratio $\omega/\omega_n \equiv r_f$, which is also a dimensionless quantity. For both systems, the scales for μ, ω/a and ω/ω_n are logarithmic to a base of 10. Moreover, the Bode plots of second-order systems are shown for undamped, underdamped and critically damped systems; for underdamped systems, plots for different values of the damping ratio ζ are displayed.

Magnification factors are usually given in *decibels*, abbreviated db. A nondimensional variable x is given in a decibel scale according with the definition

$$x_{db} = 20\log_{10}(x) \quad db \tag{2.161}$$

That is, x increases by an order of magnitude every 20 db. Likewise, the interval in which the frequency ratio increases by one order of magnitude is called a *decade*, abbreviated dec; sometimes, the *octave* is used instead, which is defined as the interval over which the frequency ratio is doubled, and abbreviated oct. For most engineering applications, the decade is the most useful unit, for which reason, this is what we will use here.

Note that, from the Bode plots displayed in Figs. 2.37 and 2.38, it is apparent that for very low values of the frequency ratio, the plot of the magnification ratio flattens out at a constant value of unity. Likewise, for very high values of the same ratio, the same plot flattens out at a constant slope. Moreover, while this slope is of -20 db/dec for first-order systems, it is of -40 db/dec for second-order systems. Note, for example, the similarity between the magnification Bode plots for first-order and critically damped second-order systems. The only difference in these plots is their slope at high frequency ratios. Instruments with Bode plots of the forms of Figs. 2.37 and 2.38 are said to be *low-pass filters*, as they reject frequencies above the reference frequency, a or ω_n.

In general, n-order systems, which occur in multi-dof systems, have Bode plots that flatten out at high frequency ratios at values of $-20n$ db/dec.

2.7.7 Applications of the Harmonic Response

From the Bode plots of damped second-order systems, it is apparent that, at excitation frequencies ω that are very low with respect to the natural frequency ω_n of the system, the magnitude of the harmonic response of these systems is virtually the same as that of the excitation. By the same token, at very high frequencies, the magnitude of the harmonic response becomes negligibly low. In the subsections below we show how we can apply these observations to practical design problems. We may, for example, want to design a pneumatic hammer so that, at the tool-end, it will transmit a very high force, capable of breaking hard rock while, at the handle end, it will transmit a gentle force to the operator. Alternatively, we may want to design instruments to measure accurately displacement, velocity or acceleration. In these cases, we want to obtain a signal transmission with the highest fidelity, and so, we should make sure that the natural frequency of our instrument is tuned with the frequency range of the signal that we want to measure.

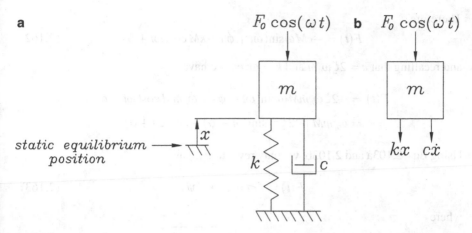

Fig. 2.39 A pneumatic hammer under harmonic excitation: (**a**) Iconic model and (**b**) its tool FBD

2.7.7.1 Vibration Isolation

Consider the model of a pneumatic rock-breaking hammer depicted in Fig. 2.39a, in which the force $F_0 \cos \omega t$ is applied by the rock on the cutting tool and the support, depicted as an inertial frame, is the handle, which is firmly held by the operator. The latter is now regarded as an inertial frame. We want to determine the relation between the force transmitted to the handle and the force applied onto the tool, for design purposes. Moreover, **the static force due to gravity in this and the examples below is irrelevant, as it does not vary harmonically, and hence, does not appear in our analysis**.

First we need to derive an expression for the transmitted force $F(t)$. From the free-body diagram of Fig. 2.39b, it is apparent that this force is the resultant of the forces acting at the ends of the spring and the dashpot, respectively. Hence,

$$F(t) = c\dot{x}_{CS} + kx_{CS}$$

where x_{CS} is the steady-state component of the displacement under a cosine excitation. Thus, the transmitted force is a linear combination of the steady-state displacement and velocity of the mass. Now, since the input is a harmonic function, so is the steady-state displacement, and hence, the velocity. The transmitted force is, then, harmonic as well, and we can fully determine it by its magnitude F_T and its phase ψ. For ease of manipulation, we represent the steady-state displacement in the form given by Eq. 2.102, and hence, the steady-state velocity is obtained by simple differentiation of the foregoing expression. For quick reference, we include expressions for these two items below:

$$x_{CS}(t) = M\cos(\omega t + \phi), \quad \dot{x}_{CS}(t) = -M\omega\sin(\omega t + \phi)$$

Therefore,

$$F(t) = -cM\omega \sin(\omega t + \phi) + kM \cos(\omega t + \phi) \tag{2.162}$$

and recalling that $c = 2\zeta \omega_n m$ and $k = \omega_n^2 m$, we have

$$F(t) = -2\zeta \omega_n m M\omega \sin(\omega t + \phi) + \omega_n^2 m M \cos(\omega t + \phi)$$
$$= \omega_n^2 m M \left[-2\zeta r_f \sin(\omega t + \phi) + \cos(\omega t + \phi) \right]$$

Using Eqs. 2.103a and 2.103b, we may rewrite $F(t)$ as

$$F(t) = F_T \cos(\omega t + \psi) \tag{2.163}$$

where

$$F_T \equiv \omega_n^2 M m \sqrt{1 + (2\zeta r_f)^2} \tag{2.164}$$

and, as the reader is invited to verify, M is given in this case as

$$M = \frac{F_0/m}{\omega_n^2 \sqrt{(1 - r_f^2)^2 + (2\zeta r_f)^2}} \tag{2.165}$$

We have, therefore,

$$F_T \equiv \frac{F_0 \sqrt{1 + (2\zeta r_f)^2}}{\sqrt{(1 - r_f^2)^2 + (2\zeta r_f)^2}} \tag{2.166}$$

Moreover, using Eq. 2.105b, we have

$$\psi \equiv \tan^{-1}(2\zeta r_f) + \phi = \tan^{-1}(2\zeta r_f) - \tan^{-1}\left(\frac{2\zeta r_f}{1 - r_f^2} \right) \tag{2.167}$$

The magnification factor μ_F of the transmitted force is thus defined as the ratio

$$\mu_F \equiv \frac{F_T}{F_0} = \sqrt{\frac{1 + (2\zeta r_f)^2}{(1 - r_f^2)^2 + (2\zeta r_f)^2}} \tag{2.168}$$

Plots of both the magnification factor μ_F and the phase angle ψ are shown in Fig. 2.40.

The similarity between the Bode plots of Figs. 2.38 and 2.40 is to be highlighted. Note that both magnification plots converge to the same asymptotes for very low frequency ratios, namely, at a constant value of unity. Furthermore, at very high frequency ratios, the Bode plots of Fig. 2.40 flatten out at a constant slope of -40 db/dec, as $\mu_F \to 0$. Moreover, the magnification plots of the same figure, for

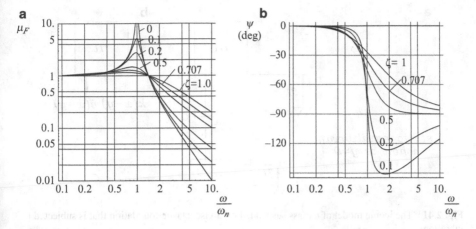

Fig. 2.40 The Bode plots of the force transmitted by a damped second-order system to its base: (a) magnification factor; and (b) phase

all values of damping coefficient ζ, cross the $\mu_F = 1$ axis at the same point. A quick calculation shows that the crossing point is located at $r_f = \sqrt{2}$, a fact that the reader is encouraged to verify.

With regard to the design of pneumatic hammers as described above, we can use the Bode plots of Fig. 2.40. From these plots it is apparent that the force transmitted to the operator will be negligibly small as long as the operation frequency ω of the rock-breaking force is substantially above $\sqrt{2}$ times the natural frequency of the tool-pneumatic cylinder of the hammer, when the cylinder is modeled as a spring-dashpot array in parallel, as shown in Fig. 2.39.

An alternative application of the above-derived Bode plots is in the design of foundations for equipment that is to remain stationary in the presence of parasitical vibration of the ground. Such vibration is usually present in environments where various machines are in operation. Equipment that should remain undisturbed by this vibration include instruments and precision machine tools. To illustrate this idea, let us assume that the equipment to be isolated from parasitical vibration can be safely modeled as a rigid body of mass m and the foundation as a spring-dashpot array in parallel, exactly as the model shown in Fig. 2.39a. Furthermore, we assume that the ground undergoes a vertical displacement $y(t) = Y \cos \omega t$. We want to relate the harmonic response of the system, $x(t)$, to this input. To this end, we sketch first the iconic model of the system in Fig. 2.41, and then derive its mathematical model.

The free-body diagram of the mass of Fig. 2.41 leads to

$$m\ddot{x} + c\dot{x} + kx = c\dot{y} + ky$$

and hence, in normal form,

$$\ddot{x} + 2\zeta\omega_n\dot{x} + \omega_n^2 x = 2\zeta\omega_n\dot{y} + \omega_n^2 y$$

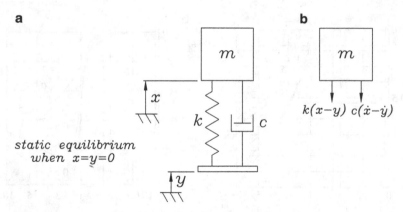

Fig. 2.41 The iconic model of a mass suspended on a viscoelastic foundation that is subjected to vibration

Further, by virtue of the harmonic form of the signal $y(t)$, we have

$$\ddot{x} + 2\zeta\omega_n\dot{x} + \omega_n^2 x = -2\zeta\omega_n\omega Y \sin\omega t + \omega_n^2 Y \cos\omega t$$

That is, the system is subject to a linear combination of a sine and a cosine signal. Using Eq. 2.103a, we rewrite the right-hand side of the above expression in cosine form, thereby obtaining

$$\ddot{x} + 2\zeta\omega_n\dot{x} + \omega_n^2 x = A\cos(\omega t + \sigma) \tag{2.169}$$

where

$$A \equiv \omega_n^2\sqrt{1 + (2\zeta r_f)^2}\,Y, \qquad \sigma \equiv \arctan(2\zeta r_f)$$

Using Eq. 2.132a, the steady–state response of the system is given by

$$x_{CS}(t) = M\cos(\omega t + \sigma + \phi) \tag{2.170a}$$

where

$$M = \frac{A/\omega_n^2}{\sqrt{(1 - r_f^2)^2 + (2\zeta r_f)^2}} \tag{2.170b}$$

or

$$M = \frac{Y\sqrt{1 + (2\zeta r_f)^2}}{\sqrt{(1 - r_f^2)^2 + (2\zeta r_f)^2}} \tag{2.170c}$$

and

$$\phi = -\tan^{-1}\left(\frac{2\zeta r_f}{1 - r_f^2}\right) \tag{2.170d}$$

Hence, the magnification factor μ_Y of the transmitted motion is given by

$$\mu_Y \equiv \frac{M}{Y} \equiv \sqrt{\frac{1 + (2\zeta r_f)^2}{(1 - r_f^2)^2 + (2\zeta r_f)^2}} \tag{2.171}$$

which is identical to the magnification factor found above for the transmitted force μ_F, and so, a Bode plot for this item is not needed.

For design purposes, then, we aim to design a foundation so that it will isolate the vibration $y(t)$ by giving the foundation a natural frequency that will be substantially below all parasitical frequencies. In this way, the parasitical frequency ratios will be substantially to the right of the value $r_f = \sqrt{2}$ of the Bode plot of Fig. 2.40.

One more application of the Bode plots of Fig. 2.40 is in the explanation of a phenomenon worrying the civil engineers in charge of road maintenance. Transport-regulating bodies specify the maximum allowable static load (mass) per axle in terrestrial vehicles, whether these transport people or goods. Thus, buses and trucks are subjected to the same regulations; transportation companies, with the aim of maximizing their profit per trip, load their units to the limit. However, as it turns out, buses produce much less damage to the road than trucks. The explanation can be found in the difference in the suspension of the two types of vehicles: those of buses are designed to give the passengers a comfortable ride; those of trucks are designed to require the least maintenance, and hence, the only damping that the latter have is the one that comes within the leaf springs that they use. In order to explain the difference that the damping makes, let us model the load-suspension system pertaining to one axle with the iconic model of Fig. 2.13a. Furthermore, we model the irregular road profile as a sinusoidal track, thereby ending up with the iconic model of Fig. 2.41, where $y = Y \cos \omega t$, and the Bode plots of Fig. 2.40 apply. Moreover, we can safely assume that trucks and buses, like any other terrestrial vehicle, have suspensions that produce vertical vibrations with a natural frequency of $\omega_n = 1$ Hz, the only difference in the two types of suspension then being in the damping. Now, the damping in trucks is very low, say, around 0.1, while that of buses can be safely assumed to be around 0.5. From Fig. 2.40a, it is apparent that μ for $\zeta = 0.1$ reaches a peak that is around four times that occurring at $\zeta = 0.5$. As a consequence, the truck transmits to the road a dynamic force that is around four times as big as that transmitted by the bus.

2.7.7.2 Velocity Meter Design

Contrary to the second example of the preceding section, we may want in some instances, such as when designing an instrument to measure the velocity of a moving body, to be able to pick up the desired velocity within its frequency range with the highest fidelity. What we mean by the latter is that we want to be able to subject a mass m to a harmonic displacement that will follow the velocity signal to be measured with the highest possible magnitude and the smallest possible phase angle.

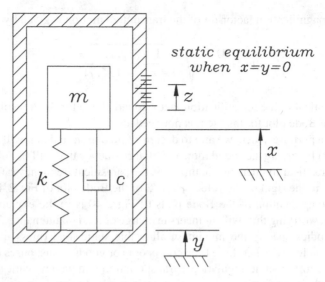

Fig. 2.42 An iconic model of the velocity meter

In order to attain this goal, obviously, the two signals, displacement of m and measured velocity, must have the same frequency. The purpose of this design task is to determine the physical parameters m, c and k of a velocity meter, modeled as a mass-spring-dashpot system, so that it will exhibit the harmonic response described above. Our iconic model in this case is shown in Fig. 2.42, where $y(t)$ is the displacement of the object whose velocity is to be measured, onto which the base of the instrument is firmly attached; moreover, x is the displacement of the mass m and the pick-up signal $z(t)$ is defined as $z \equiv y - x$. The mathematical model is now readily derived from Fig. 2.42 in the form

$$m\ddot{x} = -c(\dot{x} - \dot{y}) - k(x - y) \tag{2.172}$$

or

$$\ddot{x} + 2\zeta\omega_n(\dot{x} - \dot{y}) + \omega_n^2(x - y) = 0 \tag{2.173}$$

or, in terms of z, the measured signal,

$$\ddot{z} + 2\zeta\omega_n\dot{z} + \omega_n^2 z = \ddot{y} \tag{2.174}$$

We now assume that the velocity signal $\dot{y}$ of interest is harmonic, of the form

$$\dot{y} = V\cos\omega t \tag{2.175}$$

and hence, the foregoing model takes the form

$$\ddot{z} + 2\zeta\omega_n\dot{z} + \omega_n^2 z = -\omega V\sin\omega t$$

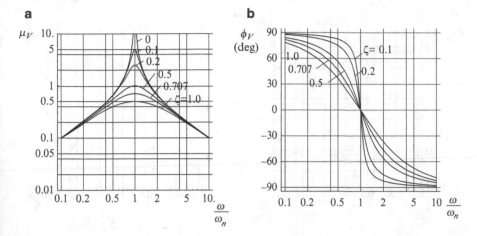

Fig. 2.43 The Bode plots of the velocity transmission of a velocity meter

However, since we want our instrument to follow the velocity signal, which is cosinusoidal, we have to express the foregoing model with a cosinusoidal excitation as well, which is done by simply adding a phase of 90° to the above sinusoidal signal, thus obtaining

$$\ddot{z} + 2\zeta\omega_n\dot{z} + \omega_n^2 z = \omega V \cos\left(\omega t + \frac{\pi}{2}\right) \qquad (2.176)$$

The steady-state response of the system, then, can be written as

$$z = \omega V M \cos(\omega t + \phi_V) \qquad (2.177)$$

where M and ϕ_V take the forms of Eqs. 2.142a, b, except that the latter is augmented by a phase angle of $\pi/2$. Now, the magnification factor that we need is one relating $\omega V M$ with the amplitude of the excitation signal V, as a dimensionless quantity. We thus define the magnification factor of interest, along with the corresponding phase angle, as shown below:

$$\mu_V \equiv \frac{\omega_n\omega V M}{V} = \frac{r_f}{\sqrt{(1-r_f^2)^2 + (2\zeta r_f)^2}}, \qquad \phi_V = \frac{\pi}{2} - \arctan\left(\frac{2\zeta r_f}{1-r_f^2}\right) \qquad (2.178)$$

which are plotted in Fig. 2.43.

An inspection of the Bode plots of Fig. 2.43 reveals that the frequency range in which the magnification factor of this instrument attains its peak value is remarkably narrow. Such an instrument is said to have a narrow *bandwidth*, which makes it useless for our envisioned application. A narrow bandwidth means that the magnification factor attains values above unity in a correspondingly narrow range

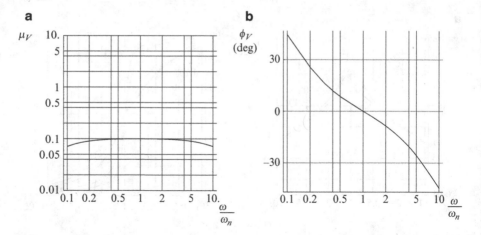

Fig. 2.44 The Bode plots of the velocity transmission of a velocity meter with a high damping ratio $\zeta = 5$

of values of r_f. A means of enhancing the bandwidth of this instrument is to design it with a rather high damping ratio, say $\zeta = 5$, as made apparent by the plot of Fig. 2.44a. From this plot, it is apparent that the velocity meter under study is useful for the measurement of velocity signals comprising a frequency ranging from $0.2\omega_n$ to about $10\omega_n$. The corresponding phase angle is shown in Fig. 2.44b.

2.7.7.3 Accelerometer Design

Accelerometers are designed to provide a displacement signal that follows as closely as possible a harmonic acceleration signal with a high amplitude and a low phase angle. We thus assume that the accelerometer is modeled as the system of Fig. 2.41, the acceleration signal $\ddot{y}(t)$ that we want to measure being harmonic, namely,

$$\ddot{y}(t) = A \cos \omega t$$

the underlying mathematical model thus taking the form

$$\ddot{z} + 2\zeta \omega_n \dot{z} + \omega_n^2 z = \ddot{y}$$

with z defined as

$$z \equiv y - x$$

The steady-state response thus becoming

$$z(t) = M_A \cos(\omega t + \phi_A)$$

with M_A given as

$$M_A = \frac{A}{\omega_n^2 \sqrt{(1 - r_f^2)^2 + (2\zeta r_f)^2}}$$

the phase being given as in Eq. 2.160, namely, as

$$\phi_A = -\arctan\left(\frac{2\zeta r_f}{1 - r_f^2}\right)$$

Now, we define the magnification factor μ_A as

$$\mu_A \equiv \frac{\omega_n^2 M_A}{A}$$

thereby obtaining

$$\mu_A = \frac{1}{\sqrt{(1 - r_f^2)^2 + (2\zeta r_f)^2}}$$

which is identical to the magnification factor of Eq. 2.160 and plotted in Fig. 2.38.

2.7.7.4   Seismograph Design

The instrument of Fig. 2.42 can be used as a *seismogragh* if we want to measure a harmonic displacement $y(t)$ of the base with the pick-up signal $z \equiv y - x$. The mathematical model we have is identical to that derived for the accelerometer, i.e.,

$$\ddot{z} + 2\zeta\omega_n\dot{z} + \omega_n^2 z = \ddot{y}$$

Upon the assumption that the displacement $y(t)$ is harmonic, we can write

$$y(t) = Y\cos\omega t$$

and hence,

$$\ddot{y} = -\omega^2 Y\cos\omega t$$

the mathematical model thus becoming

$$\ddot{z} + 2\zeta\omega_n\dot{z} + \omega_n^2 z = -\omega^2 Y\cos\omega t \equiv \omega^2 Y\cos(\omega t + \pi)$$

The steady-state response of the system is then of the form

$$z(t) = M_Y\cos(\omega t + \phi_Y)$$

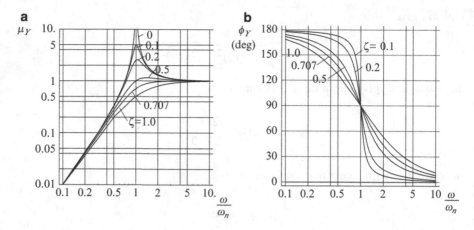

Fig. 2.45 The Bode plots of a seismograph

with M_Y and ϕ_Y given as

$$M_Y = \frac{r_f^2 Y}{\sqrt{(1-r_f^2)^2 + (2\zeta r_f)^2}}, \qquad \phi_Y = \pi - \arctan\left(\frac{2\zeta r_f}{1-r_f^2}\right)$$

The magnification factor μ_Y is thus defined as

$$\mu_Y \equiv \frac{M_Y}{Y} = \frac{r_f^2}{\sqrt{(1-r_f^2)^2 + (2\zeta r_f)^2}}$$

Plots of the magnification factor and the phase angle are shown in Fig. 2.45.

Note that the Bode plots of this instrument reveal that the seismograph is insensitive to low frequencies but very sensitive to high frequencies. The instrument is thus said to be a *high-pass filter*.

In summary, the Bode plots obtained here have a common feature: the difference between the high-frequency asymptote slope and the low-frequency asymptote slope is always $+20$ db/dec. This feature is common to all second-order systems.

2.7.8 Further Applications of Superposition

In the two examples below, we include further applications of superposition, while resorting to the time responses derived for harmonic inputs. We do this to analyze the response of a second-order system to a pulse-like input, which is not periodic.

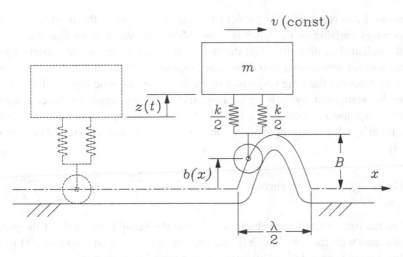

Fig. 2.46 The iconic model of a terrestrial vehicle upon encountering a bump

Example 2.7.1 (An Undamped Terrestrial Vehicle Hitting a Bump). A highly simplified iconic model of a terrestrial vehicle traveling at a constant speed v, upon hitting a bump of height B, is shown in Fig. 2.46.

In this model, we neglect the damping in the suspension and consider only its stiffness. Moreover, we assume that the bump does not affect the horizontal, uniform motion of the vehicle, the bump being modeled via the function $b(x)$, defined below:

$$b(x) \equiv \begin{cases} B\sin(2\pi x/\lambda), & \text{for} \quad 0 \le x \le \lambda/2; \\ 0 & \text{otherwise} \end{cases}$$

Here, we have $x = vt$ and hence, the period of the sine function, in terms of the wavelength λ, is $T = \lambda/v$. Determine the motion of the suspension-mass system after the vehicle has hit the bump.

Solution: In order to ease our task, we start by realizing that $b(x)$ can be synthesized as

$$b(x) = B\left(\sin\frac{2\pi x}{\lambda}\right)u(x) + B\left(\sin\frac{2\pi(x-\lambda/2)}{\lambda}\right)u\left(x-\frac{\lambda}{2}\right)$$

and so, we can obtain the desired response by superposition, i.e., as the sum of the responses to each of the two foregoing unilateral sinusoidal functions. The mathematical model of the system at hand, then, takes the form

$$\ddot{z} + \omega_n^2 z = \omega_n^2 B\left[\left(\sin\frac{2\pi vt}{\lambda}\right)u(t) + \left(\sin\frac{2\pi v(t-\lambda/(2v))}{\lambda}\right)u\left(t-\frac{\lambda}{2v}\right)\right]$$

where we have replaced $b(x)$ by $b(t)$ for consistency with the model, since the independent variable of the latter is time. Moreover, we assume that the vehicle travels undisturbed up until it hits the bump, at which time we start observing the system and set arbitrarily $t = 0$. The time response of interest can now be derived by superimposing the time responses of each of the two sine inputs of the above model. We start, then, by deriving the response to the first input. To this end, we use the total response $x_S(t)$ derived in Eq. 2.143 for an undamped second-order system acted upon by a harmonic input of the form $(\sin \omega t)u(t)$, with zero initial conditions, namely,

$$x_S(t) = \frac{1/\omega_n^2}{1-r_f^2} \left(\sin \omega t - r_f \sin \omega_n t \right), \qquad \omega \equiv \frac{2\pi v}{\lambda} \equiv \frac{2\pi}{T}, \qquad r_f = \frac{\omega}{\omega_n} = \frac{2\pi v}{\lambda}\sqrt{\frac{m}{k}}$$

That is, the time it takes the vehicle to traverse the bump is one-half of the period T associated with the wavelength λ and the speed v. The total response $z(t)$ of the system at hand is now derived by linearity and time-invariance as

$$z(t) = \frac{B}{1-r_f^2} \left\{ \left[\sin \omega t - r_f \sin \omega_n t \right] u(t) \right.$$

$$\left. + \left[\sin \omega \left(t - \frac{T}{2} \right) - r_f \sin \omega_n \left(t - \frac{T}{2} \right) \right] u\left(t - \frac{T}{2} \right) \right\}$$

or, after some rearrangement of terms,

$$z(t) = \frac{B}{1-r_f^2} \left\{ (\sin \omega t)u(t) + \left[\sin \omega \left(t - \frac{T}{2} \right) \right] u\left(t - \frac{T}{2} \right) \right\}$$

$$- \frac{Br_f}{1-r_f^2} \left\{ (\sin \omega_n t)u(t) + \left[\sin \omega_n \left(t - \frac{T}{2} \right) \right] u\left(t - \frac{T}{2} \right) \right\}$$

Note that the first term in the foregoing expression is nothing but the bump function $b(t)$ divided by $(1 - r_f^2)$. Now, as to the second term, this looks like a bump as well, but it is not. Indeed, this term is a linear combination of two harmonics of the same kind, i.e., of sines, with one delayed with respect to the other by half a period of the bump signal. Since the natural frequency ω_n of the system is, in general, different from the bump frequency ω, the two sinusoidal terms of the above expression do no cancel each other. We thus have, in summary,

$$z(t) = \frac{1}{1-r_f^2} \left[b(t) - r_f \beta(t) \right]$$

where

$$\beta(t) \equiv B \left\{ (\sin \omega_n t)u(t) + \left[\sin \omega_n \left(t - \frac{T}{2} \right) \right] u\left(t - \frac{T}{2} \right) \right\}$$

Fig. 2.47 The plot of $\beta(t)/B$ appearing in the response of the vehicle with an undamped suspension upon hitting a bump, for $\omega_n = 1\,\mathrm{Hz} = 2\pi\,\mathrm{rad/s}$ and $T = 2\,\mathrm{s}$

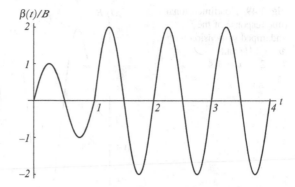

Fig. 2.48 Nondimensional time response of the undamped suspension to a bump with $\omega_n = 1\,\mathrm{Hz} = 2\pi\,\mathrm{rad/s}$ and $T = 2\,\mathrm{s}$

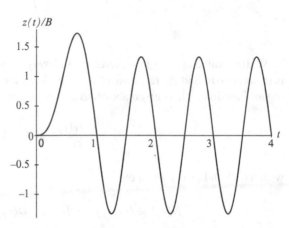

The ratio $\beta(t)/B$ is plotted in Fig. 2.47 for $\omega_n = 1\,\mathrm{Hz} = 2\pi\,\mathrm{rad/s}$ and $T = 2$ s, the total time response $z(t)$ being plotted in Fig. 2.48 in nondimensional form.

That is, the response of the undamped system to a bump is a linear combination of the bump and the function $\beta(t)$, the amplitude of this response increasing as the difference $1 - r_f^2$ decreases. Note that this difference vanishes when the frequency of the bump equals that of the spring-mass system, which occurs, in terms of the given parameters, when

$$\frac{\lambda}{v} = \frac{2\pi}{\omega_n}$$

Most terrestrial vehicles have a natural frequency of about 1 Hz, i.e., of $\omega_n = 2\pi\,\mathrm{s}^{-1}$. If we were going to encounter a sinusoidal bump with a wavelength λ obeying the foregoing relation, for a given traveling speed v, we would like to have an idea of the order of magnitude of the parameters involved, for realistic situations. For example, if the traveling speed is around 72 km/h, then $v = 20$ m/s, which thus would yield a value $\lambda = 20$ m, a rather huge wavelength, and hence, not realistic. A more realistic value of λ is 1 m, which, for the same natural frequency of the vehicle, gives a velocity $v = 1$ m/s or 3.6 km/h, a rather unusually small value.

Fig. 2.49 Nondimensional time response of the undamped suspension with $\omega_n = 1$ Hz and $T = 2\pi/\omega_n = 1$ s

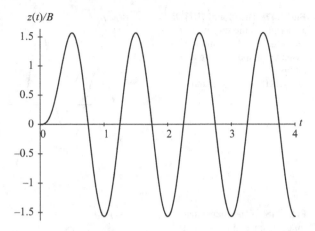

While a value of $r_f = 1$ is, therefore, unlikely to happen in real-life situations, it is instructive to find the response of the vehicle under this condition. To this end, we regard function $z(t)$, as given above, as a function of r_f, i.e.,

$$z(r_f) \equiv \frac{N(r_f)}{D(r_f)} \equiv F(r_f)$$

with $N(r_f)$ and $D(r_f)$ given by

$$N(r_f) \equiv b(t; r_f) - r_f \beta(t), \quad D(r_f) \equiv 1 - r_f^2 \tag{2.179}$$

By application of L'Hospital's rule, we have

$$F(1) = \lim_{r_f \to 1} \frac{N'(r_f)}{D'(r_f)}$$

and hence, as the reader is invited to verify,

$$z(t) = -\frac{1}{2} \left\{ t\dot{b}(t) - b(t) - \pi B[\cos(\omega_n t - \pi)]u(t - \frac{\pi}{\omega_n}) \right\}, \text{ for } r_f = 1 \text{ or } \omega_n \lambda = 2\pi v$$

Therefore, the vertical oscillations of the vehicle under study remain bounded, even under these apparently resonant conditions. The explanation is simple: r_f refers to the ratio ω/ω_n, where $\omega = 2\pi r/\lambda$ for the system of Fig. 2.46, with r denoting the radius of the wheel. However, notice that the input to this system is not harmonic, but a bump. The foregoing response is plotted in Fig. 2.49 for $\omega_n = 1$ Hz and $r_f = 1$.

It is pointed out here that, the bump being zero for $t > \lambda/v$, it disappears from the time response after the vehicle has traversed it. Hence, the remainder of $z(t)$, namely, the term containing the function $\beta(t)$, is the steady-state response of the system.

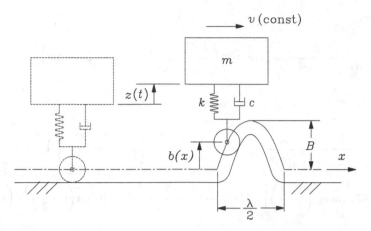

Fig. 2.50 The iconic model of a terrestrial vehicle upon encountering a bump

Example 2.7.2 (A Damped Terrestrial Vehicle Encountering a Bump). Find the time response of the vehicle introduced in the foregoing example, if we add "shocks" to its suspension, as shown in Fig. 2.50, under the assumption that the system at hand becomes underdamped.

Solution: The model now takes the form

$$\ddot{z} + 2\zeta\omega_n\dot{z} + \omega_n^2 z = 2\zeta\omega_n b(t) + \omega_n^2 b(t)$$

and hence, the response can be synthesized as the sum of the two responses $z_b(t)$ and $z_{\dot{b}}(t)$ to the first and the second terms of the right-hand side, respectively. We start by calculating $\dot{b}(t)$, namely,

$$\dot{b}(t) = \frac{db(x)}{dx}\frac{dx}{dt} \equiv \frac{db(x)}{dx}v = b'(x)v$$

and then find $b'(x)$. From the expression derived in the previous example for $b(x)$, it is now a simple matter to show that

$$b'(x) = \frac{2\pi B}{\lambda}\left[\left(\cos\frac{2\pi x}{\lambda}\right)u(x) + \left(\cos\frac{2\pi(x-\lambda/2)}{\lambda}\right)u\left(x-\frac{\lambda}{2}\right)\right]$$

2.7.9 Derivation of $z_b(t)$

To derive $z_b(t)$ we use the $x_S(t)$ response for the second-order underdamped system derived in Sect. 2.7.3, and invoke linearity and time-invariance, thereby obtaining

$$z_b(t) = \omega_n^2 Bx_S(t) + \omega_n^2 Bx_S\left(t - \frac{T}{2}\right)$$

and hence,

$$z_b(t) = -\frac{B}{(1-r_f^2)^2+4\zeta^2 r_f^2}\left[r_f e^{-\zeta\omega_n t}\left(\frac{1-2\zeta^2-r_f^2}{\sqrt{1-\zeta^2}}\sin\omega_d t - 2\zeta\cos\omega_d t\right)\right.$$

$$\left.+2\zeta r_f\cos\omega t + (r_f^2-1)\sin\omega t\right]u(t)$$

$$-\frac{B}{\left(1-r_f^2\right)^2+4\zeta^2 r_f^2}\left\{r_f e^{-\zeta\omega_n\left(t-\frac{T}{2}\right)}\left[\frac{1-2\zeta^2-r_f^2}{\sqrt{1-\zeta^2}}\sin\omega_d\left(t-\frac{T}{2}\right)\right.\right.$$

$$\left.\left.-2\zeta\cos\omega_d\left(t-\frac{T}{2}\right)\right]+2\zeta r_f\cos\omega\left(t-\frac{T}{2}\right)+(r_f^2-1)\sin\omega\left(t-\frac{T}{2}\right)\right\}u\left(t-\frac{T}{2}\right)$$

which can be recast in the form

$$z_b(t) = \frac{-1}{(1-r_f^2)^2+4\zeta^2 r_f^2}\left\{Br_f e^{-\zeta\omega_n t}\left(\frac{1-2\zeta^2-r_f^2}{\sqrt{1-\zeta^2}}\sin\omega_d t - 2\zeta\cos\omega_d t\right)u(t)\right.$$

$$+Br_f e^{-\zeta\omega_n\left(t-\frac{T}{2}\right)}\left[\frac{1-2\zeta^2-r_f^2}{\sqrt{1-\zeta^2}}\sin\omega_d\left(t-\frac{T}{2}\right)-2\zeta\cos\omega_d\left(t-\frac{T}{2}\right)\right]u\left(t-\frac{T}{2}\right)$$

$$+2\zeta r_f B\left[(\cos\omega t)\,u(t)+\cos\omega\left(t-\frac{T}{2}\right)u\left(t-\frac{T}{2}\right)\right]$$

$$\left.+(r_f^2-1)B\left[(\sin\omega t)\,u(t)+\sin\omega\left(t-\frac{T}{2}\right)u\left(t-\frac{T}{2}\right)\right]\right\}$$

2.7.9.1 Derivation of $z_{\dot b}(t)$

This response is derived likewise, i.e., by means of the response $x_C(t)$ of an underdamped system to a unilateral cosine input, namely,

$$z_{\dot b}(t) = (2\zeta\omega_n)\omega Bx_C(t)u(t)+(2\zeta\omega_n)\omega Bx_C\left(t-\frac{T}{2}\right)u\left(t-\frac{T}{2}\right)$$

Hence, upon substituting both $x_C(t)$ and $x_C(t-T/2)$ into the foregoing expression, we obtain

$$z_{\dot b}(t) = \frac{2\zeta r_f B}{(1-r_f^2)^2+4\zeta^2 r_f^2}\left\{e^{-\zeta\omega_n t}\left[\frac{-\zeta(1+r_f^2)}{\sqrt{1-\zeta^2}}\sin\omega_d t - (1-r_f^2)\cos\omega_d t\right]\right.$$

$$\left.+2\zeta r_f\sin\omega t + (1-r_f^2)\cos\omega t\right\}u(t)$$

$$+\frac{2\zeta r_f B}{\left(1-r_f^2\right)^2+4\zeta^2 r_f^2}\left\{e^{-\zeta\omega_n\left(t-\frac{T}{2}\right)}\left[\frac{-\zeta(1+r_f^2)}{\sqrt{1-\zeta^2}}\sin\omega_d\left(t-\frac{T}{2}\right)\right.\right.$$

$$\left.-(1-r_f^2)\cos\omega_d\left(t-\frac{T}{2}\right)\right]+2\zeta r_f\sin\omega\left(t-\frac{T}{2}\right)$$

$$\left.+(1-r_f^2)\cos\omega\left(t-\frac{T}{2}\right)\right\}u\left(t-\frac{T}{2}\right)$$

which can be rewritten as

$$z_b(t)=\frac{2\zeta r_f}{(1-r_f^2)^2+4\zeta^2 r_f^2}\left\{Be^{-\zeta\omega_n t}\left[\frac{-\zeta(1+r_f^2)}{\sqrt{1-\zeta^2}}\sin\omega_d t-(1-r_f^2)\cos\omega_d t\right]u(t)\right.$$

$$+Be^{-\zeta\omega_n\left(t-\frac{T}{2}\right)}\left[\frac{-\zeta(1+r_f^2)}{\sqrt{1-\zeta^2}}\sin\omega_d\left(t-\frac{T}{2}\right)\right.$$

$$\left.-(1-r_f^2)\cos\omega_d\left(t-\frac{T}{2}\right)\right]u\left(t-\frac{T}{2}\right)$$

$$+2\zeta r_f B\left[(\sin\omega t)u(t)+\left(\sin\omega\left(t-\frac{T}{2}\right)\right)u\left(t-\frac{T}{2}\right)\right]$$

$$\left.+(1-r_f^2)B\left[(\cos\omega t)u(t)+\left(\cos\omega\left(t-\frac{T}{2}\right)\right)u\left(t-\frac{T}{2}\right)\right]\right\}$$

The total response $z(t)$ of the vehicle is, then,

$$z(t)=z_b(t)+z_{\dot b}(t)$$

Shown in Fig. 2.51 are the individual responses $z_b(t)$, $z_{\dot b}(t)$ and their sum $z(t)$, for $\omega_n=1\,\text{Hz}=2\pi\,\text{rad/s}$, $\zeta=0.707$ and $T=2$ s.
Animation of this time response is available in 3-DampedBump1dof.mw.

2.8 The Periodic Response

In this section, we will assume that the systems under study are subjected to an arbitrary, though periodic, input; otherwise, conditions are similar to those of the harmonic response. Thus, we assume that a very long time has elapsed since the system under analysis was first perturbed. This is the case in many real-life situations, like aircraft fuselage under vibrations produced by the turbines at cruising speed or bodies of terrestrial vehicles under vibrations produced by the engine running at constant r.p.m., and so forth. In these examples, the source of vibration is either a turbine or an IC engine, both running at uniform angular velocity, thus producing a periodic, although non-harmonic, input.

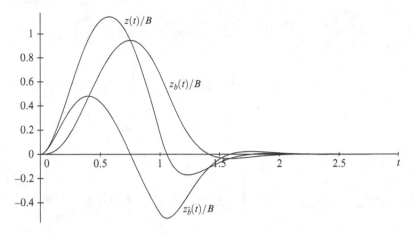

Fig. 2.51 Time response of an underdamped suspension encountering a bump

By means of *Fourier analysis*, we will decompose any periodic input into an infinite summation, i.e., a series, of harmonics of multiples of a fundamental frequency. To do this, we need some background, as described below:

2.8.1   Background on Fourier Analysis

We begin with a few definitions that will prove useful in the ensuing analysis. A function $f(t)$ is *periodic* of period T if, for any real value of t,

$$f(t) = f(t + T) \tag{2.180}$$

An *even function* $f(t)$ is a function with the property

$$f(t) = f(-t) \tag{2.181}$$

while an *odd function* $f(t)$ obeys

$$f(t) = -f(-t) \tag{2.182}$$

Now, a function need not be even or odd *everywhere*, i.e., for any value of t, to be of interest to us. Since we are interested in periodic functions, we will look at the functions under study only within a period of length T, and so, we may add that a certain function is even or odd *within the interval* $[-T/2, +T/2]$ if the above definitions hold not necessarily everywhere, but within the said interval. Moreover, we recall below a few facts that will prove to be useful in this discussion. The pertinent proofs are available in any book on Fourier analysis or partial differential equations.

Fact 1: Let a function $f(t)$ be even in the interval $[-T/2, +T/2]$ and a function $g(t)$ be odd in the same interval. The integral of the product of these functions over the same interval vanishes.

An example of even function is the cosine function, one of odd function is the sine function. Note that the cosine and the sine functions are even or, correspondingly, odd, everywhere.

It is now apparent that the sum of an arbitrary number of even functions is even as well, a similar statement holding for odd functions. Here, we must realize that, if a function is not even, it need not be odd, while keeping in mind the result below:

Fact 2: Any function $f(t)$ can always be decomposed into the sum of an even and an odd component, $f_E(t)$ and $f_O(t)$, respectively, which are given by

$$f_E(t) \equiv \frac{1}{2}[f(t) + f(-t)], \quad f_O(t) \equiv \frac{1}{2}[f(t) - f(-t)] \qquad (2.183)$$

Given *any periodic* function $f(t)$ of period T, its *Fourier expansion* is given below:

$$f(t) = a_0 + \sum_{1}^{\infty} a_k \cos\left(\frac{2\pi k}{T}t\right) + \sum_{1}^{\infty} b_k \sin\left(\frac{2\pi k}{T}t\right) \qquad (2.184a)$$

where

$$a_0 = \frac{1}{T}\int_{\mathscr{I}} f(t)\,dt \qquad (2.184b)$$

$$a_k = \frac{2}{T}\int_{\mathscr{I}} f(t)\cos\left(\frac{2\pi kt}{T}\right)dt, \quad k = 1,2,\dots,\text{etc.} \qquad (2.184c)$$

$$b_k = \frac{2}{T}\int_{\mathscr{I}} f(t)\sin\left(\frac{2\pi kt}{T}\right)dt, \quad k = 1,2,\dots,\text{etc.} \qquad (2.184d)$$

with $\mathscr{I}$ denoting *any* interval of length T, such as $[-T/2, T/2]$ or $[0, T]$.

We note that a_0 is nothing but the *mean value* of $f(t)$ in the interval of integration. Moreover, $\omega_0 = 2\pi/T$ is known as the *fundamental frequency* of $f(t)$.

The process under which the foregoing coefficients a_k and b_k are determined, for a given periodic function $f(t)$, is known as *Fourier analysis* or *spectral analysis*. Needless to say, the Fourier analysis of a given function can be conducted with paper and pencil by quadrature of the underlying integrals or by table-lookup only in special textbook-type of cases. Otherwise, the analyst has to resort to numerical methods. Note that the table-lookup procedure can be greatly eased and broadened if symbolic computation software is used. In many instances, integrals that are not available in tables can be found with the aid of this type of software.

Example 2.8.1 (A Simple Example). Find the Fourier expansion of the function

$$f(t) = \sin^2 \omega_0 t$$

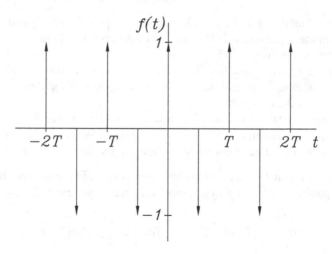

Fig. 2.52 A train of impulses

Solution: This is a periodic function of fundamental period $T = 2\pi/\omega_0$. Its Fourier expansion is particularly simple because we do not have to go through the calculation of its Fourier coefficients from integrals. Indeed, our task will be greatly eased if we recall the identity below:

$$\sin^2 \omega_0 t \equiv \frac{1}{2} - \frac{1}{2}\cos 2\omega_0 t$$

which is already the Fourier expansion of the given function, and hence,

$$a_0 = \frac{1}{2}, \quad a_2 = -\frac{1}{2}$$

all other coefficients vanishing.

Example 2.8.2 (Fourier Analysis of a Train of Impulses). Obtain the Fourier expansion of the train of impulses given below:

$$f(t) = \sum_{i=-\infty}^{\infty} (-1)^i \delta\left(t - \frac{iT}{2}\right)$$

which is plotted in Fig. 2.52.

Solution: From symmetry considerations it is apparent that the mean value of the function at hand is zero. It is also apparent that this function is even, and hence,

$$a_0 = 0, \quad b_k = 0, \quad k = 1, 2, \dots$$

The remaining Fourier coefficients are calculated below:

$$a_k = \frac{2}{T} \int_{-T^-/2}^{+T^-/2} f(t)[\cos(k\omega_0 t)]dt$$

$$= \frac{2}{T} \int_{-T^-/2}^{+T^-/2} \left[\sum_{i=-\infty}^{\infty} (-1)^i \delta\left(t - \frac{iT}{2}\right) \right] \cos\left(\frac{2\pi kt}{T}\right) dt$$

$$= \frac{2}{T} \sum_{i=-\infty}^{\infty} (-1)^i \int_{-T^-/2}^{+T^-/2} \delta\left(t - \frac{iT}{2}\right) \cos\left(\frac{2\pi kt}{T}\right) dt$$

where the integration extremes have been set so as to take into account the discontinuities in the impulse functions in the integrand.

It is apparent that the interval of integration comprises only two impulses, those at $t = -T/2$ and at $t = 0$, the impulse at $t = +T/2$ being left out of the interval. Hence, the expression for a_k reduces to

$$a_k = \frac{2}{T} \int_{-T^-/2}^{+T^-/2} \left[-\delta\left(t + \frac{T}{2}\right) + \delta(t) \right] \cos\left(\frac{2\pi kt}{T}\right) dt$$

$$= \frac{2}{T} \left[-\int_{-T^-/2}^{+T^-/2} \delta\left(t + \frac{T}{2}\right) \cos\left(\frac{2\pi kt}{T}\right) dt + \int_{-T^-/2}^{+T^-/2} \delta(t) \cos\left(\frac{2\pi kt}{T}\right) dt \right]$$

In light of the identity appearing in Eq. 2.47, the above integrands are evaluated as

$$\delta\left(t + \frac{T}{2}\right) \cos\left(\frac{2\pi kt}{T}\right) = \delta\left(t + \frac{T}{2}\right) \cos\left[\frac{2\pi k}{T}\left(-\frac{T}{2}\right)\right]$$

$$= \delta\left(t + \frac{T}{2}\right) \cos(k\pi) = \delta\left(t + \frac{T}{2}\right)(-1)^k$$

$$\delta(t) \cos\left(\frac{2\pi kt}{T}\right) = \delta(t) \cos\left[\frac{2\pi k}{T}(0)\right] = \delta(t)$$

Therefore,

$$a_k = \frac{2}{T} \left[-\int_{-T^-/2}^{+T^-/2} \delta\left(t + \frac{T}{2}\right)(-1)^k dt + \int_{-T^-/2}^{+T^-/2} \delta(t)dt \right] = \frac{2}{T} \left[-(-1)^k + 1 \right]$$

and hence,

$$a_{2k-1} = \frac{4}{T}, \quad a_{2k} = 0$$

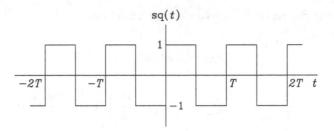

Fig. 2.53   A square wave

i.e., the Fourier expansion of the train of impulses contains only cosine terms of odd-numbered frequencies, all these terms with identical coefficients, namely,

$$f(t) \equiv \sum_{i=0}^{\infty} (-1)^i \delta \left(t - \frac{iT}{2} \right) = \frac{4}{T} \sum_{k=1}^{\infty} \cos \left[\frac{2(2k-1)\pi}{T} t \right]$$

Example 2.8.3 (Fourier Analysis of a Square Wave).   Determine the Fourier expansion of the square wave sq(t) displayed in Fig. 2.53.

Solution: It is not very difficult to realize that the integral of the train of impulses yields a train of pulses, a pulse having the shape depicted in Fig. 2.33b. Indeed, every negative impulse produces a sudden decrease of one unity, while every positive impulse produces an increase of one unity in the above integral. We can therefore realize that the integral of $f(t)$ is a signal that oscillates, about a given mean value $\overline{f}$, between $-1/2$ and $+1/2$, the jumps occurring every time an impulse fires. Therefore, the above integral undergoes oscillations of amplitude equal to $1/2$, while the given square wave undergoes oscillations of unit amplitude. As a consequence, we have

$$\mathrm{sq}(t) = 2 \int_t f(\theta) d\theta$$

where the integral has been left indefinite and the integration takes place in the domain of the dummy variable θ. Moreover, because of the linearity of the integral operation,[11] the Fourier series of the integral of the train of impulses is the integral of the Fourier series of the train, i.e.,

$$\mathrm{sq}(t) = \frac{8}{T} \sum_{k=1}^{\infty} \int_t \cos \left[\frac{2(2k-1)\pi}{T} \theta \right] d\theta + C$$

---

[11] The integral operation is both additive and homogeneous, and hence, linear.

where the integration constant is found from the condition that the mean value of $sq(t)$ is zero, which is apparent from Fig. 2.53. Moreover, upon evaluation of the integral,

$$sq(t) = \frac{4}{\pi} \sum_{k=1}^{\infty} \frac{1}{2k-1} \sin\left[\frac{2(2k-1)\pi}{T}t\right] + C$$

Since the mean value of each term of the series is zero, that of the whole series is zero as well, and hence, $C = 0$, which thus leads to

$$sq(t) = \frac{4}{\pi} \sum_{k=1}^{\infty} \frac{1}{2k-1} \sin\left[\frac{2(2k-1)\pi}{T}t\right]$$

Alternatively, we can proceed to evaluate the Fourier coefficients by direct integration, as we show below. To this end, we observe first that the square wave is an odd function with zero mean value and hence,

$$a_k = 0, \quad k = 0, 1, 2, \dots$$

the remaining coefficients being calculated from

$$b_k = \frac{2}{T} \int_{-T/2}^{T/2} sq(t) \sin\left(\frac{2\pi kt}{T}\right) dt$$

which, after evaluation and simplification, yields

$$b_k = \frac{2}{\pi k} [1 - \cos(\pi k)]$$

and hence,

$$b_k = \begin{cases} 4/(\pi k), & \text{for } k \text{ odd;} \\ 0, & \text{for } k \text{ even} \end{cases}$$

Therefore,

$$sq(t) = \frac{4}{\pi} \sum_{1}^{\infty} \frac{1}{2k-1} \sin\left[\frac{2\pi(2k-1)t}{T}\right] \equiv \frac{4}{\pi} \sum_{0}^{\infty} \frac{1}{2k+1} \sin\left[\frac{2\pi(2k+1)t}{T}\right]$$

$$= \frac{4}{\pi} \left[\sin\left(\frac{2\pi t}{T}\right) + \frac{1}{3} \sin\left(\frac{6\pi t}{T}\right) + \frac{1}{5} \sin\left(\frac{10\pi t}{T}\right) + \dots\right]$$

which thus confirms the result obtained above.

Example 2.8.4 (Fourier Analysis of a Monotonic Function). Shown in Fig. 2.54 is the profile of a pre-Columbian pyramid, of those found by tourists when visiting archeological sites in Mexico. This profile is labelled pyr(x) in that figure, in which the *wave length* λ now plays the role of the period T in the previous examples. In

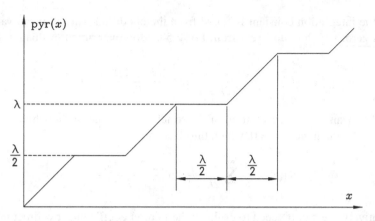

Fig. 2.54   A Pre-Columbian pyramid

thinking of tourists using wheelchairs, we would like to design the suspension of a motor-driven wheelchair to help these tourists climb up these pyramids comfortably. In order to do this, let us find the Fourier expansion of the profile.

Solution: We first note that the given profile can be obtained from the integral of the square wave of Fig. 2.53 when shifted 1 upwards. Thus, all we need to obtain the Fourier series of the given profile is (a) add 1 to the square-wave Fourier expansion derived above, (b) integrate each term of the Fourier series thus resulting, and (c) divide each integral by 2, i.e.,

$$\text{pyr}(x) = \frac{1}{2}\left[\int_x \left\{ 1 + \frac{4}{\pi}\sum_{k=1}^{\infty}\frac{1}{2k-1}\sin\left[\frac{2\pi(2k-1)\xi}{\lambda}\right]\right\}d\xi\right]$$

in which ξ is a dummy variable of integration. Hence,

$$\text{pyr}(x) = \frac{1}{2}x - \frac{\lambda}{\pi^2}\sum_{k=1}^{\infty}\frac{1}{(2k-1)^2}\cos\left[\frac{2\pi(2k-1)x}{\lambda}\right] + C$$

where C is a constant of integration that can be obtained by noting that $\text{pyr}(0) = 0$ in Fig. 2.54. Thus,

$$0 = C - \frac{\lambda}{\pi^2}\sum_{k=1}^{\infty}\frac{1}{(2k-1)^2}$$

But, as the reader can verify, e.g., using computer algebra,

$$\sum_{k=1}^{\infty}\frac{1}{(2k-1)^2} = \frac{\pi^2}{8}$$

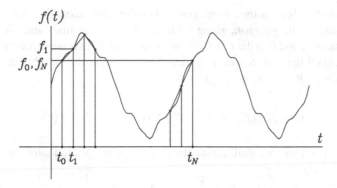

Fig. 2.55 Polygonal approximation of function $f(t)$

and hence,

$$C = \frac{\lambda}{\pi^2} \frac{\pi^2}{8} = \frac{\lambda}{8}$$

Therefore,

$$
\begin{aligned}
\text{pyr}(x) &= \frac{\lambda}{8} + \frac{1}{2}x - \frac{\lambda}{\pi^2} \sum_{k=1}^{\infty} \frac{1}{(2k-1)^2} \cos\left[\frac{2\pi(2k-1)x}{\lambda} \right] \\
&= \frac{\lambda}{8} + \frac{1}{2}x - \frac{\lambda}{\pi^2} \left[\cos\left(\frac{2\pi x}{\lambda} \right) + \frac{1}{9} \cos\left(\frac{6\pi x}{\lambda} \right) + \frac{1}{25} \cos\left(\frac{10\pi x}{\lambda} \right) + \cdots \right]
\end{aligned}
$$

Notice that the above expansion includes a linear term, which accounts for the monotonicity of $\text{pyr}(x)$.

2.8.2 The Computation of the Fourier Coefficients

Here we study the procedure under which the Fourier coefficients of a periodic function are computed using a numerical approach. We aim to compute the coefficients $\{a_k\}_0^{\infty}$ and $\{b_k\}_1^{\infty}$ of the Fourier expansion of a periodic function $f(t)$. We shall do this by simple *numerical quadrature*, as explained by Kahaner et al. [4]. Various schemes of numerical quadrature are available, the one that lends itself best to our purposes is that based on the *trapezoidal rule*. In this scheme, the interval of integration is divided into a number N of subintervals $t_{i-1} \le t \le t_i$, for $i = 1, \ldots, N$, and a set of $N+1$ points of coordinates (t_i, f_i), with $f_i \equiv f(t_i)$, for $i = 0, 1, \ldots, N$, is defined in the $f(t)$-vs.-t plane. In the next step, every pair of neighboring points is joined with a straight line, thereby obtaining a *polygonal approximation* of the given function, as illustrated in Fig. 2.55.

The desired integral is then approximated as the area below this polygon. Note that each side of the polygon, along with its two vertical lines and the segment joining points t_{i-1} and t_i in the t axis forms a trapezoid, which is the reason why this scheme is called the trapezoidal rule. Moreover, we shall refer to this trapezoid as the ith trapezoid, its area A_i being given by

$$A_i = \frac{1}{2}\Delta t_i(f_{i-1} + f_i), \quad \Delta t_i \equiv t_i - t_{i-1}, \quad f_i \equiv f(t_i) \tag{2.185}$$

Within our scheme, we shall divide the interval of integration *uniformly*, i.e.,

$$\Delta t_i \equiv \frac{T}{N}, \quad i = 1, \ldots, N \tag{2.186}$$

Now, a_0 is readily approximated as

$$a_0 \approx \frac{1}{T}\frac{T}{N}\left[\frac{1}{2}(f_0 + f_1) + \frac{1}{2}(f_1 + f_2) + \ldots + \frac{1}{2}(f_{N-2} + f_{N-1}) + \frac{1}{2}(f_{N-1} + f_N)\right]$$

$$= \frac{1}{N}\left[\frac{1}{2}(f_0 + f_N) + f_1 + f_2 + \ldots + f_{N-2} + f_{N-1}\right]$$

But, since $f(t)$ is periodic, $f_N = f_0$, and hence,

$$a_0 \approx \frac{1}{N}(f_0 + f_1 + f_2 + \ldots + f_{N-2} + f_{N-1}) \tag{2.187}$$

i.e., a_0 becomes the *mean value* of the set of sampled function values $\{f_i\}_0^{N-1}$. Moreover,

$$a_k \approx \frac{2}{T}\frac{T}{N}\left\{\frac{1}{2}[f_0\cos(k\omega_0 t_0) + f_1\cos(k\omega_0 t_1)]\right.$$

$$+ \frac{1}{2}[f_1\cos(k\omega_0 t_1) + f_2\cos(k\omega_0 t_2)] + \ldots$$

$$+ \frac{1}{2}[f_{N-2}\cos(k\omega_0 t_{N-2}) + f_{N-1}\cos(k\omega_0 t_{N-1})]$$

$$\left.+ \frac{1}{2}[f_{N-1}\cos(k\omega_0 t_{N-1}) + f_N\cos(k\omega_0 t_N)]\right\}$$

But, for $j = 1, \ldots, N$,

$$t_j \equiv t_0 + \frac{jT}{N} \quad \text{and} \quad \cos(k\omega_0 t_N) = \cos(k\omega_0 t_0)$$

and, hence, the foregoing expression becomes

$$a_k \approx \frac{2}{N} \left\{ f_0 \cos(k\omega_0 t_0) + f_1 \cos\left[k\omega_0 \left(t_0 + \frac{T}{N} \right) \right] + f_2 \cos\left[k\omega_0 \left(t_0 + \frac{2T}{N} \right) \right] + \cdots \right.$$

$$\left. + f_{N-2} \cos\left[k\omega_0 \left(t_0 + \frac{(N-2)T}{N} \right) \right] + f_{N-1} \cos\left[k\omega_0 \left(t_0 + \frac{(N-1)T}{N} \right) \right] \right\}$$

However,

$$\cos\left[k\omega_0 \left(t_0 + \frac{jT}{N} \right) \right] = \cos(k\omega_0 t_0) \cos\left(k\omega_0 \frac{jT}{N} \right) - \sin(k\omega_0 t_0) \sin\left(k\omega_0 \frac{jT}{N} \right)$$

the above expression thus becoming

$$a_k \approx \frac{2}{N} \cos(k\omega_0 t_0) \left\{ f_0 + f_1 \cos\left(k\omega_0 \frac{T}{N} \right) + f_2 \cos\left(k\omega_0 \frac{2T}{N} \right) + \cdots \right.$$

$$\left. + f_{N-2} \cos\left(k\omega_0 \frac{(N-2)T}{N} \right) + f_{N-1} \cos\left(k\omega_0 \frac{(N-1)T}{N} \right) \right\}$$

$$- \frac{2}{N} \sin(k\omega_0 t_0) \left\{ f_1 \sin\left(k\omega_0 \frac{T}{N} \right) + f_2 \sin\left(k\omega_0 \frac{2T}{N} \right) + \cdots \right.$$

$$\left. + f_{N-2} \sin\left(k\omega_0 \frac{(N-2)T}{N} \right) + f_{N-1} \sin\left(k\omega_0 \frac{(N-1)T}{N} \right) \right\}$$

and, in compact form,

$$a_k \approx \frac{2}{N} [\cos(k\omega_0 t_0)] \sum_{j=0}^{N-1} f_j \cos\left(k\omega_0 j\frac{T}{N} \right) - \frac{2}{N} [\sin(k\omega_0 t_0)] \sum_{j=1}^{N-1} f_j \sin\left(k\omega_0 j\frac{T}{N} \right)$$

$$(2.188)$$

Likewise, for the b_k coefficients, we have

$$b_k \approx \frac{2}{N} [\sin(k\omega_0 t_0)] \sum_{j=0}^{N-1} f_j \cos\left(k\omega_0 j\frac{T}{N} \right) + \frac{2}{N} [\cos(k\omega_0 t_0)] \sum_{j=1}^{N-1} f_j \sin\left(k\omega_0 j\frac{T}{N} \right)$$

$$(2.189)$$

From Eqs. 2.188 and 2.189 it is apparent that the Fourier coefficients have been approximated by a linear combination of two sums of terms, a sum C_k and a sum S_k, which are defined as

$$C_k \equiv \frac{2}{N} \sum_{j=0}^{N-1} f_j \cos\left(k\omega_0 j\frac{T}{N} \right), \quad S_k \equiv \frac{2}{N} \sum_{j=1}^{N-1} f_j \sin\left(k\omega_0 j\frac{T}{N} \right) \qquad (2.190)$$

In particular, for $k = N/2$, when N is *even*,

$$C_{N/2} = \frac{2}{N} \sum_{j=0}^{N-1} f_j \cos\left(\frac{N}{2}\omega_0 j \frac{T}{N}\right) \equiv \frac{2}{N} \sum_{j=0}^{N-1} f_j \cos\left(\omega_0 j \frac{T}{2}\right)$$

$$= \frac{2}{N} \sum_{j=0}^{N-1} f_j \cos(j\pi) = \frac{2}{N} \sum_{j=0}^{N-1} f_j (-1)^j \qquad (2.191)$$

$$S_{N/2} = \frac{2}{N} \sum_{j=1}^{N-1} f_j \sin\left(\frac{N}{2}\omega_0 j \frac{T}{N}\right) \equiv \frac{2}{N} \sum_{j=1}^{N-1} f_j \sin\left(\omega_0 j \frac{T}{2}\right)$$

$$= \frac{2}{N} \sum_{j=1}^{N-1} f_j \sin(j\pi) = 0 \qquad (2.192)$$

In summary, the Fourier coefficients can be approximated in the form

$$a_k \approx [\cos(k\omega_0 t_0)]C_k - [\sin(k\omega_0 t_0)]S_k \qquad (2.193)$$

$$b_k \approx [\sin(k\omega_0 t_0)]C_k + [\cos(k\omega_0 t_0)]S_k \qquad (2.194)$$

Interestingly, the foregoing expressions can be cast in a rather revealing vector form, namely,

$$\begin{bmatrix} a_k \\ b_k \end{bmatrix} \approx \begin{bmatrix} \cos(k\omega_0 t_0) & -\sin(k\omega_0 t_0) \\ \sin(k\omega_0 t_0) & \cos(k\omega_0 t_0) \end{bmatrix} \begin{bmatrix} C_k \\ S_k \end{bmatrix}, \quad \text{for} \quad k = 1, 2, \ldots, \text{etc.} \quad (2.195)$$

the 2×2 matrix being orthogonal; this matrix rotates vectors in the plane by an angle $k\omega_0 t_0$. Thus, the *magnitudes* of the two vectors involved in the foregoing equation are *approximately* equal, i.e.,

$$a_k^2 + b_k^2 \approx C_k^2 + S_k^2 \qquad (2.196)$$

Note that a_0 is not included in the foregoing vector expression; it appears in Eq. 2.187.

Moreover, the two foregoing sums, C_k and S_k, bear a particular significance in the context of Fourier analysis. They turn out to be the coefficients of the *discrete Fourier transform* of the sequence of discrete values $\mathcal{S} \equiv \{f_j\}_1^{N-1}$. Now, the said transform is the discrete counterpart of the *Fourier transform* of a continuous function $f(t)$, for $-\infty < t < +\infty$, and hence, finds extensive applications in engineering and economics. As a matter of fact, the discrete sequence $\mathcal{S}$ defined above, is known in economics as a *time-series*. The frequency analysis of time-series is important in many areas. For example, in management, the frequency analysis of the variations of market indicators given as time-series provides clues on trends, which helps in the decision-making process. Included in Kahaner et al. [4] is an application of time-series frequency analysis to the meteorological phenomenon known as 'El Niño'.

Because of the frequent occurrence of the C_k and S_k coefficients in many applications, software for their efficient computation is available in many scientific packages, e.g., Matlab, IMSL, Maple, Mathematica, and the like. Kahaner et al.'s book is nowadays complemented with a companion book [5] that offers down-loadable software.[12] This software package includes tools for the calculation of the foregoing coefficients in the context of a more general set of routines meant for what is called the *Fast Fourier Transform* (FFT). FFT, or FFT analysis, is used extensively in engineering, science, and economics to determine the frequency contents of either continuous or discrete functions of time, i.e., signals. Astronomer Carl Sagan, in his science fiction bestseller *Contact*, describes a group of information scientists trying to find a clue on the existence of God with the aid of the number π. What these sci-entists do is a frequency analysis of the occurrence of the digits in this number and, moreover, they try this frequency analysis in various numerical bases. The frequency map that the scientists obtain, with number 11 as a basis, turns out to be a circle!

Coming back to our problem of calculating the Fourier coefficients of a periodic signal, the natural question to ask is up to how many coefficients can we compute from a subdivision of the integration interval of length equal to one period into N subintervals? As discussed by Kahamer et al., if the interval is divided into N subintervals, then we can, at most, compute $\lfloor N/2 \rfloor$ coefficients, where $\lfloor \cdot \rfloor$ stands for the *floor* function of $(\cdot)$, defined as the greatest integer contained in the real interval $[0, N/2]$.

An interesting result, known as *Parseval's Theorem* or *Parseval's identity* [6] is now recalled, that finds extensive applications in the realm of Fourier analysis. We will use it for error estimation in our calculations. This result can be readily derived by squaring both sides of Eq. 2.184a, thus obtaining

$$f^2(t) = a_0^2 + \left[\sum_1^\infty a_k \cos\left(\frac{2\pi k}{T}t\right) \right]^2 + \left[\sum_1^\infty b_k \sin\left(\frac{2\pi k}{T}t\right) \right]^2$$

$$+ 2a_0 \sum_1^\infty a_k \cos\left(\frac{2\pi k}{T}t\right) + 2a_0 \sum_1^\infty b_k \sin\left(\frac{2\pi k}{T}t\right)$$

$$+ 2 \left[\sum_1^\infty a_k \cos\left(\frac{2\pi k}{T}t\right) \right] \left[\sum_1^\infty b_k \sin\left(\frac{2\pi k}{T}t\right) \right] \qquad (2.197)$$

In the next step, we integrate both sides of the foregoing equation within one whole period. Without going into the details, it is apparent from the above equation that there will be an infinite sum of terms of five kinds, namely,

1. $a_k^2 \cos^2\left(\frac{2\pi k}{T}t\right)$. The integral over one whole period of these terms is a_k^2 times $T/2$, for every k;

---

[12]http://www.mathworks.com/moler

2. $b_k^2 \sin^2(\frac{2\pi k}{T}t)$. The integral over one whole period of these terms is b_k^2 times $T/2$ as well, for every k;

3. $a_j b_k \cos(\frac{2\pi j}{T}t) \sin(\frac{2\pi k}{T}t)$, for $j, k = 1, 2, \ldots$, including terms for which $j = k$. The integral of every term of this kind vanishes;

4. $a_j a_k \cos(\frac{2\pi j}{T}t) \cos(\frac{2\pi k}{T}t)$, for $j, k = 1, 2, \ldots$ and $j \neq k$. The integral of every term of this kind vanishes;

5. $b_j b_k \sin(\frac{2\pi j}{T}t) \sin(\frac{2\pi k}{T}t)$, for $j, k = 1, 2, \ldots$ and $j \neq k$. The integral of every term of this kind vanishes.

As a result of the foregoing observations, the right-hand side of Eq. 2.197 reduces to the infinite sums of terms of the first two kinds described above, and hence,

$$\int_{-T/2}^{T/2} f^2(t)dt = a_0^2 T + \frac{T}{2}\sum_1^\infty a_k^2 + \frac{T}{2}\sum_1^\infty b_k^2 \qquad (2.198a)$$

or, if we divide both sides of the above equation by T,

$$\frac{1}{T}\int_{-T/2}^{T/2} f^2(t)dt = a_0^2 + \frac{1}{2}\sum_1^\infty a_k^2 + \frac{1}{2}\sum_1^\infty b_k^2 \qquad (2.198b)$$

Now, by taking the square root of both sides of the above equation, we obtain

$$[f(t)]_{\text{rms}} = \sqrt{\frac{1}{2}\left(2a_0^2 + \sum_1^\infty a_k^2 + \sum_1^\infty b_k^2\right)} \qquad (2.199a)$$

with $[f(t)]_{\text{rms}}$ defined as the *root-mean square* of $f(t)$, i.e.,

$$[f(t)]_{\text{rms}} = \sqrt{\frac{1}{T}\int_{-T/2}^{T/2} f^2(t)dt} \qquad (2.199b)$$

Therefore, the *root-mean-square value* of the periodic function $f(t)$ equals the square root of one-half the sum of the squares of its Fourier coefficients, with a_0^2 taken twice. Since root-mean square values are associated with energy—all energy functions are quadratic either in the generalized coordinates or in the generalized speeds—what Parseval's Theorem states is simply one more form of the First Law of Thermodynamics, i.e., the Principle of Conservation of Energy.

Note that the right-hand side of Eq. 2.198b is an infinite sum of squares. As a consequence, if the infinite sums are truncated after N_h harmonics, the N_h-*approximation* of the given function thus resulting leads to an *underestimation* of $[f(t)]_{\text{rms}}$, i.e.,

$$a_0^2 + \frac{1}{2}\sum_1^{N_h} a_k^2 + \frac{1}{2}\sum_1^{N_h} b_k^2 < [f(t)]_{\text{rms}}^2 \qquad (2.200a)$$

Fig. 2.56 Iconic model of an air compressor mounted on a viscoelastic foundation

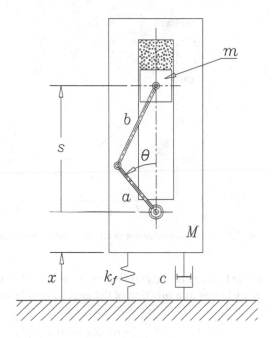

Let e_{N_h} be the error in the N_h-approximation of a given periodic function $f(t)$. The *square* of this error is defined as

$$e_{N_h}^2 \equiv [f(t)]_{\text{rms}}^2 - \left(a_0^2 + \frac{1}{2} \sum_1^{N_h} a_k^2 + \frac{1}{2} \sum_1^{N_h} b_k^2 \right) > 0 \qquad (2.200b)$$

Obviously, as more and more harmonics are included in the N_h-approximation of a given periodic function, the error decreases monotonically.

In summary, to calculate the Fourier coefficients, we have

Algorithm (Fourier):

1. If N is even, then the coefficients are calculated for $k = 1, \ldots, N/2$, with a_k being calculated for a_0 as well; otherwise, the coefficients are calculated for $k = 1, \ldots, (N-1)/2$.
2. Calculate a_0 as the mean value of the numbers $\{f_j\}_0^{N-1}$.
3. Calculate the C_k and S_k coefficients, for $k = 1, \ldots, \lfloor N/2 \rfloor$ using expressions (2.190).
4. Calculate the a_k and b_k coefficients, for $k = 1, \ldots, \lfloor N/2 \rfloor$ using expressions (2.195).

Example 2.8.5 (Spectral Analysis of the Displacement of an Air Compressor). Shown in Fig. 2.56 is the iconic model of an air compressor mounted on a viscoelastic foundation that is modeled as a spring of stiffness k_f in parallel with a dashpot of coefficient c. The *relative* displacement $s(t)$ of the piston with respect

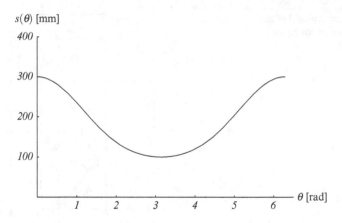

Fig. 2.57 Slider displacement vs. crank angular displacement

to the housing of the compressor is measured as indicated in the same figure and is given below as a function of the angular displacement of the crank, θ:

$$s(\theta) = a\cos\theta + b\sqrt{1 - \rho^2 \sin^2\theta}$$

where $\rho \equiv a/b$ and the crank turns at a constant angular velocity ω_0.

   For the foregoing displacement, which is plotted in Fig. 2.57 for $a = 100\,mm$ and $\rho = 0.5$, compute the Fourier coefficients of the function $s(\theta)$ numerically and give an estimate of the error incurred in the calculations. (As one can readily realize, a longhand Fourier analysis of the foregoing function would be intractable.)

Solution: It is apparent that the signal at hand is even, and hence, its sine coefficients are all zero. Therefore, we only need worry about the a_k coefficients, for $k = 0, 1, \ldots,$ N. Let us now calculate these coefficients for 1, 2, and 4 harmonics. For purposes of illustration, we obtain these coefficients below by longhand calculations for 1 and 2 harmonics, i.e., for $N = 2$ and 4; for $N_h = 4$ or $N = 8$, we resort to numerical calcualtions, but do not include the details here.

1. For $N_h = 1, N = 2$, which yields the sampled values f_0, f_1 and $f_2 = f_0$, as shown in Fig. 2.58. We thus have

$$f(t) \approx a_0 + a_1 \cos\omega_0 t$$

the coefficients a_0 and a_1 being calculated below. First, note that

$$f_0 = f(0) = 300, \qquad f_1 = f(t_1) = 100$$

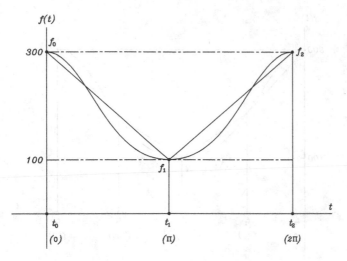

Fig. 2.58 Polygonal approximation of function $s(\theta)$ with $N = 2$ intervals

Therefore,

$$a_0 = \frac{1}{2}(f_0 + f_1) = 200$$

$$C_1 = \frac{2}{2}\sum_{j=0}^{1} f_j \cos\left(\underbrace{k}_{1}\,\omega_0 j\frac{T}{2}\right) = f_0 \underbrace{\cos(0)}_{1} + f_1 \cos\underbrace{\left(\omega_0\frac{T}{2}\right)}_{\underbrace{\pi}_{-1}} = f_0 - f_1 = 200$$

$$S_1 = \frac{2}{2}\sum_{j=0}^{1} f_j \sin\left(\omega_0 j\frac{T}{2}\right) = f_0 \underbrace{\sin(0)}_{0} + f_1 \sin\underbrace{\left(\omega_0\frac{T}{2}\right)}_{\underbrace{\pi}_{0}} = 0$$

Hence,

$$a_1 = C_1 \cos\left(\underbrace{\omega_0 t_0}_{0}\right) - S_1 \sin\left(\underbrace{\omega_0 t_0}_{0}\right) = C_1 = 200$$

Thus, with one harmonic,

$$f(t) \approx 200(1 + \cos\omega_0 t)$$

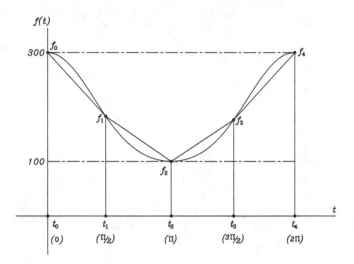

Fig. 2.59 Polygonal approximation of function $s(\theta)$ with $N = 4$ intervals

2. For $N_h = 2$, $N = 4$, which yields the sampled values f_0, f_1, f_2, f_3 and $f_4 = f_0$ of Fig. 2.59. Now we have

$$f(t) \approx a_0 + a_1 \cos \omega_0 t + a_2 \cos 2\omega_0 t$$

In order to calculate the above coefficients, we start by calculating the function values at the sampled instants:

$$f_0 = 300, \; f_1 = \underbrace{a\cos\left(\frac{\pi}{2}\right)}_{0} + b\underbrace{\sqrt{1 - \rho^2 \sin^2\left(\frac{\pi}{2}\right)}}_{1} = 200\sqrt{1 - 0.25} = 173.2051 = f_3$$

where we have made use of symmetry to find f_3. Moreover, by inspection,

$$f_2 = 100$$

and hence,

$$a_0 = \frac{1}{4}(300 + 173.2051 + 100 + 173.2051) = 186.6025$$

Furthermore,

$$C_1 = \frac{2}{4}\sum_{j=0}^{3} f_j \cos\left(\omega_0 j \frac{T}{4}\right) = \frac{1}{2}\left[f_0 \cos(0) + f_1 \cos\left(\frac{\pi}{2}\right) + f_2 \cos(\pi)\right.$$

$$\left. + f_3 \cos\left(\frac{3\pi}{2}\right)\right]\frac{1}{2}(f_0 - f_2) = 100$$

Table 2.1 Fourier coefficients of function $s(\theta)$ of the air compressor

	$N_h = 1$	$N_h = 2$	$N_h = 4$
a_0 (mm)	200.00000000000	186.60254037844	186.84270485857
a_1 (mm)	200.00000000000	100.000000000000	100.000000000000
a_2 (mm)		26.794919243112	13.397459621556
a_3 (mm)			$2.1316282072803 \times 10^{-14}$
a_4 (mm)			-0.48032896025319

$$S_1 = \frac{1}{2}\sum_{j=0}^{3} f_j \sin\left(\omega_0 j \frac{T}{4}\right) = \frac{1}{2}\left[f_0 \sin(0) + f_1 \sin\left(\frac{\pi}{2}\right) + f_2 \sin(\pi)\right.$$

$$\left. + f_3 \sin\left(\frac{3\pi}{2}\right)\right]\frac{1}{2}(f_1 - f_3) = 0$$

Likewise,

$$C_2 = \frac{1}{2}\sum_{j=0}^{3} f_j \cos\left(2\omega_0 j \frac{T}{4}\right) = \frac{1}{2}[f_0 \cos(0) + f_1 \cos(\pi) + f_2 \cos(2\pi) + f_3 \cos(3\pi)]$$

$$= \frac{1}{2}(300 - 173.2051 + 100 - 173.2051) = 26.7949$$

$$S_2 = \frac{1}{2}\sum_{j=0}^{3} f_j \sin\left(2\omega_0 j \frac{T}{4}\right) = \frac{1}{2}\left[f_0 \sin(0) + f_1 \sin(\pi) + f_2 \sin(2\pi)\right.$$

$$\left. + f_3 \sin(3\pi)\right] = 0$$

Therefore,

$$a_1 = C_1 \cos(\omega_0 t_0) - S_1 \sin(\omega_0 t_0) = 100$$

$$a_2 = C_2 \cos(2\omega_0 t_0) - S_2 \sin(2\omega_0 t_0) = 26.7949$$

and hence,

$$f(t) \approx 186.6025 + 100.0000 \cos\omega_0 t + 26.7949 \cos 2\omega_0 t$$

Note that the coefficient of the first harmonic turns out to be nothing but the crank length, which is plausible in light of the expression for $s(t)$ given above, in which this length is the coefficient of $\cos\theta$.

3. The foregoing values, as well as those for $N_h = 4$ ($N = 8$), are given with 14 digits, as obtained with the aid of Matlab, in Table 2.1. We thus have, with four harmonics and four decimals,

$$s(\theta) \approx 186.8427 + 100.0000 \cos\theta + 13.3975 \cos 2\theta - 0.4803 \cos 4\theta$$

where the third harmonic does not appear because a_3 turns out to vanish. Again, notice that the coefficient of the first harmonic is the crank length.

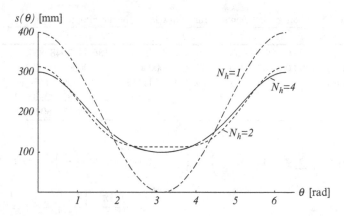

Fig. 2.60 $s(\theta)$-approximation with 1, 2, and 4 Fourier terms

The plots of $s(\theta)$ approximated with $N_h = 1$, 2 and 4 are shown in Fig. 2.60. Note, from this figure, that the value of $s(\theta)$ approximated with one single harmonic overestimates $s(\theta)$ by 33% at its peak values; when $s(\theta)$ is approximated with two harmonics, an overestimation of 4.5% is obtained. The approximation with four harmonics, on the other hand, underestimates $s(\theta)$ by less than 0.01%, but this difference, obviously, is not visible in that figure.

Let us now try to estimate the error incurred in the foregoing calculations, which we can do with the aid of Parseval's identity. To this end, we need the integral of $s^2(\theta)$ over one whole period, which gives us the rms value of $s(\theta)$, i.e.,

$$s_{\text{rms}} = \sqrt{\frac{1}{2\pi} \int_0^{2\pi} \left(a\cos\theta + b\sqrt{1 - \rho^2 \sin^2\theta} \right)^2 d\theta}$$

Upon expansion of the above integrand, $s^2(\theta)$, we obtain

$$s^2(\theta) = a^2 \cos^2\theta + 2ab\sqrt{1 - \rho^2 \sin^2\theta} + b^2(1 - \rho^2 \sin^2\theta)$$

It is now apparent that the desired integral cannot be calculated by simple quadrature because of the presence of the square-root term. However, a close look of Table 2.1 will reveal that the crank length a is the coefficient of the first harmonic of the Fourier expansion of $s(\theta)$ with more than one harmonic, and hence, the remaining harmonics of that expansion are bound to give the Fourier expansion of the second term of $s(\theta)$, $b\sqrt{1 - \rho^2 \sin^2\theta}$. Therefore, the error incurred in approximating $s(\theta)$ by its Fourier expansion is identical to the error incurred in the approximation of

Table 2.2 Fourier coefficients of function $\sqrt{1 - \rho^2 \sin^2 \theta}$ and the estimate of its rms value, $\sqrt{\Sigma}$

	$N_h = 1$	$N_h = 2$	$N_h = 4$
a_0/b	1.0000	0.9330	0.9342
a_2/b		0.1340	0.06699
a_3/b			0.0000
a_4/b			−0.002402
$\sqrt{\Sigma}$	1.0000	0.9378	0.9354

$\sqrt{1 - \rho^2 \sin^2 \theta}$, which will prove to be much easier to compute. Indeed, the rms value of the foregoing square root is

$$\left[\sqrt{1 - \rho^2 \sin^2 \theta} \right]_{rms} = \sqrt{\frac{1}{2\pi} \int_0^{2\pi} [1 - \rho^2 \sin^2 \theta] d\theta}$$

Thus,

$$\left[\sqrt{1 - \rho^2 \sin^2 \theta} \right]_{rms} = \sqrt{\frac{1}{2\pi} \int_0^{2\pi} \left[1 - \frac{\rho^2}{2}(1 - \cos 2\theta) \right] d\theta}$$

or

$$\left[\sqrt{1 - \rho^2 \sin^2 \theta} \right]_{rms} = \sqrt{\frac{1}{2\pi} \left[\left(1 - \frac{\rho^2}{2} \right) \theta + \frac{\rho^2}{4} \sin(2\theta) \right]_0^{2\pi}}$$

Finally,

$$\left[\sqrt{1 - \rho^2 \sin^2 \theta} \right]_{rms} = \sqrt{\frac{1}{2\pi} \left[1 - \frac{\rho^2}{2} \right] 2\pi} = \sqrt{1 - \frac{\rho^2}{2}}$$

For the case at hand, $\rho = 0.5$, and hence,

$$\left[\sqrt{1 - \rho^2 \sin^2 \theta} \right]_{rms} = \sqrt{0.8750} = 0.9354$$

which is exact to four digits. Table 2.2 includes the values of the Fourier coefficients of the expansion of the foregoing square root, which are identical to those of $[s(\theta) - a]/b$.

An approximation to the Fourier expansion of $s(\theta)$, frequently invoked, is obtained by approximating the square root, for "small" values of ρ, in the form

$$\sqrt{1 - \rho^2 \sin^2 \theta} \approx 1 - \frac{\rho^2}{2} \sin^2 \theta \equiv 1 - \frac{\rho^2}{4}[1 - \cos(2\theta)]$$

which thus yields the *approximate Fourier expansion*

$$\sqrt{1 - \rho^2 \sin^2 \theta} \approx 1 - \frac{\rho^2}{4} + \frac{\rho^2}{4} \cos(2\theta)$$

whose *approximate Fourier coefficients* are

$$\bar{a}_0 = 1 - \frac{\rho^2}{4} = 0.9375, \quad \bar{a}_1 = 0, \quad \bar{a}_2 = \frac{\rho^2}{4} = 0.0625$$

Therefore, according to Parseval's identity,

$$\left[\left(1 - \frac{\rho^2}{4} \right) + \rho \cos\theta + \frac{\rho^2}{4} \cos(2\theta) \right]^2_{\text{rms}} = \bar{a}_0^2 + \frac{1}{2}\bar{a}_1^2 + \frac{1}{2}\bar{a}_2^2$$

$$= 0.9375^2 + 0.5 \times 0.0625^2 = 0.8809$$

and hence,

$$\left[\left(1 - \frac{\rho^2}{4} \right) + \rho \cos\theta + \frac{\rho^2}{4} \cos(2\theta) \right]_{\text{rms}} = 0.9385$$

Displayed in Table 2.2 are the values of the coefficients of the exact Fourier expansion of $\sqrt{1 - \rho^2 \sin^2\theta}$, item $\sqrt{\Sigma}$ appearing in the same table being defined as

$$\sqrt{\Sigma} \equiv \sqrt{ \frac{1}{2} \left[\left(\frac{a_0}{b} \right)^2 + \sum_1^{N_h} \left(\frac{a_k}{b} \right)^2 \right] }$$

As compared with the exact value obtained above for the rms value of the square root, 0.9354, the approximation formula gives a fairly acceptable error of about 0.3%. Note that the *exact* Fourier expansion of the same function, as given by Table 2.2, gives an error of about 0.2% when the first two harmonics are considered. Thus, the proposed approximation, which is limited to "small" values of ρ, gives acceptable errors even for a relatively large value of ρ, namely, 0.5.

Finally, note that the exact Fourier expansion of the above square-root function matches the function itself, up to at least the first four digits, when the expansion includes the first four harmonics. The moral of the story is, then, that while simple approximations can give acceptable results, an even higher accuracy can be achieved at a rather low computational overhead.

2.8.3   The Periodic Response of First- and Second-order LTIS

In this section we shall determine the steady-state response of first- and second-order systems to periodic inputs. Such a response will be termed the *periodic response* of the system at hand. As we have seen, such an input can be represented as an infinite series of sine and cosine terms using the Fourier expansion of Eqs. 2.184a, b. Now, Eq. 2.184a can be rewritten as

$$f(t) = a_0 + \sum_1^\infty a_k \cos \omega_k t + \sum_1^\infty b_k \sin \omega_k t \qquad (2.201)$$

where

$$\omega_k = \frac{2\pi k}{T} \qquad (2.202)$$

Focusing on the terms under the summation signs, we see that they involve harmonic functions of certain frequencies ω_k. The steady-state response of the system to any one of these terms alone has the form

$$M_k^C \cos(\omega_k t + \phi_k) \quad \text{or} \quad M_k^S \sin(\omega_k t + \phi_k) \qquad (2.203)$$

where M_k^C, M_k^S and ϕ_k can be determined using the results of Sect. 2.7 with $A = a_k$, $B = b_k$, and $\omega = \omega_k$. Since a_0 can be regarded as a harmonic function of magnitude a_0 and frequency $\omega = 0$, the steady-state response M_0 of the system to the input a_0, can be determined using the results of Sect. 2.7 with $A = a_0$ and $\omega = 0$. The steady-state response of the system to the periodic input $f(t)$ can then be obtained via *superposition*, so that

$$x_P(t) = M_0 + \sum_1^\infty M_k^C \cos(\omega_k t + \phi_k) + \sum_1^\infty M_k^S \sin(\omega_k t + \phi_k) \qquad (2.204)$$

or

$$x_P(t) = M_0 + \sum_1^\infty M_k^C \cos\left(\frac{2\pi k}{T} t + \phi_k\right) + \sum_1^\infty M_k^S \sin\left(\frac{2\pi k}{T} t + \phi_k\right) \qquad (2.205)$$

In the balance of this section we will determine expressions for M_k^C, M_k^S and ϕ_k for various types of systems.

2.8.3.1 First-order LTI Dynamical Systems

If a first-order LTI dynamical system is driven by a periodic function $f(t)$, then we can model the system dynamics in the form

$$\dot{x} + ax = a_0 + \sum_1^\infty a_k \cos \frac{2\pi k}{T} t + \sum_1^\infty b_k \sin \frac{2\pi k}{T} t \qquad (2.206)$$

The term M_0 is determined by setting $\omega = 0$ and multiplying M by a_0 in Eqs. 2.118a, b, to obtain

$$M_0 = \frac{a_0}{a}, \quad \phi_0 = 0 \qquad (2.207)$$

Now, M_k^C and ϕ_k are obtained by setting $\omega = 2\pi k/T$ and multiplying M by a_k in Eqs. 2.118a, b, so that

$$M_k^C = \frac{a_k}{\sqrt{a^2 + (2\pi k/T)^2}} \qquad \phi_k = -\arctan\left(\frac{2\pi k}{Ta}\right) \qquad (2.208)$$

M_k^S is obtained likewise, by setting $\omega = 2\pi k/T$ and multiplying M by b_k in Eqs. 2.124a, b, which yields

$$M_k^S = \frac{b_k}{\sqrt{a^2 + (2\pi k/T)^2}}, \qquad (2.209)$$

2.8.3.2 Second-order Undamped LTI Dynamical Systems

Below we show the model of an undamped second-order system driven by a periodic function expressed in Fourier-expansion form, namely,

$$\ddot{x} + \omega_n^2 x = a_0 + \sum_1^\infty a_k \cos \frac{2\pi k}{T} t + \sum_1^\infty b_k \sin \frac{2\pi k}{T} t \qquad (2.210)$$

We derived the frequency response of these systems in Sect. 2.7.3 for harmonic inputs, the associated magnitude and phase expressions being displayed in Eqs. 2.135a, b and 2.144a, b. From these responses, we obtain the frequency response expressions for the system at hand, i.e.,

$$M_0 = \frac{a_0}{\omega_n^2}, \qquad \phi_0 = 0 \qquad (2.211a)$$

$$M_k^C = \frac{a_k/(\omega_n^2)}{|1 - k^2 r_0^2|}, \qquad \phi_k^C = \begin{cases} 0° & \text{if } kr_0 < 1; \\ -180° & \text{if } kr_0 > 1 \end{cases}. \qquad (2.211b)$$

$$M_k^S = \frac{b_k/(\omega_n^2)}{|1 - k^2 r_0^2|} \qquad (2.211c)$$

where r_0 is defined as the ratio

$$r_0 \equiv \frac{\omega_0}{\omega_n} \equiv \frac{2\pi}{T\omega_n} \qquad (2.212)$$

2.8.3.3 Second-order LTI Dynamical Systems

A damped system acted upon by a periodic input leads to the model below:

$$\ddot{x} + 2\zeta\omega_n\dot{x} + \omega_n^2 x = a_0 + \sum_1^\infty a_k \cos \frac{2\pi k}{T} t + \sum_1^\infty b_k \sin \frac{2\pi k}{T} t \qquad (2.213)$$

The frequency response expressions for the underdamped case were obtained in Sect. 2.7.3. We also pointed out in that subsection that the same amplitude and phase relations hold for critically damped and overdamped systems. From the two sets of magnitude and phase expressions, Eqs. 2.132a, b and 2.142a, b, we obtain the desired frequency-response expressions for the system at hand, as shown below:

$$M_0 = \frac{a_0}{\omega_n^2}, \quad \phi_0 = 0 \tag{2.214a}$$

$$M_k^C = \frac{a_k/\omega_n^2}{\sqrt{(1-k^2 r_0^2)^2 + (2\zeta k r_0)^2}} \tag{2.214b}$$

$$\phi_k = -\arctan\left[\frac{2\zeta k r_0}{1-(k r_0)^2}\right] \tag{2.214c}$$

$$M_k^S = \frac{b_k/(\omega_n^2)}{\sqrt{(1-k^2 r_0^2)^2 + (2\zeta k r_0)^2}} \tag{2.214d}$$

$$r_0 = \frac{\omega_0}{\omega_n} = \frac{2\pi}{T\omega_n} \tag{2.214e}$$

Example 2.8.6 (The Periodic Response of an Air Compressor). For the numerical values given below, find the steady-state dynamic force transmitted to the ground by the compressor-foundation system of Sect. 2.8.2, with the crank turning at a constant ω_0.

$$a = 100 \text{ mm}; \quad \rho = 0.5; \quad m = 10 \text{ kg}; \quad M = 200 \text{ kg};$$

$$\omega_n = 26.1799 \text{ s}^{-1}; \quad \omega_0 = 100 \text{ rpm}; \quad \zeta = 0.2$$

Solution: As the reader can readily verify, the mathematical model of the compressor-foundation system takes the form:

$$\ddot{x} + 2\zeta\omega_n\dot{x} + \omega_n^2 x = -\frac{m}{M+m}\ddot{s}, \quad \omega_n \equiv \sqrt{\frac{k_f}{M+m}}$$

where $s(\theta)$ is given in Example 2.8.5. Now, we need the Fourier expansion of the right-hand side of the foregoing equation. To this end, we can proceed in two ways, namely, (i) by finding $\ddot{s}(t)$ explicitly from the second derivative of $s(\theta)$ given above, and then computing the Fourier coefficients of this function; alternatively, (ii) by differentiation of the terms of the Fourier expansion of $s(\theta)$ found in Example 2.8.5. Since we already have the latter, it is apparent that the second approach should be more expeditious.

Note that $s(\theta)$ was found to be accurately approximated with its first four harmonics, namely,

$$s(\theta) = a_0 + \sum_1^4 a_k \cos(k\theta)$$

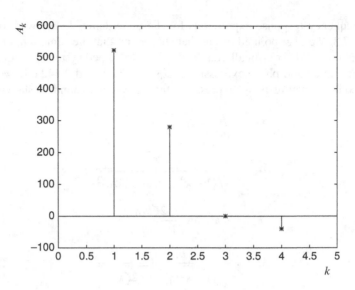

Fig. 2.61   The four Fourier coefficients A_k of the force-transmission analysis

and hence,

$$\dot{s}(\theta) = -\omega_0 \sum_1^4 k a_k \sin(k\theta), \quad \ddot{s}(\theta) = -\omega_0^2 \sum_1^4 k^2 a_k \cos(k\theta)$$

the governing equation thus taking the form

$$\ddot{x} + 2\zeta \omega_n \dot{x} + \omega_n^2 x = \sum_1^4 A_k \cos(k\theta)$$

with A_k defined as

$$A_k = \omega_0^2 \frac{m}{M+m} k^2 a_k, \quad k = 1,2,3,4$$

Shown in Fig. 2.61 are the four coefficients A_k, for the given numerical values.

Now, the force transmitted to the foundation, $F(t)$, can be expressed as a sum of harmonics having the form of the expression appearing in Eq. 2.163, i.e.,

$$F(t) = \sum_1^4 F_k \cos(k\omega_0 t + \psi_k)$$

with F_k and ψ_k derived from the expressions given in Eqs. 2.166 and 2.167, with $F_0 = A_k(M+m)$ or $F_0 = \omega_0^2 m k^2 a_k$, for $k = 1,2,3,4$. Therefore,

$$F_k \equiv \frac{A_k(M+m)\sqrt{1+(2\zeta r_k)^2}}{\sqrt{(1-r_k^2)^2+(2\zeta r_k)^2}}, \quad \psi_k \equiv \tan^{-1}(2\zeta r_k) - \tan^{-1}\left(\frac{2\zeta r_k}{1-r_k^2}\right)$$

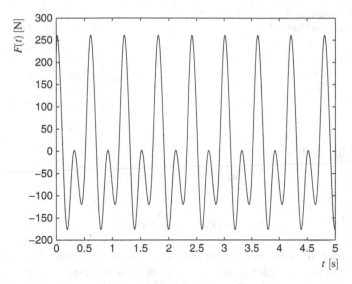

Fig. 2.62   Steady-state force transmitted to the foundation

where the counter k in the foregoing expansion is not to be confused with the stiffness of the foundation, k_f. Moreover, the counter is not needed, for the value of ω_n is given instead. Additionally,

$$r_k \equiv \frac{k\omega_0}{\omega_n}, \quad k = 1,2,3,4$$

The steady-state transmitted force, computed with four harmonics, is plotted in Fig. 2.62.

2.9   The Time Response of Systems with Coulomb Friction

Systems with Coulomb friction are more difficult to handle than those with viscous friction that we studied so far. One reason is that these systems lead to nonlinear models, as we will show presently.

We consider here a very simple model, namely, a mass coupled to an inertial frame through a linear spring and Coulomb friction, as depicted in Fig. 2.63. Here, we model friction as in Chap. 1, namely, we do not distinguish between static and dynamic friction coefficients and take the friction force as proportional to the normal force, the proportionality factor μ, i.e., the friction coefficient, being constant.

Under the foregoing conditions, it is now a simple matter to derive the governing equation of interest, namely,

$$m\ddot{x} + kx = \begin{cases} -\mu N, \text{ if } \dot{x} > 0; \\ \mu N, \quad \text{ if } \dot{x} < 0; \end{cases}$$

Fig. 2.63 Mass-spring
system subjected to Coulomb
friction

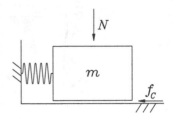

which readily leads to

$$m\ddot{x} + kx = -\mu N \text{sgn}(\dot{x}) \qquad (2.215)$$

or

$$\ddot{x} + \omega_n^2 x = -\frac{\mu N}{m}\text{sgn}(\dot{x}) \qquad (2.216)$$

Such a model is, then, said to be *piecewise linear*. We have thus, piecewise,
an undamped second-order system acted upon by a constant force of magnitude
$f_0 = \mu N/m$. Clearly, as the velocity changes sign, so does the friction force, always
opposing the motion. It is then expected that, within a period in which the velocity
does not change sign, the friction force will slow down the mass, which will
eventually come to a standstill. Now, because of the potential energy stored in the
spring upon reaching the standstill position, the mass will resume motion as long
as the force supplied by the spring is large enough to overcome the friction force.
Motion will eventually stop when the spring force is not large enough to overcome
the friction force. Thus, in order to find the desired time response, we can assume,
without loss of generality, that we start observing the system while the mass is
stationary, and so, we can assume the initial conditions $x(0) = x_0 > 0$ and $\dot{x}(0) = 0$.
We then have

$$\ddot{x} + \omega_n^2 x = \mu \frac{N}{m}, \qquad x(0) = x_0 > 0, \quad \dot{x}(0) = 0 \qquad (2.217)$$

The response of the above system is readily derived from previous results. Indeed,
this response has the form given in Eq. 2.88, with $v_0 = 0$ and $f(t) = \mu N/m =$
const, i.e.,

$$x(t) = (\cos \omega_n t)x_0 + \int_0^t \mu \frac{N}{m} \frac{1}{\omega_n}(\sin \omega_n(t - \tau))d\tau \qquad (2.218)$$

and hence,

$$x(t) = \frac{\mu N}{m\omega_n^2} + (\cos \omega_n t)\left(x_0 - \frac{\mu N}{m\omega_n^2}\right) \qquad (2.219)$$

while the velocity is

$$\dot{x}(t) = -\omega_n(\sin \omega_n t)\left(x_0 - \frac{\mu N}{m\omega_n^2}\right) \qquad (2.220)$$

Therefore, the mass will reach a standstill at a time $t = t_1 = \pi/\omega_n$. Furthermore, upon substitution of t_1 in the expression for $x(t)$, Eq. 2.219, an expression for $x_1 \equiv x(t_1)$ is readily derived, namely,

$$x_1 = \frac{\mu N}{m\omega_n^2} - \left(x_0 - \mu\frac{N}{m\omega_n^2}\right) = -x_0 + 2\frac{\mu N}{m\omega_n^2}$$

Now we regard $t = t_1$ as the initial time for the ensuing motion. The governing equation from which this motion is derived is identical to the former, Eq. 2.217, with the difference that now the non-homogeneous term changes sign, the response thus becoming

$$x(t) = -\frac{\mu N}{m\omega_n^2} + (\cos\omega_n t)\left(x_0 - 3\mu\frac{N}{m\omega_n^2}\right), \quad t \geq t_1$$

Likewise,

$$\dot{x}(t) = -\omega_n(\sin\omega_n t)\left(x_0 - 3\mu\frac{N}{m\omega_n^2}\right)$$

Therefore, the mass reaches a standstill at a time $t_2 > t_1$ given by $t_2 = 2\pi/\omega_n$, at which $x(t)$ attains the value $x_2 \equiv x(t_2)$, given by

$$x_2 = x_0 - 4\frac{\mu N}{m\omega_n^2}$$

The motion for $t > t_2$ is derived likewise. It is apparent by now that the mass will attain a standstill at $t = t_3 = 3\pi/\omega_n$, at which

$$x_3 \equiv x(t_3) = -x_0 + 6\frac{\mu N}{m\omega_n^2}$$

and hence, the mass attains rest at a sequence of positions $\{x_i\}_1^F$ given by

$$x_i = (-1)^i\left(x_0 - 2i\frac{\mu N}{m\omega_n^2}\right), \quad i = 1,\ldots,F$$

x_F, the final position, being reached when the friction force is greater than the force provided by the spring while the mass is at rest, the mass thereby stopping for good. The value F is the lowest integer F for which

$$k|x_F| < \mu N < k|x_{F-1}|$$

or

$$|x_F| < \frac{\mu N}{k} < |x_{F-1}|$$

A plot of the above time response is shown in Fig. 2.64.

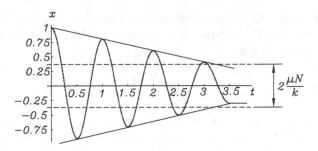

Fig. 2.64 Time response of mass-spring system subjected to Coulomb friction

Note that the time response of the system at hand is piecewise harmonic, with decreasing amplitude. However, contrary to viscous damping, which produces an exponential decrement of the motion amplitude, Coulomb damping produces a linear decrement. Hence, the nature of a damped system can be identified by inspection, i.e., by looking at its decrement.

Finally, it can be readily proven that the sequence of values $\{x_i\}_1^F$ is governed by the recursive relation given below:

$$x_{i+1} = -x_i + 2\frac{\mu N}{m\omega_n^2}, \quad x_0 = x_0, \quad i = 1,\ldots,F \tag{2.221}$$

with x_F observing the constraint derived above.

2.10   Exercises

2.1. We refer to the tugboat-barge system of Example 2.2.1 under the failure conditions described therein, but with the numerical data given below: $\tau_B = 10$ s, $\tau_T = 8$ s, $v_0 = 5$ m/s, the length of the tugging cable being 8 m. Determine the time elapsed until the occurrence of a collision. *Hint: A graphical solution is suggested here.*

2.2. The iconic model of an overhead material-transport system is sketched in Fig. 2.65. The system consists of a massless, rigid wheel driven at a controlled constant speed w along a bent track, in such a way that the load, of mass m, can be safely assumed to undergo pure translation. Note that the load is attached to the wheel by means of an undamped suspension. Find the value of w beyond which the wheel will jump off the track after reaching the horizontal track section.

2.3. Refer to the aircraft modeled upon landing as shown in Fig. 2.13. At touchdown, the fuselage lies a height $h_0 = 1/(2\pi)$ m above its static-equilibrium level (SEL), while the approach velocity has a horizontal component u_0 that can safely

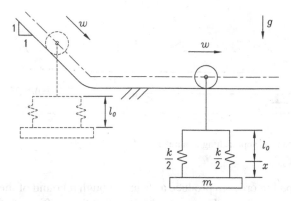

Fig. 2.65 An overhead material-transport system with an undamped suspension

be assumed constant during touchdown and a vertical component v_0 of $4\sqrt{2}$ m/s. If $\omega_n = 1$ Hz and $\zeta = \sqrt{2}/2$,

(a) Find the maximum height h_1 reached by the fuselage from the SEL right after landing
(b) Find the maximum force transmitted to the ground

2.4. Consider the overhead crane of Example 1.6.5. We want to stabilize the unstable rod-up equilibrium configuration with *analog feedback* applied by a torsional spring of stiffness k at the pin of attachment of the rod with the slider. Find the minimum value k_{min} of k that will make the feedback system stable, under the assumption that the spring is unloaded in the rod-up equilibrium configuration. Now, for a value of $k = 2k_{min}$ and constant $\ddot{u} = 0.5g$, sketch the time response of the system under the perturbations (1) $\delta\theta_0 = 0.02$ rad, $\delta\dot{\theta}_0 = 0$; (2) $\delta\theta_0 = 0$, $\delta\dot{\theta}_0 = 0.02$ rad/s; and (3) $\delta\theta_0 = 0.01$ rad, $\delta\dot{\theta}_0 = 0.06$ rad/s. Comment on the relationship among the three results. Assume that the pin provides a light damping of 10%, i.e., the damping ratio is 0.1, and $g/l = 4$ s^{-2}.

2.5. We refer to the system of Exercise 1.2. An experiment is conducted to determine the stiffness of the shaft. The experiment consists of letting the bar fall freely from its top position, i.e., with the motor exerting a zero torque on the pinion. As the bar reaches its bottom position, the motor is blocked, thereby fixing the left end of the shaft to an inertial frame. It is then noted that the link undergoes 80 small-amplitude oscillations, behaving as a rigid body, about its lowest position in 10 s. Moreover, the light damping always present, in the form of air drag, material hysteresis, and the like, produces a decrement of the amplitude of the oscillations such that, after 80 oscillations, the amplitude is only 5% of its original value. With this information, and knowing that the bar is 1 m long and weighs 117.72 N, find the stiffness of the shaft and the coefficient of damping, under the assumption that damping is linear.

2.6. The system of Example 2.4.2 is revisited here. Under the assumption that $\omega_n = 3.1451$ rad/s and $\zeta = 0.04711$, find the minimum height h from which the mass m_2

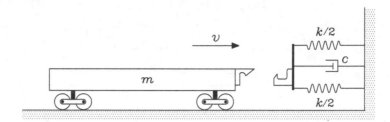

Fig. 2.66 A railroad car approaching a bumper

should be dropped in order to produce a strong-enough rebound of the pad that will exert a pull of the springs on the ground. Assume that the coefficient of restitution of the ball with the pad is $e = 0.5$. Find the instant t_{min} at which the pull is expected to occur. *Note: Because of the change in conditions, the plot of Fig. 2.25 has nothing to do with this problem.*

2.7. Boxes are transported on a horizontal track of rollers. Upon placement of a box of mass m on the rollers, the box is given a push that moves it at a speed v_0. Find the value of v_0 that will allow the box to travel a given distance d before its speed drops below $0.5 v_0$, under the assumption that the lubricant of the rollers produces linearly viscous damping of a known coefficient $c(Ns/m)$.

If a batch of different boxes, ordered by weight, is to be placed on the rollers at equal time-intervals T, with equal initial speed v_0, how would you order the placement of the boxes, in order to prevent collisions between boxes, i.e., from heavier to lighter, alternating them, etc.? Explain your rationale.

2.8. The railroad car shown in Fig. 2.66 is released with a speed of 20 km/h a distance of 50 m from the bumper to the right. The car weighs 100 kN, and its axles turn on bearings providing a linearly viscous damping that decreases its speed by 80% just before hitting the bumper. Once the bumper has been hit, assume that the car engages the bumper without backlash and without energy losses. Then, the compression of the springs is recorded. It is found that the spring compression reaches a first peak of 250 mm and a second one of only 35 mm. Find:

(a) The damping coefficient of the car axles
(b) The damping ratio of the car-bumper system
(c) The natural frequency of the same system
(d) The stiffness of the bumper
(e) The damping coefficient of the bumper

2.9. Assume that the railroad car of Fig. 2.66 weighs 98.10 kN, and travels at a velocity $v(t)$ to the right. Upon hitting the bumper, coupling is assumed to occur instantly and without backlash. Under these ideal conditions, assume further that the wheels roll without slipping on the railway, the only sink of energy being the lubricant in the axles, which can be assumed to produce linearly viscous damping

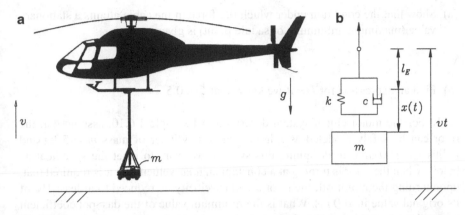

Fig. 2.67 A helicopter lifting a load: (a) the physical system (silhouette of the Aérospatiale—Eurocopter—AS 355 Écureil Helicopter); (b) the iconic model

of coefficient $c_W = 0.7$ kNs/m. The car is released with a velocity v_R and, 10 s afterwards, it is observed to hit the bumper with a velocity of 10 m/s.

(a) Find the value of v_R.
(b) The bumper is to be designed so that it will provide linearly viscous damping with a damping ratio of $\zeta = 0.5$ and undamped oscillations of $1/(2\pi)$ Hz. Specify the spring stiffness and the dashpot coefficient.
(c) Find the maximum force transmitted to the bumper, when the car approaches it as described above, i.e., with a velocity of 10 m/s.
(d) Sketch the displacement of the car as a function of time after coupling has taken place, and determine the maximum deflection experienced by the bumper.
(e) What design changes (i.e., new values of k and c) would you recommend in order to bring the maximum deflection of item (d) down to 10 % of the value found above, without changing the damping ratio? Under these changes, what is the maximum force transmitted for the given approach velocity?

2.10. Shown in Fig. 2.67a[13] is a helicopter lifting a load of mass m at a constant lifting speed v, l_E being the length of the spring when the load is airborne and under static equilibrium. The lifting mechanism is modeled in Fig. 2.67b as a linear mass-spring-dashpot system. Moreover, the force f exerted by the cable on the mass can be expressed in the form

$$f(t) = kx + c\dot{x}$$

We define the *dynamic factor* D_f as the dimensionless quantity by which the weight of the load has to be multiplied, when designing the cable, to account for the dynamic effects of lifting. We want to determine this factor by finding the maximum force exerted on the cable once the load is airborne. To this end,

---

[13]Reproduced   with   authorization   from   http://commons.wikimedia.org/wiki/   File: Helicopter_silhouette_AS-355.svg

(a) Show that the condition under which the force in the cable attains a stationary value (maximum, minimum, or saddle point) is given by

$$(4\zeta^2 - 1)\dot{x} + 2\zeta\omega_n x = 0$$

(b) Find an expression for D_f, if we know that $\zeta = 0.5$

2.11. For the motor-clutch system described in Example 1.6.10, assume that the rotor can be safely modeled as a homogeneous cylinder of mass $m = 5$ kg and radius $r = 100$ mm. In designing this system, we want to meet the specification below: when the rotor is turning at a constant angular velocity ω, it is required that, upon turning the motor off, the rotor angular velocity be reduced to at least 1% of its original value in 100 ms. What is the minimum value of the dashpot coefficient that will produce this behavior?

2.12. (To be assigned only if Problem 1.8 was previously assigned.) In designing the suspension of Problem 1.8, a test is conducted on the suspension such that the wheel is displaced from its equilibrium state by a small amount y_0 and is then released at rest. It is required that the wheel return to within 1% of its equilibrium configuration after the second oscillation. If the natural frequency of the suspension is 1 Hz, what is the required value of the damping ratio? Under these conditions, for $y_0 = 50$ mm and $\dot{y}_0 = 0$, plot the time response of the system for the first three oscillations.

2.13. A fluid clutch connects a load, e.g., the whole inertia of an automobile, to an engine, as shown in Fig. 1.23. Assume that the torque transmitted through the clutch, when it is engaged, is proportional to the difference in speed of the input and output shafts (proportionality constant c). Assume also that the speed ω of the engine shaft is constant and unaffected by the load. If the load is a pure inertia J and the system is underdamped, find the time response of the load angular velocity $\omega_R = \dot{\theta}$, after the clutch is suddenly engaged.[14]

2.14. A wheelchair crossing a ditch is modeled as the underdamped mass-spring-dashpot system shown in Fig. 2.68.

(a) Determine the time response $x(t)$ of its vertical motion, upon traversing the ditch at a constant speed v, in terms of the physical parameters of the system and the geometry of the ditch.
(b) Plot $x(t)$ vs. time at a suitable scale, for values of

$$\omega_n = 1 \text{ Hz}, \quad \zeta = \frac{\sqrt{2}}{2}, \quad \frac{v}{\lambda\omega_n} = 0.01, \frac{1}{2\pi}, 10.0$$

Comment on the results.

---

[14]This exercise is drawn from a similar one in [3].

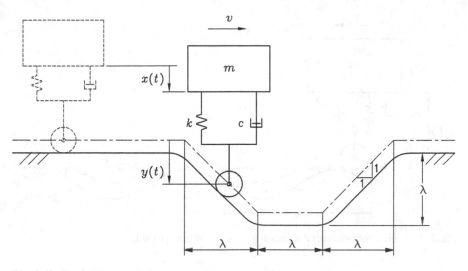

Fig. 2.68 The iconic model of a wheelchair crossing a ditch

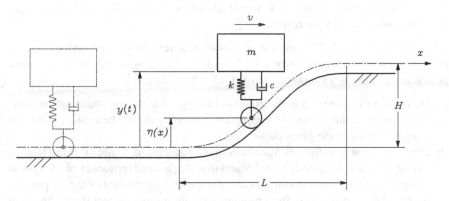

Fig. 2.69 A terrestrial vehicle climbing a cycloidal ramp

2.15. A terrestrial vehicle is modeled as the underdamped mass-spring-dashpot system shown in Fig. 2.69. Determine the time response $y(t)$ of its vertical motion, upon overcoming a *cycloidal* slope at a constant speed v, in terms of the physical parameters of the system and the geometry of the slope. The cycloidal slope is modeled via the function $\eta(x)$, defined as

$$\eta(x) \equiv \begin{cases} 0, & \text{for } x \leq 0 \\ H[x/L - 1/(2\pi)\sin(2\pi x/L)], & \text{for } 0 \leq x \leq L \\ H, & \text{for } x \geq L \end{cases}$$

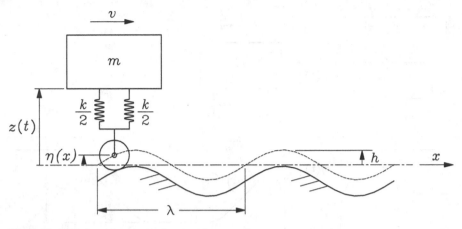

Fig. 2.70 Vehicle with undamped suspension traveling on a wavy road

Furthermore, plot the response of the system under the conditions: $c = 0$ and $\omega_n = 1$ Hz, for three values of v, $v = 0.01\omega_n L$, $v = \omega_n L/(2\pi)$, and $v = 10\omega_n L$, $L = 1$m. Comment on your results.

2.16. Shown in Fig. 2.70 is the iconic model of a vehicle with an undamped suspension—intended to model worn-out "shocks"—traveling on a wavy road whose profile is modeled as a sinusoidal wave $\eta(x) = h\sin(2\pi x/\lambda)$.

(a) In order to prevent the mass from oscillating with an amplitude greater than $5h$, a range of the values of the constant velocity v should be avoided; find this range in terms of the given parameters.

(b) Now, we want to add shock absorbers to the suspension, while knowing that the weight of the vehicle body is 10 kN and that the damped frequency of the system should be 1 Hz. Find the stiffness and the dashpot coefficient of the suspension that will prevent the mass from moving with amplitudes greater than $2.5h$ under *any* constant velocity v.

2.17. An aircraft turbine mounted on a testbed is modeled as shown in Fig. 2.71a. The turbine provides a compressive force $f(t)$ of the saturation type,[15] as plotted in Fig. 2.71b. Derive the time response of the system, $x(t)$, under zero initial conditions.

2.18. A Baja vehicle weighing 5,000 N is being designed with a suspension that has to provide a natural frequency of 1 Hz. Moreover, it is required that, upon letting the vehicle fall from a certain height h_0, the vehicle body bounce back to successive height peaks h_1 and h_2, so that h_2 be 20% of h_1. Find the dashpot coefficient and the spring stiffness of the suspension.

---

[15] The saturation function sat(x) was introduced in Eq. 1.37 and plotted in Fig. 1.26b.

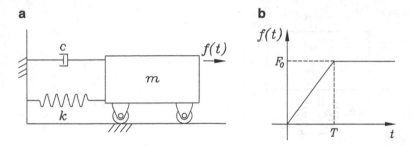

Fig. 2.71 Model for an aircraft turbine

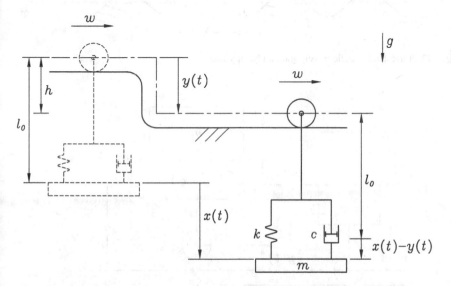

Fig. 2.72 An overhead material-transport system with an underdamped suspension

2.19. We study here the behavior of the *underdamped* material-handling system of Fig. 2.72 as the wheel goes down a step of height h along its track at $t = 0$.

(a) Derive the mathematical model of the system.
(b) Find the time response of the system.
(c) Find the values of $x(0^+)$ and $\dot{x}(0^+)$.

2.20. Shown in Fig. 2.73 is the model of a terrestrial vehicle with an undamped suspension, traveling at a constant speed v_0 on a bumpy road that is known to have a periodic profile. For purposes of our analysis, we model the road profile in the form

$$\eta(x) = h\left|\sin\left(\frac{\pi}{\lambda}x\right)\right|$$

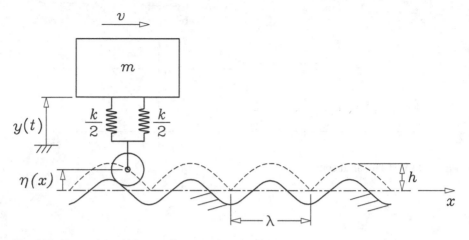

Fig. 2.73 Terrestrial vehicle traveling on an bumpy road

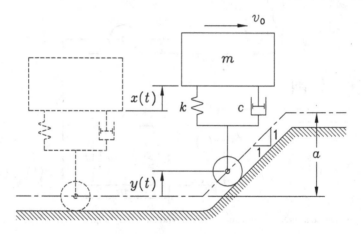

Fig. 2.74 Terrestrial vehicle overcoming a 45° ramp

for a bump height h and a wavelength λ. Upon introducing the coordinate y measuring the vertical position of the mass with respect to its equilibrium configuration, the governing equation takes the form

$$\ddot{y} + \omega_n^2 y = \omega_n^2 \eta$$

Find the steady-state response $y(t)$ of the system in terms of the parameters of the system and of the road profile. *Hint:* $2 \sin a \cos b = \sin(a+b) + \sin(a-b)$.

2.21. An *all-terrain terrestrial vehicle* with an underdamped suspension is modeled as in Fig. 2.74. It is meant to overcome a 45° slope as shown in the same figure, while traveling at a constant speed v_0. The variable x measures the vertical displacement

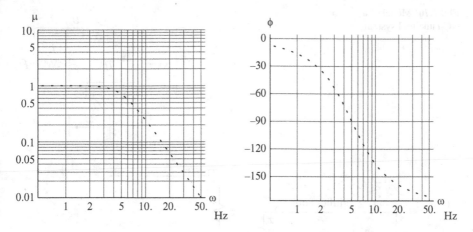

Fig. 2.75 Magnitude (μ) and phase (ϕ) of the tests conducted on a vehicle suspension

of the vehicle with respect to an inertial frame. This frame is located so that the origin corresponds to the position of static equilibrium of the vehicle when it is at rest on the ground before hitting the ramp. The variable y represents the height of the ramp from the ground. Derive the time response of the vehicle.

2.22. We refer here to the suspension system of Problem 1.8. The linearized model about the equilibrium configuration $y = 0$ takes the form

$$m_E \ddot{y} + c_E \dot{y} + k_E y = f(t)$$

where y is a "small-amplitude" displacement from equilibrium, while m_E, c_E, and k_E are the equivalent mass, equivalent dashpot coefficient, and equivalent stiffness of the linearized system, and $f(t)$ is an external vertical force. Expressions for m_E, c_E, and k_E are given as

$$m_E = 19m, \quad c_E = \frac{3}{4}c, \quad k_E = \frac{3}{4}k$$

The suspension system is tested on a machine that applies a vertical force $f(t) = F_0 \cos \omega t$ on the wheel with ω changing in a broad spectrum of frequencies, thus obtaining the plots shown in Fig. 2.75. For the displacement $y(t)$ determine the values of c and k, if $m = 5$ kg.

2.23. The tool of a NC lathe is modeled as a linearly elastic, massless spring, while the positioning mechanism is modeled as a mass-dashpot system, as shown in Fig. 2.76. When the workpiece is turning at a constant angular velocity ω_f, the workpiece exerts a displacement $y(t) = A \cos \omega_f t$ on the tip of the tool due to the irregularities in the workpiece profile, where A is assumed to be "small." Find the frequency ω_f for which the force transmitted to the machine-tool frame

Fig. 2.76 Model of a
NC-lathe-tool system

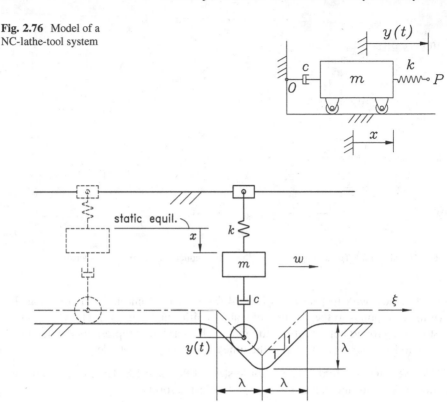

Fig. 2.77 Iconic model of a coordinate-measuring machine

is a maximum. In this model, note that the workpiece cannot, properly speaking, exert a pull on the tool. However, the spring is assumed to be under compression all the time, the displacement $y(t)$ becoming positive when the contact point of the workpiece is to the right of its nominal position; the latter is defined, in turn, as the one that the contact point would have if the workpiece were perfectly circular.[16]

2.24. Shown in Fig. 2.77 is a rough iconic model of a coordinate-measuring machine, as it is probing a nominally flat surface at a constant velocity w, while approaching a groove that can be safely modeled as a *triangular ditch*. Note that the slider is massless and can move freely and without friction on its horizontal guideway. Also note that the probe is always in contact with the surface. Under the assumption that the system at hand is underdamped, and for a set of numerical values that yield the model

$$\ddot{x} + 2\dot{x} + 2x = 200[u(t) - 2u(t-T) + u(t-2T)]$$

---

[16]This exercise is drawn from a similar one in [3].

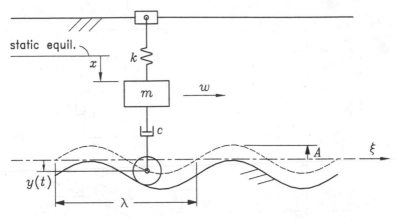

Fig. 2.78 Iconic model of a coordinate-measuring machine

find the time response of the system in terms of the corresponding responses to abrupt and impulsive excitations.

2.25. Shown in Fig. 2.78 is the same model of the coordinate-measuring machine of Exercise 2.24 as it probes a nominally flat surface at a constant velocity w. The surface shows small irregularities that can be approximated as a harmonic wave of amplitude A. Note that: (1) the slider is massless; (2) the slider can move freely and without friction on its horizontal guideway; and (3) the probe is always in contact with the surface. Under the foregoing assumptions,

(a) show that the mathematical model of the system can be cast in the form

$$\ddot{x} + 2\zeta\omega_n\dot{x} + \omega_n^2 x = f(t)$$

and find an expression for $f(t)$.
(b) Under the assumption that the system at hand is underdamped, and for a set of numerical values that yield the model

$$\ddot{x} + 2\dot{x} + 2x = 200\cos\omega t \quad [\mu m/s^2]$$

find an expression for ω in terms of the parameters of the problem.
(c) Find the value of the steady-state peak force exerted by the probe on the surface, assuming $m = 1$ kg.

2.26. Find an expression for the time response of the underdamped system of Fig. 2.79 traveling at a constant velocity w upon encountering a ramp of angle α, in terms of the response(s) of the same system to abrupt and impulsive excitations.

2.27. Shown in Fig. 2.76 is the iconic model of the tool-carrying mechanism of a lathe, with point P indicating the contact point between tool and workpiece.

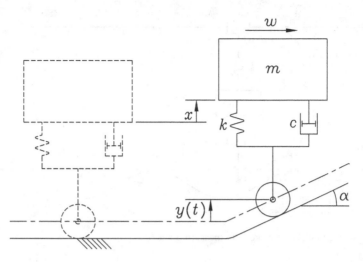

Fig. 2.79 An underdamped system encountering a ramp

The latter subjects the tip of the tool to a harmonic displacement $y(t) = A \cos \omega t$, with $A = 500$ μm, that is due to the irregularities on the machined surface. The mathematical model of this system takes the form

$$\ddot{x} + 2\zeta \omega_n \dot{x} + \omega_n^2 x = \omega_n^2 y(t)$$

If we know that $m = 1$ kg, $\zeta = 0.5$, and $\omega_n = 200/\pi$ Hz for the model at hand, find the range of cutting speeds ω that will produce a steady-state force exerted at point O higher than 48 N, with O denoting the mounting of the tool carrier on the frame of the machine.

Hint: The use of Bode plots is highly recommended here.

2.28. Shown in Fig. 2.76 is the iconic model of the tool-carrying mechanism of a lathe, with point P indicating the contact point between tool and workpiece. The presence of an irregularity on the latter subjects the tip of the tool to a *triangular bump*, as shown in Fig. 2.80. The mathematical model of this system takes the form

$$\ddot{x} + 2\zeta \omega_n \dot{x} + \omega_n^2 x = f(t)$$

(a) Find an expression for $f(t)$ in terms of the parameters of the system.
(b) Now, for a given set of numerical values of the natural frequency and the damping ratio, the foregoing system is acted upon by a function $f(t)$—not necessarily the same as that of item (a)—of the form

$$f(t) = r(at) - 2r(at - aT)$$

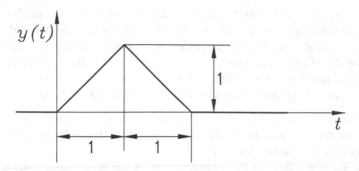

Fig. 2.80 Triangular bump profile

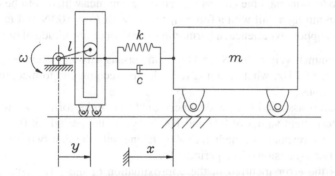

Fig. 2.81 A simplified model of a press mechanism

where $r(t)$ is the unit ramp, and T is a constant. Sketch the foregoing excitation and find the time response of the system at hand to this excitation.

2.29. A simple model of the mechanism of a press is shown in Fig. 2.81. In this model, the Scotch yoke of the left is driven by a crank that turns at a constant angular speed ω, while the spring is undeformed when $x = 0$ and $y = 0$.

(a) Under the current design it has been found that, when $\omega = \sqrt{k/m}$, the mass oscillates with an amplitude $M = 5l$. From this observation, can you estimate the damping ratio ζ?
(b) A modification is being proposed, that consists of changing the spring to one with a new stiffness k_{new}. Choose k_{new}, without modifying c, so that the amplitudes of the oscillations of the mass and the Scotch yoke coincide.

2.30. Derive the ramp response of (1) first-order systems and (2) *undamped* second-order systems. What is the steady-state component of these responses? *Skip (1) if first-order systems were not covered by your instructor.*

2.31. Derive the ramp response of (1) underdamped, (2) *critically damped* and (3) *overdamped* second-order systems. What is the steady-state component of these responses? *Skip (1) if first-order systems were not covered by your instructor.*

2.32. A wheelchair designed for climbers of pre-Columbian pyramids is to be analyzed to verify whether it meets its design specifications. To this end, we use the Fourier expansion of the pyramid profile derived in Example 2.8.4 and base our analysis on the first two harmonics. The forward velocity of the wheelchair is to be kept at a constant $3\,\mathrm{m/s}$; the length of the pyramid steps is assumed to be $\lambda/2 = 0.5\,\mathrm{m}$; the natural frequency of the chair is $1\,\mathrm{Hz}$; and the damping ratio is 0.7071.

(a) Find the magnitude and the phase of the harmonics of the steady-state response.
(b) What is the error incurred in the approximation of the pyramid profile when taking only the first two harmonics?
(c) What about the error in the approximation of the steady-state response with only two harmonics?

2.33. A production machine carries a pneumatic hammer which is to be designed so that, upon hitting a nail with a force $F_0 \cos \omega t$, with $F_0 = 100\,\mathrm{N}$ and $\omega = 10\,\mathrm{Hz}$, the machine support experiences a harmonic force with an amplitude of only $10\,\mathrm{N}$.

(a) If the hammer weighs $200\,\mathrm{N}$, and the hammer-suspension system has a natural frequency of $1\,\mathrm{Hz}$, what damping ratio do you recommend to meet the design specifications?
(b) If the nail is assumed to provide a periodic force in the form of a square wave with intermittent values of $0\,\mathrm{N}$ and $100\,\mathrm{N}$ over equal periods of 0.1 s, find the steady-state response of the hammer by taking only the first two harmonics of the Fourier expansion of the periodic force.
(c) What is the error incurred in the approximation of the force with only two harmonics?
(d) What about the error in the approximation of the force transmitted to the user?

2.34. The air compressor of Example 2.8.6 is revisited here. The velocity $\dot{s}$ of the piston with respect to the housing is given below as a function of the angular displacement θ of the crank, in dimensionless form:

$$\frac{\dot{s}}{\omega a} = \left[1 + \rho \frac{\cos \theta}{\sqrt{1 - \rho^2 \cos^2 \theta}} \right] \sin \theta$$

where ω denotes the constant angular velocity of the crank, and $\rho \equiv a/b$. Moreover, the foregoing dimensionless velocity is plotted for $\rho = 0.5$ in Fig. 2.82a. The acceleration $\ddot{s}$ appearing in the right-hand side of the mathematical model of that example can be found from the above expression by straightforward differentiation. However, as one can readily realize, a Fourier analysis of the foregoing velocity, not to speak of its time-derivative, would be untractable with longhand calculations. Here we have a typical example of an algebraically difficult problem that can be rendered tractable with a reasonable approximation. Indeed, we can approximate the velocity plot of Fig. 2.82a by a triangular wave, as depicted in Fig. 2.82b.

Express the right-hand side of the corresponding mathematical model as a series of harmonic functions, under the assumption that the crank turns at a constant rate ω_0 rad/s, for a piston velocity approximated as in Fig. 2.82b.

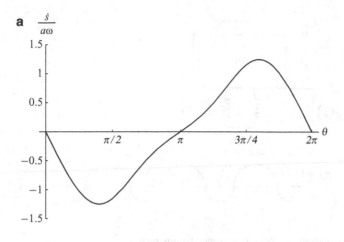

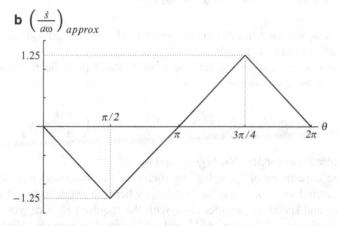

Fig. 2.82 Plot of (**a**) nondimensional piston velocity and (**b**) its piecewise linear approximation

2.35. A vehicle with worn-out shock absorbers is represented in Fig. 2.83 as it rolls through a bumpy road whose bumps follow a circular pattern of radius r. The vehicle travels with an unperturbed constant speed $v = 20\omega_n r$, where $\omega_n = \sqrt{k/m}$. We want to estimate how much the body of the vehicle will "jump" as it traverses the road. To this end, we conduct an *approximate* Fourier analysis of the road profile using a trapezoidal integration of the profile, which yields the expansion below:

$$\eta(x) \approx \tilde{a}_0 + \tilde{a}_1 \cos\left(\frac{\pi x}{r}\right) + \tilde{a}_2 \cos\left(\frac{2\pi x}{r}\right)$$

where $\tilde{a}_k$, for $k = 0, 1, 2$ are the *approximate* Fourier coefficients of the road profile, computed using the trapezoidal rule. Our computer-algebra calculations yield

$$\tilde{a}_0 = \frac{1 + \sqrt{3}}{4} r, \quad \tilde{a}_1 = -\frac{1}{2} r, \quad \tilde{a}_2 = \frac{1 - \sqrt{3}}{2} r$$

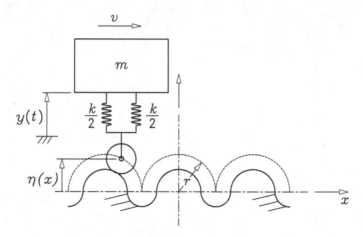

Fig. 2.83 An undamped vehicle traversing a bumpy road

(a) For starters, we want to estimate the error of the above approximation of the road profile. Find an expression for that error.
(b) Under the above approximation, the (steady-state) periodic response of the vehicle can be approximated as

$$y(t) \approx M_0 + M_1 \cos\left(\frac{\pi vt}{r} + \phi_1\right) + M_2 \cos\left(\frac{2\pi vt}{r} + \phi_2\right)$$

Find expressions for M_0, M_1, M_2, ϕ_1, and ϕ_2.
(c) Estimate the amount of "jumping" by means of the root-mean square value of $y(t)$, i.e., find an expression for $[y(t)]_{\text{rms}}$, where, in order to avoid algebraic mistakes and spending precious time with the required algebra, you need not substitute the expression for $\{M_i\}_0^2$ and $\{\phi_i\}_1^2$ found in item (b). *Hint: Express $y(t)$ as a truncated Fourier series, i.e., as a sum of (1) a constant term, (2) two cosine terms, and (3) two sine terms, none of these with a phase angle.*

2.36. A fluid clutch, similar to that of Fig. 1.23, is considered here, with a Geneva wheel—see Fig. 4.13—mounted between the motor M and the driving disk rotating at an angular velocity ω. Moreover, the load is driven by the above-mentioned disk through a viscous fluid that transmits a moment proportional to the difference in angular velocities of the two disks, the constant of proportionality being the viscous friction coefficient c. The system can be modeled as shown below:

$$\dot{\omega}_R + \frac{1}{\tau}\omega_R = \frac{1}{\tau}\omega$$

where $\omega \equiv \dot{\phi}$, $\omega_R \equiv \dot{\theta}$, and τ is the time constant of the system, defined as $\tau \equiv J/c$. Moreover, the profile of the angular displacement ϕ is approximated as indicated in Fig. 2.84.

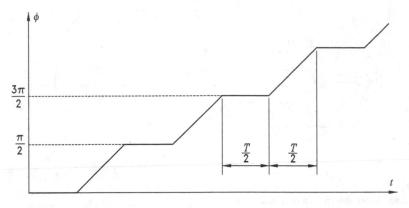

Fig. 2.84 Displacement produced by a Geneva wheel

Find an expression for the time response $\omega_R(t)$ after a long-enough time has elapsed so as to allow for the decay of all transients, i.e., at steady-state.

2.37. A fluid clutch connects a load, e.g., the whole inertia of an automobile, to an engine, as shown in Fig. 1.23. Assume that the torque transmitted through the clutch, when engaged, is proportional to the difference in speed of the input and output shafts (proportionality constant c). Assume also that the speed $\omega(t)$ of the input shaft is unaffected by the load. If the load is a pure inertia J, find the steady-state angular velocity $\omega_R = \dot{\theta}$ of the load, if the input angular velocity $\omega(t)$ is given by[17]

$$\omega(t) = \frac{c}{J} \left| \sin \frac{2\pi b t}{J} \right|$$

2.38. Shown in Fig. 2.85 is a rotor of moment of inertia J that turns at a rate $\dot{\theta}$, as it is driven by a shaft via a universal joint. If the input axis of the universal joint is driven, in turn, at a constant angular velocity ω_0, the output axis delivers a periodic angular velocity σ, which is transmitted to the left end of an elastic shaft of stiffness k. That is, if the constant angle between input and output axes is labelled α, and the angular displacement of the input shaft is denoted by ψ, then,

$$\dot{\phi} = \sigma = \frac{\cos \alpha}{1 - \sin^2 \alpha \sin^2 \psi} \omega_0, \quad \omega_0 \equiv \dot{\psi}$$

Moreover, the relation between the output angle ϕ and the input angle ψ is given by

$$\tan \phi = -\frac{\cot \psi}{\cos \alpha}$$

---

[17]This exercise is drawn from a similar one in [3].

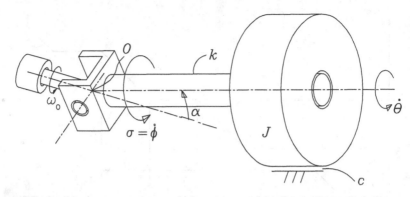

Fig. 2.85 Rotor driven by a universal joint

Using *Parseval's Formula*, we can estimate the error in the approximation of the function $\sigma(t)$ by comparing the rms value of $\sigma(t)$, $\tilde{\sigma}$, with the sum of the squares of the coefficients of the series, and hence, the square of the error in the approximation takes the form

$$e^2 = a_0^2 + \frac{1}{2}\sum_1^{N_h} a_k^2 + \frac{1}{2}\sum_1^{N_h} b_k^2 - \tilde{\sigma}^2$$

where $\tilde{\sigma}$ is calculated as

$$\tilde{\sigma} = \sqrt{\frac{1}{T}\int_0^T \sigma^2(t)dt}$$

It is highly recommended that you calculate this integral using numerical quadrature, as implemented in a Matlab routine (quad or quad8), or that you use another reliable commercial routine *with truncation-error control*, which is indicated via a user-prescribed tolerance. Assign your tolerance judiciously: if you choose a very loose tolerance, your computed integral will contain an inadmissible high error; if too tight, the procedure may take too long to finish. As well, it is recommended that you calculate the Fourier coefficients using the computational scheme given in Sect. 2.8.2.

Produce a table of error e vs. N_h, for $N_h = 1, 2, 3, 4, 5, 10$, and 20. Comment on your tabulated results.

Now, find the steady-state response $\dot{\theta}(t)$ for N_h harmonics of the Fourier expansion of $\sigma(t)$, using a suitable value of N_h, that you should choose based on the table that you produced above. Plot time response vs. time in an interval $[0, 4T]$. For this part of the problem, use the model

$$\ddot{\theta} + 2\zeta\omega_n\dot{\theta} + \omega_n^2\theta = \omega_n^2\phi$$

For the calculation, use the numerical values

$$\alpha = 30°, \quad \zeta = 0.1, \quad \omega_n = 10p, \quad p = 0.3\,\mathrm{s}^{-1}$$

Fig. 2.86 Follower motion program of quick-return cam mechanism

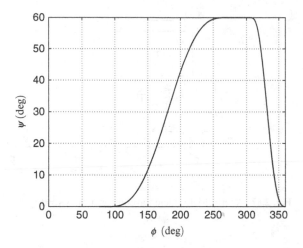

2.39. The oscillating follower of a quick-return cam mechanism goes through a *lower dwell* in the first 25% of its cycle; then, it performs a *working stroke* of 60° of amplitude for 50% of its cycle; it goes further through an *upper dwell* for 10% of its cycle, and returns to its lower dwell in the remaining 15% of its cycle. One period of the periodic angular displacement of the follower be represented by $\phi = \phi(\psi)$, where ψ is the angular displacement of the cam, which rotates at a constant rate $\dot{\psi} = \omega = $ const, the *motion program* $\phi(\psi)$ being given below. Moreover, the follower drives an elastic shaft connected to a load of moment of inertia J mounted on bearings that provide a linearly viscous dissipative torque of coefficient c, the displacement of the load being denoted by θ. The follower motion program, displayed in Fig. 2.86, is given by

$$\phi(\psi) = \begin{cases} 0, & \text{for } 0 \leq \psi \leq \dfrac{\pi}{2}; \\ \dfrac{\pi}{3}\left[\dfrac{\psi}{\pi} - \dfrac{1}{2} - \dfrac{1}{2\pi}\sin(2\psi - \pi)\right], & \text{for } \dfrac{\pi}{2} \leq \psi \leq \dfrac{3\pi}{2}; \\ \dfrac{\pi}{3}, & \text{for } \dfrac{3\pi}{2} \leq \psi \leq \dfrac{17\pi}{10}; \\ \dfrac{\pi}{3}\left[\dfrac{20}{3} - \dfrac{10\psi}{3\pi} - \dfrac{1}{2\pi}\sin\left(\dfrac{40\pi - 20\psi}{3}\right)\right], & \text{for } \dfrac{17\pi}{10} \leq \psi \leq 2\pi \end{cases}$$

while the mathematical model of the follower-load system takes the form

$$\ddot{\theta} + 2\zeta\omega_n\dot{\theta} + \omega_n^2\theta = \omega_n^2\phi$$

and the torque experienced by the shaft, $\tau(t)$, is

$$\tau(t) = k[\theta(t) - \phi(t)]$$

The numerical values of the variables involved are given as

$$\zeta = 0.1, \quad \omega_n = 10\omega, \quad \omega = 300\,\text{rpm}$$

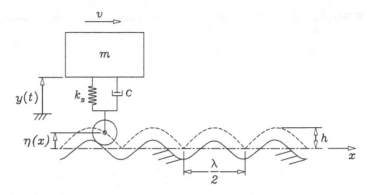

Fig. 2.87 A terrestrial vehicle traveling on a wavy road

(a) Using *Parseval's Formula*, estimate the error in the Fourier approximation of the function $\phi(t)$ by comparing the rms value $\tilde{\phi}$ of $\phi(t)$ with the sum of the squares of the coefficients of the truncated Fourier series. The difference of the two quantities produces the square of the error, e^2. Now, produce a table of error e vs. N_h, for $N_h = 1, 2, 3, 4, 5, 10$, and 20. Comment on your tabulated results.

(b) Then, find the steady-state response $\theta(t)$ for N_h harmonics of the Fourier expansion of $\phi(t)$, using a suitable value of N_h, which you should choose based on the table that you produced above. Plot the steady-state response vs. time in the interval $[0, 4T]$, where T is the duration of one cycle.

(c) Now plot the steady-state value of the torque experienced by the elastic shaft in dimensionless form, i.e., plot $\tau(t)/(J\omega_n^2)$ vs. time, in the time-interval $[0, 4T]$.

2.40. A terrestrial vehicle traveling at a constant velocity v on a wavy road with a profile that can be approximated fairly well by $h|\sin(2\pi x/\lambda)|$ is shown in Fig. 2.87, its mathematical model taking the form,

$$\ddot{y} + 2\zeta\omega_n\dot{y} + \omega_n^2 y = 2\zeta\omega_n\dot{\eta}(t) + \omega_n^2\eta(t)$$

where $\omega_n \equiv \sqrt{k_s/m}$, $\zeta \equiv c/(2m\omega_n)$, and $0 < \zeta < 1$, while

$$\eta = h|\sin(2\pi x/\lambda)|.$$

(a) With $\omega_0 \equiv 2\pi v/\lambda$, express the right-hand side of the mathematical model in the form

$$2\zeta\omega_n\dot{\eta}(t) + \omega_n^2\eta(t) = \sum_{k=0}^{\infty} C_k\cos(k\omega_0 t + \psi_k)$$

and give expressions for C_k and ψ_k.

(b) Using only the first two harmonics of the Fourier expansion of the right-hand side of the mathematical model, find the **steady-state** response of the system.

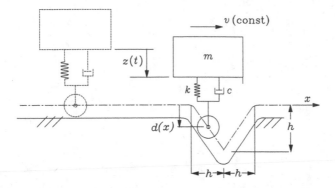

Fig. 2.88 The model of an automobile suspension encountering a triangular ditch

2.41. The model of an automobile suspension is depicted in Fig. 2.88. Find the time response of the model as the vehicle approaches a ditch that is modeled by a triangular pulse. Assume that the traveling speed v of the vehicle is constant, and that this speed is not affected by the ditch. Use the relations

$$\zeta = \frac{\sqrt{2}}{2}, \quad \omega_n \equiv \sqrt{\frac{k}{m}} = \frac{v}{h}.$$

References

1. Moler CB, Van Loan C (1978) Nineteen dubious ways to compute the exponential of a matrix. SIAM Rev 20(4):801–836
2. Chicurel-Uziel E (2007) Dirac delta representation by parametric equations. Applications to impulsive vibration systems. J Sound Vibr 305:134–150
3. Cannon RH (1967) Dynamics of physical systems. McGraw-Hill Book Co., New York
4. Kahaner D, Moler C, Nash S (1989) Numerical methods and software. Prentice Hall, Inc., Englewood Cliffs, NJ
5. Moler C (2004) Numerical computing with MATLAB, Electronic edition. The MathWorks, Inc., Nantick
6. Strang G (1986) Introduction to applied mathematics. Wessley-Cambridge Press, Wessley, pp 274–276

Chapter 3
Simulation of Single-dof Systems

Solving a problem no longer means writing down an infinite series or finding a formula like Cramer's rule, but constructing an effective algorithm.

Strang, G., 1988, *Linear Algebra*, Third Edition, Harcourt Brace Jovanovich College Publishers, Fort Worth, TX.

3.1 Preamble

Simulation consists in producing the time response of a dynamical system to a certain input and certain initial conditions with the aid of a model of the system. In some instances, like in flight simulators, the model is a piece of hardware that is wired so as to behave like the actual system. Flight simulators have attained such a degree of development that, nowadays, commercial pilots can obtain certification without ever having flown a real aircraft, but rather, by accumulating a certain number of hours at the cockpit of a flight simulator. The simulation we will study here takes place not on a hardware model, but rather on a piece of software. Thus, the simulation of interest is based on the mathematical model of the system at hand. However, as we shall see, the time-response analysis conducted in Chap. 2 is not suitable for numerical simulation, some further work being needed, as explained in this chapter.

The advantages of simulation are obvious: simulations spare us dealing with the actual system, which implies that we don't need expensive investments to conduct tests, with the inherent risk of damaging the equipment or even the operator using it. Below we give an introduction to the main items behind simulation, while focusing on single-degree-of-freedom systems. Simulation of multi-dof systems is studied in Chap. 7

J. Angeles, *Dynamic Response of Linear Mechanical Systems: Modeling, Analysis and Simulation*, Mechanical Engineering Series, DOI 10.1007/978-1-4419-1027-1_3, © Springer Science+Business Media, LLC 2011

3.2   The Zero-Order Hold (ZOH)

The time response of first- and second-order mechanical systems obtained in
Chap. 2, although general enough, is not suitable for computer implementation.
Indeed, those expressions involve a reevaluation of both the zero-input and the
zero-state responses at every instant. Obviously, a straightforward evaluation of
these responses is not advisable, and hence, alternatives that are more suitable
for numerical implementation should be explored. Below we derive relations for
the digital simulation of the systems under study, based on the *zero-order hold*
(ZOH) [1]. The ZOH allows us to convert linear dynamical systems of the type at
hand, termed *continuous-time* systems, into a *discrete* form, suitable for numerical
simulation. The discrete model thus obtained is termed, correspondingly, a *discrete-
time* linear dynamical system. Due to the discrete form of this system, it is also
called *digital*, as opposed to its original counterpart, which is also termed *analog*.
The zero-order hold can thus be regarded as an analog-to-digital (A/D) converter.

The zero-order hold is a device that transforms a given input $f(t)$ into a *sampled*
signal $\bar{f}(t)$. This is done by defining $\bar{f}(t)$ in the form

$$\bar{f}(t) \equiv f(t_k), \quad \text{for} \quad t_k \le t < t_{k+1} \tag{3.1}$$

and hence, $\bar{f}(t)$ is constant in the interval $t_k \le t < t_{k+1}$, which is called the
sampling interval. The A/D conversion of the original analog function into its digital
counterpart is illustrated in Fig. 3.1. The ZOH box of Fig. 3.1, thus, represents a
device that responds with a constant value f_k of the input signal, when a circuit

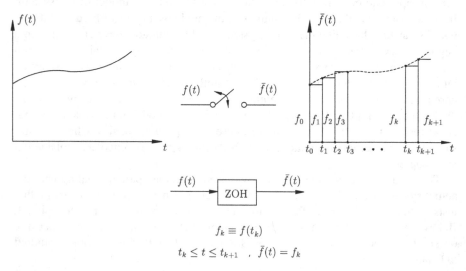

Fig. 3.1 A/D conversion by means of the zero-order hold

is closed instantaneously. The output of the ZOH is kept at f_k until the next sampling instant t_{k+1}, at which time the circuit is closed again, and the output is kept henceforth at f_{k+1}, until the next sampling instant.

Now we apply this concept to obtain the digital version of the governing equations of first- and second-order systems.

3.3 First-Order Systems

The system at hand takes the form

$$\dot{x} = -ax + f(t), \quad x(0) = x_0, \quad t \geq 0 \tag{3.2}$$

The time response of this system was found in Chap. 2 to be

$$x(t) = e^{-at}x_0 + \int_0^t e^{-a(t-\tau)} f(\tau) d\tau \tag{3.3}$$

Now, for simulation we cannot evaluate the foregoing response *continuously* in time, for simulation is intended to be conducted on a computer; therefore, the simulation interval $0 \leq t \leq t_N$ should be divided into N subintervals defining a set of sampling instants $\{t_k\}_0^N$. Moreover, we would like to be able to calculate the response at $t = t_{k+1}$ based on that at $t = t_k$. To this end, we assume that we have evaluated the foregoing response up until an instant $t = t_k$, and let $x_k \equiv x(t_k)$ be the initial condition of the system at hand for the new interval. Thus, the response at $t = t_{k+1}$, x_{k+1}, is given by

$$x_{k+1} = e^{-a(t_{k+1}-t_k)} x_k + \int_{t_k}^{t_{k+1}} e^{-a(t_{k+1}-\tau)} f(\tau) d\tau$$

In order to evaluate the above integral, we resort now to the ZOH and replace $f(t)$ by its staircase approximation $\overline{f}(t)$. Moreover, we let

$$f_k \equiv f(t_k)$$

The expression for x_{k+1} obtained above reduces now to

$$x_{k+1} = e^{-a(t_{k+1}-t_k)} x_k + e^{-a(t_{k+1})} f_k \int_{t_k}^{t_{k+1}} e^{a\tau} d\tau$$

Furthermore, we assume that all subintervals are of equal length, i.e., $\Delta t_k \equiv t_{k+1} - t_k = h$, for all k, and so, the foregoing expression reduces to

$$x_{k+1} = e^{-ah} x_k + \frac{1}{a}(1 - e^{-ah}) f_k$$

Hence, the response at time t_{k+1} is calculated simply as a linear combination of the response and the forcing term or input at the same instant t_k. This response, then, takes the form

$$x_{k+1} = Fx_k + Gf_k \qquad (3.4a)$$

with the constant coefficients F and G defined as

$$F = e^{-ah}, \quad G \equiv \frac{1}{a}(1 - e^{-ah}) \qquad (3.4b)$$

The system appearing in Eq. 3.4a is a *discrete-time* system. Notice that this is a dynamical system itself that evolves at discrete intervals of time, such as a savings account, that gains interest at finite intervals of, say, one month. In a savings account, x_k is the monthly balance, F is $1 + r$, where r is the monthly interest in decimal form, G is unity, and f_k is the amount deposited or withdrawn during the same period. The evolution of the savings account is a dynamical system because its state, i.e., its balance at any time, is not just a function of the amount deposited or withdrawn in that month, but rather a *functional*[1] of the total history of deposits and withdrawals. As well, the savings account is termed *causal*, for its balance does not depend on future deposits or withdrawals, but only on past ones. *Causality* is a charateristic of dynamical systems.

Note that Eq. 3.4a gives the response at instant t_{k+1} based on that at instant t_k. Sometimes one may be interested in the response at a particular time and not in the whole past history. In this case an expression for x_n in terms of the initial condition x_0 and the history of the input, $\{ f_k \}_0^n$ is needed. This expression is derived below by first evaluating x_k for successive values of k, namely,

$$x_1 = Fx_0 + Gf_0$$
$$x_2 = Fx_1 + Gf_1 = F(Fx_0 + Gf_0) + Gf_1$$
$$\quad = F^2x_0 + FGf_0 + Gf_1$$
$$x_3 = Fx_2 + Gf_2 = F(F^2x_0 + FGf_0 + Gf_1) + Gf_2$$
$$\quad = F^3x_0 + F^2Gf_0 + FGf_1 + Gf_2$$

and so on. Hence, the general expression for x_n is

$$x_n = F^n x_0 + \sum_{k=0}^{n-1} F^k G f_{n-1-k} \qquad (3.5)$$

---

[1] The zero-state part of the total response of the system, i.e., the second term of the RHS—right-hand side—of Eq. 2.86 and the third term of Eq. 2.88, make it apparent that the response of the system at time $t > 0$ is determined by the whole past history of the input, $f(t - \tau)$, for $0 \leq \tau \leq t$. This is what is called a *functional*.

which is the *discrete-time response* of the first-order system, the counterpart of the continuous-time response of Eq. 3.3. More specifically, the discrete-time counterpart of the exponential e^{-at} is the nth power F^n, i.e., e^{-nah}, while the summation above is the discrete-time counterpart of the continuous-time convolution. Notice that the sum of the exponent of F in the summation appearing in Eq. 3.5 and the subscript of the input is exactly $n-1$, the instant previous to that at which the response is calculated. This relation parallels that of the convolution, in which the instant at which the impulse response and that at which the input appear in the integrand add up to the current instant t.

Example 3.3.1 (Clutch Tests). The clutch of Example 1.6.11 is to be tested by driving it with a velocity of the motor that varies harmonically with a frequency ω and an amplitude A. Thus, the motor delivers an input angular velocity to the clutch disk of the form $A\cos\omega t$. Moreover, the clutch disk is assumed to be turning at a constant 1,500 rpm when it is engaged by the motor turning with an amplitude of 300 rpm at a frequency of 1 Hz. Furthermore, the time constant of the system is 2 s. In particular, we want to know the velocity of the rotor at time $t = 10\,s$. What would be a suitable sampling interval h?

Solution: In our case the system is governed by the continuous-time model

$$\dot{p} = -\frac{1}{\tau}p + \frac{1}{\tau}A\cos\omega t, \quad p(0) = p_0, \quad t \geq 0$$

where p denotes the angular velocity of the rotor, to distinguish it from the frequency of oscillations of the angular velocity of the motor, ω. The discrete-time response of the above system takes the form

$$p_{k+1} = Fp_k + Gf_k, \quad p(0) = p_0$$

whose coefficients are calculated as

$$F = e^{-h/\tau}, \quad G = \tau(1 - e^{-h/\tau})$$

from which it is apparent that we need a *sufficiently small* h/τ ratio. We can choose, for example, $h/\tau = 0.01$, which thus yields $h = 0.02\,s$, and hence,

$$F = 0.990, \quad G = 0.0199\,s$$

while

$$f_k = \frac{1}{\tau}A\cos\omega t_k$$

where A is the given amplitude and ω is the frequency at which the angular velocity of the motor is varying harmonically, i.e., $A = 300$ rpm $\equiv 31.416\,s^{-1}$ and $\omega = 1$ Hz $\equiv 2\pi\,s^{-1}$. Hence,

$$f_k = (31.416/2)\cos(2\pi t_k) = 15.708\cos(6.283kh) = 15.708\cos(0.126k)$$

We thus have

$$p_{k+1} = 0.990p_k + 0.0199\,[15.708\cos(0.126k)] = 0.990p_k + 0.313\cos(0.126k)$$

Now, what we need in order to find p at $t = 10\,\text{s}$ is the *solution* of the *difference equation* 3.4a, i.e., Eq. 3.5. We thus have to determine n first, which is done from the relation $nh = 10\,\text{s}$, i.e., $n = 500$. Upon substitution of the foregoing numerical values into the time response given in the form of Eq. 3.5, we have

$$p_{500} = 0.990^{500}(50\pi) + \sum_{k=0}^{499} 0.990^k (0.313)\cos[0.126(499-k)]$$

$$= 0.990^{500}(50\pi) + 0.313 \sum_{k=0}^{499} 0.9900^k \cos[0.126(499-k)] \qquad (3.6)$$

where 50π is the initial angular velocity of the rotor in radians per second.[2] Since the numerical value of F, 0.9900, is smaller than unity, its integer powers are also smaller than unity. As a matter of fact, the numerical values of these decrease monotonically with the exponent, as this increases, to the point that, for large enough exponents, this numerical value virtually vanishes. In our case,

$$0.990^{500} = 0.00674$$

and hence, the zero-input response at $t = 10\,\text{s}$ has diminished to less than 1% of its initial value. This part of the response thus belongs to the transient response of the system. The value of the first term in Eq. 3.6 is

$$0.990^{500}(50\pi) = 1.058$$

The second term, using 16 digits in all calculations, is

$$\sum_{k=0}^{499} 0.990^k \cos[0.126(499-k)] = 0.0408$$

and hence, the required value is

$$p(10) \equiv p_{500} = 1.099\,\text{s}^{-1}$$

---

[2] The above calculations have been conducted with 16 digits, but only three significant digits are displayed.

3.4 Second-Order Systems

Here, we will study, as usual, undamped, underdamped, critically-damped and overdamped systems.

3.4.1 Undamped Systems

The model at hand is the usual

$$\ddot{x} + \omega_n^2 x = f(t), \quad x(0) = x_0, \quad \dot{x}(0) = v_0, \quad t \geq 0 \tag{3.7}$$

its time response having been obtained in Chap. 2 as

$$x(t) = (\cos \omega_n t)x_0 + \frac{1}{\omega_n}(\sin \omega_n t)v_0 + \int_0^t \frac{1}{\omega_n}[\sin \omega_n(t - \tau)]f(\tau)d\tau \tag{3.8}$$

What we need now is an expression for $x_{k+1} \equiv x(t_{k+1})$ in terms of $x_k \equiv x(t_k)$, $\dot{x}(t_k) \equiv v_k$ and $f_k \equiv f(t_k)$. Thus, we regard instant t_k as the initial instant and compute x_{k+1} from instant t_k, i.e.,

$$x_{k+1} = \cos(\omega_n h)x_k + \frac{1}{\omega_n}\sin(\omega_n h)v_k + \frac{1}{\omega_n}\int_{t_k}^{t_k+h}\sin[\omega_n(t_k + h - \tau)]f(\tau)d\tau \tag{3.9}$$

where we have assumed that $t_{k+1} - t_k = h$, for all k.

To calculate the foregoing integral, we let $u \equiv t_k + h - \tau$, which yields

$$\int_{t_k}^{t_k+h}\sin[\omega_n(t_k + h - \tau)]d\tau = -\int_h^0 \sin \omega_n u\, du \tag{3.10}$$

and hence,

$$\int_{t_k}^{t_k+h}\sin[\omega_n(t_k + h - \tau)]d\tau = \frac{1}{\omega_n}(1 - \cos \omega_n h) \tag{3.11}$$

thereby showing that the said integral is a constant as long as the sampling takes place at equal intervals of length h. Thus, x_{k+1} takes the form

$$x_{k+1} = (\cos \omega_n h)x_k + \frac{1}{\omega_n}(\sin \omega_n h)v_k + \frac{1}{\omega_n^2}(1 - \cos \omega_n h)f_k \tag{3.12}$$

However, Eq. 3.12 requires the updating of v_k at every sampling instant, and hence, an expression for v_{k+1} in terms of variables evaluated at instant t_k is needed.

This expression is obtained from the time derivative of $x(t)$ as given by Eq. 3.8, namely,

$$\dot{x}(t) \equiv v(t) = -\omega_n(\sin \omega_n t)x_0 + (\cos \omega_n t)v_0 + \int_0^t [\cos \omega_n(t-\tau)]f(\tau)d\tau \quad (3.13)$$

To obtain the desired expression, we have to evaluate, between t_k and t_{k+1}, the integral appearing in Eq. 3.13. The foregoing value is termed here L, i.e.,

$$L \equiv \int_{t_k}^{t_k+h} \cos[\omega_n(t_k+h-\tau)]f_k\,d\tau \quad (3.14)$$

which is calculated by resorting to a substitution like that used in Eq. 3.10, the integral thus reducing to

$$L = \frac{1}{\omega_n}(\sin \omega_n h)f_k \quad (3.15)$$

Then, the final expression for v_{k+1} takes the form

$$v_{k+1} = -(\omega_n \sin \omega_n h)x_k + (\cos \omega_n h)v_k + \frac{1}{\omega_n}(\sin \omega_n h)f_k \quad (3.16)$$

Thus, whereas Eq. 3.12 allows the updating of the displacement, the velocity is updated using Eq. 3.16.

Moreover, we can show that the simulation of undamped second-order systems is formally identical to that of first-order systems, which can be made apparent by writing Eqs. 3.12 and 3.16 in vector form. If we let $\mathbf{z}_k \equiv [x_k, v_k]^T$ denote the vector of state variables at $t = t_k$, then

$$\mathbf{z}_{k+1} = \mathbf{F}\mathbf{z}_k + \mathbf{g}f_k \quad (3.17a)$$

with $\mathbf{F}$ and $\mathbf{g}$ defined now as a 2×2 matrix and a two-dimensional vector, namely,

$$\mathbf{F} \equiv \begin{bmatrix} \cos \omega_n h & (\sin \omega_n h)/\omega_n \\ -\omega_n \sin \omega_n h & \cos \omega_n h \end{bmatrix}, \quad \mathbf{g} \equiv \begin{bmatrix} (1-\cos \omega_n h)/\omega_n^2 \\ (\sin \omega_n h)/\omega_n \end{bmatrix} \quad (3.17b)$$

thereby showing that, indeed, the simulation scheme for undamped second-order systems is formally identical to that for first-order systems. The obvious difference is, of course, that the scheme developed above appears in terms of two-dimensional vectors. Hence, the equivalent expression for $\mathbf{z}_n$ in terms of $\mathbf{z}_0$ is readily derived as

$$\mathbf{z}_N = \mathbf{F}^N\mathbf{z}_0 + \sum_{k=0}^{N-1} \mathbf{F}^k\mathbf{g}f_{N-1-k} \quad (3.18)$$

Here, again, notice that the discrete counterpart of the matrix exponential is the $\mathbf{F}$ matrix raised to the nth power and that the summation is the discrete counterpart of the continuous-time convolution. Moreover, notice that the calculation of the nth power of $\mathbf{F}$, for values of $n > 2$, when powers lower that the nth are not needed, can be computed with the aid of the Cayley-Hamilton Theorem, introduced in Appendix A

So far we have not considered the effect of roundoff errors that are unavoidable when computing with floating-point arithmetic. If we look closely at matrix $\mathbf{F}$ as given by Eq. 3.17b, we will note that its diagonal entries have their absolute values comprised between zero and unity, while its off-diagonal entries are unbounded, and can take any real value, depending on the value of ω_n. For very large values of ω_n, for example, a correspondingly large roundoff error will be incurred when computing the $(2, 1)$ entry of $\mathbf{F}$. Moreover, this roundoff error will be magnified as the simulation proceeds, for exponent n takes larger and larger values. As a means to alleviate the buildup of roundoff error that can lead to catastrophic results, we rewrite Eqs. 3.8 and 3.13, if in terms of two new variables, $w(t) \equiv \dot{x}(t)/\omega_n$ and $\phi(t) \equiv f(t)/\omega_n^2$, thereby obtaining

$$x(t) = (\cos \omega_n t)x_0 + (\sin \omega_n t)w_0 + \int_0^t [\sin \omega_n(t - \tau)]\,\omega_n \phi(\tau)d\tau \qquad (3.19a)$$

$$\frac{\dot{x}(t)}{\omega_n} \equiv w(t) = -(\sin \omega_n t)x_0 + (\cos \omega_n t)w_0 + \int_0^t [\cos \omega_n(t - \tau)]\,\phi(\tau)d\tau \qquad (3.19b)$$

The corresponding simulation scheme, i.e., the discrete-time response of the undamped second-order system, is thus the discrete-time version of Eq. 3.19a and b, namely,

$$\zeta_{k+1} = \mathbf{H}\zeta_k + \mathbf{h}\phi_k \qquad (3.20a)$$

where

$$\zeta_k = \begin{bmatrix} x_k \\ w_k \end{bmatrix}, \quad \mathbf{H} \equiv \begin{bmatrix} \cos(\omega_n h) & \sin(\omega_n h) \\ -\sin(\omega_n h) & \cos(\omega_n h) \end{bmatrix}, \quad \mathbf{h} \equiv \begin{bmatrix} 1 - \cos(\omega_n h) \\ \sin(\omega_n h) \end{bmatrix} \qquad (3.20b)$$

Notice that the foregoing simulation scheme needs to be complemented with the relation giving explicitly v_k, namely,

$$\dot{x} \equiv v_k = \omega_n w_k \qquad (3.20c)$$

Now, matrix $\mathbf{H}$, and vector $\mathbf{h}$ by the way, are better behaved numerically. Note that, in the foregoing scheme, $\mathbf{H}$ is an orthogonal matrix, i.e.,

$$\mathbf{H}\mathbf{H}^T = \mathbf{1} \qquad (3.21)$$

with $\mathbf{1}$ denoting the 2×2 identity matrix. Moreover, $\det(\mathbf{H}) = +1$, which makes $\mathbf{H}$ *proper orthogonal*.[3] Proper orthogonal matrices represent rotations. An important

---

[3] Orthogonal matrices whose determinant is -1 are termed *improper*; they represent reflections.

property of orthogonal matrices, whether proper or not, is that the product of any two orthogonal matrices remains orthogonal. As a consequence, the powers of any orthogonal matrix are also orthogonal. One more important property of orthogonal matrices is that, upon multiplying a vector, the vector only changes its direction, but its magnitude—or, more precisely, its *Euclidean norm*—remains unchanged.

The simulation scheme of Eq. 3.18 now takes an alternative, numerically robust[4] form:

$$\zeta_N = \mathbf{H}^N \zeta_0 + \sum_{k=0}^{N-1} \mathbf{H}^k \mathbf{h} \phi_{N-1-k} \tag{3.22}$$

As a consequence of the orthogonality of $\mathbf{H}$, the foregoing zero-input response $\zeta_n = \mathbf{H}^n \zeta_0$ is *conservative*, i.e., te total energy of the system at instant t_n is necessarily identical to that at $t_0 = 0$. Indeed, $\mathbf{H}$ represents a rotation in the x-w plane through an angle $\theta = \omega_n h$, and hence, $\mathbf{H}^2$ represents a rotation through an angle 2θ and, in general, $\mathbf{H}^n$ represents a rotation through an angle $n\theta$. That is

$$\mathbf{H}^n = \begin{bmatrix} \cos n\omega_n h & \sin n\omega_n h \\ -\sin n\omega_n h & \cos n\omega_n h \end{bmatrix} \tag{3.23}$$

Given that rotations preserve the vector magnitude,

$$\|\zeta_n\| \equiv \|\mathbf{H}^n \zeta_0\| = \|\zeta_n\|$$

i.e.,

$$x_n^2 + w_n^2 = x_0^2 + w_0^2 \quad \Rightarrow \quad x_n^2 + \frac{\dot{x}_n^2}{\omega_n^2} = x_0^2 + \frac{\dot{x}_0^2}{\omega_n^2}$$

Upon clearing denominators, and dividing the both sides by 2, the foregoing equation leads to

$$\frac{1}{2} m \dot{x}_n^2 + \frac{1}{2} k x_n^2 = \frac{1}{2} m \dot{x}_0^2 + \frac{1}{2} k x_0^2$$

thereby showing that the energy of the system at instant t_n is identical to that at instant t_0.

Example 3.4.1 (Discrete-time Response of an Undamped Suspension). Here we set up the simulation model for the system analyzed in Example 2.7.1. We assume that the suspension under study has a natural frequency of 1 Hz and that the bump has a height of 500 mm and a wavelength λ of 1,000 mm. Moreover, the vehicle is speeding at $v_0 = 100$ km/h when it hits the bump, and we assume that the bump does not affect its uniform horizontal speed. We want to calculate the value of the

---

[4]Robustness means here insensitivity to roundoff errors.

vertical displacement and velocity of the car body at the instant just after the bump has been overcome, as well as the amount of energy transferred to the suspension by the bump. Finally, we want to calculate the value of the vertical displacement and the vertical velocity at $t \approx 5$ s after the vehicle hit the bump.

Solution: The model has the form

$$\ddot{x} + \omega_n x^2 = \omega_n^2 b(t), \quad x(0) = 0, \quad \dot{x}(0) = 0$$

with $b(t)$ defined as the 'bump function', already introduced in Example 2.7.1. We now have to determine the length h of The sampling interval. What we need is a small-enough value of h in order to yield a product $\omega_n h$ small enough so as not to miss the bump. From the given information, the vehicle traverses the bump in a time $\Delta t = \lambda/(2v_0)$, i.e., $\Delta t = 0.018$ s. We will thus choose h so as to be able to sample the bump, say, five times, and hence, $h = \Delta t/4 = 0.0045$ s. Now, the product $\omega_n h$ takes the value $\omega_n h = 0.0283$ rad $= 1.620°$. Moreover, the sampling instants t_k are given by

$$t_k = kh = k\frac{\Delta t}{4} = k\frac{\lambda}{8v_0} = 0.0045k\,\text{s}$$

Hence, the bump is traversed in five sampling instants $t_0, t_1, \ldots, t_4$, the sampled bump, b_k, now taking the form

$$b_k \equiv b(t_k) = \begin{cases} 500\sin(\pi k/4) \text{ mm}, & \text{for} \quad 0 \le k \le 4 \\ 0 & \text{for} \quad k > 4 \end{cases}$$

Now, let $x_k \equiv x(t_k)$, $v_k \equiv \dot{x}(t_k)$ and $w_k \equiv v_k/\omega_n$, with the state-variable vector ζ_k defined as

$$\zeta_k \equiv \begin{bmatrix} x_k \\ w_k \end{bmatrix}$$

Moreover, function ϕ_k of Eq. 3.20a turns out to be, in this case, b_k. The foregoing definitions permit us to write the discrete-time model in the form of Eq. 3.20a, i.e.,

$$\zeta_{k+1} = H\zeta_k + hb_k$$

with H and h given as in Eq. 3.20b, which require the values below:

$$\cos(\omega_n h) = 0.9996, \quad \sin(\omega_n h) = 0.0283$$

Therefore,

$$H = \begin{bmatrix} 0.9996 & 0.0283 \\ -0.0283 & 0.9996 \end{bmatrix}, \quad h = \begin{bmatrix} 0.0004 \\ 0.0283 \end{bmatrix}$$

We thus have, for the given numerical values, $b_0 = 0$, $b_1 = 70.72$, $b_2 = 100$, $b_3 = 70.72$, and $b_4 = 0$, all of them in mm.

The calculation of the response at t_5 is now straightforward:

$$\zeta_1 = \mathbf{H}\zeta_0 + \mathbf{h}b_0 = \mathbf{0}$$

$$\zeta_2 = \mathbf{H}\zeta_1 + \mathbf{h}b_1 = \begin{bmatrix} 0.0283 \\ 2.002 \end{bmatrix}$$

$$\zeta_3 = \mathbf{H}\zeta_2 + \mathbf{h}b_2 = \begin{bmatrix} 0.2149 \\ 1.002 \end{bmatrix}$$

$$\zeta_4 = \mathbf{H}\zeta_3 + \mathbf{h}b_3 = \begin{bmatrix} 28.5 \\ 1004 \end{bmatrix}$$

$$\zeta_5 = \mathbf{H}\zeta_4 + \mathbf{h}b_4 = \begin{bmatrix} 56.88 \\ 1002 \end{bmatrix}$$

where all values are given in mm. Therefore,

$$x_5 = 56.88 \text{ mm}, \quad \dot{x}_5 = \omega_n w_5 = 2\pi \times 1002 = 6298 \text{ mm/s} = 6.298 \text{ m/s} = 22.67 \text{ km/h}$$

the vehicle body thus attaining a vertical velocity compared to the traveling speed, a result of the undamped suspension. Now, the energy E_u transferred to the suspension is readily calculated as

$$E_u = \frac{1}{2}m\dot{x}_5^2 + \frac{1}{2}kx_5^2 = \frac{1}{2}m\left(\dot{x}_5^2 + \omega_n^2 x_5^2\right) = 19.90m \text{ Nm}$$

Further, in order to calculate the state-variable vector at $t \approx 5$ s, we first determine the sampling instant closest to 5 s, namely, $k = \lfloor 5/h \rfloor = 1111$, with the 'floor function' $\lfloor \cdot \rfloor$ defined in Sect. 2.8.2. Therefore, what we need is ζ_{1111}, which is readily obtained by regarding $t = 4h$ as the initial time and ζ_5 as the initial value of ζ, the input to the system thus vanishing. All we need is thus the zero-input response of the system, i.e.,

$$\zeta_{1111} = \mathbf{H}^{1106}\zeta_5$$

for the sampled bump—just as its analog counterpart—vanishes for $k > 4$. The value of $\mathbf{H}^{1106}$ can now be obtained with the aid of the Cayley-Hamilton Theorem, as explained in Appendix A. However, this is not needed, given the special form of matrix $\mathbf{H}$. Since this matrix is orthogonal in the case of undamped systems, the argument of its trigonometric functions represents simply an angle of rotation, say θ. Therefore, the nth power of $\mathbf{H}$ is nothing but a rotation through an angle $n\theta$, and hence,

$$\mathbf{H}^{1106} = \begin{bmatrix} \cos(2\pi \times 1106 \times 0.0045) & \sin(2\pi \times 1106 \times 0.0045) \\ -\sin(2\pi \times 1106 \times 0.0045) & \cos(2\pi \times 1106 \times 0.0045) \end{bmatrix}$$

$$= \begin{bmatrix} 0.9896 & -0.1440 \\ 0.1440 & 0.9896 \end{bmatrix} \tag{3.24}$$

Therefore,

$$\zeta_{1111} = \begin{bmatrix} -88.06 \\ 1000 \end{bmatrix} \text{mm}$$

or

$$x_{1111} = -88.06 \text{ mm}, \quad \dot{x}_{1111} = 2\pi \times 1000 = 6283 \text{ mm/s}$$

As the reader can verify, the energy of the suspension at t_{1111} is identical to that at t_5, to four digits of accuracy, as it should be, for the system is conservative.

3.4.2 Damped Systems

Our model here is the usual

$$\ddot{x} + 2\zeta\omega_n\dot{x} + \omega^2 x = f(t), \quad x(0) = x_0, \quad \dot{x}(0) = v_0 \tag{3.25}$$

We will derive a suitable simulation procedure for this system based on the ZOH introduced in Sect. 3.2. To this end, we now write the above model in state-variable form, as done in Chap. 2, thereby obtaining a system of two first-order linear ODEs, namely,

$$\dot{x} = v \tag{3.26a}$$

$$\dot{v} = -\omega_n^2 x - 2\zeta\omega_n\dot{x} + f(t) \tag{3.26b}$$

with the initial conditions $x(0) = x_0$ and $v(0) = v_0$. We can now write Eq. 3.26a and b in state-variable form, namely,

$$\dot{z} = Az + bf(t), \quad z(0) = z_0 \tag{3.27}$$

where

$$A \equiv \begin{bmatrix} 0 & 1 \\ -\omega_n^2 & -2\zeta\omega_n \end{bmatrix}, \quad b \equiv \begin{bmatrix} 0 \\ 1 \end{bmatrix}, \quad z \equiv \begin{bmatrix} x \\ v \end{bmatrix} \tag{3.28}$$

In the above definitions, A is a 2×2 matrix, while b and z are two-dimensional vectors.

The total response of the above system was obtained in Chap. 2 in the form

$$z(t) = e^{At}z_0 + \int_0^t e^{A(t-\tau)}bf(\tau)\,d\tau \tag{3.29}$$

Similar to the case of undamped systems, Eq. 3.29 is not suitable for simulation, for it requires computing a matrix exponential and a convolution every sampled instant t_k. Thus, a more suitable model for simulation is required, which is derived below with the aid of the ZOH.

We begin by *sampling* the input function $f(t)$ at equal intervals $h \equiv t_{k+1} - t_k$, thereby obtaining its *staircase approximation* $\overline{f}(t)$ defined earlier. Now, assuming that we have computed a value $\mathbf{z}_k \equiv \mathbf{z}(t_k)$, we shall derive an expression for $\mathbf{z}_{k+1} \equiv \mathbf{z}(t_{k+1})$. We do this by invoking the time-invariance of the system at hand, while regarding t_k as the initial time and calculate $\mathbf{z}_{k+1}$ from the total response, Eq. 3.29, if with $\overline{f}(t)$ instead of $f(t)$. Obviously, the shorter the sampling interval h, the closer the former approaches the latter. Thus, we have

$$\mathbf{z}_{k+1} = e^{\mathbf{A}(t_{k+1}-t_k)}\mathbf{z}_k + \int_{t_k}^{t_{k+1}} e^{\mathbf{A}(t_{k+1}-\tau)}\mathbf{b}\overline{f}(\tau)\,d\tau \tag{3.30}$$

where the first exponential readily reduces to a constant, namely,

$$\mathbf{F} \equiv e^{\mathbf{A}(t_{k+1}-t_k)} \equiv e^{\mathbf{A}h} \tag{3.31}$$

The integral is evaluated as we describe below: We let $\theta \equiv t_{k+1} - \tau$ and rewrite that integral in terms of θ. To this end, we need the extremes of the integration interval, which are readily determined by noticing that $\theta = h$ when $\tau = t_k$ and $\theta = 0$ when $\tau = t_{k+1}$. Moreover, $d\tau = -d\theta$, and hence,

$$\int_{t_k}^{t_{k+1}} e^{\mathbf{A}(t_{k+1}-\tau)}\mathbf{b}\overline{f}(\tau)\,d\tau = -\int_h^0 e^{\mathbf{A}\theta}\mathbf{b}f_k\,d\theta \tag{3.32}$$

where $f_k \equiv f(t_k)$, and so, it is a constant. Since vector $\mathbf{b}$ is a constant as well, the product $\mathbf{b}f_k$ can be taken out of the integration sign, thereby obtaining

$$\int_{t_k}^{t_{k+1}} e^{\mathbf{A}(t_{k+1}-\tau)}\mathbf{b}\overline{f}(\tau)\,d\tau = \mathbf{g}f_k \tag{3.33}$$

where $\mathbf{g}$ is defined below:

$$\mathbf{g} \equiv \left(\int_0^h e^{\mathbf{A}\theta}\,d\theta\right)\mathbf{b} \tag{3.34}$$

Thus, all we need in the expressions above is the integral of the exponential of $\mathbf{A}$, from $\theta = 0$ to $\theta = h$. If we recall the integral of the scalar $e^{a\theta}$ in the same interval, we can write the above integral in an explicit form. The said scalar integral is reproduced below for quick reference:

$$\int_0^h e^{a\theta}\,d\theta = \frac{e^{ah} - 1}{a} \tag{3.35}$$

We can now mimic the above expression to obtain the matrix integral of interest, by *properly* replacing the scalar a by matrix $\mathbf{A}$. What we mean by properly replacing a scalar by a matrix refers, obviously, to the care with which we must handle matrix algebra, since a division by a matrix is meaningless and the product of matrices is, in general, not commutative. Obviously, then, the division by the scalar a in the aforementioned expression is to be replaced by the inverse of $\mathbf{A}$, which requires that $\mathbf{A}$ be *invertible*. Additionally, the numerator of the same expression becomes the matrix $e^{\mathbf{A}h} - \mathbf{1}$, where $\mathbf{1}$ is defined here as the 2×2 identity matrix. The question that remains, then, is where to write $\mathbf{A}^{-1}$, to the right or to the left of the foregoing difference. This question can be readily answered if we recall that any square matrix *commutes* with its analytic functions.[5] By the same token, then, any two analytic functions of the same matrix commute, while $\mathbf{A}^{-1}$, the exponential of $\mathbf{A}h$ and the identity matrix are all analytic functions of $\mathbf{A}$. Thus, where we place the inverse of $\mathbf{A}$, whether to the right or to the left of the above difference, becomes immaterial, and hence,

$$\int_0^h e^{\mathbf{A}\theta} d\theta = \mathbf{A}^{-1}(e^{\mathbf{A}h} - \mathbf{1}) \equiv (e^{\mathbf{A}h} - \mathbf{1})\mathbf{A}^{-1} \tag{3.36}$$

Therefore,

$$\mathbf{g} \equiv \mathbf{A}^{-1}(e^{\mathbf{A}h} - \mathbf{1})\mathbf{b} \equiv (e^{\mathbf{A}h} - \mathbf{1})\mathbf{A}^{-1}\mathbf{b} \tag{3.37}$$

Notice that the second expression of Eq. 3.37 is *slightly* more convenient than the first one. Indeed, if we let $\mathbf{c} = \mathbf{A}^{-1}\mathbf{b}$, then

$$\mathbf{g} = (\mathbf{F} - \mathbf{1})\mathbf{c} \tag{3.38a}$$

and $\mathbf{c}$ is computed as the solution of a system of two linear equations in two unknowns, namely,

$$\mathbf{A}\mathbf{c} = \mathbf{b} \tag{3.38b}$$

The simulation scheme of the system at hand takes the form

$$\mathbf{z}_{k+1} = \mathbf{F}\mathbf{z}_k + \mathbf{g}f_k \tag{3.39}$$

with the initial condition $\mathbf{z}_0 = \mathbf{z}(0)$. The simulation algorithm, then, can be summarized as

Algorithm Damped-1dof

1. Calculate the exponential of $\mathbf{A}h$: $\mathbf{F} \leftarrow e^{\mathbf{A}h}$
2. Calculate $\mathbf{c}$ from Eq. 3.38b
3. $\quad \mathbf{g} \leftarrow (\mathbf{F} - \mathbf{1})\mathbf{c}$;
4. Compute $\mathbf{z}_{k+1}$ as indicated in Eq. 3.39

---

[5]Fact 4 of Appendix A.

Note that the constant matrices involved in the foregoing algorithm can be computed by longhand calculations. Indeed, all we do is replace t by h in the exponential formula we found for $\mathbf{A}t$ in Appendix A. For underdamped systems, e.g.,

$$\mathbf{F} = \frac{e^{-\zeta\omega_n h}}{\sqrt{1-\zeta^2}} \begin{bmatrix} \sqrt{1-\zeta^2}\cos\omega_d h + \zeta\sin\omega_d h & \dfrac{1}{\omega_n}\sin\omega_d h \\ -\omega_n\sin\omega_d h & \sqrt{1-\zeta^2}\cos\omega_d h - \zeta\sin\omega_d h \end{bmatrix}$$
$$(3.40)$$

while, for damped and overdamped systems, Eqs. A.31 and A.32 apply. Moreover, still for undamped systems,

$$\mathbf{F} - \mathbf{1} = C \begin{bmatrix} R\cos\omega_d h + \zeta\sin\omega_d h - \dfrac{1}{C} & \dfrac{1}{\omega_n}\sin\omega_d h \\ -\omega_n\sin\omega_d h & R\cos\omega_d h - \zeta\sin\omega_d h - \dfrac{1}{C} \end{bmatrix} \quad (3.41a)$$

with C and R defined as

$$C = \frac{e^{-\zeta\omega_n h}}{R}, \quad R = \sqrt{1-\zeta^2} \qquad (3.41b)$$

Furthermore, the calculation of $\mathbf{c}$ requires solving a system of two linear equations in two unknowns. In the case at hand, however, this solution is not necessary, as the inverse of $\mathbf{A}$ can be readily calculated *symbolically*, i.e.,[6]

$$\mathbf{A}^{-1} = \frac{1}{\omega_n^2} \begin{bmatrix} -2\zeta\omega_n & -1 \\ \omega_n^2 & 0 \end{bmatrix} \qquad (3.42)$$

and hence, $\mathbf{g}$ becomes, for underdamped systems,

$$\mathbf{g} = \frac{1}{\omega_n^2} \begin{bmatrix} 1 - e^{-\zeta\omega_n h}\left(\dfrac{\zeta}{\sqrt{1-\zeta^2}}\sin\omega_d h + \cos\omega_d h\right) \\ \dfrac{\omega_n e^{-\zeta\omega_n h}}{\sqrt{1-\zeta^2}}\sin\omega_d h \end{bmatrix} \qquad (3.43)$$

thereby completing the required calculations.

The system at hand is now modeled by the discrete-time system given below.

$$\mathbf{z}_{k+1} = \mathbf{F}\mathbf{z}_k + \mathbf{g}f_k \qquad (3.44)$$

---

[6]Equation 3.42 is valid for all three kinds of damped systems.

which, in component form becomes, still for underdamped systems,

$$x_{k+1} = \frac{e^{-\zeta\omega_n h}}{\sqrt{1-\zeta^2}} \left[\left(\sqrt{1-\zeta^2}\cos(\omega_d h) + \zeta\sin(\omega_d h) \right) x_k + \frac{1}{\omega_n}\sin(\omega_d h)v_k \right.$$

$$\left. - \left(\frac{\zeta}{\omega_n^2}\sin(\omega_d h) + \frac{\sqrt{1-\zeta^2}}{\omega_n^2}\cos(\omega_d h) \right) f_k \right] + \frac{1}{\omega_n^2}f_k \quad (3.45a)$$

and

$$v_{k+1} = \frac{e^{-\zeta\omega_n h}}{\sqrt{1-\zeta^2}} \left[-\omega_n\sin(\omega_d h)x_k + \left(\sqrt{1-\zeta^2}\cos\omega_d h - \zeta\sin\omega_d h \right)v_k \right.$$

$$\left. + \frac{1}{\omega_n}\sin(\omega_d h)f_k \right] \quad (3.45b)$$

As in the undamped case, the matrix $\mathbf{F}$ derived above is not dimensionally homogeneous and hence, is prone to numerical instability. We solve this problem as in the undamped case, i.e., by expressing the foregoing simulation scheme in terms of the new variables $w_k \equiv v_k/\omega_n$ and $\phi_k \equiv f_k/\omega_n^2$, and hence, $\mathbf{F}$, $\mathbf{z}_k$ and $\mathbf{g}$ are replaced by $\mathbf{H}$, ζ_k and $\mathbf{h}$, respectively, as given below:

$$\mathbf{H} = C \begin{bmatrix} R\cos\omega_d h + \zeta\sin\omega_d h & \sin\omega_d h \\ -\sin\omega_d h & R\cos\omega_d h - \zeta\sin\omega_d h \end{bmatrix} \quad (3.46a)$$

$$\zeta_k \equiv \begin{bmatrix} x_k \\ w_k \end{bmatrix}, \quad \mathbf{h} = \begin{bmatrix} 1 - e^{-\zeta\omega_n h}\left(\frac{\zeta}{R}\sin\omega_d h + \cos\omega_d h \right) \\ C\sin\omega_d h \end{bmatrix} \quad (3.46b)$$

with C and R defined in Eq. 3.41b. The simulation scheme that provides the discrete-time response of the second-order damped system now takes the form

$$\zeta_{k+1} = \mathbf{H}\zeta_k + \mathbf{h}\phi_k \quad (3.47a)$$

which, again, has to be complemented with

$$v_k = \omega_n w_k \quad (3.47b)$$

Similar to the discrete-time response of the undamped system, Eq. 3.18, that of the damped system takes the form

$$\mathbf{z}_N = \mathbf{F}^N \mathbf{z}_0 + \sum_{k=0}^{N-1} \mathbf{F}^k \mathbf{g} f_{N-1-k} \quad (3.48)$$

Example 3.4.2 (Discrete-time Response of a Damped Suspension). The discrete-time response of the suspension analyzed in the previous example, if with 'shocks' added to it, is required here. To this end, assume that the natural frequency of the system is still 1 Hz and that the damping ratio is now 0.7071, while the bump has the same features as above. Again, assume that the vehicle is speeding at 100 km/h and that the bump does not affect its uniform horizontal speed. Calculate: (a) the vertical displacement and velocity of the vehicle body at the instant just after it has overcome the bump; (b) the energy transferred by the bump to the suspension; (c) the energy dissipated by the shocks; and (d) the vertical displacement and velocity of the vehicle body at a time $t = 10h$, where h is the sampling interval, that is taken as in the undamped case.

Solution: First, we note that the model takes the form

$$\ddot{x} + 2\zeta\omega_n\dot{x} + \omega_n^2 x = f(t)$$

with $f(t)$ defined now as

$$f(t) \equiv 2\zeta\omega_n\dot{b}(t) + \omega_n^2 b(t)$$

$b(t)$ denoting the bump function defined in Example 3.4.1, and $\dot{b}(t)$ its time derivative. The ϕ_k function of Eq. 3.47a thus takes the form

$$\phi_k \equiv \frac{2\zeta}{\omega_n}\dot{b}_k + b_k$$

where b_k denotes, as in Example 3.4.1, $b(t_k)$, while $\dot{b}_k$ denotes, correspondingly, $\dot{b}(t_k)$, and hence,

$$\phi_k \equiv \begin{cases} 3928\cos(0.7854k) + 100\sin(0.7854k) \text{ mm}, & \text{for} \quad 0 \le k \le 4 \\ 0, & \text{for} \quad k > 4 \end{cases}$$

That is,

$$\phi_0 = 3928, \quad \phi_1 = 2848, \quad \phi_2 = 100, \quad \phi_3 = -270.8, \quad \phi_4 = -392.8$$

with all foregoing values in mm, all other values of ϕ_k being zero. Furthermore, we determine the damped frequency of the system at hand, namely,

$$\omega_d = \sqrt{1 - \zeta^2}\omega_n = 0.7071(2\pi) = 4.4429\,\text{s}^{-1}$$

and hence,

$$\omega_d h = 0.0200\,\text{rad} = 1.1455°$$

Moreover,

$$\cos \omega_d h = 0.9998, \quad \sin \omega_d h = 0.0200, \quad e^{-\zeta \omega_n h} = 0.9802$$

Hence, for $\mathbf{H}$, $\mathbf{h}$ and ζ_k given as in Eq. 3.50a and b),

$$\mathbf{H} = \begin{bmatrix} 0.9996 & 0.0277 \\ -0.0277 & 0.9604 \end{bmatrix}, \quad \mathbf{h} = \begin{bmatrix} 0.0004 \\ 0.0277 \end{bmatrix}$$

What we need now is ζ_5, as in Example 3.4.1, which is calculated below:

$$\zeta_1 = \mathbf{H}\zeta_0 + \mathbf{h}\phi_0 = \begin{bmatrix} 1.549 \\ 112.5 \end{bmatrix} \text{ mm}$$

$$\zeta_2 = \mathbf{H}\zeta_1 + \mathbf{h}\phi_1 = \begin{bmatrix} 5.690 \\ 183.5 \end{bmatrix} \text{ mm}$$

$$\zeta_3 = \mathbf{H}\zeta_2 + \mathbf{h}\phi_2 = \begin{bmatrix} 10.81 \\ 178.8 \end{bmatrix} \text{ mm}$$

$$\zeta_4 = \mathbf{H}\zeta_3 + \mathbf{h}\phi_3 = \begin{bmatrix} 14.69 \\ 96.40 \end{bmatrix} \text{ mm}$$

$$\zeta_5 = \mathbf{H}\zeta_4 + \mathbf{h}\phi_4 = \begin{bmatrix} 15.81 \\ -16.68 \end{bmatrix} \text{ mm}$$

and hence,

$$x_5 = 15.81 \text{ mm}, \quad \dot{x}_5 = -2\pi \times 16.68 = -104.8 \text{ mm/s}$$

Therefore, the energy E_d transmitted to the suspension upon crossing the bump is given by

$$E_d = \frac{1}{2}m\dot{x}_5^2 + \frac{1}{2}kx_5^2 = \frac{1}{2}m\left(\dot{x}_5^2 + \omega_n^2 x_5^2\right) = 0.0104m \text{ Nm}$$

Obviously, the energy dissipated by the shocks is simply the difference between the energy transferred to the undamped suspension minus the above figure, i.e.,

$$\Delta E = E_u - E_d = 19.89m \text{ Nm}$$

That is, the shocks absorbed 99.95% of the energy E_u transferred by the bump. In order to gain more insight into the role of the shocks, let us assume a rather light vehicle, with a mass of 500 kg, which gives a ΔE of 9,940 Nm, or 2.761 watt h, which is the amount of energy consumed by a 60-watt light bulb when left lit for 2.761 min.

Now we calculate ζ_{10}. We do this simply as in Example 3.4.1, namely,

$$\zeta_{10} = \mathbf{H}^5 \mathbf{z}_5$$

with $\mathbf{H}^5$ obtained using the procedure introduced in Appendix A. Note that, in this case, we cannot do the shortcut that we did in Example 3.4.1, for matrix $\mathbf{H}$ is no longer orthogonal. Thus,

$$\mathbf{H}^5 = f_0 \mathbf{1} + f_1 \mathbf{H}$$

with coefficients f_0 and f_1 calculated from the formulas derived in Appendix A, namely,

$$f_0 = \frac{\lambda_2^5 \lambda_1 - \lambda_1^5 \lambda_2}{\lambda_2 - \lambda_1}, \quad f_1 = \frac{\lambda_1^5 - \lambda_2^5}{\lambda_1 - \lambda_2}$$

in which λ_1 and λ_2 are the eigenvalues of $\mathbf{H}$:

$$\lambda_{1,2} = 0.9800 \pm j0.0196 = 0.9802 e^{\pm j0.0200}$$

The foregoing expressions simplify to

$$f_0 = -\lambda_1 \lambda_2 (\lambda_1 + \lambda_2)(\lambda_1^2 + \lambda_2^2)$$
$$f_1 = \lambda_1^4 + \lambda_1^3 \lambda_2 + \lambda_1^2 \lambda_2^2 + \lambda_1 \lambda_2^3 + \lambda_2^4$$

and hence, upon substituting numerical values,

$$f_0 = -3.761 \quad f_1 = 4.607$$

Therefore,

$$\mathbf{H}^5 = \begin{bmatrix} 0.8440 & 0.1271 \\ -0.1271 & 0.6640 \end{bmatrix}$$

whence,

$$\zeta_{10} = \begin{bmatrix} 56.12 \\ -65.44 \end{bmatrix} \text{ mm}$$

i.e.,

$$x_{10} = 11.22 \text{ mm}, \quad \dot{x}_{10} = -2\pi \times 13.01 \text{ mm/s} = -82.24 \text{ mm/s}$$

The energy of the suspension, then, at the end of $10h = 0.045$ s, has reduced to

$$E_s = \frac{1}{2}m\dot{x}_{10}^2 + \frac{1}{2}kx_{10}^2 = \frac{1}{2}m\left(\dot{x}_{10}^2 + \omega_n^2 x_{10}^2\right) = 0.0059 \text{ Nm}$$

Note that the shocks have now absorbed only 56.73% of the energy that the suspension had since it overcame the bump. See 3-DampedBump1dof.mw.

Example 3.4.3 (A Mass Acted Upon by a Time-varying Force). The system under study now consists of a particle of mass m acted upon by a force $f(t)$. We want to derive a scheme for the simulation of the ensuing motion when the mass starts from initial conditions $x(0) = x_0$ and $\dot{x}(0) = v_0$.

Solution: The model at hand takes the form

$$m\ddot{x} = f(t), \quad x(0) = x_0, \quad \dot{x}(0) = v_0, \quad t \geq 0$$

We start by writing the governing equations in state-variable form, namely,

$$\dot{x} = v$$
$$\dot{v} = \frac{1}{m} f(t)$$

and hence, matrix $\mathbf{A}$ and vector $\mathbf{b}$ can be readily identified as

$$\mathbf{A} = \begin{bmatrix} 0 & 1 \\ 0 & 0 \end{bmatrix}, \quad \mathbf{b} = \begin{bmatrix} 0 \\ 1/m \end{bmatrix}$$

Because of the absence of a spring, no natural frequency occurs in the model, and hence, the simulation scheme of Eq. 3.39 suffices in this case, the discrete-time model now taking the form

$$\mathbf{z}_{k+1} = \mathbf{F}\mathbf{z}_k + \mathbf{g}f_k$$

where $\mathbf{z}_k$ is defined as $\mathbf{z}_k \equiv \mathbf{z}(t_k)$ and $\mathbf{z}(t) \equiv [x(t), v(t)]^T$ is the state-variable vector. Now we need the exponential of $\mathbf{A}$, but this is derived in Appendix A

$$\mathbf{F} = e^{\mathbf{A}h} = \begin{bmatrix} 1 & h \\ 0 & 1 \end{bmatrix}$$

where h is the sampling interval. Moreover, $\mathbf{g}$ is derived as

$$\mathbf{g} = \left(\int_0^h e^{\mathbf{A}t} dt \right) \mathbf{b} = \begin{bmatrix} h & h^2/2 \\ 0 & h \end{bmatrix} \begin{bmatrix} 0 \\ 1/m \end{bmatrix} = \begin{bmatrix} h^2/(2m) \\ h/m \end{bmatrix}$$

and hence, the simulation scheme reduces to

$$x_{k+1} = x_k + hv_k + \frac{h^2}{2m} f_k$$
$$v_{k+1} = v_k + \frac{h}{m} f_k \tag{3.49}$$

Note that the foregoing scheme is nothing but the motion of a particle of mass m acted upon by a constant force f_k in the time interval $t_k \leq t \leq t_{k+1}$, at the beginning of which the state of the particle is given by $[x_k, v_k]^T$, a result to be expected.

Fig. 3.2 A Geneva wheel
driving a rotor

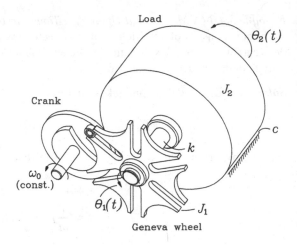

Example 3.4.4 (Discrete-time Response of a Geneva Wheel). Shown in Fig. 3.2 is
the model of the driving mechanism of a movie camera, consisting of a crank
that turns at a constant angular velocity ω_0, and drives a *Geneva wheel* with an
intermittent motion. This motion consists, in turn, of alternating dwell-and-forward
phases of equal duration T each, with T defined as the time it takes the crank to go
through one full rotation, that is,

$$T = \frac{2\pi}{\omega_0}$$

with ω_0 measured in rad/s. Moreover, the motion transmitted to the wheel is given by

$$\theta_1(t) = \frac{\pi}{3T}t + \theta_p(t)$$

and displayed in Fig. 3.3, with $\theta_p(t)$ defined as a periodic component of the motion,
that is given by

$$\theta_p(t) = \begin{cases} -\dfrac{\pi}{3T}t, & \text{for} \quad 0 \le t \le T/2; \\ \dfrac{\pi}{3T}(t-T) + \arctan[f(t)], & \text{for} \quad T/2 \le t \le T \end{cases}$$

while $f(t)$ is given, in turn, as

$$f(t) = -\frac{\sin \omega_0 t}{2 + \cos \omega_0 t}$$

Under the assumption that the load J_2 is originally at rest and θ_2 is set to
a value $\theta_2(0) = 0$, the wheel starts being driven by the crank, which turns at a
constant ω_0. By simulation, find the periodic and the nonperiodic parts of the steady-
state response of the load, $\theta_{2p}(t)$, and $\theta_{2n}(t)$, respectively. For the calculation,

Fig. 3.3 Angular diplacement of the Geneva wheel when driven at a uniform rate $\omega_0 = 2\pi/T$

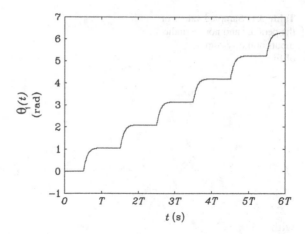

use the numerical values

$$\zeta = 0.1, \quad \omega_n = 10\omega_0, \quad \omega_0 = 0.3\,\mathrm{s}^{-1}$$

Solution: The model has the form

$$\ddot{\theta}_2 + 2\zeta\omega_n\dot{\theta}_2 + \omega_n^2\theta_2 = \omega_n^2\left(\frac{\pi}{3T}t + \theta_p(t)\right)$$

By linearity, the time response $\theta_2(t)$ can be decomposed into a periodic part, $\theta_{2p}(t)$, and a nonperiodic part, $\theta_{2n}(t)$. Therefore,

$$\ddot{\theta}_{2p} + 2\zeta\omega_n\dot{\theta}_{2p} + \omega_n^2\theta_{2p} = \omega_n^2\theta_p(t) \tag{3.50a}$$

$$\ddot{\theta}_{2n} + 2\zeta\omega_n\dot{\theta}_{2n} + \omega_n^2\theta_{2n} = \omega_n^2\frac{\pi}{3T}t \tag{3.50b}$$

We now have to determine the length h of the sampling interval. The periodic motion has a frequency of ω_0 rad/s. We can choose to sample this period in, say, 10 intervals, and hence, $h = T/10 = \pi/5\omega_0$. The sampling instants t_k are thus given by

$$t_k = kh = \frac{\pi}{5\omega_0}k = 2.0944k\,\mathrm{s}$$

Since Eq. 3.50a and b are of the usual form, Eqs. 3.25, 3.46a, b and 3.47a can be used directly. Here, ϕ_{pk} and ϕ_{nk}, for the periodic and nonperiodic part, respectively, are given by

$$\phi_{pk} = \theta_p(t_k) = \begin{cases} -\dfrac{\pi}{30}k, & \text{for} \quad 0 \le k \le 5; \\[2ex] \dfrac{\pi}{3}\left(\dfrac{k}{10} - 1\right) + \arctan f_k, & \text{for} \quad 5 \le k \le 10 \end{cases}$$

Table 3.1 Sampled values of
the periodic and non-periodic
inputs of the system
of Fig. 3.2

k	ϕ_{pk}	ϕ_{nk}
0	0	0
1	−0.1047	0.1047
2	−0.2094	0.2094
3	−0.3142	0.3142
4	−0.4189	0.4189
5	−0.5236	0.5236
6	0.0396	0.6283
7	0.1982	0.7330
8	0.1813	0.8378
9	0.1016	0.9425
10	0	1.0472

with

$$f_k = f(t_k) = -\frac{\sin(\pi k/5)}{2 + \cos(\pi k/5)}$$

and

$$\phi_{nk} = \frac{\pi}{30}k$$

For the first period T, ϕ_{p_k} and ϕ_{n_k} are given in Table 3.1.
Furthermore, the damped frequency of the system is

$$\omega_d = \sqrt{1 - \zeta^2}\omega_n = 2.9850\,\text{s}^{-1}$$

and hence,

$$\omega_n h = 6.2832\,\text{rad}, \quad \omega_d h = 6.2517\,\text{rad}$$

Moreover,

$$\cos \omega_d h = 0.9995, \quad \sin \omega_d h = -0.0315, \quad e^{-\zeta \omega_n h} = 0.5335$$

Hence,

$$\mathbf{H} = \begin{bmatrix} 0.5315 & -0.0169 \\ 0.0169 & 0.5349 \end{bmatrix}, \quad \mathbf{h} = \begin{bmatrix} 0.4685 \\ -0.0169 \end{bmatrix}$$

Now, using Eq. 3.47a, the periodic part θ_{2p} and the nonperiodic part θ_{2n} of the
time response are readily obtained as the first component of ζ_p and ζ_n respectively,

$$\zeta_{p(k+1)} = \mathbf{H}\zeta_{pk} + \mathbf{h}\phi_{pk}$$
$$\zeta_{n(k+1)} = \mathbf{H}\zeta_{nk} + \mathbf{h}\phi_{nk}$$

Table 3.2 Discrete-time periodic and non-periodic responses of the system of Fig. 3.2

k	θ_{2p}	θ_{2n}	θ_2
0	0	0	0
1	0	0	0
2	−0.0491	0.0491	0
3	−0.1242	0.1242	0
4	−0.2133	0.2133	0
5	−0.3097	0.3097	0
6	−0.4100	0.4100	0
7	−0.1995	0.5124	0.3129
8	−0.0131	0.6159	0.6028
9	0.0781	0.7200	0.7981
10	0.0892	0.8243	0.9136

Using the zero initial conditions, we have $\zeta_{p0} = \zeta_{n0} = [0,\ 0]^T$, and thus,

$$\zeta_{p1} = \mathbf{H}\zeta_{p0} + \mathbf{h}\phi_{p0} = \begin{bmatrix} 0 \\ 0 \end{bmatrix} \text{ rad}, \quad \zeta_{n1} = \mathbf{H}\zeta_{n0} + \mathbf{h}\phi_{n0} = \begin{bmatrix} 0 \\ 0 \end{bmatrix} \text{ rad}$$

$$\zeta_{p2} = \mathbf{H}\zeta_{p1} + \mathbf{h}\phi_{p1} = \begin{bmatrix} -0.0491 \\ 0.0018 \end{bmatrix} \text{ rad}, \quad \zeta_{n2} = \mathbf{H}\zeta_{n1} + \mathbf{h}\phi_{n1} = \begin{bmatrix} 0.0491 \\ -0.0018 \end{bmatrix} \text{ rad}$$

and so on.

Finally, the total time response $\theta_2(t)$ is simply

$$\theta_2(t) = \theta_{2p}(t) + \theta_{2n}(t)$$

For the first period T, we obtain the values displayed in Table 3.2.

The periodic and nonperiodic parts of the time response as well as the total response are plotted in Fig. 3.4, for the interval $[0,\ T]$. We can see that we have a very rough approximation of the time response of the Geneva wheel. The same time responses are plotted for an interval $[0,\ 4T]$ in Fig. 3.5 when using a smaller sampling interval h, say $h = T/100$.

3.5 Exercises

3.1. Find all equilibrium configurations of the mechanical system given in Example 1.6.5, where $a = zg$ and $mg/(kl) = z \times 10^{-7}$, with z denoting your seven-digit telephone number. If you don't have one, then use a friend's but make sure that your friend is not taking the same vibrations/dynamics course as you are.

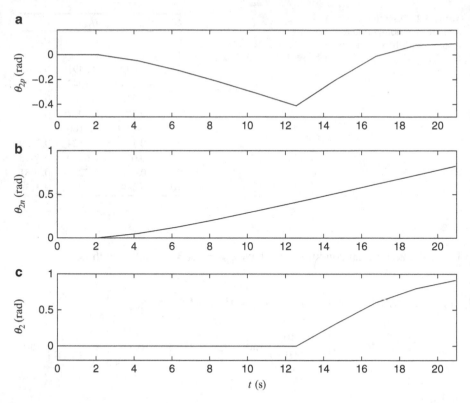

Fig. 3.4 Discrete-time response of the system of Fig. 3.2 over one period T, with a sample time $h = T/10$: (**a**) periodic part; (**b**) non-periodic part; and (**c**) total response

Now determine which equilibrium configurations are stable, unstable or marginally stable. Upon choosing a stable configuration, obtain the steady-state response of the system when this is acted upon by a square-wave force of the shape of that displayed in Fig. 2.53, with an amplitude of $(M + m)g/10$, for $M = m$, and acting on the cart in the horizontal direction. Assume the period $T = 1/(2\omega_n)$, where ω_n is the natural frequency of the linearized model.

3.2. Shown in Fig. 2.56 is an air compressor mounted on a viscoelastic foundation that is modeled as a spring of stiffness k in parallel with a dashpot of coefficient c. Furthermore, the moving parts of the compressor produce negligible inertia forces, except for the piston of mass m. Additionally, the housing of the compressor, as well as all other components besides the slider-crank mechanism, are lumped in one single mass M. Under these conditions, the mathematical model of the compressor-foundation system takes the form

$$(M + m)\ddot{x} + c\dot{x} + kx = -m\ddot{s}$$

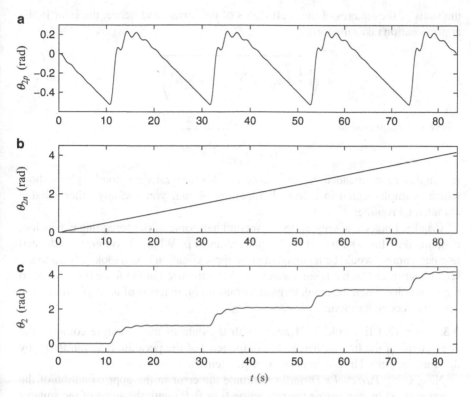

Fig. 3.5 Discrete-time response of the system of Fig. 3.2 over four periods T, with a sample time $h = T/100$: (a) periodic part; (b) non-periodic part; and (c) total response

where $s(t)$ is the *relative* displacement of the piston with respect to the housing of the compressor, measured as indicated in the figure. The foregoing model is now cast in *normal form*, namely,

$$\ddot{x} + 2\zeta\omega_n\dot{x} + \omega_n^2 x = -\frac{m}{M+m}\ddot{s}$$

The displacement s of the piston with respect to the housing is given below as a function of the angular displacement of the crank, θ:

$$s(\theta) = a\cos\theta + b\sqrt{1 - \rho^2\sin^2\theta}$$

in which ω denotes the constant angular velocity of the crank and $\rho \equiv a/b = z \times 10^{-7}$, with z defined as in Exercise 3.1 above.

Compute the coefficients of the first twenty harmonic components of the foregoing periodic function. Now, using *Parseval's Formula*, estimate the error in the approximation of the function $x(t)$ by comparing the rms value $\tilde{x}$ of $x(t)$ with the sums of the squares of the coefficients of the series, and hence, the error in the approximation takes the form

$$e = \sqrt{\frac{1}{T}\left(a_0^2 + 4\sum_0^N a_k^2 + 4\sum_1^N b_k^2\right) - \tilde{x}}$$

where $\tilde{x}$ is calculated as

$$\tilde{x} = \sqrt{\frac{1}{T}\int_{-T/2}^{T/2} x^2(t)\,dt}$$

It is highly recommended that you calculate this integral with Romberg's method, which is implemented in a Matlab function, or that you use any other reliable commercial routine.

Usually, books take only up to the second harmonic of the foregoing expansion, claiming that this is sufficient for 'small' values of ρ. What relative error (indicated as a percentage) would be incurred in your approximation if you took only up to the second harmonic? What relative error would be incurred in the force transmitted to the foundation when the crank turns at a constant ω, in terms of ω, if you took only up to the second harmonic?

3.3. We revisit Example 3.4.4 here. With the data of this example compute the coefficients of the first N_h harmonic components of the periodic function $\theta_p(t)$, by dividing the period in $2N_h$ intervals of equal length.

Now, using *Parseval's Formula*, estimate the error in the approximation of the function $\theta_p(t)$ by comparing the rms value $\tilde{\theta}$ of $\theta_p(t)$ with the sums of the squares of the coefficients of the series. The error in the approximation can be expressed as

$$e = \sqrt{a_0^2 + \frac{1}{2}\sum_1^{N_h} a_k^2 + \frac{1}{2}\sum_1^{N_h} b_k^2 - \tilde{\theta}^2}$$

where $\tilde{\theta}$ is calculated as

$$\tilde{\theta} = \sqrt{\frac{1}{T}\int_0^T \theta_p^2(t)\,dt}$$

It is highly recommended that you calculate this integral using numerical quadrature, as implemented in a Matlab function (quad or quad8), or that you use another reliable commercial routine, *with truncation-error control*, which is indicated via a user-prescribed tolerance. Assign your tolerance judiciously: if you choose a very loose tolerance, your computed integral will contain an inadmissibly high error; if too tight, the procedure may take too long for a negligible gain. As well, it is recommended that you calculate the Fourier coefficients using the computational scheme given in Sect. 2.8.2.

Produce a table of error e vs. N_h, for $N_h = 1, 2, 3, 4, 5, 10$, and 20. Comment on your tabulated results.

Now, find the periodic part $\theta_{2p}(t)$ of the steady-state response $\theta_2(t)$ for N_h harmonics of the Fourier expansion of $\theta_p(t)$, using a suitable value of N_h, that you should choose based on the table that you produced above. Plot this periodic response vs. time in an interval $[0, 2T]$. For this part of the exercise, use the model

$$\ddot{\theta}_{2p} + 2\zeta\omega_n\dot{\theta}_{2p} + \omega_n^2\theta_{2p} = \omega_n^2\theta_p(t)$$

Here, you should use the Fourier series of $\theta_p(t)$ with N_h harmonics, as found above. For the calculation, use the numerical values

$$\zeta = 0.1, \quad \omega_n = 10\omega_0, \quad \omega_0 = 0.3\,\text{s}^{-1}$$

Let the non-periodic part of the steady-state response be $\theta_{2n}(t)$. Find an expression for this response.

3.4. The oscillating follower of a quick-return cam mechanism goes through a *lower dwell* in the first 25% of its cycle, then performs a *working phase* of 60° of amplitude for 50% of its cycle, goes then through an *upper dwell* for 10% of its cycle, and returns to its lower dwell in the remaining 15% of its cycle. Let the angular displacement of the follower be represented by $\phi = \phi(\psi)$, where ψ is the angular displacement of the cam, which rotates at a constant rate $\dot{\psi} = \omega = \text{const}$, the *motion program* $\phi(\psi)$ being given below. Moreover, the follower drives an elastic shaft connected to a load of moment of inertia J mounted on bearings that provide a linearly viscous dissipative torque of coefficient c, the displacement of the load being denoted by θ. We thus have the follower motion program given by

$$\phi(\psi) = \begin{cases} 0, & \text{for} \quad 0 \le \phi \le \pi/2; \\ h\sigma(x), & \text{for} \quad \pi/2 \le \phi \le 3\pi/2; \\ h, & \text{for} \quad 3\pi/2 \le \phi \le 17\pi/10; \\ h\sigma(1-x), & \text{for} \quad 17\pi/10 \le \phi \le 2\pi; \end{cases}$$

with h as the maximum value for $\phi(\psi)$ and $\sigma(x)$ defined as

$$\sigma(x) = -20x^7 + 70x^6 - 84x^5 + 35x^4, \quad 0 \le x \le 1$$

while the mathematical model of the follower-load system takes the form

$$\ddot{\theta} + 2\zeta\omega_n\dot{\theta} + \omega_n^2\theta = \omega_n^2\phi$$

and the torque experienced by the shaft, $\tau(t)$, is

$$\tau(t) = k[\theta(t) - \phi(t)]$$

The numerical values of the variables involved are given as

$$\zeta = 0.1, \quad \omega_n = 10\omega, \quad \omega = 300 \, \text{rpm}$$

(a) Using *Parseval's Formula*, estimate the error in the approximation of the function $\phi(t)$ by comparing the rms value $\tilde{\phi}$ of $\phi(t)$ with the sums of the squares of the coefficients of the series. The difference of the two quantities is the error, e. Now, produce a table of error e vs. N_h, for $N_h = 1, 2, 3, 4, 5, 10$, and 20. Comment on your tabulated results.

(b) Then, find the steady-state response $\theta(t)$ for N_h harmonics of the Fourier expansion of $\phi(t)$, using a suitable value of N_h, that you should choose based on the table that you produced above. Plot time response vs. time in the interval $[0, 4T]$, where T is the duration of one cycle.

(c) Now plot the steady-state value of the torque experienced by the elastic shaft in dimensionless form, i.e., plot $\tau(t)/(J\omega_n^2)$ vs. time, in the time-interval $[0, 4T]$

Reference

1. Åström KJ, Wittenmark B (1990) Computer-controlled systems: theory and design. Prentice-Hall Inc, Englewood Cliffs, NJ

Chapter 4
Modeling of Multi-dof Mechanical Systems

How dare we, creatures of length, height, and width,
speak of four-dimensional space?
Is it possible by using all our three-dimensional intelligence
to imagine a superspace of four dimensions?
And how would a four-dimensional cube or sphere look like?

Gamow, G., 1961, *One, Two, Three ... Infinity,*
Dover Publications, Inc., New York.

4.1 Introduction

The modeling of multi-dof systems calls for various concepts that were not required when studying single-dof systems. For example, the concept of *degree of freedom*, to begin with, plays a major role here. Furthermore, the concept of *vector* of generalized coordinates, and the corresponding *vector* of generalized speeds arise only in multi-dof systems. Associated with these vectors, we have the *vector* of generalized force, while the concepts of *generalized mass matrix*, *generalized damping matrix* and *generalized stiffness matrix* enter in this context as natural extensions of their scalar counterparts encountered in single-dof systems. We shall introduce these concepts via the *vector* form of the Lagrange governing equations. One more concept that pertains to multi-dof systems is that of *rigid modes*, namely, non-trivial motions under which the potential energy of the whole system does not undergo any change.

The modeling techniques that we will introduce in this chapter allow us to model a variety of structures and machines. As a matter of fact, these mechanical systems are made of beams, plates, shells and other elements that can be accurately modeled as continua. We will be mostly concerned with *linear* mechanical systems composed of a finite set of mechanical elements. If the system at hand is representable by a set of n independent generalized coordinates, we say that the system has n degrees of freedom. Chapter 8 gives an overview of the simplest instances of continuous systems: bars under tension/compression or torsion; strings, and beams.

J. Angeles, *Dynamic Response of Linear Mechanical Systems: Modeling, Analysis and Simulation*, Mechanical Engineering Series, DOI 10.1007/978-1-4419-1027-1_4,
© Springer Science+Business Media, LLC 2011

4.2 The Derivation of the Governing Equations

A mechanical system is available to us through the *configuration* it attains at a given instant. A configuration consists, in turn, of the *relative* layout of the components of the system and of the *absolute* layout of the overall system with respect to an inertial frame. Such as in the case of single-dof systems, both the relative layout of the system components and that of the whole system with respect to an inertial frame are described by *signed* distances between landmark points in each component, whether rigid body, particle, massless spring or massless dashpot, and *signed* angles between landmark lines in the same components. Again, such as in the case of single-dof systems, these distances and angles constitute the *generalized coordinates* of the system.

Now, if we assume that the dof of a system is known, say *n*, it is always possible to find *n* independent generalized coordinates that fully describe the system configuration. However, an arbitrary set of *n* generalized coordinates may not necessarily be independent. In the examples that we will study here, the independent generalized coordinates will be either apparent or given as part of the problem statement.

We recall, moreover, that the time derivatives of the generalized coordinates are the generalized speeds of the system at hand. Furthermore, if the system has *n* dof, then it has a set of *n* independent generalized coordinates $\{q_i\}_1^n$ and *n* independent generalized speeds $\{\dot{q}_i\}_1^n$, that are conveniently stored in the *n*-dimensional arrays $\mathbf{q}$ and $\dot{\mathbf{q}}$, respectively, defined as

$$\mathbf{q} \equiv \begin{bmatrix} q_1 \\ q_2 \\ \vdots \\ q_n \end{bmatrix}, \quad \dot{\mathbf{q}} \equiv \begin{bmatrix} \dot{q}_1 \\ \dot{q}_2 \\ \vdots \\ \dot{q}_n \end{bmatrix} \tag{4.1}$$

Additionally, the whole set of both independent generalized coordinates and independent generalized speeds constitutes the set of $2n$ *state variables* of the system. If we denote by $\mathbf{x}$ the *state-variable vector*, we then have

$$\mathbf{x} \equiv \begin{bmatrix} \mathbf{q} \\ \dot{\mathbf{q}} \end{bmatrix} \tag{4.2}$$

Once a set of independent generalized coordinates and speeds has been decided on, the generalized forces are determined *uniquely*. Indeed, let the sum of all powers developed by driving devices supplying controlled forces be denoted by Π and that of all dissipation functions associated with dashpots and other sinks of energy be denoted by Δ. Then, the generalized force ϕ_{fi} associated with the *i*th generalized coordinate q_i is derived as

$$\phi_{fi} = \frac{\partial \Pi}{\partial \dot{q}_i} \tag{4.3}$$

Likewise, the generalized force ϕ_{di} stemming from dissipation and associated with the ith generalized coordinate q_i is derived as

$$\phi_{di} = -\frac{\partial \Delta}{\partial \dot{q}_i} \qquad (4.4)$$

The associated n-dimensional vectors of generalized driving force contributed by force-controlled drivers, ϕ_f, and generalized dissipative force, ϕ_d, are defined as

$$\phi_f \equiv \begin{bmatrix} \phi_{f1} \\ \phi_{f2} \\ \vdots \\ \phi_{fn} \end{bmatrix}, \quad \phi_d \equiv \begin{bmatrix} \phi_{d1} \\ \phi_{d2} \\ \vdots \\ \phi_{dn} \end{bmatrix} \qquad (4.5)$$

We have, therefore, in vector form,

$$\phi_f = \frac{\partial \Pi}{\partial \dot{\mathbf{q}}} \equiv \begin{bmatrix} \partial\Pi/\partial\dot{q}_1 \\ \partial\Pi/\partial\dot{q}_2 \\ \vdots \\ \partial\Pi/\partial\dot{q}_n \end{bmatrix}, \quad \phi_d = -\frac{\partial \Delta}{\partial \dot{\mathbf{q}}} \equiv \begin{bmatrix} -\partial\Delta/\partial\dot{q}_1 \\ -\partial\Delta/\partial\dot{q}_2 \\ \vdots \\ -\partial\Delta/\partial\dot{q}_n \end{bmatrix} \qquad (4.6)$$

Henceforth, the derivative, whether partial or total, of a scalar quantity with respect to a vector is understood as the vector of the corresponding derivatives with respect to each of the components of the vector. Likewise, we will come across derivatives of vectors with respect to vectors. These are matrices, as discussed in more detail in Sect. 4.5.

While the Lagrangian formalism can be applied to mechanical systems of particles and both rigid and flexible bodies, and to other physical systems as well, we will focus henceforth on mechanical systems composed of rigid bodies undergoing *planar motion*. Furthermore, we recall the notation introduced in Chap. 1, namely, we will denote the mass of the ith body as m_i, its moment of inertia about an axis perpendicular to the plane of motion, with respect to the center of mass, being denoted by I_i. We denote, moreover, the position and the velocity vectors of the center of mass of the ith rigid body by $\mathbf{c}_i$ and $\dot{\mathbf{c}}_i$, respectively, its angular velocity being denoted by the scalar ω_i. The kinetic energy T_i of the ith body can then be written in the form

$$T_i = \frac{1}{2}m_i\|\dot{\mathbf{c}}_i\|^2 + \frac{1}{2}I_i\omega_i^2 \qquad (4.7)$$

Thus, if the system under analysis is formed of r rigid bodies, its kinetic energy becomes

$$T = \frac{1}{2}\sum_1^r m_i\|\dot{\mathbf{c}}_i\|^2 + \frac{1}{2}\sum_1^r I_i\omega_i^2 \qquad (4.8)$$

As we will show with examples, both the square of the magnitude of the velocity vector of the center of mass of each body and the square of its scalar angular velocity can always be written as *quadratic* expressions of the independent generalized speeds.

When the system is driven by controlled motions independent of the generalized speeds, the foregoing quantities become quadratic expressions not only of the generalized speeds, but also of the *controlled rates*, which are defined as the time derivatives of the *controlled variables*. Moreover, the coefficients associated with the above quadratic expressions are in many instances functions of the independent generalized coordinates.

By conducting a kinematic analysis similar to that introduced in Chap. 1, it is always possible to express the kinetic energy of the overall system as a quadratic function of the generalized speeds. This transformation of the kinetic energy from a quadratic form of the magnitudes of the centers of mass and the angular velocities into a quadratic form in the generalized speeds is best understood with the examples ahead. We shall then derive an expression for the kinetic energy of the system in the form

$$T = \frac{1}{2} \sum_{i,j=1}^{n} m_{ij} \dot{q}_i \dot{q}_j + \sum_{i=1}^{n} p_i \dot{q}_i + T_0(\mathbf{q}, t) \tag{4.9}$$

or, in compact form,

$$T = \frac{1}{2} \dot{\mathbf{q}}^T \mathbf{M} \dot{\mathbf{q}} + \mathbf{p}^T \dot{\mathbf{q}} + T_0(\mathbf{q}, t) \tag{4.10}$$

where the *mass matrix* $\mathbf{M}$ plays the role of the scalar generalized mass of single-dof systems, its entries being the coefficients m_{ij} appearing in Eq. 4.9; these entries are, in general, functions of the generalized coordinates, but not of the generalized speeds. If we now group all p_i coefficients of Eq. 4.9 into the n-dimensional array $\mathbf{p}$, we end up with the definitions below:

$$\mathbf{M} = \mathbf{M}(\mathbf{q}) \equiv \begin{bmatrix} m_{11} & m_{12} & \cdots & m_{1n} \\ m_{12} & m_{22} & \cdots & m_{2n} \\ \vdots & \cdots & \ddots & \vdots \\ m_{1n} & m_{2n} & \cdots & m_{nn} \end{bmatrix}, \quad \mathbf{p} = \mathbf{p}(\mathbf{q}, t) \equiv \begin{bmatrix} p_1 \\ p_2 \\ \vdots \\ p_n \end{bmatrix} \tag{4.11}$$

where it is noteworthy that $\mathbf{M}$ is *symmetric*. Note that each product $p_i \dot{q}_i$ has units of energy, while $\dot{q}_i$ has units of generalized speed. It is then apparent that the coefficients p_i of Eq. 4.9 have units of generalized momentum, which could be either momentum or angular momentum.

In setting up the Lagrange equations of the system under study, we need an expression for its potential energy, which, as in the single-dof case, can be either elastic or gravitational. The former appears because of the presence of springs, the latter because of the gravity field. The potential energy due to the gravity field is, in general, a nonlinear function of the generalized coordinates, *but does not depend on the generalized speeds*. Thus, we will assume that, in general, the potential energy

V_g due to gravity is a nonlinear function of the generalized coordinates. Likewise, the elastic potential energy V_e is a nonlinear function of the generalized coordinates, and hence, the total potential energy V, the sum of the two, is also a function solely of the generalized coordinates, i.e.,

$$V = V(\mathbf{q}) = V_g(\mathbf{q}) + V_e(\mathbf{q}) \tag{4.12}$$

Now we turn to the formulation of the governing equations using a Lagrangian approach. Here, we will need the *Lagrangian L* of the system, which is defined exactly as in the case of single-dof systems, namely, as

$$L \equiv T - V \tag{4.13}$$

the Lagrange equations taking the form

$$\frac{d}{dt}\left(\frac{\partial L}{\partial \dot{\mathbf{q}}}\right) - \frac{\partial L}{\partial \mathbf{q}} = \phi_f + \phi_d \tag{4.14}$$

It will become apparent, with the aid of the examples, that the foregoing equations take a general form resembling that of the single-dof case, except that now the generalized mass becomes the $n \times n$ mass matrix and all other quantities become n-dimensional vectors, namely,

$$\mathbf{M}\ddot{\mathbf{q}} + \mathbf{h}(\mathbf{q}, \dot{\mathbf{q}}) = \phi_p + \phi_m + \phi_f + \phi_d \tag{4.15}$$

where the vectors of generalized force that arise from a potential and from motion-controlled sources, ϕ_p and ϕ_m, respectively, are contributed by the Lagrangian. The above mathematical model is thus a system of n *nonlinear* ordinary differential equations of the second order. The integration of these equations, with prescribed initial values $\mathbf{q}(0) = \mathbf{q}_0$ and $\dot{\mathbf{q}}(0) = \mathbf{r}_0$, and given *generalized active forces* $\phi_m(t)$ and $\phi_f(t)$, lies beyond the scope of the book. As pertaining to linear, constant-coefficient systems, the integration of the underlying ODEs for two-dof systems in *symbolic form* will be discussed in Chap. 5, for n-dof systems *numerically* in Chap. 7.

If we recall Eq. 1.39b, $h(q, \dot{q})$ for the single-dof case is *quadratic* in $\dot{q}$ and nonlinear in q. Likewise, the n-dimensional term $\mathbf{h}(\mathbf{q}, \dot{\mathbf{q}})$ of Eq. 4.15 is quadratic in the generalized speeds $\dot{\mathbf{q}}$ and nonlinear in the generalized coordinates $\mathbf{q}$. This term thus plays the role of the generalized Coriolis and centrifugal forces of the single-dof case. For this reason, this term is also called the vector of generalized Coriolis and centrifugal forces. Moreover, terms ϕ_g and ϕ_m of the right-hand side of Eq. 4.15 stem from the left-hand side of Eq. 4.14, the former appearing in the Lagrangian because it represents generalized forces due to gravity, the latter from controlled-motion actuators, the remaining two terms of the right-hand side of that equation having been already explained. The whole right-hand side of that equation is called the *vector of generalized forces* and is represented by ϕ. Note that $\phi = \phi(\mathbf{q}, \dot{\mathbf{q}}, t)$.

From Eq. 4.15 it is apparent that matrix $\mathbf{M}$ plays a key role in the dynamics of mechanical systems. Moreover, this matrix has properties that impact on the behavior of these systems. First and foremost, given the form of the kinetic energy of Eq. 4.9, it is apparent that $\mathbf{M}$ is symmetric and hence, all its eigenvalues are real. Furthermore, since the kinetic energy cannot be negative, it turns out that $\mathbf{M}$ is positive-semidefinite, and hence, its eigenvalues are not only real, but also non-negative and, in fact, in many instances, they are all positive, $\mathbf{M}$ then being positive-definite. Moreover, the term $\mathbf{h}(\mathbf{q}, \dot{\mathbf{q}})$ does not have any particular structure, except that $\mathbf{h}$ is, in general, a nonlinear function of $\mathbf{q}$ and a quadratic function of $\dot{\mathbf{q}}$.

Just like in Chap. 1, we now have the seven-step procedure outlined below to set up the governing equations of n-dof mechanical systems:

1. Introduce a set of generalized coordinates $q_1, \ldots, q_n$ and their time rates of change $\dot{q}_1, \ldots, \dot{q}_n$, defining the state of the system. Group them in the n-dimensional vectors $\mathbf{q}$ and $\dot{\mathbf{q}}$, i.e.,

$$\mathbf{q} = [q_1, \ldots, q_n]^T : \quad \text{vector of generalized coordinates}$$
$$\dot{\mathbf{q}} = [\dot{q}_1, \ldots, \dot{q}_n]^T : \quad \text{vector of generalized speeds}$$

   and, by means of a kinematic analysis, express the squares of the magnitudes of the c.o.m. velocities and the squares of the angular velocities as quadratic expressions in the generalized speeds.
2. Evaluate $T = T(\mathbf{q}, \dot{\mathbf{q}})$, the kinetic energy of the whole system, as the sum of the individual kinetic energy expressions.
3. Evaluate $V = V(\mathbf{q})$, the potential energy of the whole system, as the sum of the individual expressions, for every potential-energy-storing element.
4. Evaluate $L \equiv T - V$, the Lagrangian of the whole system: $L = L(\mathbf{q}, \dot{\mathbf{q}})$.
5. Evaluate $\Pi = \Pi(\mathbf{q}, \dot{\mathbf{q}})$, the power supplied to the system from force/torque-controlled sources ($\Pi \geq 0$), then its partial derivative $\partial \Pi / \partial \dot{\mathbf{q}}$, to obtain ϕ_f:

$$\phi_f = \partial \Pi / \partial \dot{\mathbf{q}} \tag{4.16}$$

6. Evaluate $\Delta = \Delta(\mathbf{q}, \dot{\mathbf{q}})$, the sum of the dissipation functions of all dissipative elements of the system ($\Delta \geq 0$), as well as its partial derivative $\partial \Delta / \partial \dot{\mathbf{q}}$, to obtain ϕ_d:

$$\phi_d = \partial \Delta / \partial \dot{\mathbf{q}} \tag{4.17}$$

7. Write the governing equations using the foregoing partial derivatives:

$$\frac{d}{dt}\left(\frac{\partial L}{\partial \dot{\mathbf{q}}}\right) - \frac{\partial L}{\partial \mathbf{q}} = \phi_f - \phi_d \tag{4.18}$$

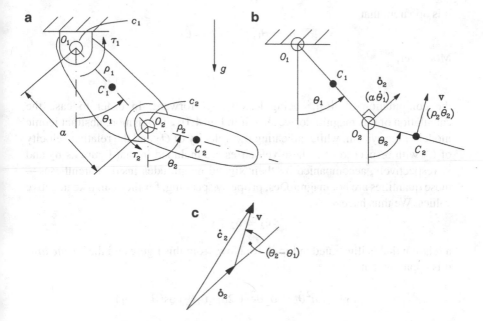

Fig. 4.1 A two-link robotic arm

Example 4.2.1 (A Two-Link Robotic Arm with Viscous Damping) A two-link
robotic arm suspended from the ceiling is shown in the iconic model of Fig. 4.1.
In this model we have assumed that the two links are rigid bodies pinned at O_1 and
O_2, the moment of inertia of body 1 with respect to O_1 being denoted by J_1, that
of body 2 with respect to its center of mass being denoted by J_2, while the masses
of these bodies are labelled m_1 amd m_2, respectively. Moreover, we assume that the
joints are lubricated with a fluid that provides linearly viscous damping, while the
motors at the joints provide torques τ_1 and τ_2. As indicated in the figure, the effect of
gravity is considered and the first link is assumed to be *inertially symmetric*. What
we mean by this is that its center of mass C_1 is aligned with the two joint centers.
Derive the Lagrange equations of motion of this system.

Solution: The seven steps leading to the mathematical model of Eq. 4.18 are now
implemented.

1. **Kinematics**. We start by introducing the generalized coordinates of the system.
   Apparently, the system has two dof, and hence, two generalized coordinates are
   needed to describe it in an arbitrary configuration. We thus define the generalized
   coordinates as angles θ_k, for $k = 1, 2$, as shown in Fig. 4.1, that measure the
   orientation of the two links with respect to the vertical. In setting up the kinetic
   energy, we shall resort to Eq. 4.8, where four kinematic variables are needed,
   namely, $\dot{c}_1$, $\dot{c}_2$, ω_1 and ω_2. As a matter of fact, the variables as such are not
   needed, but rather $||\dot{c}_1||^2$, $||\dot{c}_2||^2$, ω_1^2 and ω_2^2. From the definitions of θ_1 and θ_2,

it is apparent that

$$\omega_1 = \dot{\theta}_1, \qquad \omega_2 = \dot{\theta}_2$$

Moreover,

$$\|\dot{\mathbf{c}}_1\|^2 = \rho_1^2 \dot{\theta}_1^2$$

the computation of $\|\dot{\mathbf{c}}_2\|^2$ being less straightforward. In order to ease the calculation of this magnitude, we sketch in Fig. 4.1b an even more abstract iconic model of the system, while indicating the velocity of O_2 and the relative velocity of C_2 with respect to O_2. These velocities are included in that figure as $\dot{\mathbf{o}}_2$ and $\mathbf{v}$, respectively, accompanied by their signed magnitudes inside parentheses—these quantities are not magnitudes, properly speaking, for they can take negative values. We thus have

$$\dot{\mathbf{c}}_2 = \dot{\mathbf{o}}_2 + \mathbf{v}$$

a relation that is illustrated in Fig. 4.1c. Now, from this figure and the *cosine law*, it is apparent that

$$\|\dot{\mathbf{c}}_2\|^2 = a^2\dot{\theta}_1^2 + \rho_2^2\dot{\theta}_2^2 + 2a\rho_2\dot{\theta}_1\dot{\theta}_2\cos(\theta_2 - \theta_1)$$

2. **Kinetic energy**. We first recall that J_1 denotes the moment of inertia of body 1 with respect to O_1, while J_2 that of body 2 with respect to its c.o.m., whence,

$$T_1 = \frac{1}{2}J_1\dot{\theta}_1^2, \quad T_2 = \frac{1}{2}J_2\dot{\theta}_2^2 + \frac{1}{2}m_2[a^2\dot{\theta}_1^2 + \rho_2^2\dot{\theta}_2^2 + 2a\rho_2\dot{\theta}_1\dot{\theta}_2\cos(\theta_2 - \theta_1)]$$

Therefore, the kinetic energy of the overall system is

$$T = \frac{1}{2}\left(J_1\dot{\theta}_1^2 + J_2\dot{\theta}_2^2\right) + \frac{1}{2}m_2\left[a^2\dot{\theta}_1^2 + \rho_2^2\dot{\theta}_2^2 + 2a\rho_2\dot{\theta}_1\dot{\theta}_2\cos(\theta_2 - \theta_1)\right]$$

3. **Potential energy**. This is readily obtained as

$$V = -m_1g\rho_1\cos\theta_1 - m_2g(a\cos\theta_1 + \rho_2\cos\theta_2)$$

where m_1 and m_2 are the masses of the first and second links, respectively.
4. **Lagrangian**. The Lagrangian becomes

$$L = \frac{1}{2}\left(J_1\dot{\theta}_1^2 + J_2\dot{\theta}_2^2\right) + \frac{1}{2}m_2\left[a^2\dot{\theta}_1^2 + \rho_2^2\dot{\theta}_2^2 + 2a\rho_2\dot{\theta}_1\dot{\theta}_2\cos(\theta_2 - \theta_1)\right]$$

$$+ m_1g\rho_1\cos\theta_1 + m_2g(a\cos\theta_1 + \rho_2\cos\theta_2)$$

The partial derivatives of the Lagrangian needed to set up the Lagrange equations of the system are now computed as

$$\frac{\partial L}{\partial \dot{\theta}_1} = (J_1 + m_2 a^2)\dot{\theta}_1 + m_2 a \rho_2 \dot{\theta}_2 \cos(\theta_2 - \theta_1)$$

$$\frac{d}{dt}\left(\frac{\partial L}{\partial \dot{\theta}_1}\right) = (J_1 + m_2 a^2)\ddot{\theta}_1 + m_2 a \rho_2 \ddot{\theta}_2 \cos(\theta_2 - \theta_1)$$

$$-m_2 a \rho_2 \dot{\theta}_2 (\dot{\theta}_2 - \dot{\theta}_1)\sin(\theta_2 - \theta_1)$$

$$\frac{\partial L}{\partial \theta_1} = m_2 a \rho_2 \dot{\theta}_1 \dot{\theta}_2 \sin(\theta_2 - \theta_1) - m_1 g \rho_1 \sin \theta_1 - m_2 g a \sin \theta_1$$

$$\frac{\partial L}{\partial \dot{\theta}_2} = (J_2 + m_2 \rho_2^2)\dot{\theta}_2 + m_2 a \rho_2 \dot{\theta}_1 \cos(\theta_2 - \theta_1)$$

$$\frac{d}{dt}\left(\frac{\partial L}{\partial \dot{\theta}_2}\right) = (J_2 + m_2 \rho_2^2)\ddot{\theta}_2 + m_2 a \rho_2 \ddot{\theta}_1 \cos(\theta_2 - \theta_1)$$

$$-m_2 a \rho_2 \dot{\theta}_1 (\dot{\theta}_2 - \dot{\theta}_1)\sin(\theta_2 - \theta_1)$$

$$\frac{\partial L}{\partial \theta_2} = -m_2 a \rho_2 \dot{\theta}_1 \dot{\theta}_2 \sin(\theta_2 - \theta_1) - m_2 g \rho_2 \sin \theta_2$$

5. **Power supplied**. The total power supplied from force-controlled sources is the sum of the powers delivered by the two motors, i.e.,

$$\Pi = \tau_1 \dot{\theta}_1 + \tau_2(\dot{\theta}_2 - \dot{\theta}_1)$$

its partial derivatives being

$$\phi_{f1} = \frac{\partial \Pi}{\partial \dot{\theta}_1} = \tau_1 - \tau_2, \quad \phi_{f2} = \frac{\partial \Pi}{\partial \dot{\theta}_2} = \tau_2$$

6. **Dissipation function.** This item is the sum of the dissipation functions associated with the individual joints, i.e.,

$$\Delta = \frac{1}{2}c_1 \dot{\theta}_1^2 + \frac{1}{2}c_2(\dot{\theta}_2 - \dot{\theta}_1)^2$$

which is quadratic in the *relative* angular velocity of the second link with respect to the former, not in the absolute angular velocity of this link, its partial derivatives being

$$\phi_{d1} = \frac{\partial \Delta}{\partial \dot{\theta}_1} = (c_1 + c_2)\dot{\theta}_1 - c_2 \dot{\theta}_2, \quad \phi_{d2} = \frac{\partial \Delta}{\partial \dot{\theta}_2} = c_2(\dot{\theta}_2 - \dot{\theta}_1)$$

7. **Governing equations**. After simplifications, the Lagrange equations reduce to

$$(J_1 + m_2 a^2)\ddot{\theta}_1 + m_2 a \rho_2[\cos(\theta_2 - \theta_1)]\ddot{\theta}_2 - m_2 a \rho_2 \dot{\theta}_2^2 \sin(\theta_2 - \theta_1)$$

$$= -g(m_1 \rho_1 + m_2 a)\sin \theta_1 + \tau_1 - \tau_2 - (c_1 + c_2)\dot{\theta}_1 + c_2 \dot{\theta}_2$$

$$m_2 a \rho_2 [\cos(\theta_2 - \theta_1)] \ddot{\theta}_1 + (J_2 + m_2 \rho_2^2) \ddot{\theta}_2 + m_2 a \rho_2 \dot{\theta}_1^2 \sin(\theta_2 - \theta_1)$$

$$= -m_2 g \rho_2 \sin \theta_2 + \tau_2 - c_2 (\dot{\theta}_2 - \dot{\theta}_1)$$

It is now a simple matter to identify the mass matrix via the coefficients of $\ddot{\theta}_1$ and $\ddot{\theta}_2$, which yields

$$\mathbf{M} = \begin{bmatrix} J_1 + m_2 a^2 & m_2 a \rho_2 \cos(\theta_2 - \theta_1) \\ m_2 a \rho_2 \cos(\theta_2 - \theta_1) & J_2 + m_2 \rho_2^2 \end{bmatrix}$$

a 2×2 matrix that is apparently symmetric, as it should. Moreover, $\mathbf{M}$ is *positive-definite*, as shown below: Indeed, to show that $\mathbf{M}$ is positive-definite, all we need is show that its two eigenvalues are positive. Furthermore, to prove that the two eigenvalues of a 2×2 matrix are positive, or negative, for that matter, we do not need to actually compute its two eigenvalues, if we recall from Appendix A that the trace of the matrix is the sum of its eigenvalues,[1] while its determinant is the product of these. Hence, the two eigenvalues of a 2×2 matrix are positive if and only if both the trace and the determinant of the matrix are positive. Now we have

$$\operatorname{tr}(\mathbf{M}) = J_1 + J_2 + m_2 (a^2 + \rho_2^2) > 0$$

Moreover,

$$\det(\mathbf{M}) = (J_1 + m_2 a^2)(J_2 + m_2 \rho_2^2) - m_2^2 a^2 \rho_2^2 \cos^2(\theta_2 - \theta_1)$$

$$= J_1 J_2 + m_2 (J_1 \rho_2^2 + J_2 a^2) + m_2^2 a^2 \rho_2^2 \sin^2(\theta_2 - \theta_1) > 0$$

thereby showing that the mass matrix of this system is positive-definite.

Further, the $\mathbf{h}(\mathbf{q}, \dot{\mathbf{q}})$ vector and the vectors of generalized force are readily identified, namely,

$$\mathbf{h}(\mathbf{q}, \dot{\mathbf{q}}) = \begin{bmatrix} -m_2 a \rho_2 \dot{\theta}_2^2 \sin(\theta_2 - \theta_1) \\ m_2 a \rho_2 \dot{\theta}_1^2 \sin(\theta_2 - \theta_1) \end{bmatrix}$$

$$\boldsymbol{\phi}_p = - \begin{bmatrix} g(m_1 \rho_1 + m_2 a) \sin \theta_1 \\ m_2 g \rho_2 \sin \theta_2 \end{bmatrix}, \boldsymbol{\phi}_f = \begin{bmatrix} \tau_1 - \tau_2 \\ \tau_2 \end{bmatrix}$$

$$\boldsymbol{\phi}_d = \begin{bmatrix} -(c_1 + c_2) \dot{\theta}_1 + c_2 \dot{\theta}_2 \\ -c_2 (\dot{\theta}_2 - \dot{\theta}_1) \end{bmatrix}$$

---

[1] In the case at hand, the trace of $\mathbf{M}$ has a physical meaning, as the diagonal entries of $\mathbf{M}$ bear the same units, those of moment of inertia. Depending on the choice of the generalized coordinates, these entries may bear distinct units, in which case the trace is meaningless. In such cases, the eigenvalues *must* be computed to prove positive-definiteness.

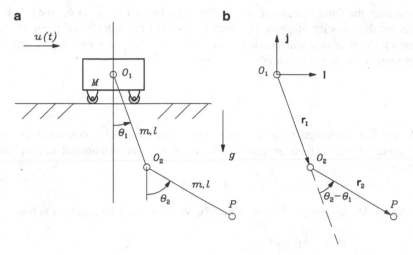

Fig. 4.2 A two-link gantry robot

The system being apparently driven under controlled forces supplied by actuators, it contains a corresponding term ϕ_f. As well, since no apparent controlled motions are present, the corresponding term ϕ_m does not appear above. Finally, note that the vector of generalized gravity force, ϕ_p, does not involve the generalized speeds. This is the case in general, for this force stems from a potential energy that does not depend on the velocities of the system elements.

Example 4.2.2 (A Two-dof Gantry Robot). Now we analyze a similar system, but with an added complexity. The system at hand is the gantry robot of Fig. 4.2. It consists of two identical links that can be modeled as slender rods of lengths l and mass m, with the base joint O_1 now mounted on a trolley that can slide under controlled motion along a horizontal track. The motion of the trolley is thus given as a function $u(t)$. If the two joints are lubricated with a fluid that produces linearly viscous torques and the same joints are acted upon by motors producing driving torques τ_k, for $k = 1, 2$, find the mathematical model of the system under the action of gravity.

Solution: We proceed as in the foregoing example, i.e., in seven steps.

1. **Kinematics.** The system at hand involves one more independent motion when compared to the system of Example 4.2.1, namely, the motion of the base. However, the latter is prescribed and does not count toward the dof of the system, which is still two, and hence, fully described by θ_1 and θ_2, that are, again, regarded as the generalized coordinates of the system. Nevertheless, because of the base motion, the procedure used in Example 4.2.1 to find the square of the magnitude of the velocity of the center of mass C_2 is no longer practical. Another approach to derive this expression is now introduced, that finds frequent applications in robotics. Let $\mathbf{r}_1$ and $\mathbf{r}_2$ denote the vectors directed from O_1 to O_2

and from the latter to P, as shown in Fig. 4.2b. Furthermore, let $\dot{\mathbf{o}}_k$, for $k = 1, 2$, denote the velocity of point O_k, with $\dot{\mathbf{p}}$ denoting the velocity of point P. The ensuing derivations will become more handleable if we now recall matrix $\mathbf{E}$ introduced in Sect. 1.6, namely,

$$\mathbf{E} \equiv \begin{bmatrix} 0 & -1 \\ 1 & 0 \end{bmatrix}$$

Matrix $\mathbf{E}$, it is recalled, rotates vectors in the plane through $90°$ counterclockwise, without changing their magnitude. $\mathbf{E}$ is in fact an orthogonal matrix and, moreover,

$$\mathbf{E}^2 = -\mathbf{1}$$

where $\mathbf{1}$ denotes the 2×2 identity matrix. We then have the relations below[2]

$$\dot{\mathbf{o}}_1 = \dot{u}\mathbf{i}$$

$$\dot{\mathbf{o}}_2 = \dot{\mathbf{o}}_1 + \dot{\theta}_1 \mathbf{E}\mathbf{r}_1 = \dot{u}\mathbf{i} + \dot{\theta}_1 \mathbf{E}\mathbf{r}_1$$

$$\dot{\mathbf{p}} = \dot{\mathbf{o}}_2 + \dot{\theta}_2 \mathbf{E}\mathbf{r}_2 = \dot{u}\mathbf{i} + \dot{\theta}_1 \mathbf{E}\mathbf{r}_1 + \dot{\theta}_2 \mathbf{E}\mathbf{r}_2$$

Furthermore, from the modeling assumptions,

$$\dot{\mathbf{c}}_1 = \dot{u}\mathbf{i} + \frac{1}{2}\dot{\theta}_1 \mathbf{E}\mathbf{r}_1$$

and hence,

$$\|\dot{\mathbf{c}}_1\|^2 = \left(\dot{u}\mathbf{i} + \frac{1}{2}\dot{\theta}_1 \mathbf{E}\mathbf{r}_1 \right)^T \left(\dot{u}\mathbf{i} + \frac{1}{2}\dot{\theta}_1 \mathbf{E}\mathbf{r}_1 \right) = \dot{u}^2 + \mathbf{i} \cdot (\mathbf{E}\mathbf{r}_1)\dot{u}\dot{\theta}_1 + \frac{1}{4}\|\mathbf{E}\mathbf{r}_1\|^2 \dot{\theta}_1^2$$

But $\mathbf{E}\mathbf{r}_1$ is simply $\mathbf{r}_1$ with its first and second components interchanged, the first one having, additionally, its sign reversed. From Fig. 4.2a, b,

$$\mathbf{r}_1 = \begin{bmatrix} l \sin \theta_1 \\ -l \cos \theta_1 \end{bmatrix} \Rightarrow \mathbf{E}\mathbf{r}_1 = \begin{bmatrix} l \cos \theta_1 \\ l \sin \theta_1 \end{bmatrix}$$

and hence, $\mathbf{i} \cdot (\mathbf{E}\mathbf{r}_1)$ is just the first component of $\mathbf{E}\mathbf{r}_1$, i.e.,

$$\mathbf{i} \cdot (\mathbf{E}\mathbf{r}_1) = l \cos \theta_1$$

Moreover, since $\mathbf{E}$ preserves the magnitude of vectors, $\|\mathbf{E}\mathbf{v}\| \equiv \|\mathbf{v}\|$, for *any* vector $\mathbf{v}$, and hence,

$$\|\mathbf{E}\mathbf{r}_1\| \equiv \|\mathbf{r}_1\| = l$$

---

[2]Notice the *recursive* nature of these relations: the second is based on the first, as the third on the second. Recursion is at the core of computational algorithms in multibody system dynamics.

Therefore,

$$\|\dot{c}_1\|^2 = \dot{u}^2 + l(\cos\theta_1)\dot{u}\dot{\theta}_1 + \frac{1}{4}l^2\dot{\theta}_1^2$$

Likewise,

$$\dot{c}_2 = \dot{u}\mathbf{i} + \dot{\theta}_1\mathbf{E}\mathbf{r}_1 + \frac{1}{2}\dot{\theta}_2\mathbf{E}\mathbf{r}_2$$

whence,

$$\|\dot{c}_2\|^2 = \dot{u}^2 + \|\mathbf{E}\mathbf{r}_1\|^2\dot{\theta}_1^2 + \frac{1}{4}\|\mathbf{E}\mathbf{r}_2\|^2\dot{\theta}_2^2 + 2\mathbf{i}\cdot(\mathbf{E}\mathbf{r}_1)\dot{u}\dot{\theta}_1 + \mathbf{i}\cdot(\mathbf{E}\mathbf{r}_2)\dot{u}\dot{\theta}_2$$

$$+ (\mathbf{E}\mathbf{r}_1)\cdot(\mathbf{E}\mathbf{r}_2)\dot{\theta}_1\dot{\theta}_2$$

where

$$\mathbf{r}_2 = \begin{bmatrix} l\sin\theta_2 \\ -l\cos\theta_2 \end{bmatrix} \Rightarrow \mathbf{E}\mathbf{r}_2 = \begin{bmatrix} l\cos\theta_2 \\ l\sin\theta_2 \end{bmatrix}$$

Now we can readily write

$$\mathbf{i}\cdot(\mathbf{E}\mathbf{r}_2) = l\cos\theta_2, \quad \|\mathbf{E}\mathbf{r}_2\| = l$$

Moreover,

$$(\mathbf{E}\mathbf{r}_1)\cdot(\mathbf{E}\mathbf{r}_2) = (\mathbf{E}\mathbf{r}_1)^T(\mathbf{E}\mathbf{r}_2) \equiv \mathbf{r}_1^T\mathbf{E}^T\mathbf{E}\mathbf{r}_2$$

and if we recall further that $\mathbf{E}$ is an orthogonal matrix, then $\mathbf{E}^T\mathbf{E} = \mathbf{1}$, where $\mathbf{1}$ denotes the 2×2 identity matrix. Hence,

$$(\mathbf{E}\mathbf{r}_1)\cdot(\mathbf{E}\mathbf{r}_2) = \mathbf{r}_1^T\mathbf{r}_2 = \mathbf{r}_1\cdot\mathbf{r}_2 = l^2\cos(\theta_2 - \theta_1)$$

Therefore,

$$\|\dot{c}_2\|^2 = \dot{u}^2 + l^2\left[\dot{\theta}_1^2 + \frac{1}{4}\dot{\theta}_2^2 + \dot{\theta}_1\dot{\theta}_2\cos(\theta_2 - \theta_1)\right] + l\dot{u}\left(2\dot{\theta}_1\cos\theta_1 + \dot{\theta}_2\cos\theta_2\right)$$

2. **Kinetic energy**. The kinetic energy of each of the two links is labeled T_1 and T_2, respectively, that of the trolley T_3. The corresponding expressions are

$$T_1 = \frac{1}{2}m\|\dot{c}_1\|^2 + \frac{1}{2}\frac{1}{12}ml^2\dot{\theta}_1^2, \quad T_1 = \frac{1}{2}m\|\dot{c}_2\|^2 + \frac{1}{2}\frac{1}{12}ml^2\dot{\theta}_2^2, \quad T_3 = \frac{1}{2}M\dot{u}^2$$

the kinetic energy of the overall system thus being

$$T = \frac{1}{2}\left(m\|\dot{c}_1\|^2 + \frac{1}{12}ml^2\dot{\theta}_1^2 + m\|\dot{c}_2\|^2 + \frac{1}{12}ml^2\dot{\theta}_2^2\right) + \frac{1}{2}M\dot{u}^2$$

Upon introduction of the expressions for $\|\dot{\mathbf{c}}_1\|$ and $\|\dot{\mathbf{c}}_2\|$ derived in item 1 of this example into the above expression for T, we obtain

$$T = \frac{1}{2}m\left[\frac{4}{3}l^2\dot{\theta}_1^2 + \frac{1}{3}l^2\dot{\theta}_2^2 + l^2[\cos(\theta_2 - \theta_1)]\dot{\theta}_1\dot{\theta}_2 + l\dot{u}(3\dot{\theta}_1\cos\theta_1 + \dot{\theta}_2\cos\theta_2)\right.$$

$$\left. +2\dot{u}^2\right] + \frac{1}{2}M\dot{u}^2$$

3. **Potential energy**. The derivation of the potential energy is more straightforward:

$$V = -\frac{3}{2}mgl\cos\theta_1 - \frac{1}{2}mgl\cos\theta_2$$

where we have used as a reference level that of O_1.

4. **Lagrangian**. This is simply the difference of the two foregoing items, i.e.,

$$L = \frac{1}{2}m\left[\frac{4}{3}l^2\dot{\theta}_1^2 + \frac{1}{3}l^2\dot{\theta}_2^2 + l^2[\cos(\theta_2 - \theta_1)]\dot{\theta}_1\dot{\theta}_2 + l\dot{u}(3\dot{\theta}_1\cos\theta_1 + \dot{\theta}_2\cos\theta_2)\right.$$

$$\left. +2\dot{u}^2\right] + \frac{3}{2}mgl\cos\theta_1 + \frac{1}{2}mgl\cos\theta_2$$

its partial derivatives being listed below:

$$\frac{\partial L}{\partial \dot{\theta}_1} = \frac{1}{2}m\left\{\frac{8}{3}l^2\dot{\theta}_1 + l^2[\cos(\theta_2 - \theta_1)]\dot{\theta}_2 + 3l\dot{u}\cos\theta_1\right\}$$

$$\frac{d}{dt}\left(\frac{\partial L}{\partial \dot{\theta}_1}\right) = \frac{1}{2}m\left\{\frac{8}{3}l^2\ddot{\theta}_1 + l^2[\cos(\theta_2 - \theta_1)]\ddot{\theta}_2 - l^2[\sin(\theta_2 - \theta_1)](\dot{\theta}_2 - \dot{\theta}_1)\dot{\theta}_2\right.$$

$$\left. +3l\ddot{u}\cos\theta_1 - 3l\dot{u}\dot{\theta}_1\sin\theta_1\right\}$$

$$\frac{\partial L}{\partial \theta_1} = \frac{1}{2}m\left\{l^2[\sin(\theta_2 - \theta_1)]\dot{\theta}_1\dot{\theta}_2 - 3l\dot{u}\dot{\theta}_1\sin\theta_1 - 3gl\sin\theta_1\right\}$$

$$\frac{\partial L}{\partial \dot{\theta}_2} = \frac{1}{2}m\left\{\frac{2}{3}l^2\dot{\theta}_2 + l^2[\cos(\theta_2 - \theta_1)]\dot{\theta}_1 + l\dot{u}\cos\theta_2\right\}$$

$$\frac{d}{dt}\left(\frac{\partial L}{\partial \dot{\theta}_2}\right) = \frac{1}{2}m\left\{\frac{2}{3}l^2\ddot{\theta}_2 + l^2[\cos(\theta_2 - \theta_1)]\ddot{\theta}_1 - l^2[\sin(\theta_2 - \theta_1)](\dot{\theta}_2 - \dot{\theta}_1)\dot{\theta}_1\right.$$

$$\left. +l\ddot{u}\cos\theta_2 - l\dot{u}\dot{\theta}_2\sin\theta_2\right\}$$

$$\frac{\partial L}{\partial \theta_2} = \frac{1}{2}m\left\{-l^2[\sin(\theta_2 - \theta_1)]\dot{\theta}_1\dot{\theta}_2 - l\dot{u}\dot{\theta}_2\sin\theta_2 - gl\sin\theta_2\right\}$$

5. **Power supplied**. The power supplied to the system is readily derived, as in the previous example, namely,

$$\Pi = \tau_1 \dot{\theta}_1 + \tau_2 (\dot{\theta}_2 - \dot{\theta}_1)$$

its partial derivatives yielding the generalized forces of ϕ_{f1} and ϕ_{f2}:

$$\phi_{f1} = \frac{\partial \Pi}{\partial \dot{\theta}_1} = \tau_1 - \tau_2, \quad \phi_{f2} = \frac{\partial \Pi}{\partial \dot{\theta}_2} = \tau_2$$

6. **Dissipation function**. Now we have

$$\Delta = \frac{1}{2} c_1 \dot{\theta}_1^2 + \frac{1}{2} c_2 (\dot{\theta}_2 - \dot{\theta}_1)^2$$

its partial derivatives yielding the dissipative forces of ϕ_{d1} and ϕ_{d2}:

$$\phi_{d1} = \frac{\partial \Delta}{\partial \dot{\theta}_1} = (c_1 + c_2) \dot{\theta}_1 - c_2 \dot{\theta}_2, \quad \phi_{d2} = \frac{\partial \Delta}{\partial \dot{\theta}_2} = c_2 (\dot{\theta}_2 - \dot{\theta}_1)$$

7. **Lagrange equations**. The Lagrange equations of the system at hand become

$$\frac{4}{3} ml^2 \ddot{\theta}_1 + \frac{1}{2} ml^2 [\cos(\theta_2 - \theta_1)] \ddot{\theta}_2 - \frac{1}{2} ml^2 [\sin(\theta_2 - \theta_1)] \dot{\theta}_2^2 = \frac{3}{2} mgl \sin \theta_1$$

$$- \frac{3}{2} ml \ddot{u} \cos \theta_1 + \tau_1 - \tau_2 - (c_1 + c_2) \dot{\theta}_1 + c_2 \dot{\theta}_2$$

$$\frac{1}{2} ml^2 [\cos(\theta_2 - \theta_1)] \ddot{\theta}_1 + \frac{1}{3} ml^2 \ddot{\theta}_2 + \frac{1}{2} ml^2 [\sin(\theta_2 - \theta_1)] \dot{\theta}_1^2 = \frac{1}{2} mgl \sin \theta_2$$

$$- \frac{1}{2} ml \ddot{u} \cos \theta_2 + \tau_2 - c_2 (\dot{\theta}_2 - \dot{\theta}_1)$$

where, interestingly, the terms in $\ddot{u}$ vanished.

The mass matrix, the $\mathbf{h}(\mathbf{q}, \dot{\mathbf{q}})$ vector and the generalized force vectors are readily identified as

$$\mathbf{M} = ml^2 \begin{bmatrix} 4/3 & (1/2) \cos(\theta_2 - \theta_1) \\ (1/2) \cos(\theta_2 - \theta_1) & 1/3 \end{bmatrix}$$

$$\mathbf{h}(\mathbf{q}, \dot{\mathbf{q}}) = \frac{1}{2} ml^2 [\sin(\theta_2 - \theta_1)] \begin{bmatrix} -\dot{\theta}_2^2 \\ \dot{\theta}_1^2 \end{bmatrix}$$

$$\phi_p = \frac{1}{2}mgl \begin{bmatrix} 3\sin\theta_1 \\ \sin\theta_2 \end{bmatrix}, \quad \phi_f = \begin{bmatrix} \tau_1 - \tau_2 \\ \tau_2 \end{bmatrix}$$

$$\phi_m = \frac{1}{2}ml\ddot{u}\cos\theta_1 \begin{bmatrix} 3 \\ 1 \end{bmatrix}, \quad \phi_d = \begin{bmatrix} -(c_1+c_2)\dot{\theta}_1 + c_2\dot{\theta}_2 \\ -c_2(\dot{\theta}_2 - \dot{\theta}_1) \end{bmatrix}$$

Note, again, that the vector of generalized gravity forces does not involve the generalized speeds, and that the matrix of generalized mass is positive-definite.

4.3 Equilibrium States

In this section we will study the equilibrium states of n-dof mechanical systems. Moreover, for the sake of simplicity, we will assume that the system is described by a set of n independent generalized coordinates. The system at hand can thus be assumed to be governed by Eq. 4.15. This system is in an equilibrium state if $\dot{q} = 0$ and $\ddot{q} = 0$, which implies that, at equilibrium, q attains a constant value q_E, determined from the equilibrium equations. These are then derived from Eq. 4.15, by simply setting $\dot{q} = 0$ and $\ddot{q} = 0$ in this equation, thereby obtaining an equation formally identical to Eq. 1.46, namely,

$$\phi(q_E, 0) = 0 \tag{4.19}$$

because $h(q, \dot{q})$, being quadratic homogeneous in $\dot{q}$, vanishes at equilibrium.

The equilibrium equation is thus obtained upon solving the system of n algebraic[3] equations (4.19) in the n unknown components of vector q_E. Note that this system of equations is, in general, nonlinear, and hence, may or may not admit a real solution. Moreover, if it admits a real solution, this is, in general, not unique. The solution of systems of nonlinear algebraic equations can be quite challenging, but the examples with which we will be concerned lend themselves to rather straightforward, if ad-hoc solutions, as will be illustrated with the examples below.

Example 4.3.1 (Equilibrium Configurations of the Two-Link Robot). Obtain the equilibrium configurations of Example 4.2.1, when the motors are turned off and assumed to apply no braking torques.

Solution: Upon setting $\dot{\theta}_k = 0$, $\ddot{\theta}_k = 0$ and $\tau_k = 0$, for $k = 1, 2$, in the Lagrange equations of Example 4.2.1, we obtain the equilibrium equations in the form

$$g(m_1\rho_1 + m_2a)\sin\theta_1 = 0$$

$$m_2g\rho_2\sin\theta_2 = 0$$

---

[3] As opposed to differential.

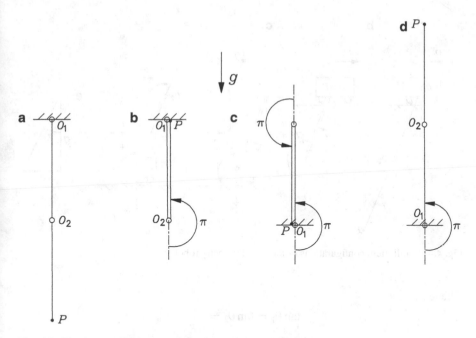

Fig. 4.3 The four equilibrium configurations of a two-link robot

whence it is apparent that, at an equilibrium state, we have

$$\sin \theta_1 = 0, \quad \sin \theta_2 = 0$$

Thus, the system is at equilibrium whenever the joints are at rest and the center of mass C_2 is aligned with the two joint centers, all these points lying on a vertical. It is apparent, moreover, that four such equilibrium configurations are possible. Indeed, let us call joint O_2 the *elbow*, a common terminology in robotics. There are two equilibrium configurations of the elbow-down type and two of the elbow-up. Each of these configurations corresponds to the arm either fully extended or fully retracted, as displayed in Fig. 4.3.

Example 4.3.2 (Equilibrium Configurations of the Gantry Robot). Determine the equilibrium configurations of the system shown in Fig. 4.2 under $\ddot{u} \equiv a = \text{const}$, when the motors are turned off and exert no braking torque.

Solution: We start by setting $\dot{\theta}_k = 0$, $\ddot{\theta}_k = 0$ and $\tau_k = 0$, for $k = 1, 2$, in the governing equations derived in Example 4.2.2 above. The equilibrium equations now take the form

$$\frac{3}{2} mgl \sin \theta_1 = -\frac{3}{2} mla \cos \theta_1$$

$$\frac{1}{2} mgl \sin \theta_2 = -\frac{1}{2} mla \cos \theta_2$$

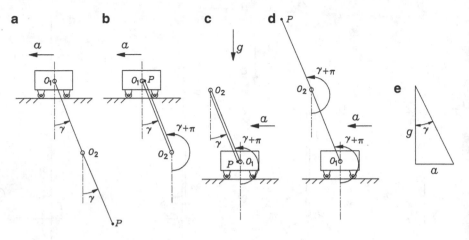

Fig. 4.4 Equilibrium configurations of the two-dof gantry robot

whence

$$\tan \theta_1 = \tan \theta_2 = -\frac{a}{g}$$

Let us define γ as

$$\gamma = \arctan\left(-\frac{a}{g}\right)$$

Here, we recall that, $\tan(x) = \tan(x + \pi)$; as a consequence, there are four possible solutions to the equilibrium equations, and thus, four equilibrium states, given by $(\theta_1, \theta_2) = (\gamma, \gamma)$, $(\gamma, \gamma + \pi)$, $(\gamma + \pi, \gamma)$, $(\gamma + \pi, \gamma + \pi)$, which are shown in Fig. 4.4a–d, respectively, the relation among a, g and γ appearing in Fig. 4.4e. Here, note that the negative sign in the argument of the $\arctan(\cdot)$ function indicates that an acceleration a to the left, i.e., negative according to the positive direction of $u(t)$ assumed in Fig. 4.2, produces an angle γ ccw from the vertical.

Example 4.3.3 (A Time-Varying Equilibrium State). Shown in Fig. 4.5 is the overhead crane of Example 1.6.5, but now with the motor that drives the cart providing not a controlled motion, but rather a controlled force $F(t)$; likewise, the motion of the rod is now controlled by a motor providing a torque $\tau(t)$. Find the Lagrange equations of the system under the above conditions, and then determine the equilibrium states of this system under the assumption that damping is negligible and that $F = 0$ and $\tau = 0$.

Solution: The translation of the cart becomes now a generalized coordinate, that we label x. We thus have a two-dof system with vector of generalized coordinates defined as $\mathbf{q} \equiv [x, \theta]^T$, its kinetic energy taking a form similar to that derived in Example 1.6.5, namely,

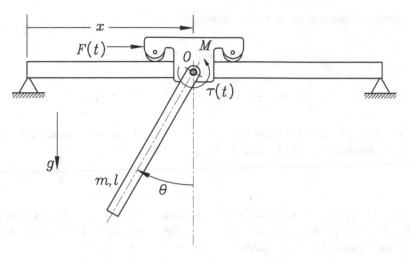

Fig. 4.5   A two-dof overhead crane

$$T = \frac{1}{2}(M+m)\dot{x}^2 + \frac{1}{6}ml^2\dot{\theta}^2 - \frac{1}{2}ml(\cos\theta)\dot{x}\dot{\theta}$$

while the potential energy is the same as in that example, i.e., $V = -(1/2)mgl\cos\theta$, and hence,

$$L = \frac{1}{2}(M+m)\dot{x}^2 + \frac{1}{6}ml^2\dot{\theta}^2 - \frac{1}{2}ml(\cos\theta)\dot{x}\dot{\theta} + \frac{1}{2}mgl\cos\theta$$

The partial derivatives needed to set up the Lagrange equations of the system are now

$$\frac{\partial L}{\partial \dot{\mathbf{q}}} = \begin{bmatrix} (M+m)\dot{x} - \frac{1}{2}ml\dot{\theta}\cos\theta \\ \frac{1}{3}ml^2\dot{\theta} - \frac{1}{2}ml\dot{x}\cos\theta \end{bmatrix}$$

$$\frac{d}{dt}\left(\frac{\partial L}{\partial \dot{\mathbf{q}}}\right) = \begin{bmatrix} (M+m)\ddot{x} - \frac{1}{2}ml(\ddot{\theta}\cos\theta - \dot{\theta}^2\sin\theta) \\ \frac{1}{3}ml^2\ddot{\theta} - \frac{1}{2}ml\ddot{x}\cos\theta + \frac{1}{2}ml\dot{x}\dot{\theta}\sin\theta \end{bmatrix}$$

$$\frac{\partial L}{\partial \mathbf{q}} = \begin{bmatrix} 0 \\ \frac{1}{2}ml\dot{x}\dot{\theta}\sin\theta - \frac{1}{2}mgl\sin\theta \end{bmatrix}$$

Moreover,

$$\Pi = F\dot{x} + \tau\dot{\theta}, \quad \Delta = 0$$

the Lagrange equations of the system thus being

$$(M+m)\ddot{x} - \frac{1}{2}ml(\ddot{\theta}\cos\theta - \dot{\theta}^2\sin\theta) = F(t)$$

$$\frac{1}{3}ml^2\ddot{\theta} - \frac{1}{2}ml\ddot{x}\cos\theta + \frac{1}{2}mgl\sin\theta = \tau(t)$$

Upon setting $\dot{x}$, $\dot{\theta}$, $\ddot{x}$, $\ddot{\theta}$, $F(t)$ and $\tau(t)$ equal to zero in the above equations, the first becomes identically satisfied, while the second yields

$$\sin\theta = 0$$

and hence, equilibrium is possible for any constant value $\dot{x}_E = v_0$, as long as $\sin\theta = 0$, which yields the two equilibrium configurations $\theta_E = 0$ or π, corresponding to rod-down and rod-up, exactly as in Example 1.6.5.

Note that, such as in the case of single-dof systems, the above time-varying equilibrium configuration occurs concurrently with $\partial L/\partial x = 0$, and the governing equation associated with x is conservative and autonomous. Thus, we have here a similar condition for the occurrence of time-varying equilibrium states to that derived in the single-dof case, i.e., *a time-varying equilibrium state of an autonomous system is possible if (1) the governing equation associated with the generalized coordinate q_k is free of dissipative terms, and (2) $\partial L/\partial q_k = 0$.*

In this example we have two generalized coordinates that are of different dimensions, one having units of length, the other being dimensionless. This special feature brings about interesting consequences. In fact, the mass matrix of this system takes the form

$$\mathbf{M} = \begin{bmatrix} M+m & -\frac{1}{2}ml\cos\theta \\ -\frac{1}{2}ml\cos\theta & \frac{1}{2}ml^2 \end{bmatrix}$$

Apparently this matrix is symmetric, but we would be interested in verifying whether it is positive-definite as well. In order to do this, we recall our well-known rule of verifying positive definiteness of 2×2 matrices based upon the signs of the trace and the determinant of the matrix under study, which give two necessary and sufficient conditions for our purpose. So, let us find the trace of this matrix:

$$\mathrm{tr}(\mathbf{M}) = M + m + \frac{1}{3}ml^2$$

which is *physically meaningless* because we cannot compute the sum in the right-hand side. Note that the third term has units of moment of inertia, while the first two have units of mass.

So as to verify the positive-definiteness of $\mathbf{M}$, the reader is invited to compute its eigenvalues and verify that both, although bearing different units, are positive.

Alternatively, let us make a change of variables, by redefining the vector of generalized coordinates in dimensionally homogeneous form, namely, as

$$\mathbf{q} \equiv \begin{bmatrix} x/l \\ \theta \end{bmatrix} \equiv \begin{bmatrix} \xi \\ \theta \end{bmatrix}$$

Now we rewrite the governing equations in terms of the new set of generalized coordinates, thereby obtaining

$$(M+m)l\ddot{\xi} - \frac{1}{2}ml(\ddot{\theta}\cos\theta - \dot{\theta}^2\sin\theta) = F(t)$$

$$\frac{1}{3}ml\ddot{\theta} - \frac{1}{2}ml\ddot{\xi}\cos\theta + \frac{1}{2}mg\sin\theta = \frac{\tau(t)}{l}$$

where we have divided the two sides of the second equation by l. Now the mass matrix becomes

$$\mathbf{M} = \begin{bmatrix} (M+m)l & -\frac{1}{2}ml\cos\theta \\ -\frac{1}{2}ml\cos\theta & \frac{1}{2}ml \end{bmatrix}$$

which is now dimensionally homogeneous, its trace being

$$\text{tr}(\mathbf{M}) = (M+m)l + \frac{1}{2}ml = Ml + \frac{3}{2}ml$$

and has units of Kg · m. We might as well have multiplied the two sides of the first of the last two equations by l, while leaving the second unchanged. The mass matrix thus resulting would have had units of moment of inertia. In any case, the mass matrix can be proven to be positive-definite.

4.4 Linearization of the Governing Equations About Equilibrium States

We now derive the governing equations upon 'small' perturbations of the equilibrium states. To this end, we introduce perturbations $\delta\mathbf{q}$, $\delta\dot{\mathbf{q}}$ and $\delta\ddot{\mathbf{q}}$, which lead to

$$\mathbf{q} \equiv \mathbf{q}_E + \delta\mathbf{q}, \quad \dot{\mathbf{q}} \equiv \delta\dot{\mathbf{q}}, \quad \ddot{\mathbf{q}} \equiv \delta\ddot{\mathbf{q}} \tag{4.20}$$

Upon conducting an analysis similar to that of Sect. 1.10, we end up with the n-dof counterpart of Eq. 1.53a, namely,

$$\mathbf{M}_E\delta\ddot{\mathbf{q}} + \mathbf{C}_E\delta\dot{\mathbf{q}} + \mathbf{K}_E\delta\mathbf{q} = \delta\boldsymbol{\phi}(t)$$

which is a system of n second-order ODEs in $\delta\mathbf{q}$, with the three $n \times n$ matrices $\mathbf{M}_E$, $\mathbf{C}_E$ and $\mathbf{K}_E$ defined as the $n \times n$ matrix counterparts of the scalar coefficients m_E,

c_E and k_E, respectively, of Eq. 1.53a, with the definitions below that parallel those of Eq. 1.53b, namely,

$$\mathbf{M}_E \equiv \mathbf{M}(\mathbf{q}_E), \quad \mathbf{C}_E \equiv -\left.\frac{\partial \boldsymbol{\phi}}{\partial \dot{\mathbf{q}}}\right|_{(\mathbf{q}_E, \mathbf{0})}, \quad \mathbf{K}_E \equiv -\left.\frac{\partial \boldsymbol{\phi}}{\partial \mathbf{q}}\right|_{(\mathbf{q}_E, \mathbf{0})} \tag{4.21}$$

For brevity, when the foregoing linearization process is self-understood, we shall delete the subscript E and the symbol δ from the above equation, which thus takes a simplified form, namely,

$$\mathbf{M}\ddot{\mathbf{q}} + \mathbf{C}\dot{\mathbf{q}} + \mathbf{K}\mathbf{q} = \boldsymbol{\phi}(t) \tag{4.22}$$

The above system of equations is the vector counterpart of the scalar Eq. 1.57. Just as in the scalar case, we can now determine the nature of the equilibrium states based on the matrix coefficients of Eq. 4.22. For quick reference, we recall here that a system is said to be autonomous when all its motors are turned off and no external perturbations act upon it. As well, we recall that the scalar linearized equation of an autonomous system is asymptotically stable if and only if all its coefficients are positive; it is marginally stable when all its coefficients are positive, except for that of the velocity term, which vanishes. Without entering in the details of the proof, which we can skip at an introductory level, we can state similar conditions for the stability of Eq. 4.22, but here care must be taken in that the coefficients of this equation are now matrices and a matrix cannot be called, properly speaking, positive or negative. However, if all the eigenvalues of a matrix are positive, then the matrix is termed *positive-definite*; if none is negative, then the matrix is termed *positive-semidefinite*. Similar definitions apply to *negative-definite* and *negative-semidefinite* matrices. Note, however, that, if a matrix is neither positive-definite nor semidefinite, it is not necessarily negative-definite or semidefinite. Matrices with positive and negative eigenvalues are termed *sign-indefinite*.

The stability conditions for Eq. 4.22, *with its right-hand side equated to zero*, are thus summarized as:

1. Equation 4.22 is asymptotically stable if all its coefficients are positive-definite, except for $\mathbf{K}$, which can be positive-semidefinite
2. Equation 4.22 is marginally stable if coefficient $\mathbf{C}$ is positive-semidefinite, but all others are positive-definite

Note that the above conditions are sufficient, but not necessary. In fact, under condition 1, $\mathbf{C}$ should be positive-definite, and hence, symmetric, for $\mathbf{C}$ is supposed to have real eigenvalues. However, when *gyroscopic* forces are present, like in systems mounted on platforms turning under controlled motion, then $\mathbf{C}$ is the sum of a symmetric part and a skew-symmetric part. Moreover, if the symmetric part is positive-definite, then the system is asymptotically stable, for the gyroscopic forces do not contribute to destabilizing the system. Gyroscopic forces lie beyond the scope of this book.

When all matrix coefficients of Eq. 4.22 are at least positive-semidefinite, they are directly associated with our basic mechanical elements introduced in Chap. 1. In fact, it is apparent that the mass matrix $\mathbf{M}$ is associated with mass, which is the reason why it is called the mass matrix. Likewise, when $\mathbf{C}$ and $\mathbf{K}$ are at least positive-semidefinite, they are called the *damping matrix* and the *stiffness matrix*. Moreover, because $\mathbf{M}$ is associated with mass in any instance, it is always at least positive-semidefinite. In fact, if every generalized coordinate is associated with a non-negligible inertia, then $\mathbf{M}$ is positive-definite. This matrix becomes semidefinite only when some generalized coordinates do not represent the displacement, whether angular or translational, of a rotor or, correspondingly, of a mass. In this book we will discuss only systems with a positive-definite mass matrix.

Example 4.4.1 (Linearized Model of a Two-link Robot). Out of the four possible equilibrium configurations of the system of Example 4.3.1, analyze that at which $\theta_k = 0$, for $k = 1, 2$, with regard to stability.

Solution: In order to proceed with the linearization of the mathematical model of the system at hand, let us perturb the generalized variables—coordinates, speeds and accelerations—at the equilibrium configuration. For compactness, we shall group the generalized coordinates θ_k, for $k = 1, 2$, into vector θ, the generalized speeds being grouped into vector $\dot{\theta}$, while the generalized accelerations are grouped into vector $\ddot{\theta}$, and the generalized torques are grouped into vector τ. The perturbed variables are now

$$\theta = \delta\theta, \quad \dot{\theta} = \delta\dot{\theta}, \quad \ddot{\theta} = \delta\ddot{\theta}, \quad \tau = \delta\tau$$

In the linearization process we shall need the trigonometric functions of the perturbed variables. These are readily computed below for quick reference:

$$\cos(\delta\theta_k) \approx 1, \quad \sin(\delta\theta_k) \approx \delta\theta_k$$

$$\cos(\delta\theta_2 - \delta\theta_1) \approx 1 + \delta\theta_1\delta\theta_2 \approx 1, \quad \sin(\delta\theta_2 - \delta\theta_1) \approx \delta\theta_2 - \delta\theta_1$$

where we have considered that $\delta\theta_1\delta\theta_2 \approx 0$ because it is a second-order term. Upon substituting the foregoing values into the governing equations, the linearized equations are derived as

$$(J_1 + m_2 a^2)\delta\ddot{\theta}_1 + m_2 a\rho_2\delta\ddot{\theta}_2 + (c_1 + c_2)\delta\dot{\theta}_1 - c_2\delta\dot{\theta}_2 + g(m_1\rho_1 + m_2 a)\delta\theta_1$$
$$= \delta\tau_1 - \delta\tau_2$$

$$m_2 a\rho_2\delta\ddot{\theta}_1 + (J_2 + m_2\rho_2^2)\delta\ddot{\theta}_2 + c_2(\delta\dot{\theta}_2 - \delta\dot{\theta}_1) + m_2 g\rho_2\delta\theta_2 = \delta\tau_2$$

From the above linearized model it is now a simple matter to identify the mass, damping and stiffness matrices, as well as the generalized force. These are

$$
\mathbf{M} = \begin{bmatrix} J_1 + m_2 a^2 & m_2 a \rho_2 \\ m_2 a \rho_2 & J_2 + m_2 \rho_2^2 \end{bmatrix}, \quad
\mathbf{C} = \begin{bmatrix} c_1 + c_2 & -c_2 \\ -c_2 & c_2 \end{bmatrix}
$$

$$
\mathbf{K} = g \begin{bmatrix} m_1 \rho_1 + m_2 a & 0 \\ 0 & m_2 \rho_2 \end{bmatrix}, \quad
\phi = \begin{bmatrix} \delta \tau_1 - \delta \tau_2 \\ \delta \tau_2 \end{bmatrix}
$$

Note that here gravity plays the role of springs, the stiffness matrix thus being proportional to g. Moreover, all three matrices above are positive-definite, and hence, in the absence of external torques τ_k, for $k = 1, 2$, the system is asymptotically stable. Since real-life joints always carry a certain amount of damping, robotic manipulators of this kind are asymptotically stable under small-amplitude motions, a rather comforting result for robotics engineers. The positive-definiteness of the three above matrices may not be apparent at first glance, but it is there. Indeed, all it takes to prove positive-definiteness is to calculate the trace and the determinant of these matrices, which are both positive, for each matrix.

Example 4.4.2 (Linearized Model of a Gantry Robot). For the system of Fig. 4.2, derive the linearized equations of at the equilibrium configuration of Fig. 4.4a. Moreover, in the absence of perturbations, i.e., for $\delta \ddot{u} = 0$, is this configuration stable? unstable? marginally stable?

Solution: First, we rewrite the governing equations in a simpler form, by multiplying their two sides by 2 and dividing them by ml^2, which thus leads to

$$
\frac{8}{3} \ddot{\theta}_1 + [\cos(\theta_2 - \theta_1)] \ddot{\theta}_2 - [\sin(\theta_2 - \theta_1)] \dot{\theta}_2^2 + \frac{3g}{l} \sin \theta_1
$$

$$
+ \frac{2(c_1 + c_2)}{ml^2} \dot{\theta}_1 - \frac{2c_2}{ml^2} \dot{\theta}_2 = \frac{2}{ml^2} (\tau_1 - \tau_2) - \frac{3}{l} \ddot{u} \cos \theta_1
$$

$$
[\cos(\theta_2 - \theta_1)] \ddot{\theta}_1 + \frac{2}{3} \ddot{\theta}_2 + [\sin(\theta_2 - \theta_1)] \dot{\theta}_1^2 + \frac{g}{l} \sin \theta_2
$$

$$
+ \frac{2c_2}{l^2} (\dot{\theta}_2 - \dot{\theta}_1) = \frac{2}{ml^2} \tau_2 - \frac{1}{l} \ddot{u} \cos \theta_2
$$

Next, we substitute the generalized variables of the above equations by their perturbed values at the equilibrium states found in Example 4.3.2. Of these states, we choose to linearize about the *elbow-down* configuration. Thus, we have now, for $k = 1, 2$,

$$
\theta_k \leftarrow \gamma + \delta \theta_k, \quad \dot{\theta}_k \leftarrow \delta \dot{\theta}_k, \quad \ddot{\theta}_k \leftarrow \delta \ddot{\theta}_k
$$

Moreover, we set $\ddot{u} = a + \delta\ddot{u}$, where a is a prescribed constant value of $\ddot{u}$ at equilibrium, and hence, the linearized equations take the form

$$\frac{8}{3}\delta\ddot{\theta}_1 + \cos(\delta\theta_2 - \delta\theta_1)\delta\ddot{\theta}_2 - \sin(\delta\theta_2 - \delta\theta_1)(\delta\dot{\theta}_2)^2 + \frac{3g}{l}\sin(\gamma + \delta\theta_1)$$

$$+\frac{2(c_1 + c_2)}{ml^2}\delta\dot{\theta}_1 - \frac{2c_2}{ml^2}\delta\dot{\theta}_2 = -3\frac{a + \delta\ddot{u}}{l}\cos(\gamma + \delta\theta_1)$$

$$\cos(\delta\theta_2 - \delta\theta_1)\delta\ddot{\theta}_1 + \frac{2}{3}\delta\ddot{\theta}_2 + \sin(\delta\theta_2 - \delta\theta_1)(\delta\dot{\theta}_1)^2 + \frac{g}{l}\sin(\gamma + \delta\theta_2)$$

$$+\frac{2c_2}{ml^2}(\delta\dot{\theta}_2 - \delta\dot{\theta}_1) = -\frac{a + \delta\ddot{u}}{l}\cos(\gamma + \delta\theta_2)$$

We note that

$$\cos(\delta\theta_2 - \delta\theta_1) \approx 1, \qquad \sin(\delta\theta_2 - \delta\theta_1) \approx \delta\theta_2 - \delta\theta_1$$

Moreover, at the elbow-down configuration

$$\cos\gamma = \frac{g}{\sqrt{a^2 + g^2}}, \qquad \sin\gamma = \frac{-a}{\sqrt{a^2 + g^2}}$$

Hence,

$$\sin(\gamma + \delta\theta_k) = \sin\gamma\cos\delta\theta_k + \cos\gamma\sin\delta\theta_k \approx \sin\gamma + (\cos\gamma)\delta\theta_k$$

$$= \frac{-a + g\delta\theta_k}{\sqrt{a^2 + g^2}}$$

$$\cos(\gamma + \delta\theta_k) = \cos\gamma\cos\delta\theta_k - \sin\gamma\sin\delta\theta_k \approx \cos\gamma - (\sin\gamma\delta)\theta_k = \frac{g + a\delta\theta_k}{\sqrt{a^2 + g^2}}$$

After retaining only constant and linear terms, and taking into account the equilibrium equations, the linearized model becomes

$$\frac{8}{3}\delta\ddot{\theta}_1 + \delta\ddot{\theta}_2 + \frac{2(c_1 + c_2)}{ml^2}\delta\dot{\theta}_1 - \frac{2c_2}{ml^2}\delta\dot{\theta}_2 + \frac{3}{l}\sqrt{a^2 + g^2}\delta\theta_1 = -\frac{3g}{l\sqrt{a^2 + g^2}}\delta\ddot{u}$$

$$\delta\ddot{\theta}_1 + \frac{2}{3}\delta\ddot{\theta}_2 + \frac{2c_2}{ml^2}\delta\dot{\theta}_2 - \frac{2c_2}{ml^2}\delta\dot{\theta}_1 + \frac{\sqrt{a^2 + g^2}}{l}\delta\theta_2 = -\frac{g}{l\sqrt{a^2 + g^2}}\delta\ddot{u}$$

Hence, the linearized equations can be cast in the form of Eq. 4.22, with the two-dimensional vectors of generalized coordinates, generalized speeds, generalized accelerations and generalized forces $\mathbf{q}$, $\dot{\mathbf{q}}$, $\ddot{\mathbf{q}}$ and ϕ, respectively, defined below, along with the mass, damping and stiffness matrices:

$$\mathbf{q} \equiv \begin{bmatrix} \delta\theta_1 \\ \delta\theta_2 \end{bmatrix}, \quad \dot{\mathbf{q}} \equiv \begin{bmatrix} \delta\dot{\theta}_1, \\ \delta\dot{\theta}_2 \end{bmatrix}, \quad \ddot{\mathbf{q}} \equiv \begin{bmatrix} \delta\ddot{\theta}_1 \\ \delta\ddot{\theta}_2 \end{bmatrix}, \quad \phi \equiv \begin{bmatrix} -3B \\ -B \end{bmatrix} \frac{\delta\ddot{u}}{l}$$

$$\mathbf{M} \equiv \begin{bmatrix} 8/3 & 1 \\ 1 & 2/3 \end{bmatrix}, \quad \mathbf{C} \equiv \frac{2}{ml^2} \begin{bmatrix} c_1 + c_2 & -c_2 \\ -c_2 & c_2 \end{bmatrix}, \quad \mathbf{K} \equiv \begin{bmatrix} 3A & 0 \\ 0 & A \end{bmatrix}$$

where

$$A = \frac{\sqrt{a^2 + g^2}}{l}, \qquad B = \frac{g}{l\sqrt{a^2 + g^2}}$$

Now it is apparent that the stiffness matrix is composed of both g and the prescribed constant acceleration a.

Finally, a quick calculation of the trace and the determinant of each of $\mathbf{M}$, $\mathbf{C}$ and $\mathbf{K}$ shows that these three matrices are positive-definite. Therefore, in the absence of perturbations, the equilibrium configuration of Fig. 4.4a is asymptotically stable.

4.5   Lagrange Equations of Linear Mechanical Systems

When we know beforehand that a mechanical system is linear, its Lagrange equations can be derived with much less effort than in the general case. In particular, we will be concerned only with *constant-coefficient* linear mechanical systems of the form of Eq. 4.22, in which all matrix coefficients are constant. Henceforth, these systems will be referred to simply as *linear*, for the sake of brevity, the constancy of their matrix coefficients being implicit. An important property of systems of this class is that their kinetic energy is both *independent* of the generalized coordinates and *quadratic* in the generalized speeds, while their potential energy is *quadratic* in the generalized coordinates and, as usual, *independent* of the generalized speeds.

A few preliminary definitions are now introduced. A *quadratic form* Q of n variables x_1, x_2, $\ldots$, x_n takes the general form

$$Q = \sum_{i=1}^{n} a_{ii}x_i^2 + 2\sum_{i=1}^{n}\sum_{j=i+1}^{n} a_{ij}x_ix_j + \sum_{i=1}^{n} b_i x_i + Q_0 \qquad (4.23a)$$

which can be rewritten alternatively in terms of the two n-dimensional vectors $\mathbf{x}$ and $\mathbf{b}$, and a *symmetric* matrix $\mathbf{A}$, all of which are defined as

$$\mathbf{x} \equiv \begin{bmatrix} x_1 \\ x_2 \\ \vdots \\ x_n \end{bmatrix}, \quad \mathbf{b} \equiv \begin{bmatrix} b_1 \\ b_2 \\ \vdots \\ b_n \end{bmatrix}, \quad \mathbf{A} \equiv \begin{bmatrix} a_{11} & a_{12} & \cdots & a_{1n} \\ a_{12} & a_{22} & \cdots & a_{2n} \\ \vdots & \vdots & \ddots & \vdots \\ a_{1n} & a_{2n} & \cdots & a_{nn} \end{bmatrix} \qquad (4.23b)$$

and hence,

$$Q = \mathbf{x}^T \mathbf{A} \mathbf{x} + \mathbf{b}^T \mathbf{x} + Q_0 \tag{4.23c}$$

What is interesting in this context is that the potential and kinetic energies of linear mechanical systems are *quadratic* in the generalized coordinates and the generalized speeds, respectively. Moreover, the power supplied to the system by external actions is *linear* in the generalized speeds, and the power dissipation is *independent* of the generalized coordinates and *quadratic* in the generalized speeds.

Now we turn to the energy functions of the system at hand. Although these can take a more general form, we will assume rather simple forms, namely,

$$T = \frac{1}{2}\dot{\mathbf{q}}^T \mathbf{M}\dot{\mathbf{q}} + \mathbf{p}^T\dot{\mathbf{q}} + T_0 \tag{4.24a}$$

$$V = \frac{1}{2}\mathbf{q}^T \mathbf{K}\mathbf{q} \tag{4.24b}$$

$$\Pi = \mathbf{f}_f^T\dot{\mathbf{q}} \tag{4.24c}$$

$$\Delta = \frac{1}{2}\dot{\mathbf{q}}^T \mathbf{C}\dot{\mathbf{q}} \tag{4.24d}$$

where the $n \times n$ constant matrices $\mathbf{M}$, $\mathbf{C}$ and $\mathbf{K}$ are the mass, damping and stiffness matrices of the system, as defined in Eq. 4.11 of Sect. 4.2. As well, $\mathbf{f}_f$ is the vector of generalized forces, as provided by actuators, while $\mathbf{p}$ is, as in the scalar case, a vector with units of generalized momentum. This vector thus contains inputs from controlled motions, and can be a function of $\mathbf{q}$, a function of time, or even a constant. For the systems under study, only the case in which $\mathbf{p}$ is independent of $\mathbf{q}$ and $\dot{\mathbf{q}}$ will be considered. Finally, T_0 is, in general, a function of $\mathbf{q}$ and time, but **not** of $\dot{\mathbf{q}}$. Again, for the systems of interest here, only the case in which T_0 is independent of $\mathbf{q}$ and $\dot{\mathbf{q}}$ will be considered. From Eq. 4.24a, b, d, it is apparent that the mass, damping and stiffness matrices can be calculated as the *Hessian* matrices of the kinetic energy, the dissipation function and the potential energy, respectively, the first two with respect to the generalized speeds, the third one with respect to the generalized coordinates. Here, we recall that a Hessian matrix is a matrix of second derivatives. In our case, since the order of differentiation is immaterial, the above Hessian matrices are all symmetric. We thus have

$$\mathbf{M} = \frac{\partial^2 T}{\partial \dot{\mathbf{q}}^2} \equiv \begin{bmatrix} \partial^2 T/\partial \dot{q}_1^2 & \partial^2 T/\partial \dot{q}_1 \partial \dot{q}_2 & \cdots & \partial^2 T/\partial \dot{q}_1 \partial \dot{q}_n \\ \partial^2 T/\partial \dot{q}_2 \partial \dot{q}_1 & \partial^2 T/\partial \dot{q}_2^2 & \cdots & \partial^2 T/\partial \dot{q}_2 \partial \dot{q}_n \\ \vdots & \vdots & \ddots & \vdots \\ \partial^2 T/\partial \dot{q}_n \partial \dot{q}_1 & \partial^2 T/\partial \dot{q}_n \partial \dot{q}_2 & \cdots & \partial^2 T/\partial \dot{q}_n^2 \end{bmatrix} \tag{4.25a}$$

$$
C = \frac{\partial^2 \Delta}{\partial \dot{q}^2} \equiv
\begin{bmatrix}
\partial^2 \Delta/\partial \dot{q}_1^2 & \partial^2 \Delta/\partial \dot{q}_1 \partial \dot{q}_2 & \cdots & \partial^2 \Delta/\partial \dot{q}_1 \partial \dot{q}_n \\
\partial^2 \Delta/\partial \dot{q}_2 \partial \dot{q}_1 & \partial^2 \Delta/\partial \dot{q}_2^2 & \cdots & \partial^2 \Delta/\partial \dot{q}_2 \partial \dot{q}_n \\
\vdots & \vdots & \ddots & \vdots \\
\partial^2 \Delta/\partial \dot{q}_n \partial \dot{q}_1 & \partial^2 \Delta/\partial \dot{q}_n \partial \dot{q}_2 & \cdots & \partial^2 \Delta/\partial \dot{q}_n^2
\end{bmatrix}
\tag{4.25b}
$$

$$
K = \frac{\partial^2 V}{\partial q^2} \equiv
\begin{bmatrix}
\partial^2 V/\partial q_1^2 & \partial^2 V/\partial q_1 \partial q_2 & \cdots & \partial^2 V/\partial q_1 \partial q_n \\
\partial^2 V/\partial q_2 \partial q_1 & \partial^2 V/\partial q_2^2 & \cdots & \partial^2 V/\partial q_2 \partial q_n \\
\vdots & \vdots & \ddots & \vdots \\
\partial^2 V/\partial q_n \partial q_1 & \partial^2 V/\partial q_n \partial q_2 & \cdots & \partial^2 V/\partial q_n^2
\end{bmatrix}
\tag{4.25c}
$$

While we have assumed at the outset that matrix M is *positive-definite*, C and K can be either *positive-semidefinite* or *positive-definite*.

Henceforth, a mechanical system will be characterized as linear if its energy functions take the forms given in Eq. 4.24a–d, with p and T_0 independent of q and $\dot{q}$. Moreover, deriving expressions for the above energy functions is straightforward, for this can be done separately, for each element at a time. The total amounts are then calculated by simple addition.

Given the simple forms of the energy functions of linear mechanical systems, their associated Lagrange equations take an equally simple form, namely,

$$
M\ddot{q} + C\dot{q} + Kq = -\dot{p} + f_f
\tag{4.26}
$$

where the independence of p and T_0 from q and $\dot{q}$ has been taken into account. In the above equation we can recognize the generalized force supplied by motion-controlled sources, f_m, as

$$
f_m = -\dot{p}
\tag{4.27}
$$

Example 4.5.1 (A Simple Model for the Whirling of Shafts). When a massive rotor, like in power-generation equipment, is mounted on a shaft, a slight offset of its c.o.m. can induce inertia forces that are large enough to produce a motion of the c.o.m. around the axis of the shaft—in the undeformed configuration of the shaft. This occurs because of the flexibility of the shaft and the bearings on which it is mounted. The same phenomenon can occur in moderately massive rotors when these spin at high-enough rates, as in grinding wheels. This phenomenon, known as *whirling of shafts*, is illustrated in Fig. 4.6a. A simple model of this situation considers that the rotor is a homogeneous rigid cylinder whose center of mass is offset with respect to the axis of the shaft. Moreover, the inertia forces and the flexibility of the mounting (shaft and bearings) produce vibration of the c.o.m. in the X-Y plane. The rotor, furthermore, is assumed to rotate only about an axis parallel to the neutral axis of the shaft, and hence, no gyroscopic effects are considered.

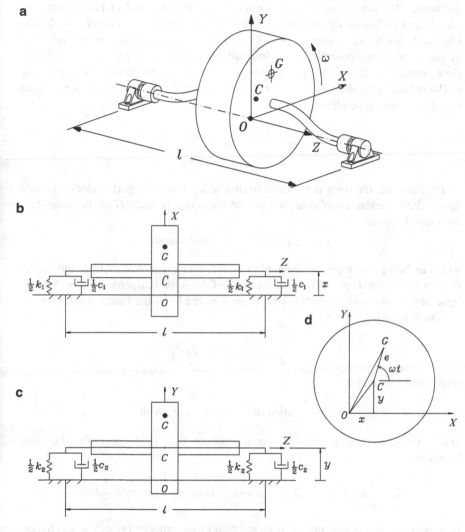

Fig. 4.6 (a) Rotor-shaft system under whirling (b–c) plant and (c–d) side views of rotor-shaft system

In reality, especially when high natural frequencies of the mounting are excited, the rotor shows precession and nutation, but we will not be concerned with these complex motions in this simple model. Thus, we can model the whole system as illustrated in Fig. 4.6b–d. In these figures, we assume that all the flexibility of shaft and bearings is lumped in the springs on which the shaft is mounted. Moreover, the shaft is now modeled as a rigid body that is rigidly coupled to the rotor. Derive the mathematical model of this system.

Solution: The system under study apparently has two dof, and so, two generalized coordinates suffice to describe its motion. Let the generalized coordinates be the Cartesian coordinates x and y of point C of the shaft axis contained in the mid plane of the rotor. These coordinates, additionally, are measured from the *equilibrium configuration* of the system; as a consequence, then, gravity forces will not appear in the model to be derived below. We can thus define the vectors of generalized coordinates and generalized speeds as

$$\mathbf{q} \equiv \begin{bmatrix} x \\ y \end{bmatrix}, \quad \dot{\mathbf{q}} \equiv \begin{bmatrix} \dot{x} \\ \dot{y} \end{bmatrix}$$

Furthermore, the rotor is assumed to turn at a constant angular velocity ω, and hence, the Cartesian coordinates u and v of the center of mass G of the rotor-shaft ensemble become

$$u = x + e\cos\omega t, \quad v = y + e\sin\omega t$$

where ω, being a constant, is assumed to stem from a controlled motion, and hence, *is not a generalized speed.* Now, let the mass of the rotor-shaft ensemble be denoted by m and its moment of inertia about G be denoted by J, the kinetic energy of the system thus taking the form

$$T = \frac{1}{2}\left[m(\dot{u}^2 + \dot{v}^2) + J\omega^2\right]$$

with $\dot{u}$ and $\dot{v}$ given by

$$\dot{u} = \dot{x} - e\omega\sin\omega t, \quad \dot{v} = \dot{y} + e\omega\cos\omega t$$

and so, the kinetic energy, expressed as a quadratic form of the generalized speeds, becomes

$$T = \frac{1}{2}m\left[\dot{x}^2 + \dot{y}^2 + 2\omega e(\dot{y}\cos\omega t - \dot{x}\sin\omega t) + \omega^2 e^2\right] + \frac{1}{2}J\omega^2$$

from which it is a simple matter to identify all three terms of Eq. 4.24a, and hence, $\mathbf{M}$ and $\mathbf{p}$. Indeed,

$$\frac{1}{2}\dot{\mathbf{q}}^T\mathbf{M}\dot{\mathbf{q}} = \frac{1}{2}m(\dot{x}^2 + \dot{y}^2), \quad \mathbf{p}^T\dot{\mathbf{q}} = m\omega e(-\dot{x}\sin\omega t + \dot{y}\cos\omega t), \quad T_0 = \frac{1}{2}(J + me^2)\omega^2$$

Therefore,

$$\mathbf{M} \equiv m\begin{bmatrix} 1 & 0 \\ 0 & 1 \end{bmatrix}, \quad \mathbf{p} \equiv m\omega e\begin{bmatrix} -\sin\omega t \\ \cos\omega t \end{bmatrix} \equiv m\omega\mathbf{E}(\mathbf{g} - \mathbf{c})$$

with $\mathbf{E}$ defined already in Sect. 1.6 and $\mathbf{c}$ and $\mathbf{g}$ defined as the position vectors of points C and G of Fig. 4.6d, respectively. Therefore, $\mathbf{p}$ is the momentum of a particle of mass m placed at point G and moving with a velocity ωe in a direction perpendicular to line CG.

On the other hand, the potential energy is simply

$$V = \frac{1}{2}k_1 x^2 + \frac{1}{2}k_2 y^2$$

while the power supplied to the system by controlled forces is zero and the dissipation function takes the form

$$\Delta = \frac{1}{2}c_1 \dot{x}^2 + \frac{1}{2}c_2 \dot{y}^2$$

Now it is a simple matter to determine the damping and stiffness matrices. This is done by application of the Hessian formulas, thereby obtaining

$$\mathbf{C} = \begin{bmatrix} c_1 & 0 \\ 0 & c_2 \end{bmatrix}, \quad \mathbf{K} = \begin{bmatrix} k_1 & 0 \\ 0 & k_2 \end{bmatrix}$$

Moreover, the generalized force arises solely from a motion-controlled source, i.e., $\mathbf{f} = \mathbf{f}_m$, the latter being given by Eq. 4.27. Note that, from the above expressions, $\mathbf{p}$ is a function solely of time, independent of the generalized coordinates and the generalized sapped, while T_0 is a constant. Hence,

$$\mathbf{f} = \mathbf{f}_m = -\dot{\mathbf{p}} = m\omega^2 e \begin{bmatrix} \cos \omega t \\ \sin \omega t \end{bmatrix}$$

That is, the generalized force of the system stems from a controlled motion, that of the rotor about its axis. It has become apparent that the two equations are decoupled, for all three matrices obtained above are diagonal. Hence, the system leads, in fact, to two independent single-dof systems. Upon dividing each of these by m, the said systems become

$$\ddot{x} + 2\zeta_1 \omega_1 \dot{x} + \omega_1^2 x = \omega^2 e \cos \omega t$$
$$\ddot{y} + 2\zeta_2 \omega_2 \dot{y} + \omega_2^2 y = \omega^2 e \sin \omega t$$

where ω_k and ζ_k, for $k = 1, 2$, are the natural frequencies and the damping ratios of the x and y equations, as defined in Eq. 1.56a, namely,

$$\omega_k \equiv \sqrt{\frac{k_k}{m}}, \quad \zeta_k \equiv \frac{c_k}{2m\omega_k}, \quad k = 1, 2$$

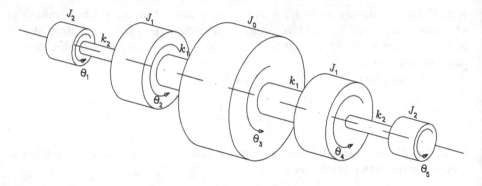

Fig. 4.7 A five-dof model of an aircraft wing

Example 4.5.2 (A Model for the Torsional Vibrations of Aircraft Wings). A simple model for the torsional-vibration analysis of aircraft wings is shown in Fig. 4.7, which consists of a flexible shaft of piecewise constant stiffness and five rotors. Note that the layout is symmetric about the mid-plane of the aircraft, and the shaft consists of four sections with different stiffnesses. The central rotor accounts for the aircraft body, while the other two rotors for the engines on each side. The generalized coordinates of the system are the angular displacements θ_k, for $k = 1, \ldots, 5$. Derive the mass and the stiffness matrices of this model.

Solution: We calculate the required matrices as the Hessians of the kinetic energy and the potential energy, and hence, we need expressions for these items. These expressions, with all angles measured from an inertial frame, are readily derived below:

$$T = \frac{1}{2}\left[J_2(\dot{\theta}_1^2 + \dot{\theta}_5^2) + J_1(\dot{\theta}_2^2 + \dot{\theta}_4^2) + J_0\dot{\theta}_3^2\right]$$

$$V = \frac{1}{2}\left[k_2[(\theta_2 - \theta_1)^2 + (\theta_5 - \theta_4)^2] + k_1[(\theta_3 - \theta_2)^2 + (\theta_4 - \theta_3)^2]\right]$$

Hence, the mass matrix is readily shown to be diagonal, namely,

$$\mathbf{M} = \mathrm{diag}(J_2, J_1, J_0, J_1, J_2) \equiv \begin{bmatrix} J_2 & 0 & 0 & 0 & 0 \\ 0 & J_1 & 0 & 0 & 0 \\ 0 & 0 & J_0 & 0 & 0 \\ 0 & 0 & 0 & J_1 & 0 \\ 0 & 0 & 0 & 0 & J_2 \end{bmatrix}$$

while the stiffness matrix takes the form

$$
\mathbf{K} = \begin{bmatrix}
k_2 & -k_2 & 0 & 0 & 0 \\
-k_2 & k_2 + k_1 & -k_1 & 0 & 0 \\
0 & -k_1 & 2k_1 & -k_1 & 0 \\
0 & 0 & -k_1 & k_2 + k_1 & -k_2 \\
0 & 0 & 0 & -k_2 & k_2
\end{bmatrix},
$$

It is apparent that matrix $\mathbf{M}$ is positive-definite, its eigenvalues being its diagonal entries, which are all positive. However, matrix $\mathbf{K}$ not being diagonal, its sign-definition via its eigenvalues is less apparent. Nevertheless, $\mathbf{K}$ is *tridiagonal* and exhibits the symmetries corresponding to the symmetry of the iconic model. This feature makes it possible to obtain its eigenvalues in closed form by means of computer algebra, namely,

$$
\lambda_1 = 0, \quad \lambda_2 = \frac{3}{2}k_1 + k_2 + \frac{1}{2}r_1, \quad \lambda_3 = \frac{3}{2}k_1 + k_2 - \frac{1}{2}r_1, \quad \lambda_4 = \frac{1}{2}k_1 + k_2 + \frac{1}{2}r_2,
$$
$$
\lambda_5 = \frac{1}{2}k_1 + k_2 - \frac{1}{2}r_2
$$

with r_1 and r_2 given by

$$
r_1 = \sqrt{9k_1{}^2 - 8k_1k_2 + 4k_2^2} \equiv \sqrt{5k_1^2 + 4(k_1 - k_2)^2}, \quad r_2 = \sqrt{k_1^2 + 4k_2^2}
$$

and λ_i, for $i = 1, \ldots, 5$ can be proven to be all positive. As $\mathbf{K}$ contains one zero eigenvalue and four positive eigenvalues, the matrix is positive-definite. The physical significance of this result is that, when all rotors turn by the same amount, the shaft stores no potential energy, which is apparent because, in this case, the wing and the aircraft body rotate as a rigid body. The existence of a semidefinite stiffness matrix thus leads to a motion that is called a *rigid mode*. These systems occur in other instances, less obvious than this example. More on these systems is discussed in Sect. 4.6.

Example 4.5.3 (A Model for the Vertical Vibration of Mass-transit Cars). Derive the mathematical model of the car shown in Fig. 1.1, which is mounted on two *bogies*, each carrying two wheel axles, the cars running on pneumatic tires. This design is seen in operation in subway systems of cities like Paris, Montreal, Mexico City and Sao Paolo. Because of the type of tires, each of the wheels of an axle must be allowed to rotate independently; otherwise, the tires would wear too often. Now, there are two types of cars, namely, *tractors*, with powered wheels, and *trailers*, with idle wheels. The axles of bogies of tractors are thus provided with a *differential gear train* to allow for power transmission to each of the two sections of the axle at different speeds. Furthermore, each axle is driven by a DC motor, that is integrated to the differential gear train, thereby forming a *motor-differential bridge*.

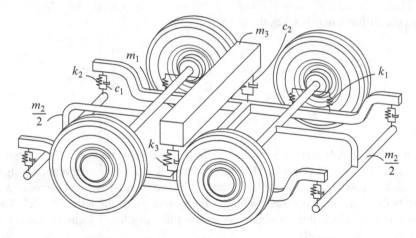

Fig. 4.8 Mechanical model of the suspension system

Solution: The *mechanical model* of the array described above is shown in Fig. 4.8, which is essentially that of Fig. 1.1, if with more detail. This model consists of an H-shaped structural element, which is for this reason termed *the H* in the subway jargon. Moreover, the suspension itself consists of two parts, the *primary* and the *secondary* suspension. The primary suspension is composed, in turn, of eight identical springs of stiffness k_1 and four more of stiffness k_2, where k_1 accounts for the coupling of the H to the axle and k_2 for the support of the motor-differential bridge. The car body is coupled to the H via a secondary suspension, composed of two identical springs of stiffness k_3. Furthermore, the spring stiffness of the rubber wheels is k_4. The model includes dashpots of damping coefficient c_1 and c_2, to account for either the natural damping of the primary and secondary suspensions or for that of shock absorbers. Referring to Fig. 4.8, we have the definitions below:

- m_1: mass of the chassis
- $m_2/2$: mass of each motor-differential bridge
- m_3: half the mass of the car body

The iconic model corresponding to the mechanical model of Fig. 4.8 appears in Fig. 4.9. In order to derive the mathematical model of the system appearing in Fig. 4.9, we define now the three-dimensional vector of generalized coordinates $\mathbf{x}$ as

$$\mathbf{x} = \begin{bmatrix} x_1, \, x_2, \, x_3 \end{bmatrix}^T$$

where all three components are measured from their equilibrium configurations, thereby doing away with gravity terms, as per the discussion of Example 1.10.4.

The mathematical model corresponding to Fig. 4.9 takes the form

$$\mathbf{M\ddot{x}} + \mathbf{C\dot{x}} + \mathbf{Kx} = \mathbf{0}$$

Fig. 4.9 Iconic model of
suspension system

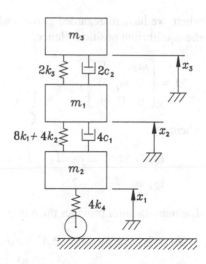

In the mathematical model displayed above, **M**, **C** and **K** are the 3×3 mass, damping and stiffness matrices, respectively, where the mass and the stiffness of the elements involved are supposed to be constant. Moreover, the matrices appearing in the above model can be obtained as the Hessian matrices of: the kinetic energy with respect to the vector $\dot{\mathbf{x}}$ of generalized velocities; the dissipation function with respect to the same; and of the potential energy with respect to the vector $\mathbf{x}$ of generalized coordinates, respectively, i.e.,

$$\mathbf{M} = \frac{\partial^2 T}{\partial \dot{\mathbf{x}}^2}, \quad \mathbf{C} = \frac{\partial^2 \Delta}{\partial \dot{\mathbf{x}}^2}, \quad \mathbf{K} = \frac{\partial^2 V}{\partial \mathbf{x}^2}$$

where T and V are the kinetic and the potential energies, respectively, while Δ is the dissipation function of the system. These are readily derived from Fig. 4.9, namely,

$$T = \frac{1}{2}m_1\dot{x}_2^2 + \frac{1}{2}m_2\dot{x}_1^2 + \frac{1}{2}m_3\dot{x}_3^2$$

$$\Delta = \frac{1}{2}(4c_1)(\dot{x}_2 - \dot{x}_1)^2 + \frac{1}{2}(2c_2)(\dot{x}_3 - \dot{x}_2)^2$$

and

$$V = \frac{1}{2}(8k_1 + 4k_2 + 4k_4)x_1^2 + \frac{1}{2}(8k_1 + 4k_2 + 2k_3)x_2^2$$

$$+ \frac{1}{2}2k_3x_3^2 - (8k_1 + 4k_2)x_1x_2 - 2k_3x_2x_3$$

where we have disregarded gravity, which is plausible if we measure each x_i from the equilibrium position. Hence,

$$\mathbf{M} = \begin{bmatrix} m_2 & 0 & 0 \\ 0 & m_1 & 0 \\ 0 & 0 & m_3 \end{bmatrix}, \quad \mathbf{C} = \begin{bmatrix} 4c_1 & -4c_1 & 0 \\ -4c_1 & 4c_1 + 2c_2 & -2c_2 \\ 0 & -2c_2 & 2c_2 \end{bmatrix}, \quad \mathbf{K} = \begin{bmatrix} k_{11} & k_{12} & 0 \\ k_{12} & k_{22} & k_{23} \\ 0 & k_{23} & k_{33} \end{bmatrix}$$

where

$$k_{11} = 8k_1 + 4k_2 + 4k_4, \quad k_{12} = -8k_1 - 4k_2, \quad k_{22} = 8k_1 + 4k_2 + 2k_3,$$

$$k_{23} = -k_{33} = -2k_3 \tag{4.28}$$

The manufacturer provides the numerical values given below:

$$k_1 = 4.9 \times 10^6, \quad k_2 = 3.43 \times 10^6, \quad k_3 = 8.37 \times 10^5, \quad k_4 = 1.783 \times 10^6$$

$$m_1 = 1.971 \times 10^3, \quad m_2 = 3.256 \times 10^3, \quad m_3 = 1.578 \times 10^4$$

where stiffness units are N/m and mass units are kg. In the latter, m_3 is half the mass of the car body under full load, i.e., when the cars are fully occupied by passengers. Dashpots are included in the model for completeness, although the suspension bears no shock absorbers. A study was conducted for the Mexico City *Sistema de Transporte Colectivo* with the purpose of deciding on the addition of shock absorbers to attenuate the vertical vibration that exhibited the cars of the time in the longest distances between stations [1]. According to this study, feasible values found are

$$c_1 = 19.61 \text{ kN s/m}, \quad c_2 = 123.6 \text{ kN s/m}$$

Moreover, for an angular velocity ω rad/s of the wheels, under no slipping, the velocity v of the train is given by

$$v = 0.5 \, d\omega$$

where $d = 0.960$ m is the diameter of the wheels, thereby completing the modeling of the undamped system.

Note that $\mathbf{M}$ and $\mathbf{K}$ are positive-definite, but $\mathbf{C}$ is positive-semidefinite. Indeed, under motions with $x_1 = x_2 = x_3$, the system dissipates no energy, which explains the semidefiniteness of $\mathbf{C}$.

4.6 Systems with Rigid Modes

Generally speaking, systems with rigid modes are those exhibiting non-trivial motions where no changes in the potential energy are experienced by the system. For example, in conservative systems, i.e., undamped systems, rigid modes occur

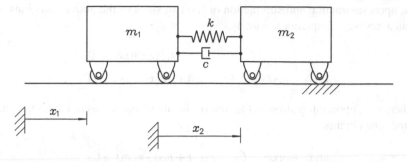

Fig. 4.10 A two-dof semidefinite system

whenever their stiffness matrix is positive-semidefinite, just as in Example 4.5.2 above. For this reason, these systems are also termed *semidefinite*. Note that semidefinite matrices always have at least one vanishing eigenvalue. Associated with this eigenvalue, semidefinite matrices have an eigenvector that is mapped into the zero vector when multiplied by the semidefinite matrix. Furthermore, a damped system can also be semidefinite if both its damping and stiffness matrices are semidefinite *and* they both share the same eigenvectors associated with their zero eigenvalues.

A damped semidefinite system is shown in Fig. 4.10, consisting of two masses coupled by a spring-dashpot parallel array, that can move otherwise freely on a flat surface.

As the reader can readily verify, the damping and stiffness matrices of this system are

$$\mathbf{C} = c\mathbf{A}, \quad \mathbf{K} = k\mathbf{A}, \quad \mathbf{A} = \begin{bmatrix} 1 & -1 \\ -1 & 1 \end{bmatrix}$$

Now, these two matrices are semidefinite, as both are proportional to matrix $\mathbf{A}$, whose eigenvalues are $\lambda_1 = 0$ and $\lambda_2 = 2$, its corresponding eigenvectors being

$$\mathbf{e}_1 = \frac{\sqrt{2}}{2} \begin{bmatrix} 1 \\ 1 \end{bmatrix}, \quad \mathbf{e}_2 = \frac{\sqrt{2}}{2} \begin{bmatrix} -1 \\ 1 \end{bmatrix}$$

the eigenvalues of $\mathbf{C}$ being $c\lambda_j$, those of $\mathbf{K}$ being $k\lambda_j$, for $j = 1, 2$. The mechanical significance of the eigenvalues of these matrices is better understood if we write the governing equations of the system in the absence of external forces:

$$m_1 \ddot{x}_1 + c(\dot{x}_1 - \dot{x}_2) + k(x_1 - x_2) = 0$$
$$m_2 \ddot{x}_2 - c(\dot{x}_1 - \dot{x}_2) - k(x_1 - x_2) = 0$$

Now, upon summation and subtraction of the two sides of the above equations, we obtain a new set of equations, namely,

$$m_1\ddot{x}_1 + m_2\ddot{x}_2 = 0$$

$$m_1\ddot{x}_1 - m_2\ddot{x}_2 + 2c(\dot{x}_1 - \dot{x}_2) + 2k(x_1 - x_2) = 0$$

Furthermore, upon integration of the first of the above equations twice with respect to time, one obtains

$$m_1\dot{x}_1 + m_2\dot{x}_2 = C_1, \quad m_1 x_1 + m_2 x_2 = c_1 t + C_2$$

where C_1 and C_2 are integration constants whose values depend on the given initial conditions. Now, from the above equations we can solve for, say x_2, in terms of x_1 and for $\dot{x}_2$ in terms of $\dot{x}_1$, namely,

$$\dot{x}_2 = \frac{C_1 - m_1\dot{x}_1}{m_2}, \quad x_2 = \frac{C_1 t + C_2 - m_1 x_1}{m_2}$$

Furthermore, we substitute the two above expressions in the second of the new set of governing equations derived above, which, after simplifications, leads to

$$\ddot{x}_1 + c\frac{m_1 + m_2}{m_1 m_2}\dot{x}_1 + k\frac{m_1 + m_2}{m_1 m_2}x_1 = \frac{cC_1 + kC_2}{m_2} + \frac{k}{m_2}C_1 t$$

and can be rewritten in normal form, namely, as

$$\ddot{x}_1 + 2\zeta_{eq}\omega_{eq}\dot{x}_1 + \omega_{eq}^2 x_1 = A + Bt$$

with the *equivalent* natural frequency and the *equivalent* damping ratio defined as

$$\omega_{eq} \equiv \sqrt{\frac{k}{m_{eq}}}, \quad \zeta_{eq} \equiv \frac{c}{2m_{eq}\omega_{eq}}$$

and the *equivalent mass* m_{eq} and constants A and B defined, in turn, as

$$m_{eq} \equiv \frac{m_1 m_2}{m_1 + m_2}, \quad A \equiv \frac{cC_1 + kC_2}{m_2}, \quad B \equiv \frac{k}{m_2}C_1$$

We have thus decoupled the *motion of the center of mass* from the *motion about the center of mass*. The former leads to a stationary c.o.m., the latter to the motion of mass m_{eq}, coupled to an inertial frame via a spring-dashpot parallel array with damping coefficient c and spring stiffness k.

The zero eigenvalue of the mass and damping matrices, thus, corresponds to the rigid mode, i.e., a harmonic motion with zero frequency, while the second eigenvalue leads to the motion of the equivalent mass-spring-dashpot system.

Fig. 4.11 A landing gear modeled as a double pendulum

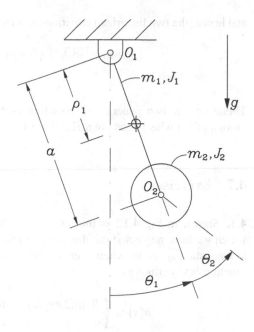

While the rigid mode of the foregoing example was straightforward to identify, other instances exist where the rigid mode is more elusive. We illustrate this idea with an example below.

Example 4.6.1 (A Landing Gear). Introduced in Example 4.2.1 is a system that can model a landing gear, as shown in Fig. 4.11. We assume that the wheel is statically balanced and hence, its center of mass coincides with its centroid O_2. Derive the governing equations of this system in the absence of driving torques, and show that this system admits a rigid mode. For simplicity, neglect viscous friction.

Solution: To simplify matters, we focus on the linearized model. For quick reference, we rewrite below the stiffness matrix derived in Example 4.4.1, with $\rho_2 = 0$, namely,

$$\mathbf{K} = g \begin{bmatrix} m_1\rho_1 + m_2a & 0 \\ 0 & 0 \end{bmatrix}$$

Since $\mathbf{K}$ is already in diagonal form, we need not calculate its eigenvalues; these are its diagonal entries, which include the factor g. Hence, $\mathbf{K}$ has one eigenvalue equal to zero, which indicates that it is positive-semidefinite, the system thus admitting a rigid mode. Moreover, the mass matrix simplifies to

$$\mathbf{M} = \begin{bmatrix} J_1 + m_2a^2 & 0 \\ 0 & J_2 \end{bmatrix}$$

and hence, the two linearized equations reduce to

$$(J_1 + m_2 a^2)\delta\ddot\theta_1 + g(m_1\rho_1 + m_2 a)\delta\theta_1 = 0$$
$$J_2\delta\ddot\theta_2 = 0$$

Therefore, the two generalized coordinates $\delta\theta_1$ and $\delta\theta_2$ are decoupled, i.e., the motion of the wheel does not affect that of the carrier, and vice versa.

4.7   Exercises

4.1. Shown in Fig. 4.12 is the iconic two-dof model of a terrestrial vehicle, in which we have neglected the damping of the suspension. The vehicle is travelling at a constant speed w, when it encounters a bump of height B, represented by the function $b(x)$ defined as

$$b(x) \equiv \begin{cases} B\sin(2\pi x/\lambda), & \text{for}\quad 0 \le x \le \lambda/2; \\ 0 & \text{otherwise} \end{cases}$$

Derive the mathematical model of the system.

4.2. Shown in Fig. 4.13 is the model of the indexing mechanism of a production machine, consisting of a crank that turns at constant angular velocity ω_0, and

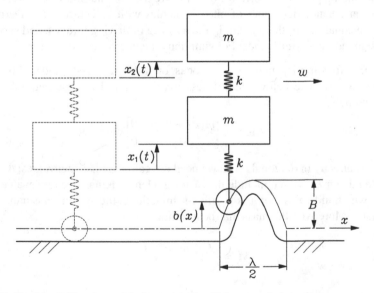

Fig. 4.12 The two-dof iconic model of a terrestrial vehicle encountering a bump

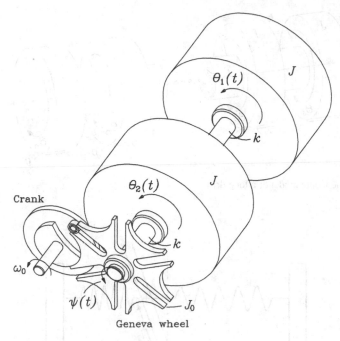

Fig. 4.13 Indexing mechanism of a production machine

drives a *Geneva wheel* with an intermittent motion. This motion consists, in turn, of alternating dwell-and-forward phases of equal duration $T/2$ each, with T defined as the time it takes the crank to go through one full rotation, that is,

$$T = \frac{2\pi}{\omega_0}$$

and ω_0 measured in rad/s. Moreover, the motion transmitted to the wheel is approximated by

$$\psi(t) = \frac{\pi}{3T}t + \psi_p(t)$$

where $\psi_p(t)$ is a periodic function of period T, that we need not specify here. Derive the mathematical model of the system.

4.3. A long drill for deep-boring is modeled as a mechanical system with n identical rotors of moments of inertia J connected by $n-1$ identical elastic shafts of torsional stiffness k and subject to linearly viscous friction of coefficient c, as indicated in Fig. 4.14. Derive expressions for the mass matrix $\mathbf{M}$, the stiffness matrix $\mathbf{K}$, and the damping matrix $\mathbf{C}$, when using $\theta_1, \theta_2, \ldots, \theta_{n-1}, \theta_n$ as generalized coordinates. These angles are measured from an inertial frame.

4.4. The frame of mass m_1 of Fig. 4.15 rides on frictionless rollers, while the uniform rod is pivoted to the frame via a pin that provides linearly viscous damping. Derive the mathematical model of the system under the assumption that l_2 is large

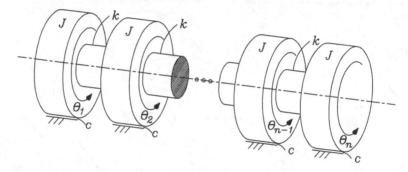

Fig. 4.14 The iconic model of a long drill

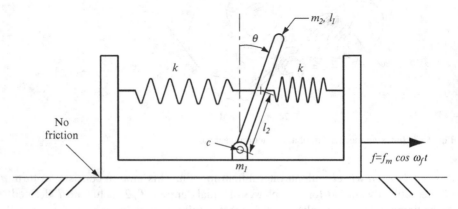

Fig. 4.15 A mechanical system mounted on an air cushion

enough with respect to the distance between the two vertical walls of the frame so as to allow us to assume that the point of attachment of the springs to the rod moves on an essentially straight path. Furthermore, under this assumption, find a relation among the system parameters so that the equilibrium state under which the frame travels at uniform speed and the bar remains vertical is stable. Linearize the model about this equilibrium state and obtain expressions for its mass, stiffness, and damping matrices.[4]

4.5. A gear transmission is modeled as indicated in Fig. 4.16. In this model, two identical disks of moment of inertia J are coupled via smaller disks of moment of inertia αJ, where $\alpha < 1$. Moreover, the two smaller disks are coupled to each other via an elastic shaft of torsional stiffness k. Under the assumption that the smaller disks roll without slipping with respect to the larger ones, we have the relations

$$\theta_3 = -r\theta_1, \quad \theta_4 = -r\theta_2$$

---

[4]Taken, with some modifications, from Cannon [2].

Fig. 4.16 Gear transmission

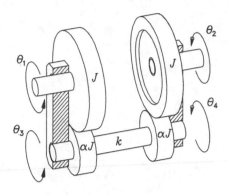

in which r denotes the transmission ratio. Find the mass and stiffness matrices of the system, using θ_1 and θ_2 as independent generalized coordinates, which are measured from an inertial frame.

4.6. Assume that the slider of the overhead crane of Fig. 1.19 has a mass $2m$, and that it is acted upon by a force $f(t)$. Derive the governing equations of the system, and find its equilibrium configurations for $f(t) = 0$. Then, linearize these equations about a marginally stable state, and determine the mass and stiffness matrices of the system. Are these matrices positive-definite or semidefinite?

4.7. Obtain the mathematical model of the system of Fig. 4.1, Example 4.2.1, with a different set of generalized coordinates, $q_1 = \theta_1$, $q_2 = \theta_2 - \theta_1$.

4.8. We revisit Example 4.2.2, but now we assume that the trolley is acted upon by a horizontal force $F(t)$, and hence, u is now a third generalized coordinate. Obtain the mathematical model of the system thus resulting.

4.9. Derive the linearized model for the system of Fig. 4.1, Example 4.2.1, using the generalized coordinates defined in Exercise 4.7.

4.10. Derive the linearized model of the three-dof system introduced in Exercise 4.8.

References

1. Angeles J, Espinosa I (1981) Suspension-system synthesis for mass transport vehicles with prescribed dynamic behavior, ASME Paper 81-DET-44. Proc. 1981 ASME Design Engineering Technical Conference, Hartford, 20–23 Sept 1981
2. Cannon RH (1967) Dynamics of physical systems, McGraw-Hill Book Co., New York

Chapter 5
Vibration Analysis of Two-dof Systems

And here the two brothers [Tweedledum' and Twedledee]
gave each other a hug, and then they held out the two hands that
were free,
to shake hands with her [Alice].

Carroll, L., 1872, *Through the Looking Glass and What Alice*
Found There.
Taken from the 1982 edition of
The Complete Illustrated Works of Lewis Carroll,
Chancellor Press, London.

5.1 Introduction

The vibration analysis of two-dof systems bears many common features with that of n-dof systems, yet it also bears simplifying features that make it amenable to longhand calculations. Moreover, as we show here, two-dof systems can be analyzed completely with graphical methods. We show that the *Mohr circle*, a rather fundamental tool in engineering analysis, finds extensive applications in understanding the behavior of the kind of systems at hand. Once the analysis of two-dof systems has been fully understood, that of n-dof systems comes virtually as a byproduct.

A fundamental difference between single-dof systems and multi-dof systems is their zero-input response. In fact, while the zero-input response of the former is always harmonic, that of the latter is seldom so. This difference will be made apparent in this chapter, which focuses on undamped, linear mechanical systems. An outline of damped systems is given at the end of the chapter.

J. Angeles, *Dynamic Response of Linear Mechanical Systems: Modeling, Analysis and Simulation*, Mechanical Engineering Series, DOI 10.1007/978-1-4419-1027-1_5,
© Springer Science+Business Media, LLC 2011

5.2   The Natural Frequencies and the Natural Modes of Two-dof Undamped Systems

A two-dof undamped linear mechanical system is shown in Fig. 5.1. It is composed of two masses m_1 and m_2, coupled by three springs of stiffness k_1, k_2 and k_3, which are assumed to be unloaded when $x_1 = x_2 = 0$. The generalized coordinates of this system are x_1 and x_2, which are grouped in vector $\mathbf{x}$. The governing equation of this system can be readily derived using the method outlined in Chap. 3. The governing equation thus resulting is given below:

$$\mathbf{M}\ddot{\mathbf{x}} + \mathbf{K}\mathbf{x} = \mathbf{0}, \quad \mathbf{x}(0) = \mathbf{x}_0, \quad \dot{\mathbf{x}}(0) = \mathbf{v}_0, \quad t \geq 0 \qquad (5.1)$$

where $\mathbf{0}$ denotes the two-dimensional zero vector, whereas the mass matrix $\mathbf{M}$ and the stiffness matrix $\mathbf{K}$ are given as

$$\mathbf{M} = \begin{bmatrix} m_1 & 0 \\ 0 & m_2 \end{bmatrix}, \quad \mathbf{K} = \begin{bmatrix} k_1 + k_2 & -k_2 \\ -k_2 & k_2 + k_3 \end{bmatrix} \qquad (5.2)$$

In the next step of our analysis we will reduce the model given in Eq. 5.1 to one with a leading coefficient equal to the identity matrix. In our example, the stiffness matrix is positive-definite, but this need not be always so, for positive-semidefinite stiffness matrices can be equally handled. However, we will assume that the mass matrix is positive-definite. Semidefinite mass matrices imply the presence of particles or rigid bodies of zero mass or zero moment of inertia about a certain axis, which will be left out in our study. The above-mentioned reduction is described below: first, since $\mathbf{M}$ is positive-definite by assumption, it can always be factored in the form

$$\mathbf{M} = \mathbf{N}^T \mathbf{N} \qquad (5.3)$$

The above factoring is not unique. Two of the most common ways of factoring a positive-definite matrix are the *Cholesky* factoring [1] and the *square-root* factoring.

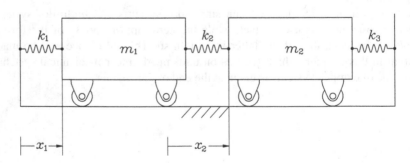

Fig. 5.1 A two-dof system

The former consists of defining $\mathbf{N}$ as upper-triangular with non-negative diagonal entries, in which case the factoring becomes *unique*. The latter consists of defining $\mathbf{N}$ as positive-definite, which is possible because $\mathbf{M}$ is positive-definite as well. In this case $\mathbf{N}$ can be assumed symmetric and, since the eigenvalues of $\mathbf{M}$ are positive, their square roots are real, $\mathbf{N}$ thus becoming *a square root* of $\mathbf{M}$. Note, however, that this square root *is not unique*. Indeed, if we imagine, without loss of generality, that $\mathbf{M}$ is in diagonal form—this is plausible because $\mathbf{M}$ is symmetric, and hence, has real eigenvalues and mutually orthogonal eigenvectors— then its square root is also diagonal and its diagonal entries are the square roots of the corresponding diagonal entries of $\mathbf{M}$. Moreover, since each diagonal entry of $\mathbf{N}$ can be either positive or negative, we conclude that $\mathbf{M}$, being $n \times n$, admits 2^n different square roots. Of these, only one is positive-definite, and only one is negative-definite, the remaining roots being sign-indefinite. Thus, matrix $\mathbf{N}$ of the factoring of Eq. 5.3 is *unique* if we define it as the positive-definite square root of $\mathbf{M}$. Numerically, the Cholesky factoring is more economical than the square-root factoring. However, the latter is adopted here because (1) in the case of two-dof systems, it lends itself to a Mohr-circle treatment, and (2) in simulation, the square-root factoring is done off-line and hence, the overhead difference in the two types of factoring—which is higher for the square-root factoring—becomes immaterial. Thus, we define $\mathbf{N}$ as *the positive-definite square root* of $\mathbf{M}$, and represent it as

$$\mathbf{N} \equiv \sqrt{\mathbf{M}} \tag{5.4}$$

If we now substitute $\mathbf{M} = \mathbf{N}^2$ into Eq. 5.1 and multiply both sides of that equation by $\mathbf{N}^{-1}$, which exists by virtue of the positive-definiteness[1] of $\mathbf{N}$, we derive

$$\mathbf{N}\ddot{\mathbf{x}} + \mathbf{N}^{-1}\mathbf{K}\mathbf{x} = \mathbf{0} \tag{5.5}$$

Now, let us introduce a change of variable:

$$\mathbf{y} = \mathbf{N}\mathbf{x} \quad \Rightarrow \quad \ddot{\mathbf{y}} = \mathbf{N}\ddot{\mathbf{x}} \tag{5.6}$$

Further, substitution of Eq. 5.6 into Eq. 5.5 yields a new form of the governing equation, with a leading coefficient equal to $\mathbf{1}$, the 2×2 identity matrix, namely,

$$\ddot{\mathbf{y}} + \mathbf{P}\mathbf{y} = \mathbf{0}, \quad \mathbf{y}(0) \equiv \mathbf{y}_0 = \mathbf{N}\mathbf{x}_0, \quad \dot{\mathbf{y}}(0) \equiv \mathbf{w}_0 = \mathbf{N}\mathbf{v}_0, \quad t \geq 0 \tag{5.7}$$

where

$$\mathbf{P} \equiv \mathbf{N}^{-1}\mathbf{K}\mathbf{N}^{-1} \tag{5.8}$$

---

[1] A positive-definite matrix is necessarily *non-singular* and hence, *invertible*.

Because the coefficient of the second-derivative term of Eq. 5.7 is the identity matrix, that equation will be henceforth termed the *normal form* of the governing equations.

Furthermore, since $\mathbf{K}$ is at least positive-semidefinite and $\mathbf{N}$ and, consequently, $\mathbf{N}^{-1}$, are symmetric and positive-definite, matrix $\mathbf{P}$ is positive-semidefinite at least. Hence, a positive-definite (or, at least, semidefinite) matrix, say $\boldsymbol{\Omega}$, exists whose square is $\mathbf{P}$. We have thus defined

$$\mathbf{P} \equiv \boldsymbol{\Omega}^2 \equiv \mathbf{N}^{-1}\mathbf{K}\mathbf{N}^{-1} \tag{5.9a}$$

Moreover, the positive-semidefinite (or definite) square root of $\mathbf{P}$, namely,

$$\boldsymbol{\Omega} \equiv \sqrt{\mathbf{N}^{-1}\mathbf{K}\mathbf{N}^{-1}} \tag{5.9b}$$

is termed the *frequency matrix*. Therefore, the normal form of the governing equations can be rewritten as

$$\ddot{\mathbf{y}} + \boldsymbol{\Omega}^2\mathbf{y} = \mathbf{0}, \quad \mathbf{y}(0) = \mathbf{y}_0, \quad \dot{\mathbf{y}}(0) = \mathbf{s}_0, \quad t \geq 0 \tag{5.10}$$

Note that the scalar equation governing the free-vibration motion of a single-dof second-order, undamped system, Eq. 2.10 with $\zeta = 0$, has the form

$$\ddot{y} + \omega_n^2 y = 0, \quad y(0) = y_0, \quad \dot{y}(0) = s_0, \quad t \geq 0 \tag{5.11}$$

where ω_n is the natural frequency of the system, defined as $\sqrt{k/m}$. The striking similarity between Eqs. 5.10 and 5.11 is to be highlighted.

Moreover, since $\boldsymbol{\Omega}$ is at least positive-semidefinite, besides being necessarily symmetric, its eigenvalues are non-negative and its eigenvectors are mutually orthogonal.

Let $\{\omega_i\}_1^2$ be the two—non-negative—eigenvalues of $\boldsymbol{\Omega}$. These are the *natural frequencies* of the system under study. Moreover, let $\{\mathbf{e}_i\}_1^2$ denote the *unit* eigenvectors of $\boldsymbol{\Omega}$, which are necessarily mutually orthogonal, i.e.,

$$\|\mathbf{e}_1\| = \|\mathbf{e}_2\| = 1, \quad \mathbf{e}_1^T\mathbf{e}_2 = 0 \tag{5.12}$$

That is, the sets $\{\omega_i\}_1^2$ and $\{\mathbf{e}_i\}_1^2$ verify the relations

$$\boldsymbol{\Omega}\mathbf{e}_i = \omega_i\mathbf{e}_i, \quad i = 1, 2 \tag{5.13}$$

Now, by virtue of Fact 1 of Appendix A, ω_i and $\mathbf{e}_i$ also verify

$$\boldsymbol{\Omega}^2\mathbf{e}_i = \omega_i^2\mathbf{e}_i, \quad i = 1, 2 \tag{5.14}$$

and hence, if we express Ω^2 in terms of $\mathbf{K}$ and $\mathbf{N}$, as given by Eq. 5.9a, we have

$$\mathbf{N}^{-1}\mathbf{K}\mathbf{N}^{-1}\mathbf{e}_i = \omega_i^2 \mathbf{e}_i$$

or

$$\mathbf{K}\mathbf{N}^{-1}\mathbf{e}_i = \omega_i^2 \mathbf{N}\mathbf{e}_i \qquad (5.15)$$

Vector $\mathbf{N}^{-1}\mathbf{e}_i$ appearing in Eq. 5.15 is called the ith *modal vector* of the given system, the reasons behind this name becoming apparent in the ensuing discussion. Let the ith modal vector be represented by $\mathbf{f}_i$, i.e.,

$$\mathbf{f}_i \equiv \mathbf{N}^{-1}\mathbf{e}_i, \quad i = 1,2 \qquad (5.16a)$$

and hence,

$$\mathbf{e}_i = \mathbf{N}\mathbf{f}_i, \quad i = 1,2 \qquad (5.16b)$$

Upon substitution of Eq. 5.16a into the left-hand side of Eq. 5.15, and of Eq. 5.16b into the right-hand side of the same equation, we have

$$\mathbf{K}\mathbf{f}_i = \omega_i^2 \mathbf{N}^2 \mathbf{f}_i \qquad (5.17a)$$

i.e.,

$$\omega_i^2 \mathbf{M}\mathbf{f}_i = \mathbf{K}\mathbf{f}_i \qquad (5.17b)$$

or

$$(\omega_i^2 \mathbf{M} - \mathbf{K})\mathbf{f}_i = \mathbf{0}, \quad i = 1,2 \qquad (5.17c)$$

Note that the eigenvectors of Ω are, by definition, of unit magnitude. However, the modal vectors are not, in general, of unit magnitude. These vectors, nevertheless, obey an interesting relationship, as made apparent below. Let us compute the product

$$p_i \equiv \mathbf{f}_i^T \mathbf{M}\mathbf{f}_i = \mathbf{e}_i^T (\mathbf{N}^{-1})^T \mathbf{M}\mathbf{N}^{-1}\mathbf{e}_i = \mathbf{e}_i^T \mathbf{N}^{-1}\mathbf{M}\mathbf{N}^{-1}\mathbf{e}_i, \quad i = 1,2$$

Now, since $\mathbf{M} = \mathbf{N}^2$, it is apparent that

$$\mathbf{N}^{-1}\mathbf{M}\mathbf{N}^{-1} = \mathbf{1}$$

with $\mathbf{1}$ defined already as the 2×2 identity matrix. Therefore, the foregoing product p_i reduces to

$$P_i = \mathbf{e}_i^T \mathbf{N}^{-1}\mathbf{N}\mathbf{N}\mathbf{N}^{-1}\mathbf{e}_i = \mathbf{e}_i^T \mathbf{e}_i = 1, \quad i = 1,2$$

which means that

$$\mathbf{f}_i^T \mathbf{M} \mathbf{f}_i = 1, \quad i = 1,2 \tag{5.18}$$

i.e., the modal vectors are of unit magnitude if we define their magnitude properly, namely, as the *weighted* magnitude

$$\|\mathbf{f}_i\|_{\mathbf{M}}^2 \equiv \mathbf{f}_i^T \mathbf{M} \mathbf{f}_i, \quad i = 1,2 \tag{5.19}$$

Thus, the modal vectors *are of unit magnitude with respect to the mass matrix*.

Now it is just natural to investigate the magnitude of the modal vectors *with respect to the stiffness matrix*. To this end, we define the product

$$q_i \equiv \mathbf{f}_i^T \mathbf{K} \mathbf{f}_i \equiv \mathbf{e}_i^T \mathbf{N}^{-1} \mathbf{K} \mathbf{N}^{-1} \mathbf{e}_i = \mathbf{e}_i^T \mathbf{\Omega}^2 \mathbf{e}_i, \quad i = 1,2$$

Since $\mathbf{e}_i$ is an eigenvector of $\mathbf{\Omega}^2$, of eigenvalue ω_i^2, we have

$$q_i = \omega_i^2$$

and hence,

$$\mathbf{f}_i^T \mathbf{K} \mathbf{f}_i = \omega_i^2, \quad i = 1,2 \tag{5.20}$$

i.e., the weighted magnitude of the ith modal vector, when the weighting matrix is the stiffness matrix, is the ith natural frequency, namely,[2]

$$\|\mathbf{f}_i\|_{\mathbf{K}} = \omega_i \tag{5.21}$$

One more property of the modal vectors follows from Eq. 5.17b, if we multiply its two sides by $\mathbf{M}^{-1}$, namely,

$$\mathbf{M}^{-1} \mathbf{K} \mathbf{f}_i = \omega_i^2 \mathbf{f}_i \tag{5.22}$$

where the product $\mathbf{M}^{-1} \mathbf{K}$ is known as the *dynamic matrix* of the system, i.e.,

$$\mathbf{D} \equiv \mathbf{M}^{-1} \mathbf{K} \tag{5.23}$$

Therefore, the eigenvalues of the dynamic matrix are the natural frequencies-squared, its eigenvectors being the modal vectors. Note, however, that these eigenvectors are not of unit magnitude.

Thus, the natural frequencies and the modal vectors can be calculated from the *eigenvalue problem* associated with the dynamic matrix. Indeed, Eq. 5.22 can be rewritten as

$$(\omega_i^2 \mathbf{1} - \mathbf{M}^{-1} \mathbf{K}) \mathbf{f}_i = \mathbf{0} \tag{5.24}$$

---

[2]Properly speaking, the magnitude displayed in Eq. 5.21 requires that $\mathbf{K}$ be positive-definite, but we need not worry about this technicality here.

However, note that the dynamic matrix not being necessarily symmetric, we cannot resort to the Mohr circle to graphically solve the foregoing eigenvalue problem. Therefore, we have to solve this problem in an entirely algebraic fashion, which we do below first in general, and then, with the aid of an example.

Since the modal vectors are nonzero, the above equation implies that the matrix difference in parentheses must be singular, i.e.,

$$\det(\omega_i^2 \mathbf{1} - \mathbf{M}^{-1} \mathbf{K}) = 0 \qquad (5.25)$$

which, upon expansion, leads to a fourth-degree polynomial in ω_i, for the determinant of an $n \times n$ matrix whose entries are polynomials of the pth degree is a p^nth-degree polynomial. In our case, $p = 2$ and $n = 2$, which thus leads to $p^n = 4$. As a matter of fact, the foregoing determinant is *quadratic* in ω_i^2, and hence, the above characteristic equation takes the form

$$\omega_i^4 + 2a\omega_i^2 + b = 0, \quad i = 1,2 \qquad (5.26)$$

which is the *characteristic polynomial* of the dynamic matrix, with coefficients a and b depending on the two matrices $\mathbf{M}$ and $\mathbf{K}$. The two roots of the foregoing characteristic equation can thus be expressed as

$$\omega_{1,2}^2 = -a \pm \sqrt{a^2 - b} \geq 0 \qquad (5.27)$$

The foregoing eigenvalues are bound to be nonnegative, for we know that these are the eigenvalues of the frequency matrix squared, which is at least positive-semidefinite. Moreover, from basic algebra we know that the coefficient of the linear term in a quadratic equation—in ω_i^2 in the case at hand—is the negative of the sum of the two roots, while the independent term is the product of the two roots. Hence, in Eq. 5.26, $a < 0$ and $b \geq 0$. Once the two natural frequencies have been calculated with the aid of Eq. 5.27, the natural modes are found from Eq. 5.24. Now, since this equation represents a homogeneous system of linear equations in $\mathbf{f}_i$, it determines this unknown up to a certain factor. This factor is determined, in turn, by imposing conditions (5.18).

As we will show in Sect. 5.3, the zero-input response of two-dof systems, in general, *is not harmonic*, as opposed to that of single-dof systems. However, harmonic motions are possible, but only if they are shaped by the modal vectors. That is, if $\mathbf{x}(t)$ has the shape

$$\mathbf{x}(t) = A_i(\cos \omega_i t)\mathbf{f}_i \qquad (5.28)$$

Upon substitution of $\mathbf{x}(t)$ and $\ddot{\mathbf{x}}(t)$, as given above, in the governing equation (5.1), we have

$$\mathbf{M}\ddot{\mathbf{x}} + \mathbf{K}\mathbf{x} = -A_i(\cos \omega_i t)(\omega_i^2 \mathbf{M} - \mathbf{K})\mathbf{f}_i$$

By virtue of Eq. 5.17c, the right-hand side of the foregoing equation vanishes, and hence, $\mathbf{x}(t)$, as given by Eq. 5.28, verifies the governing equation. Since we have two modal vectors for a two-dof system, harmonic motions of two distinct shapes are possible. Moreover, each of these two kinds of motions oscillates at one of the natural frequencies of the system. Each of the two shapes is called a *mode*; hence the name given to vectors $\mathbf{f}_i$. Furthermore, any of the three Eqs. 5.17a–c above is termed the *modal equation* of the system at hand. Now it is a simple matter to realize that the modal vectors $\{\mathbf{f}_i\}_1^2$ are *orthogonal with respect to the mass matrix*. Indeed, if we express the orthogonality of the eigenvectors $\mathbf{e}_i$ in terms of Eq. 5.16b, we obtain

$$\mathbf{f}_1^T \mathbf{N}^2 \mathbf{f}_2 = 0$$

and, by virtue of definition (5.4), the foregoing equation leads to

$$\mathbf{f}_1^T \mathbf{M} \mathbf{f}_2 = 0 \tag{5.29}$$

thereby showing the normality of the modal vectors with respect to the mass matrix. Furthermore, if both sides of Eq. 5.17b, for $i = 2$, are multiplied from the left by $\mathbf{f}_1^T$, then, by virtue of Eq. 5.29, one can readily verify that the two foregoing vectors are orthogonal with respect to the stiffness matrix as well, i.e.,

$$\mathbf{f}_1^T \mathbf{K} \mathbf{f}_2 = 0 \tag{5.30}$$

In the system of Fig. 5.1, let us assume that

$$m_1 = m, \quad m_2 = 4m, \quad k_1 = k, \quad k_2 = 8k, \quad k_3 = 4k \tag{5.31}$$

$\mathbf{N}$ thus being a diagonal matrix with diagonal entries $\sqrt{m}$ and $2\sqrt{m}$, i.e.,

$$\mathbf{N} = \sqrt{m} \begin{bmatrix} 1 & 0 \\ 0 & 2 \end{bmatrix} \tag{5.32}$$

and hence,

$$\mathbf{\Omega}^2 = \frac{k}{m} \begin{bmatrix} 1 & 0 \\ 0 & 1/2 \end{bmatrix} \begin{bmatrix} 9 & -8 \\ -8 & 12 \end{bmatrix} \begin{bmatrix} 1 & 0 \\ 0 & 1/2 \end{bmatrix} = \omega^2 \begin{bmatrix} 9 & -4 \\ -4 & 3 \end{bmatrix} \tag{5.33}$$

with ω^2 defined as $\omega^2 \equiv k/m$. The Mohr circle of $\mathbf{\Omega}^2$ is shown in Fig. 5.2a.

From Fig. 5.2a, the eigenvalues of $\mathbf{\Omega}^2$ are determined by the intersection of its Mohr circle with the horizontal axis, namely,

$$\omega_1^2 = \omega^2, \quad \omega_2^2 = 11\omega^2$$

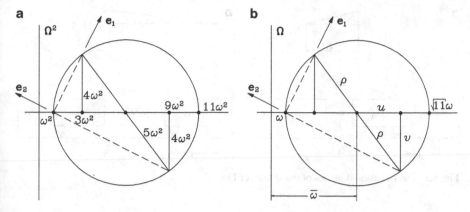

Fig. 5.2 Mohr circle of (**a**) Ω^2 of the system of Fig. 5.1, (**b**) Ω of the same system

while the corresponding eigenvectors, e_1 and e_2, are calculated using the relations derived in Appendix A, thus obtaining

$$e_1 = \frac{\sqrt{5}}{5}\begin{bmatrix}1\\2\end{bmatrix}, \quad e_2 = \frac{\sqrt{5}}{5}\begin{bmatrix}-2\\1\end{bmatrix}$$

Moreover, the natural frequencies of the system are readily found to be

$$\omega_1 = \omega, \quad \omega_2 = \sqrt{11}\,\omega$$

Furthermore, in Fig. 5.2b we introduce two new concepts, the *mean frequency* and the *frequency radius*, denoted by $\overline{\omega}$ and ρ, namely,

$$\overline{\omega} = \frac{1+\sqrt{11}}{2}\omega, \quad \rho = \frac{\sqrt{11}-1}{2}\omega$$

the frequency matrix then taking the form

$$\Omega = \begin{bmatrix}\overline{\omega}+u & -v\\-v & \overline{\omega}-u\end{bmatrix}$$

with u and v defined below in terms of the mean frequency and the frequency radius. From the similarity of the corresponding triangles in Fig. 5.2a,b, we obtain

$$u = \frac{3}{5}\rho = \frac{3(\sqrt{11}-1)}{10}\omega, \quad v = \frac{4}{5}\rho = \frac{4(\sqrt{11}-1)}{10}\omega$$

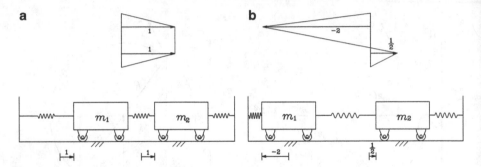

Fig. 5.3 The two natural modes of the system of Fig. 5.1

and hence,

$$\overline{\omega}+u = \frac{1+4\sqrt{11}}{5}\omega, \quad \overline{\omega}-u = \frac{4+\sqrt{11}}{5}\omega$$

Therefore, the positive-definite square root of Ω^2 is

$$\Omega = \frac{\omega}{5}\begin{bmatrix} 1+4\sqrt{11} & 2(1-\sqrt{11}) \\ 2(1-\sqrt{11}) & 4+\sqrt{11} \end{bmatrix}$$

Moreover, the modal vectors are

$$\mathbf{f}_1 = \mathbf{N}^{-1}\mathbf{e}_1 = \frac{\sqrt{5m}}{5m}\begin{bmatrix} 1 \\ 1 \end{bmatrix}, \quad \mathbf{f}_2 = \mathbf{N}^{-1}\mathbf{e}_2 = \frac{\sqrt{5m}}{5m}\begin{bmatrix} -2 \\ 1/2 \end{bmatrix},$$

The natural modes of the system, drawn from the modal vectors, are then represented as shown in Fig. 5.3, which are apparently obeyed if the vectors of initial conditions, x_0 and v_0, are both proportional to any one of the two modal vectors of the system at hand. An animation of the Mohr circles of the harmonic functions of the frequency matrix of this system is included in 5-MohrCircleAnmtn.mw.

Alternatively, we can find the natural frequencies and the natural modes of the system under study using an algebraic approach, i.e., by computing the eigenvalues and eigenvectors of the dynamic matrix, which was defined in Eq. 5.23. To do this, let us first calculate the dynamic matrix of the system under study:

$$\mathbf{D} \equiv \mathbf{M}^{-1}\mathbf{K} = \frac{k}{m}\begin{bmatrix} 9 & -8 \\ -2 & 3 \end{bmatrix}$$

which, apparently, is not symmetric, its characteristic equation being

$$\det(\omega_i^2\mathbf{1} - \mathbf{D}) \equiv \det\begin{bmatrix} \omega_i^2 - 9\omega^2 & 8\omega^2 \\ 2\omega^2 & \omega_i^2 - 3\omega^2 \end{bmatrix}$$

where

$$\omega^2 \equiv \frac{k}{m}$$

Upon expansion of the above determinant, the characteristic polynomial takes the form

$$P(\omega^2) = \omega_i^4 - 12\omega^2\omega_i^2 + 11\omega^4$$

whose roots can be readily found as

$$\omega_{1,2}^2 = \omega^2, 11\omega^2$$

The modal vectors are now calculated.

For $\omega_1 = \omega$:

$$\omega_1^2 \mathbf{1} - \mathbf{D} = \omega^2 \begin{bmatrix} -8 & 8 \\ 2 & -2 \end{bmatrix}, \quad \mathbf{f}_1 \equiv \begin{bmatrix} x_1 \\ y_1 \end{bmatrix}$$

and hence, the associated system of homogeneous equations, derived from Eq. 5.24, takes the form

$$2\omega^2 \begin{bmatrix} -4 & 4 \\ 1 & -1 \end{bmatrix} \begin{bmatrix} x_1 \\ y_1 \end{bmatrix} = \begin{bmatrix} 0 \\ 0 \end{bmatrix}$$

which leads to

$$-4x_1 + 4y_1 = 0$$

$$x_1 - y_1 = 0$$

Therefore, $y_1 = x_1$, the above system thus reducing to one single equation in one unknown, say x_1. In order to determine x_1, we apply condition (5.18), namely,

$$\begin{bmatrix} x_1 & x_1 \end{bmatrix} \begin{bmatrix} m & 0 \\ 0 & 4m \end{bmatrix} \begin{bmatrix} x_1 \\ x_1 \end{bmatrix} = 1$$

or

$$5mx_1^2 = 1$$

whence,

$$x_1 = \pm \frac{\sqrt{5m}}{5m}$$

If we take the positive sign in the above expression, then,

$$\mathbf{f}_1 = \frac{\sqrt{5m}}{5m}\begin{bmatrix} 1 \\ 1 \end{bmatrix}$$

For $\omega_2 = \sqrt{11}\omega$:

$$\omega_2^2 \mathbf{1} - \mathbf{D} = \omega^2 \begin{bmatrix} 2 & 8 \\ 2 & 8 \end{bmatrix}, \quad \mathbf{f}_2 \equiv \begin{bmatrix} x_2 \\ y_2 \end{bmatrix}$$

That is,

$$2\omega^2 \begin{bmatrix} 1 & 4 \\ 1 & 4 \end{bmatrix} \begin{bmatrix} x_2 \\ y_2 \end{bmatrix} = \begin{bmatrix} 0 \\ 0 \end{bmatrix}$$

thereby leading to

$$x_2 + 4y_2 = 0$$

$$x_2 + 4y_2 = 0$$

whence $x_2 = -4y_2$, the above system thus reducing to one single equation in one unknown, say y_2, which is found by application of condition (5.24), namely,

$$\begin{bmatrix} -4y_2 & y_2 \end{bmatrix} \begin{bmatrix} m & 0 \\ 0 & 4m \end{bmatrix} \begin{bmatrix} -4y_2 \\ y_2 \end{bmatrix} = 1$$

or

$$20my_2^2 = 1$$

Therefore,

$$y_2 = \pm\frac{\sqrt{20m}}{20m} \equiv= \pm\frac{1}{2}\frac{\sqrt{5m}}{5m}$$

and hence, upon taking the positive sign in the above expression, we obtain

$$\mathbf{f}_2 = \frac{\sqrt{5m}}{5m}\begin{bmatrix} -2 \\ 1/2 \end{bmatrix}$$

thereby completing the intended calculations, which yield the same results as when one proceeds with the Mohr circle applied to the frequency matrix.

Example 5.2.1 (Modal Analysis of a Two-dof Gantry Robot). Determine the natural frequencies and the natural modes of the two-dof gantry robot introduced in Example 4.2.2, a task that is known as *modal analysis*.

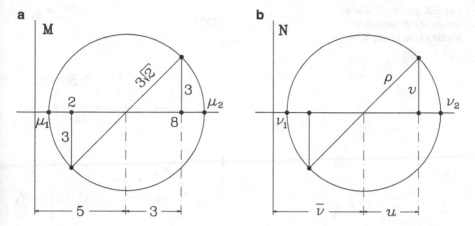

Fig. 5.4 Mohr circles of: (**a**) the mass matrix **M** of the two-dof gantry robot; and (**b**) the positive-definite square root of **M**

Solution: For quick reference, we recall below the mass and stiffness matrices of the linearized model derived in Example 4.4.2 about the stable *elbow-down* configuration. Moreover, we consider in that model that the disturbance $\ddot{u}$ vanishes and neglect damping, i.e., we assume $c_1 = c_2 = 0$. The said matrices are

$$\mathbf{M} = \frac{1}{3}\begin{bmatrix} 8 & 3 \\ 3 & 2 \end{bmatrix}, \quad \mathbf{K} = A\begin{bmatrix} 3 & 0 \\ 0 & 1 \end{bmatrix}, \quad A \equiv \frac{\sqrt{a^2 + g^2}}{l}$$

Furthermore, it is apparent that the modal analysis is not affected if both the mass and the stiffness matrices are multiplied by the same factor. So, let us redefine these matrices as

$$\mathbf{M} = \begin{bmatrix} 8 & 3 \\ 3 & 2 \end{bmatrix}, \quad \mathbf{K} = \omega^2\begin{bmatrix} 3 & 0 \\ 0 & 1 \end{bmatrix}, \quad \omega^2 \equiv 3\frac{\sqrt{a^2 + g^2}}{l}$$

The positive-definite square root of **M**, matrix **N**, is now calculated from the Mohr circle of **M**, as shown in Fig. 5.4.

From Fig. 5.4a, the eigenvalues μ_1 and μ_2 of **M** are readily calculated as

$$\mu_1 = 5 - 3\sqrt{2} = 0.7574, \quad \mu_2 = 5 + 3\sqrt{2} = 9.243$$

and hence, the eigenvalues ν_1 and ν_2 of **N** are

$$\nu_1 = \sqrt{\mu_1} = 0.8703, \quad \nu_2 = \sqrt{\mu_2} = 3.040$$

Therefore, the parameters of the Mohr circle of **N**, as shown in Fig. 5.4b, are

$$\bar{\nu} \equiv \frac{1}{2}(\nu_1 + \nu_2) = 1.955, \quad \rho \equiv \frac{1}{2}(\nu_2 - \nu_1) = 1.085$$

Fig. 5.5 Mohr circle of the
square of the frequency
matrix of the gantry robot

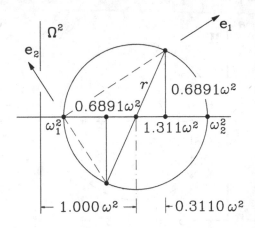

while

$$u = v = \frac{\sqrt{2}}{2}\rho = 0.7672$$

Therefore,

$$\mathbf{N} = \begin{bmatrix} \overline{v}+u & v \\ v & \overline{v}-u \end{bmatrix} = \begin{bmatrix} 2.722 & 0.7672 \\ 0.7672 & 1.188 \end{bmatrix}$$

whence,

$$\mathbf{N}^{-1} = \begin{bmatrix} 0.4491 & -0.2900 \\ -0.2900 & 1.029 \end{bmatrix}$$

the frequency matrix-squared then becoming

$$\mathbf{\Omega}^2 = \omega^2 \begin{bmatrix} 0.6891 & -0.6891 \\ -0.6891 & 1.311 \end{bmatrix}$$

The eigenvalues and eigenvectors of $\mathbf{\Omega}^2$, and hence, those of $\mathbf{\Omega}$, are now determined from the Mohr circle of $\mathbf{\Omega}^2$, displayed in Fig. 5.5. From this figure, one can readily find

$$r \equiv \omega^2 \sqrt{0.3110^2 + 0.6891^2} = 0.7560\omega^2$$

and hence,

$$\omega_1^2 = \omega^2 - r = 0.2440\omega^2, \quad \omega_2^2 = \omega^2 + r = 1.756\omega^2$$

the natural frequencies thus being

$$\omega_1 = 0.4939\omega, \quad \omega_2 = 1.325\omega$$

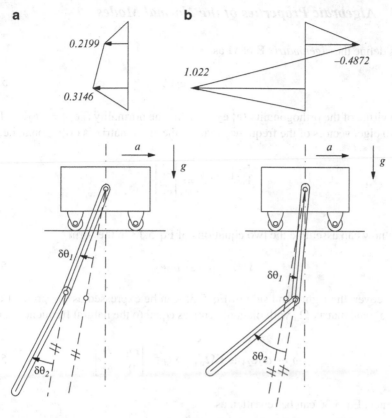

Fig. 5.6 Natural modes of the two-dof gantry robot: (**a**) first mode, at $\omega_1 = 0.4939\omega$; (**b**) second mode, at $\omega_2 = 1.325\omega$

while the corresponding eigenvectors are

$$\mathbf{e}_1 = \begin{bmatrix} 0.8401 \\ 0.5425 \end{bmatrix}, \quad \mathbf{e}_2 = \begin{bmatrix} -0.5425 \\ 0.8401 \end{bmatrix}$$

The modal vectors are now determined from simple matrix-times-vector multiplications, as indicated in Eq. 5.16a, which thus yields

$$\mathbf{f}_1 = \begin{bmatrix} 0.2199 \\ 0.3146 \end{bmatrix}, \quad \mathbf{f}_2 = \begin{bmatrix} -0.4872 \\ 1.022 \end{bmatrix}$$

The normal modes are represented in Fig. 5.6.

5.2.1 Algebraic Properties of the Normal Modes

Let us define the *eigenmatrix* $\mathbf{E}$ of $\mathbf{\Omega}$ as

$$\mathbf{E} \equiv \begin{bmatrix} \mathbf{e}_1 & \mathbf{e}_2 \end{bmatrix} \tag{5.34}$$

By virtue of the orthogonality ($\mathbf{e}_1^T \mathbf{e}_2 = 0$) and the normality ($\|\mathbf{e}_1\| = \|\mathbf{e}_2\| = 1$) of the two eigenvectors of the frequency matrix, the eigenmatrix is orthogonal, i.e.,

$$\mathbf{E}^{-1} = \mathbf{E}^T = \begin{bmatrix} \mathbf{e}_1^T \\ \mathbf{e}_2^T \end{bmatrix}$$

We now can assemble the two equations of Eq. 5.13 in the form

$$\mathbf{\Omega E} = \begin{bmatrix} \omega_1 \mathbf{e}_1 & \omega_2 \mathbf{e}_2 \end{bmatrix} \tag{5.35}$$

Moreover, the right-hand side of Eq. 5.35 can be expressed as the product of $\mathbf{E}$ by a diagonal matrix $\mathbf{\Omega}_d$ with diagonal entries equal to the natural frequencies, i.e.,

$$\begin{bmatrix} \omega_1 \mathbf{e}_1 & \omega_2 \mathbf{e}_2 \end{bmatrix} \equiv \mathbf{E}\mathbf{\Omega}_d, \quad \mathbf{\Omega}_d \equiv \begin{bmatrix} \omega_1 & 0 \\ 0 & \omega_2 \end{bmatrix} \tag{5.36}$$

and hence, Eq. 5.36 can be rewritten as

$$\mathbf{\Omega E} = \mathbf{E}\mathbf{\Omega}_d$$

Therefore,

$$\mathbf{\Omega}_d = \mathbf{E}^T \mathbf{\Omega E}, \quad \text{or} \quad \mathbf{\Omega} = \mathbf{E}\mathbf{\Omega}_d\mathbf{E}^T \tag{5.37}$$

That is, the eigenmatrix $\mathbf{E}$ diagonalizes $\mathbf{\Omega}$ in the sense of Eq. 5.37. Also note that

$$\mathbf{\Omega}_d^2 = \mathbf{E}^T \mathbf{\Omega E}\mathbf{E}^T \mathbf{\Omega E} = \mathbf{E}^T \mathbf{\Omega}^2 \mathbf{E} \tag{5.38}$$

which means that the eigenmatrix also diagonalizes $\mathbf{\Omega}^2$. Now, if we define the *modal matrix* $\mathbf{F}$ as

$$\mathbf{F} \equiv \begin{bmatrix} \mathbf{f}_1 & \mathbf{f}_2 \end{bmatrix} \equiv \mathbf{N}^{-1}\mathbf{E} \tag{5.39}$$

then it is apparent from Eqs. 5.18 and 5.29 that

$$\mathbf{F}^T \mathbf{M}\mathbf{F} = \mathbf{1} \tag{5.40}$$

Moreover, from Eqs. 5.20 and 5.30,

$$\mathbf{F}^T \mathbf{K} \mathbf{F} = \mathbf{\Omega}_d^2 \equiv \begin{bmatrix} \omega_1^2 & 0 \\ 0 & \omega_2^2 \end{bmatrix} \tag{5.41}$$

That is, $\mathbf{F}$ diagonalizes $\mathbf{M}$ in the sense of Eq. 5.40, while $\mathbf{F}$ diagonalizes $\mathbf{K}$ in the sense of Eq. 5.41.

The modal matrix of the gantry robot of Example 5.2.1 is, thus,

$$\mathbf{F} = \begin{bmatrix} 0.2199 & -0.4872 \\ 0.3146 & 1.022 \end{bmatrix}$$

Furthermore, the modal matrix also diagonalizes the dynamic matrix, but in a sense different from that in which it diagonalizes the mass and the stiffness matrices. Indeed, let us calculate

$$\mathbf{F}^{-1} \mathbf{D} \mathbf{F} = (\mathbf{N}^{-1} \mathbf{E})^{-1} \mathbf{M}^{-1} \mathbf{K} \mathbf{N}^{-1} \mathbf{E}$$

where $\mathbf{M}^{-1}$ can be written as $\mathbf{N}^{-1} \mathbf{N}^{-1}$, and hence,

$$\mathbf{F}^{-1} \mathbf{D} \mathbf{F} = \mathbf{E}^T \underbrace{\mathbf{N} \mathbf{N}^{-1}}_{1} \underbrace{\mathbf{N}^{-1} \mathbf{K} \mathbf{N}^{-1}}_{\mathbf{\Omega}^2} \mathbf{E} = \mathbf{E}^T \mathbf{\Omega}^2 \mathbf{E}$$

which is readily recognized as $\mathbf{\Omega}_d^2$. Therefore,

$$\mathbf{F}^{-1} \mathbf{D} \mathbf{F} = \mathbf{\Omega}_d^2 \tag{5.42}$$

The difference between the foregoing form of diagonalization and that displayed in Eqs. 5.40 and 5.41 is to be highlighted: While the modal matrix appears inverted in Eq. 5.42, it appears transposed in Eqs. 5.40 and 5.41.

5.3 The Zero-Input Response of Two-dof Systems

In this section we derive the zero-input response of a two-dof system. We distinguish here between those with positive-definite and systems with positive-semidefinite stiffness matrices, the latter being characterized by the presence of a rigid mode. If the system at hand contains a rigid-body mode, then the frequency matrix is positive-semidefinite, and hence, singular. In this case, however, finding the time response of the system is, in a way, a simpler task, for the problem reduces to finding the response of a single-dof system. We thus begin by discussing the zero-input response of semidefinite systems.

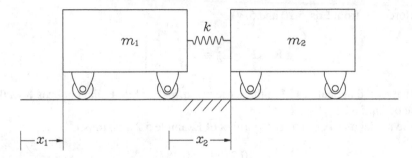

Fig. 5.7 A two-dof semidefinite system

5.3.1 Semidefinite Systems

We illustrate the general approach to handling this kind of systems with an example. Consider the system of Fig. 5.7, which, as we will show below, contains a rigid-body mode.

The governing equation of this system has the form of Eq. 5.1, with

$$\mathbf{M} = \begin{bmatrix} m_1 & 0 \\ 0 & m_2 \end{bmatrix}, \quad \mathbf{K} = \begin{bmatrix} k & -k \\ -k & k \end{bmatrix} \tag{5.43}$$

The mathematical model of the system under study thus takes on the scalar form

$$m_1 \ddot{x}_1 + k(x_1 - x_2) = 0 \tag{5.44a}$$

$$m_2 \ddot{x}_2 - k(x_1 - x_2) = 0 \tag{5.44b}$$

with initial conditions

$$x_k(0) = a_k, \quad \dot{x}_k(0) = b_k, \quad k = 1, 2 \tag{5.44c}$$

As shown in Sect. 4.6, the foregoing equations can be rewritten as

$$m_1 \ddot{x}_1 + m_2 \ddot{x}_2 = 0 \tag{5.45a}$$

$$m_1 \ddot{x}_1 - m_2 \ddot{x}_2 + 2k(x_1 - x_2) = 0 \tag{5.45b}$$

The first of the two above equations states nothing but the conservation of momentum. Dividing both sides of this equation by $m_1 + m_2$ and integrating once with respect to time gives

$$\frac{m_1 \dot{x}_1 + m_2 \dot{x}_2}{m_1 + m_2} = v_0 \tag{5.46}$$

where v_0 is a constant of integration whose mechanical interpretation is given below. Integrating this equation once more gives

$$\frac{m_1 x_1 + m_2 x_2}{m_1 + m_2} = v_0 t + x_0 \tag{5.47}$$

where x_0 is another constant of integration. Now it should be apparent that the left-hand side of this equation is nothing but the coordinate of the c.o.m. of the system, which we shall denote by $x(t)$, so that

$$x(t) = x_0 + v_0 t \tag{5.48}$$

and it is now clear that v_0 and x_0 are the initial velocity and the initial position, respectively, of the c.o.m. Moreover, Eq. 5.46 simply states the well-known fact that the velocity of the c.o.m. of the system is constant.

From Eq. 5.47, we can readily solve for x_2, namely,

$$x_2(t) = \frac{(m_1 + m_2)(x_0 + v_0 t) - m_1 x_1(t)}{m_2} \tag{5.49}$$

Furthermore, from Eq. 5.45a, it is apparent that $\ddot{x}_2 = -(m_1/m_2)\ddot{x}_1$. If, furthermore, we substitute the above expression for x_2 into Eq. 5.45b, we obtain, after simplifications,

$$m_1 \ddot{x}_1 + k\left(1 + \frac{m_1}{m_2}\right) x_1 = k(m_1 + m_2)\frac{x_0 + v_0 t}{m_2}$$

If we now divide both sides of the above equation by m_1, we obtain

$$\ddot{x}_1 + k\frac{m_1 + m_2}{m_1 m_2} x_1 = k\frac{m_1 + m_2}{m_1 m_2}(x_0 + v_0 t)$$

or

$$\ddot{x}_1 + \frac{k}{m_{eq}} x_1 = \frac{k}{m_{eq}}(x_0 + v_0 t) \tag{5.50}$$

with the *equivalent mass* m_{eq} defined as

$$m_{eq} \equiv \frac{m_1 m_2}{m_1 + m_2} \tag{5.51}$$

The normal form of the above equation is thus readily derived as

$$\ddot{x}_1 + \omega_n^2 x_1 = \omega_n^2 x_0 + \omega_n^2 v_0 t, \quad x_1(0) = a_1, \quad \dot{x}_1(0) = b_1 \tag{5.52}$$

with ω_n defined as the natural frequency of the non-rigid mode, which is given by

$$\omega_n \equiv \sqrt{\frac{k}{m_{eq}}} \tag{5.53}$$

x_0 being the initial value of the abscissa of the center of mass, i.e.,

$$x_0 \equiv \frac{m_1 x_1(0) + m_2 x_2(0)}{m_1 + m_2} \equiv \frac{m_1 a_1 + m_2 a_2}{m_1 + m_2} \tag{5.54a}$$

and v_0 is the initial value of the velocity of the center of mass, i.e.,

$$v_0 \equiv \frac{m_1 \dot{x}_1(0) + m_2 \dot{x}_2(0)}{m_1 + m_2} \equiv \frac{m_1 b_1 + m_2 b_2}{m_1 + m_2} \tag{5.54b}$$

It is now apparent that $x_1(t)$ can be computed separately from $x_2(t)$, the analysis thus leading to that of two decoupled single-dof systems, one that is governed by Eq. 5.52, determining the non-rigid mode of the system, the other one being simply the motion of the center of mass of the system. We discuss below the derivation of the time response of the non-rigid mode. To do this, we need to find the response of a system acted upon by a constant input and that of the same system acted upon by an input that is linear in time, with nonzero initial conditions, as given by Eq. 5.52. The desired response can then be found by superposition, as explained below.

First and foremost, it is essential that we identify properly the excitation terms of Eq. 5.52. At a first glance, one would be tempted to treat the first term of the right-hand side of that equation as a step function, the second as a ramp function, but they are neither. Indeed, by virtue of the Principle of Conservation of Momentum, the center of mass of the system has been moving with a velocity v_0 even before the observations started at time $t = 0$, while the value x_0 is constant. Hence, we cannot apply here either the step response nor the ramp response to find the response of the system of Eq. 5.52.

Since the system is acted upon by two inputs, $\omega_n^2 x_0$ and $\omega_n^2 v_0 t$, it seems natural to find its time response by superposition. However, superposition applies not only to the two different inputs, but also to the nonzero initial conditions. In other words, we have to superimpose the zero-input response to its two zero-state responses obtained for each of its inputs. We proceed then by finding all the individual responses; then we superimpose them all. Thus,

1. Zero-input response: The system is driven solely by its initial conditions, i.e.,

$$\ddot{x}_1 + \omega_n^2 x_1 = 0, \quad x_1(0) = a_1, \quad \dot{x}_1(0) = b_1 \tag{5.55}$$

The desired time response, derived in Sect. 2.7.3, is displayed in Eq. 2.8. Hence,

$$x_1(t) = a_1 \cos \omega_n t + \frac{b_1}{\omega_n} \sin \omega_n t \tag{5.56}$$

2. Zero-state response: We divide this response into two parts, that due to $\omega_n^2 x_0$ and that due to $\omega_n^2 v_0 t$.

   (a) First zero-state response: We have

$$\ddot{x}_1 + \omega_n^2 x_1 = \omega_n^2 x_0, \quad x_1(0) = 0, \quad \dot{x}_1(0) = 0 \tag{5.57}$$

   We can find this response using the results of Sect. 2.7.4, by setting $\zeta = 0$, and hence,

$$x_1(t) = x_0(1 - \cos \omega_n t) \tag{5.58}$$

   (b) Second zero-state response: Now the system under study takes the form

$$\ddot{x}_1 + \omega_n^2 x_1 = \omega_n^2 v_0 t, \quad x_1(0) = 0, \quad \dot{x}_1(0) = 0 \tag{5.59}$$

The desired time response can now be found by linearity, namely, from the zero-state response derived in item (a). Let us call the input of Eq. 5.57 $u_1(t)$, that of Eq. 5.59 $u_2(t)$, the corresponding time responses being denoted by $\xi_1(t)$ and $\xi_2(t)$. Now we note the relation between $u_1(t)$ and $u_2(t)$, namely,

$$u_2(t) = \frac{v_0}{x_0} \int_0^t u_1(\theta) d\theta \tag{5.60}$$

where θ is a dummy integration variable. From the above relation for the two inputs, we can readily derive the relation for the corresponding responses, i.e.,

$$\xi_2(t) = \frac{v_0}{x_0} \int_0^t \xi_1(\theta) d\theta \tag{5.61}$$

where $\xi_1(\theta)$ is nothing but $x_1(\theta)$, as given by Eq. 5.58, i.e.,

$$\xi_1(\theta) = x_0(1 - \cos \omega_n \theta) \tag{5.62}$$

and hence,

$$\xi_2(t) = v_0 \int_0^t (1 - \cos \omega_n \theta) d\theta$$

The desired response, $x_1(t)$, to the given input, $\omega_n^2 v_0 t$, is then obtained upon expansion of the foregoing integral, namely,

$$x_1(t) = \xi_2(t) = v_0 \left(t - \frac{1}{\omega_n} \sin \omega_n t \right) \tag{5.63}$$

Alternatively, expression (5.63) could have been obtained from the results of Sect. 2.7.4 for the response to a linear input, with $\zeta = 0$.

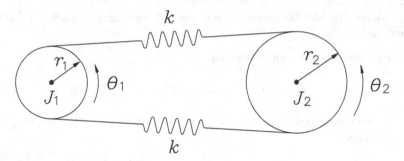

Fig. 5.8 An elastic belt-pulley transmission

Now the total time response is simply the sum of all three individual responses displayed in Eqs. 5.56, 5.58 and 5.63, namely,

$$x_1(t) = a_1 \cos \omega_n t + \frac{b_1}{\omega_n} \sin \omega_n t + x_0(1 - \cos \omega_n t) + v_0 t - \frac{v_0}{\omega_n} \sin \omega_n t$$

or, more systematically,

$$x_1(t) = x_0 + v_0 t + (a_1 - x_0) \cos \omega_n t + \frac{b_1 - v_0}{\omega_n} \sin \omega_n t \qquad (5.64)$$

i.e., the motion of the first mass is given as the superposition of the uniform-velocity motion of the center of mass and a harmonic motion about the center of mass.

The motion of the second mass is simply derived from Eqs. 5.49 and 5.64. After obvious simplifications, this response is derived as

$$x_2(t) = x_0 + v_0 t - \frac{m_1}{m_2} \left[(a_1 - x_0) \cos \omega_n t + \frac{b_1 - v_0}{\omega_n} \sin \omega_n t \right] \qquad (5.65)$$

which is, likewise, the superposition of the motion of the center of mass and a harmonic motion, thereby completing the analysis of the whole system. Animations of the zero-input response of this system, that include its rigid and flexible mode, are included in 5-Semidef2dof.mw.

Example 5.3.1 (A Belt-Pulley Transmission). Shown in Fig. 5.8 is a belt-pulley transmission composed of two pulleys that are modeled as homogeneous rigid disks of radii r_1 and r_2 and centroidal moments of inertia J_1 and J_2, respectively. Moreover, the belt is assumed elastic and weightless, while neglecting its viscosity, so that the belt can be modeled as two identical lumped springs of stiffness k. Show that this system admits a rigid mode and describe it. Moreover, the belt is assumed to be purely elastic; if, additionally, we neglect the viscous effect of the lubricant and the dry friction in the bearings, no dissipation is present.

Solution: We will study the zero-input response of this system and so, we will assume zero input power. The kinetic and potential energies of the system can be readily derived as

$$T = \frac{1}{2}J_1\dot{\theta}_1^2 + \frac{1}{2}J_2\dot{\theta}_2^2, \quad V = \frac{1}{2}2k(r_1\theta_1 - r_2\theta_2)^2$$

It is now apparent from the above expression for V that no changes of potential energy occur when the system undergoes a motion whose the generalized coordinates are related by $r_1\theta_1 - r_2\theta_2 = 0$. This means that the system has one rigid mode, which is given by the above relation, and hence, its stiffness matrix is bound to be positive-semidefinite, i.e., singular. As a consequence, the frequency matrix of the system is also positive-semidefinite. It is a simple matter to derive the mass and stiffness matrices, as shown below:

$$\mathbf{M} = \begin{bmatrix} J_1 & 0 \\ 0 & J_2 \end{bmatrix}, \quad \mathbf{K} = 2k \begin{bmatrix} r_1^2 & -r_1 r_2 \\ -r_1 r_2 & r_2^2 \end{bmatrix}$$

The reader can now compute the determinant of matrix $\mathbf{K}$ and realize that, indeed, this matrix is singular. Moreover, since the trace of the same matrix is apparently positive, so is the sum of its eigenvalues, and, since one eigenvalue is bound to be zero, the remaining eigenvalue is bound to be positive, the matrix thus being positive-semidefinite.

In deriving the time response of the system under study it will prove advantageous to introduce a change of variables. One obvious candidate is the rigid mode, and hence, we shall define one of the new variables as $r_1\theta_1 - r_2\theta_2$. Since we need one additional new variable, independent from the former, the obvious choice is to define it as $r_1\theta_1 + r_2\theta_2$. In fact, we will define the latter as the first variable, the former as the second, i.e.,

$$\psi_1 \equiv r_1\theta_1 + r_2\theta_2$$
$$\psi_2 \equiv r_1\theta_1 - r_2\theta_2$$

Hence, in the presence of a rigid mode, ψ_2 vanishes, but ψ_1 does not. Likewise, if we have a motion under which ψ_1 vanishes but ψ_2 does not, we call this a *purely flexible mode*. In general, for arbitrary initial conditions, the time response will contain a combination of both the rigid and the purely flexible mode. Below we shall obtain the time response of the two foregoing modes independent from each other, i.e., by decoupling one from the other. In order to obtain the time response in the given coordinates, we need the inverse change of coordinates, namely,

$$\theta_1 \equiv \frac{1}{2r_1}(\psi_1 + \psi_2) \quad \theta_2 \equiv \frac{1}{2r_2}(\psi_1 - \psi_2)$$

and hence,

$$\ddot{\theta}_1 \equiv \frac{1}{2r_1}(\ddot{\psi}_1 + \ddot{\psi}_2) \quad \ddot{\theta}_2 \equiv \frac{1}{2r_2}(\ddot{\psi}_1 - \ddot{\psi}_2)$$

Now, we have initial conditions on the original variables; the initial conditions for the new variables are readily found from the former, i.e.,

$$\psi_1(0) = r_1\theta_1(0) + r_2\theta_2(0), \quad \psi_2(0) = r_1\theta_1(0) - r_2\theta_2(0)$$
$$\dot{\psi}_1(0) = r_1\dot{\theta}_1(0) + r_2\dot{\theta}_2(0), \quad \dot{\psi}_2(0) = r_1\dot{\theta}_1(0) - r_2\dot{\theta}_2(0)$$

Upon the foregoing change of variable, the governing equations become

$$\ddot{\psi}_1 + \ddot{\psi}_2 + \frac{4kr_1^2}{J_1}\psi_2 = 0$$

$$\ddot{\psi}_1 - \ddot{\psi}_2 - \frac{4kr_2^2}{J_2}\psi_2 = 0$$

from which $\ddot{\psi}_1$ can be eliminated by subtracting the second equation from the first one, thus obtaining

$$\ddot{\psi}_2 + 2k\frac{J_1r_2^2 + J_2r_1^2}{J_1J_2}\psi_2 = 0$$

or, in a more familiar form,

$$\ddot{\psi}_2 + \omega_n^2\psi_2 = 0$$

with initial conditions $\psi_2(0)$ and $\dot{\psi}_2(0)$ given above and the natural frequency ω_n defined as

$$\omega_n \equiv \sqrt{2k\left(\frac{r_2^2}{J_2} + \frac{r_1^2}{J_1}\right)}$$

The time response of the system at hand is thus the familiar

$$\psi_2(t) = \psi_2(0)\cos\omega_n t + \frac{\dot{\psi}_2(0)}{\omega_n}\sin\omega_n t$$

Furthermore, a differential equation for $\psi_1(t)$ can be readily derived if the two governing equations in the new coordinates are added sidewise, thus obtaining

$$\ddot{\psi}_1 + 2k\left(\frac{r_1^2}{J_1} - \frac{r_2^2}{J_2}\right)\psi_2 = 0$$

or

$$\ddot{\psi}_1 = 2k \left(\frac{r_2^2}{J_2} - \frac{r_1^2}{J_1} \right) \left[\psi_2(0) \cos \omega_n t + \frac{\dot{\psi}_2(0)}{\omega_n} \sin \omega_n t \right]$$

from which $\psi_1(t)$ can be derived by simple quadrature. Indeed, by time-integrating the above equation, we obtain

$$\dot{\psi}_1(t) = 2k \left(\frac{r_2^2}{J_2} - \frac{r_1^2}{J_1} \right) \left[\frac{\dot{\psi}_2(0)}{\omega_n} \sin \omega_n t - \frac{\psi_2(0)}{\omega_n^2} \cos \omega_n t \right] + c_1$$

where c_1 is an integration constant, to be determined from the initial conditions. Thus, at $t = 0$,

$$\dot{\psi}_1(0) = -2k \left(\frac{r_2^2}{J_2} - \frac{r_1^2}{J_1} \right) \frac{\psi_2(0)}{\omega_n^2} + c_1$$

from which

$$c_1 = \dot{\psi}_1(0) + 2k \left(\frac{r_2^2}{J_2} - \frac{r_1^2}{J_1} \right) \frac{\psi_2(0)}{\omega_n^2}$$

and hence,

$$\dot{\psi}_1(t) = \dot{\psi}_1(0) + 2k \left(\frac{r_2^2}{J_2} - \frac{r_1^2}{J_1} \right) \left[\frac{\dot{\psi}_2(0)}{\omega_n} \sin \omega_n t + \frac{\psi_2(0)}{\omega_n^2} (1 - \cos \omega_n t) \right]$$

Upon integration of the last equation, we obtain

$$\psi_1(t) = \dot{\psi}_1(0)t + 2k \left(\frac{r_2^2}{J_2} - \frac{r_1^2}{J_1} \right) \left[-\frac{\dot{\psi}_2(0)}{\omega_n^2} \cos \omega_n t + \frac{\psi_2(0)}{\omega_n^2} \left(t - \frac{1}{\omega_n} \sin \omega_n t \right) \right] + c_2$$

where c_2 is a second integration constant, to be detrmined from the initial conditions as well. At $t = 0$, we have

$$\psi_1(0) = -2k \left(\frac{r_2^2}{J_2} - \frac{r_1^2}{J_1} \right) \frac{\dot{\psi}_2(0)}{\omega_n^2} + c_2$$

whence

$$c_2 = \psi_1(0) + 2k \left(\frac{r_2^2}{J_2} - \frac{r_1^2}{J_1} \right) \frac{\dot{\psi}_2(0)}{\omega_n^2}$$

Therefore,

$$\psi_1(t) = \psi_1(0) + \dot{\psi}_1(0)t + 2k \left(\frac{r_2^2}{J_2} - \frac{r_1^2}{J_1} \right) \left[\frac{\psi_2(0)}{\omega_n^2} (1 - \cos \omega_n t) \right.$$
$$\left. + \frac{\dot{\psi}_2(0)}{\omega_n^2} \left(t - \frac{1}{\omega_n} \sin \omega_n t \right) \right]$$

Now, in order to find the time response in the original generalized coordinates, all we need is to change back to those coordinates, thereby obtaining

$$\theta_1(t) = r_2(A + Bt) + r_1 J_2(C\cos\omega_n t + D\sin\omega_n t)$$
$$\theta_2(t) = r_1(A + Bt) + r_2 J_1(C\cos\omega_n t + D\sin\omega_n t)$$

with coefficients A, B, C, and D defined as

$$A \equiv \frac{r_2 J_1 \theta_1(0) + r_1 J_2 \theta_2(0)}{r_1^2 J_2 + r_2^2 J_1}, \quad B \equiv \frac{r_2 J_1 \dot{\theta}_1(0) + r_1 J_2 \dot{\theta}_2(0)}{r_1^2 J_2 + r_2^2 J_1}$$

$$C \equiv \frac{r_1 \theta_1(0) + r_2 \theta_2(0)}{r_1^2 J_2 + r_2^2 J_1}, \quad D \equiv \frac{r_1 \dot{\theta}_1(0) - r_2 \dot{\theta}_2(0)}{\omega_n(r_1^2 J_2 + r_2^2 J_1)}$$

From the above expressions it is apparent that the motion of the overall system is the superposition of the rigid mode and the purely-flexible mode. Indeed, the first term of each of the two above expressions represents a displacement that is linear in time, and hence, corresponds to a uniform motion starting from an initial value $r_2 A$ or, correspondingly, $r_1 A$, and moving with the constant speed $r_2 B$ or, correspondingly, $r_1 B$. The second term comprises the flexible mode.

Furthermore, if the two original governing equations are added sidewise, we obtain an interesting result, namely,

$$J_1 \ddot{\theta}_1 + J_2 \ddot{\theta}_2 = 2k(r_2 \theta_2 - r_1 \theta_1)(r_1 + r_2)$$

or, if we define $h(t)$ as the total angular momentum of the overall system, i.e.,

$$h(t) \equiv J_1 \dot{\theta}_1 + J_2 \dot{\theta}_2$$

then, the above equation can be rewritten as

$$\dot{h}(t) = 2k(r_2 \theta_2 - r_1 \theta_1)(r_1 + r_2)$$

in which the right-hand side vanishes only if the system moves under its rigid mode. As a consequence, then, the *angular momentum of the whole system is not preserved*. Can the reader give a physical explanation of this fact?

We have thus derived a procedure, described below, to obtain the zero-input response of a semidefinite two-dof system:

1. Decouple the rigid from the flexible mode, thus obtaining two new governing equations, one of which contains one single generalized coordinate
2. Find the time response of the subsystem with one single generalized coordinate
3. Find the remaining time response by substituting the time response found in step-2 above

5.3.2   Systems with a Positive-Definite Frequency Matrix

Under the assumption that the stiffness matrix is positive-definite, the frequency matrix, defined in Eq. 5.9a, is positive-definite as well. In this case, the response of the system at hand can be derived from the response of a single-dof system, as we will show presently.

The time response of a single-dof mass-spring system governed by Eq. 5.11 is now recalled for quick reference:

$$y(t) = (\cos \omega_n t)y_0 + \frac{1}{\omega_n}(\sin \omega_n t)s_0, \quad t \geq 0 \tag{5.66}$$

Further, the response of the system appearing in Eq. 5.10 can be readily derived just by mimicking the response of the scalar system of Eq. 5.11, given by Eq. 5.66. In doing so, we will rewrite Eq. 5.66 in vector form, with $y(t)$ replaced by the two-dimensional vector $\mathbf{y}(t)$ and the scalar ω_n by the 2×2 frequency matrix $\boldsymbol{\Omega}$. Here, care must be taken when mimicking the $1/\omega_n$ factor in the above equation, for the *reciprocal* of a matrix does not exist. This reciprocal must be substituted by $\boldsymbol{\Omega}^{-1}$ instead, which exists because $\boldsymbol{\Omega}$ is assumed to be positive-definite. The zero-input response of the two-dof system, then, takes the form[3]

$$\mathbf{y}(t) = (\cos \boldsymbol{\Omega} t)\mathbf{y}_0 + \boldsymbol{\Omega}^{-1}(\sin \boldsymbol{\Omega} t)\mathbf{s}_0, \quad t \geq 0 \tag{5.67}$$

where $\cos \boldsymbol{\Omega} t$ and $\sin \boldsymbol{\Omega} t$ are the analytic functions of $\boldsymbol{\Omega} t$ derived from the corresponding scalar functions $\cos \omega_i t$ and $\sin \omega_i t$, with ω_i denoting the ith eigenvalue of the frequency matrix, that is, the ith natural frequency of the system at hand, for $i = 1, 2$. Note from Eq. 5.67 that, if both $\mathbf{y}_0$ and $\mathbf{w}_0$ are proportional to the eigenvector $\mathbf{e}_i$, then $\mathbf{y}(t)$ is harmonic, with frequency ω_i. Otherwise, in general, $\mathbf{y}(t)$ *is not harmonic*.

The foregoing matrix functions bear some of the properties of their scalar counterparts, namely,

$$\cos^2 \boldsymbol{\Omega} t + \sin^2 \boldsymbol{\Omega} t = 1 \tag{5.68a}$$

$$\frac{d}{dt}(\cos \boldsymbol{\Omega} t) = -\boldsymbol{\Omega} \sin \boldsymbol{\Omega} t \equiv -(\sin \boldsymbol{\Omega} t)\boldsymbol{\Omega} \tag{5.68b}$$

$$\frac{d}{dt}(\sin \boldsymbol{\Omega} t) = \boldsymbol{\Omega} \cos \boldsymbol{\Omega} t \equiv (\cos \boldsymbol{\Omega} t)\boldsymbol{\Omega} \tag{5.68c}$$

$$\int_0^t \cos \boldsymbol{\Omega} \theta d\theta = \boldsymbol{\Omega}^{-1} \sin \boldsymbol{\Omega} t \equiv (\sin \boldsymbol{\Omega} t)\boldsymbol{\Omega}^{-1} \tag{5.68d}$$

$$\int_0^t \sin \boldsymbol{\Omega} \theta d\theta = \boldsymbol{\Omega}^{-1}(1 - \cos \boldsymbol{\Omega} t) \equiv (1 - \cos \boldsymbol{\Omega} t)\boldsymbol{\Omega}^{-1} \tag{5.68e}$$

---

[3]The second term of the right-hand side of Eq. 5.67 can be alternatively expressed with the order of $\boldsymbol{\Omega}^{-1}$ and $\sin \boldsymbol{\Omega} t$ reversed, a consequence of Fact 4 of Appendix A.

where $\mathbf{1}$ is the 2×2 identity matrix and Fact 4 of Appendix A has been invoked. The reader can readily verify the foregoing relations, for example, by assuming, without loss of generality, that $\mathbf{\Omega}$ is in diagonal form. Using Eqs. 5.68b, c, the proof that $\mathbf{y}(t)$, as given by Eq. 5.67, is the integral of the system of ODE of Eq. 5.10 is straightforward.

In mimicking the trigonometric relations for matrices, care must be taken in that, in general, for *arbitrary* 2×2 matrices $\mathbf{A}$ and $\mathbf{B}$,

$$\cos(\mathbf{A}+\mathbf{B}) \neq \cos\mathbf{A}\cos\mathbf{B} - \sin\mathbf{A}\sin\mathbf{B}$$

$$\sin(\mathbf{A}+\mathbf{B}) \neq \sin\mathbf{A}\cos\mathbf{B} + \cos\mathbf{A}\sin\mathbf{B}$$

Nevertheless, if $\mathbf{A}$ and $\mathbf{B}$ in the above relations *share* the same set of eigenvectors, then they *commute* under the operation of multiplication, and the relations above do hold with the equality sign. This is the case if, for example, $\mathbf{A}$ and $\mathbf{B}$ are analytic functions of the same matrix, or if one is an analytic function of the other. Now, since the vector of generalized coordinates is $\mathbf{x}$, rather than $\mathbf{y}$, a transformation back to the original coordinates is in order. From Eq. 5.6, we have

$$\mathbf{x}(t) = \mathbf{N}^{-1}(\cos\mathbf{\Omega}t)\mathbf{N}\mathbf{x}_0 + \mathbf{N}^{-1}\mathbf{\Omega}^{-1}(\sin\mathbf{\Omega}t)\mathbf{N}\mathbf{v}_0, \quad t \geq 0 \qquad (5.69a)$$

where $\dot{\mathbf{x}}(0) \equiv \mathbf{v}_0$ and $\mathbf{s}_0 \equiv \mathbf{N}\mathbf{v}_0$. Note from Eq. 5.69a that, if both $\mathbf{x}_0$ and $\mathbf{v}_0$ are proportional to the modal vector $\mathbf{f}_i$, then both $\mathbf{N}\mathbf{x}_0$ and $\mathbf{N}\mathbf{v}_0$ are proportional to the eigenvector $\mathbf{e}_i$. As a consequence, then, the above zero-input response is harmonic, with frequency ω_i.

In simulation studies, we also need $\dot{\mathbf{x}}$, which can be readily derived by straight-forward differentiation of Eq. 5.69a, which thus leads to

$$\dot{\mathbf{x}}(t) = -\mathbf{N}^{-1}\mathbf{\Omega}(\sin\mathbf{\Omega}t)\mathbf{N}\mathbf{x}_0 + \mathbf{N}^{-1}(\cos\mathbf{\Omega}t)\mathbf{N}\mathbf{v}_0, \quad t \geq 0 \qquad (5.69b)$$

Example 5.3.2 (Zero-Input Response of a Two-dof System). In computing the time response of the system of Fig. 5.1 with the numerical values given in Eq. 5.31, we resort to Eq. 5.69a. In this equation, the sine and cosine functions of the frequency matrix are needed. These functions are computed from the Mohr circles of Fig. 5.9a,b.

The matrices thus resulting are displayed below:

$$\cos\mathbf{\Omega}t \equiv \begin{bmatrix} \bar{c}+u_C & -v_C \\ -v_C & \bar{c}-u_C \end{bmatrix}, \quad \sin\mathbf{\Omega}t \equiv \begin{bmatrix} \bar{s}+u_S & -v_S \\ -v_S & \bar{s}-u_S \end{bmatrix}$$

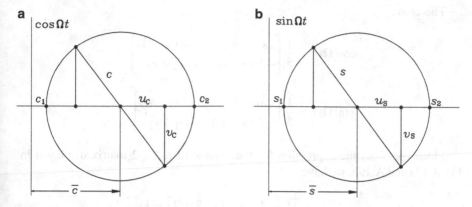

Fig. 5.9 The Mohr circle of (a) $\cos \Omega t$ and (b) $\sin \Omega t$ of the two-dof system composed of two masses coupled by three springs

where variables $\bar{c}$ and c are the abscissa of the center and the radius of the Mohr circle of $\cos(\Omega t)$, with similar definitions for $\bar{s}$ and s, i.e.,

$$\bar{c} \equiv \bar{c}(t) \equiv \frac{\cos \omega_1 t + \cos \omega_2 t}{2}, \quad c \equiv c(t) \equiv \frac{\cos \omega_2 t - \cos \omega_1 t}{2}$$

$$\bar{s} \equiv \bar{s}(t) \equiv \frac{\sin \omega_1 t + \sin \omega_2 t}{2}, \quad s \equiv s(t) \equiv \frac{\sin \omega_2 t - \sin \omega_1 t}{2}$$

In the Mohr circles of Fig. 5.9a,b, we have

$$c_k \equiv \cos \omega_k t, \quad s_k \equiv \sin \omega_k t, \quad k = 1, 2$$

Moreover, from similarity of triangles,

$$u_C = \frac{3}{5}c, \quad v_C = \frac{4}{5}c, \quad u_S = \frac{3}{5}s, \quad v_S = \frac{4}{5}s$$

and hence,

$$\bar{c} + u_C = \frac{c_1 + 4c_2}{5}, \quad \bar{c} - u_C = \frac{4c_1 + c_2}{5}$$

$$\bar{s} + u_S = \frac{s_1 + 4s_2}{5}, \quad \bar{s} - u_S = \frac{4s_1 + s_2}{5}$$

Therefore,

$$\cos(\Omega t) = \frac{1}{5}\begin{bmatrix} c_1 + 4c_2 & -2(c_2 - c_1) \\ -2(c_2 - c_1) & 4c_1 + c_2 \end{bmatrix}$$

$$\sin(\Omega t) = \frac{1}{5}\begin{bmatrix} s_1 + 4s_2 & -2(s_2 - s_1) \\ -2(s_2 - s_1) & 4s_1 + s_2 \end{bmatrix}$$

Moreover, from the expression for the inverse of a 2×2 matrix displayed in Eq. A.19a and A.19b, we have

$$\Omega^{-1} = \frac{\sqrt{11}}{55\omega}\begin{bmatrix} 4 + \sqrt{11} & 2(\sqrt{11} - 1) \\ 2(\sqrt{11} - 1) & 1 + 4\sqrt{11} \end{bmatrix}$$

Now, let the initial conditions be given as

$$\mathbf{x}_0 = \begin{bmatrix} a_1 \\ a_2 \end{bmatrix}, \quad \mathbf{v}_0 = \begin{bmatrix} b_1 \\ b_2 \end{bmatrix}$$

Hence,

$$\mathbf{y}_0 = \mathbf{N}\mathbf{x}_0 = \sqrt{m}\begin{bmatrix} a_1 \\ 2a_2 \end{bmatrix}, \quad \mathbf{s}_0 = \mathbf{N}\mathbf{v}_0 = \sqrt{m}\begin{bmatrix} b_1 \\ 2b_2 \end{bmatrix}$$

Upon expansion of the time response of Eq. 5.69a and rearrangement of terms, we obtain the said response as a linear combination of functions $\cos\omega_1 t$, $\cos\omega_2 t$, $\sin\omega_1 t$ and $\sin\omega_2 t$. The terms entering in that expression are calculated below:

$$[\cos(\Omega t)]\mathbf{y}_0 = \frac{\sqrt{m}}{5}\begin{bmatrix} (a_1 + 4a_2)c_1 + 4(a_1 - a_2)c_2 \\ 2(a_1 + 4a_2)c_1 - 2(a_1 - a_2)c_2 \end{bmatrix}$$

$$[\Omega^{-1}\sin(\Omega t)]\mathbf{s}_0 = \frac{\sqrt{m}}{55\omega}\begin{bmatrix} 11(b_1 + 4b_2)s_1 + 4\sqrt{11}(b_1 - b_2)s_2 \\ 22(b_1 + 4b_2)s_1 - 2\sqrt{11}(b_1 - b_2)s_2 \end{bmatrix}$$

Furthermore,

$$[\mathbf{N}^{-1}(\cos\Omega t)]\mathbf{y}_0 = \frac{1}{5}\begin{bmatrix} (a_1 + 4a_2)c_1 + 4(a_1 - a_2)c_2 \\ (a_1 + 4a_2)c_1 - (a_1 - a_2)c_2 \end{bmatrix}$$

$$[\mathbf{N}^{-1}\Omega^{-1}\sin(\Omega t)]\mathbf{s}_0 = \frac{1}{55\omega}\begin{bmatrix} 11(b_1 + 4b_2)s_1 + 4\sqrt{11}(b_1 - b_2)s_2 \\ 11(b_1 + 4b_2)s_1 - \sqrt{11}(b_1 - b_2)s_2 \end{bmatrix}$$

and hence,

$$x_1(t) = \frac{1}{5}[(a_1 + 4a_2)\cos\omega t + 4(a_1 - a_2)\cos\sqrt{11}\omega t]$$

$$+ \frac{1}{55\omega}[11(b_1 + 4b_2)\sin\omega t + 4\sqrt{11}(b_1 - b_2)\sin\sqrt{11}\omega t]$$

$$x_2(t) = \frac{1}{5}[(a_1 + 4a_2)\cos\omega t - (a_1 - a_2)\cos\sqrt{11}\omega t]$$

$$+ \frac{1}{55\omega}[11(b_1 + 4b_2)\sin\omega t - \sqrt{11}(b_1 - b_2)\sin\sqrt{11}\omega t]$$

In summary, then, the time response of the system is a linear combination of the harmonic functions associated with its two natural frequencies. Now a question arises, namely, *is the zero-input response of two-dof systems, like that of their single-dof counterparts, in general, harmonic?* For the answer to be positive, we need a certain time interval T at which the harmonics of the two frequencies both cover integer numbers of cycles, N_1 and N_2, namely,

$$\omega_1 T = 2\pi N_1, \quad \omega_2 T = 2\pi N_2$$

Now, if we divide the corresponding sides of the second of the two foregoing equations by those of the first, we obtain

$$\frac{\omega_2}{\omega_1} = \frac{N_2}{N_1}$$

which means that, for the zero-input response to be harmonic for arbitrary initial values, the ratio of the two natural frequencies must be a rational number. In our example we have

$$\frac{\omega_2}{\omega_1} = \frac{\sqrt{11}\omega}{\omega} = \sqrt{11}$$

which is, apparently, irrational; as a consequence, the response of the system is not, in general, harmonic. As we have seen earlier, we need a special set of initial conditions, namely, shaped by any of the two modal vectors, in order to obtain a harmonic zero-input response. Now, from the form of the response derived above, it is apparent that a harmonic response is possible if the initial conditions obey any one of the two relations below:

1. $a_1 = a_2$ and $b_1 = b_2$
2. $a_1 = -4a_2$ and $b_1 = -4b_2$

which are apparently obeyed if the vectors of initial conditions, x_0 and v_0, are both proportional to any one of the two modal vectors of the system at hand.

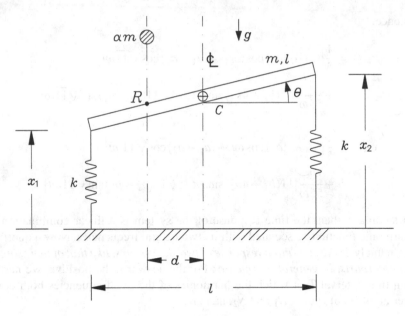

Fig. 5.10 A test pad mounted on an elastic suspension

Example 5.3.3 (A Two-dof Test Pad).   Shown in Fig. 5.10 is a test pad mounted on an elastic suspension, consisting of a slender rigid bar of mass m and length l. A small ball of mass αm is dropped from a height h at an offset $d = l/4$ with respect to the centerline of the bar, and α is the ratio of the mass of the ball to the mass of the bar, as indicated in Fig. 5.10. Find the time response of the pad under a perfectly elastic shock of the ball with the bar and under the assumption that the bar (a) is at rest prior to being hit by the ball and (b) undergoes small-amplitude rotations that allow for a linear-model analysis.

Solution: We define the generalized coordinates x_1, x_2 and θ, as shown in Fig. 5.10. In order to obtain a mathematical model with a zero input, we aim to determine the initial conditions right after the impact of the ball with the bar. In the analysis below we measure velocities the same way as we measure the displacements of the end-points of the bar, i.e., assuming that upward velocities are positive.

The initial conditions are determined from the basic Principle of Conservation of Momentum. In the analysis below, let v_B and v_R denote the velocities of the ball and of point R of the bar, where R denotes the point of the bar where collision takes place. Let v_{rel}, moreover, be defined as the relative velocity of the ball with respect to the bar, and hence, just prior to the shock,

$$v_{\text{rel}}(0^-) = v_B(0^-) - v_R(0^-)$$

while, right after the shock,

$$v_{rel}(0^+) = v_B(0^+) - v_R(0^+)$$

From the data,

$$v_{rel}(0^-) = -\sqrt{2gh}$$

while, from the assumption of a perfectly elastic shock,

$$v_{rel}(0^+) = -v_{rel}(0^-)$$

and hence,

$$v_B(0^+) - v_R(0^+) = \sqrt{2gh} \qquad (5.70)$$

Moreover, from the nature of the shock, it is apparent that no displacement jumps occur upon collision; only velocity jumps appear. Now, since the force developed in the springs is proportional to the displacements, it is apparent that no impulsive forces are developed by the springs. As a consequence, both the momentum and the angular momentum of the whole ball-pad system are preserved. The system momentum just before and just after the shock is calculated below:

$$p(0^-) = -\alpha m \sqrt{2gh}, \quad p(0^+) = m v_C(0^+) + \alpha m v_B(0^+)$$

where v_C is the velocity of the center of mass of the bar. Upon equating the two above expressions, we obtain

$$v_C(0^+) + \alpha v_B(0^+) = -\alpha \sqrt{2gh} \qquad (5.71)$$

Moreover, the angular momentum about any point O fixed on the centerline right before and right after the shock is given by

$$h_O(0^-) = \alpha m d \sqrt{2gh}, \quad h_O(0^+) = \frac{ml^2}{12}\dot{\theta}(0^+) - \alpha m d v_B(0^+)$$

Therefore, the conservation of the angular momentum, for $d = l/4$, leads to

$$l\dot{\theta}(0^+) - 3\alpha v_B(0^+) = 3\alpha \sqrt{2gh} \qquad (5.72)$$

Furthermore, from planar kinematics we have,

$$v_R(0^+) = v_C(0^+) - \frac{l}{4}\dot{\theta}(0^+) \qquad (5.73)$$

Upon substitution of Eq. 5.73 into Eq. 5.70, we obtain

$$v_B(0^+) - v_C(0^+) + \frac{l}{4}\dot{\theta}(0^+) = \sqrt{2gh} \tag{5.74}$$

In summary, then, we have derived three independent equations, (5.71), (5.72), and (5.74), to determine three unknowns, namely, $v_B(0^+)$, $v_C(0^+)$ and $\dot{\theta}(0^+)$. We proceed now to solve for these unknowns: From Eq. 5.71, we can solve for $v_C(0^+)$, namely,

$$v_C(0^+) = \alpha[-\sqrt{2gh} - v_B(0^+)] \tag{5.75}$$

Likewise, from Eq. 5.72, we can solve for $\dot{\theta}(0^+)$ as

$$\dot{\theta}(0^+) = \frac{3\alpha}{l}[\sqrt{2gh} + v_B(0^+)] \tag{5.76}$$

Now, upon substitution of Eqs. 5.75 and 5.76 into Eq. 5.74, an equation in $v_B(0^+)$ alone is derived, from which

$$v_B(0^+) = \frac{1 - 7\alpha/4}{1 + 7\alpha/4}\sqrt{2gh} \tag{5.77}$$

whence corresponding expressions for $v_C(0^+)$ and $\dot{\theta}(0^+)$ are readily derived:

$$v_C(0^+) = -\frac{2\alpha}{1 + 7\alpha/4}\sqrt{2gh}, \quad \dot{\theta}(0^+) = \frac{6\alpha}{1 + 7\alpha/4}\frac{\sqrt{2gh}}{l} \tag{5.78}$$

Now, the initial conditions for the generalized coordinates $x_1(t)$ and $x_2(t)$, and their time derivatives are derived below. Note that, since no jumps in the generalized coordinates occur, we have

$$x_1(0^+) = x_2(0^+) = 0 \tag{5.79}$$

Moreover, from planar kinematics we have, for any instant t,

$$\dot{x}_1(t) = v_C(t) - \dot{\theta}(t)\frac{l}{2}, \quad \dot{x}_2(t) = v_C(t) + \dot{\theta}(t)\frac{l}{2}$$

and hence,

$$\dot{x}_1(0^+) = v_C(0^+) - \dot{\theta}(0^+)\frac{l}{2} = -\frac{5\alpha}{1 + 7\alpha/4}\sqrt{2gh} \tag{5.80}$$

$$\dot{x}_2(0^+) = v_C(0^+) + \dot{\theta}(0^+)\frac{l}{2} = \frac{\alpha}{1 + 7\alpha/4}\sqrt{2gh} \tag{5.81}$$

Having derived the initial conditions for the generalized coordinates $x_1(t)$ and $x_2(t)$ and their time derivatives, it is now a simple matter to obtain the time response of the system under zero-input conditions. To this end, we derive expressions for the mass and stiffness matrices. The former is computed as the Hessian of the kinetic energy, which in turn takes the form

$$T = \frac{1}{2}mv_C^2 + \frac{1}{2}J_C\dot{\theta}^2$$

where J_C is the moment of inertia of the bar about the center of mass, i.e., $J_C = ml^2/12$, and hence,

$$T = \frac{1}{2}mv_C^2 + \frac{1}{24}ml^2\dot{\theta}^2$$

Upon expressing the velocity of the center of mass and the angular velocity in terms of the generalized velocities as

$$v_C = \frac{\dot{x}_1 + \dot{x}_2}{2}, \qquad \dot{\theta} = \frac{\dot{x}_2 - \dot{x}_1}{l} \tag{5.82}$$

we have

$$T = \frac{1}{6}m(\dot{x}_1^2 + \dot{x}_1\dot{x}_2 + \dot{x}_2^2)$$

Moreover, the potential energy of the system is

$$V = \frac{1}{2}k(x_1^2 + x_2^2)$$

and hence, the mass and stiffness matrices are readily derived as

$$\mathbf{M} = \frac{1}{6}m\begin{bmatrix} 2 & 1 \\ 1 & 2 \end{bmatrix}, \quad \mathbf{K} = k\begin{bmatrix} 1 & 0 \\ 0 & 1 \end{bmatrix} \equiv k\mathbf{1}$$

with $\mathbf{1}$ defined as the 2×2 identity matrix. By virtue of the particular form of the stiffness matrix, the square of the frequency matrix is obtained as

$$\mathbf{\Omega}^2 = 2k\mathbf{M}^{-1} = \omega^2\begin{bmatrix} 2 & -1 \\ -1 & 2 \end{bmatrix}, \quad \omega^2 \equiv \frac{2k}{m}$$

its Mohr circle being shown in Fig. 5.11a. From this circle, we can readily determine the eigenvalues of $\mathbf{\Omega}^2$ and hence, the natural frequencies ω_1 and ω_2, namely,

$$\omega_1 = \omega = \sqrt{\frac{k}{m}}, \quad \omega_2 = \sqrt{3}\omega = \sqrt{3\frac{k}{m}}$$

Moreover, from the same Mohr circle we can derive that of the frequency matrix, as shown in Fig. 5.11b.

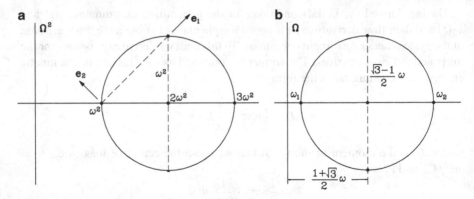

Fig. 5.11 The Mohr circles of (**a**) Ω^2, and (**b**) Ω for the test pad

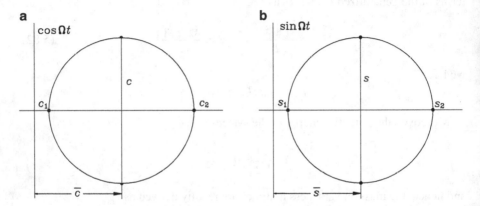

Fig. 5.12 The Mohr circles of: (**a**) $\cos \Omega t$: and (**b**) $\sin \Omega t$ for the test pad

Hence, the frequency matrix is

$$\Omega = \frac{\omega}{2} \begin{bmatrix} 1+\sqrt{3} & 1-\sqrt{3} \\ 1-\sqrt{3} & 1+\sqrt{3} \end{bmatrix}$$

In the ensuing calculations we will also need the harmonic functions of the frequency matrix multiplied by t. These can be determined from the Mohr circle of Ω, as displayed in Fig. 5.12a,b.

From the Mohr circles of the corresponding harmonic functions, one can readily obtain

$$\cos(\Omega t) = \begin{bmatrix} \bar{c} & -c \\ -c & \bar{c} \end{bmatrix}, \quad \sin(\Omega t) = \begin{bmatrix} \bar{s} & -s \\ -s & \bar{s} \end{bmatrix}$$

where $\bar{c}$, c, $\bar{s}$ and s are defined as

$$\bar{c} \equiv \bar{c}(t) \equiv \frac{1}{2}(\cos\omega_1 t + \cos\omega_2 t), \quad c \equiv c(t) \equiv \frac{1}{2}(\cos\omega_2 t - \cos\omega_1 t)$$

$$\bar{s} \equiv \bar{s}(t) \equiv \frac{1}{2}(\sin\omega_1 t + \sin\omega_2 t), \quad s \equiv c(t) \equiv \frac{1}{2}(\sin\omega_2 t - \sin\omega_1 t)$$

Now, since the stiffness matrix is proportional to the identity matrix, we note that we can write, in the given generalized coordinates,

$$\ddot{\mathbf{x}} + \Omega^2 \mathbf{x} = \mathbf{0}$$

and hence, the time response, in this particular case, takes the simple form

$$\mathbf{x}(t) = \cos(\Omega t)\mathbf{x}_0 + \Omega^{-1}\sin(\Omega t)\mathbf{v}_0$$

Moreover, in our case,

$$\mathbf{x}_0 = \mathbf{0}, \quad \mathbf{v}_0 = \begin{bmatrix} -5a \\ a \end{bmatrix}, \quad a \equiv \frac{\alpha}{1 + 7\alpha/4}\sqrt{2gh}$$

Furthermore, a simple calculation leads to

$$\Omega^{-1} = \frac{\sqrt{3}}{6\omega}\begin{bmatrix} 1+\sqrt{3} & -(1-\sqrt{3}) \\ -(1-\sqrt{3}) & 1+\sqrt{3} \end{bmatrix} = \frac{1}{6\omega}\begin{bmatrix} 3+\sqrt{3} & 3-\sqrt{3} \\ 3-\sqrt{3} & 3+\sqrt{3} \end{bmatrix}$$

and hence, upon substitution of all quantities involved, the final result turns out to be

$$\mathbf{x}(t) = -\frac{a}{\omega}\begin{bmatrix} 2\sin\omega_1 t + \sqrt{3}\sin\omega_2 t \\ 2\sin\omega_1 t - \sqrt{3}\sin\omega_2 t \end{bmatrix}$$

or, in terms of the original parameters,

$$\mathbf{x}(t) = -\frac{\alpha}{1+7\alpha/4}\sqrt{\frac{2mgh}{k}}\begin{bmatrix} 2\sin\omega_1 t + \sqrt{3}\sin\omega_2 t \\ 2\sin\omega_1 t - \sqrt{3}\sin\omega_2 t \end{bmatrix}$$

Therefore, the zero-state response of the system at hand is a linear combination of the sine functions of $\omega_1 t$ and $\omega_2 t$. Moreover, the two natural frequencies observe the ratio

$$\frac{\omega_2}{\omega_1} = \frac{\sqrt{3}\omega}{\omega} = \sqrt{3}$$

which is irrational. As a consequence, the two ends of the rod move with non-periodic motion. However, the motion of the c.o.m., given by the coordinate $x(t)$, and the rotation of the bar, $\theta(t)$, are harmonic, as we can readily verify, namely,

$$x(t) = -\frac{2\alpha}{1+7\alpha/4}\sqrt{\frac{2mgh}{k}}\sin\sqrt{\frac{k}{m}}t$$

$$\theta(t) = \frac{2\sqrt{3}\alpha}{1+7\alpha/4}\frac{1}{l}\sqrt{\frac{2mgh}{k}}\sin\sqrt{\frac{3k}{m}}t$$

Thus, the first natural frequency corresponds to the translation of the c.o.m., the second to the rotation of the bar about its center of mass.

5.3.3 The Beat Phenomenon

Mechanical systems that are only slightly coupled exhibit an interesting behavior known as *beat*. A feature of this behavior is a periodic response, which is caused by its two natural frequencies being *relatively* close to each other. What we mean by this, is that the ratio of the frequency radius to the mean frequency is much smaller than unity, thereby appearing as if the ratio of the two natural frequencies were a rational number.

In order to illustrate the beat phenomenon, let us consider again the system of Fig. 5.1, with $m_1 = m_2 = m$ but now we let $k_1 = k_3 = k$ and $k_2 = \alpha k$, and assume that $\alpha \ll 1$. Thus, the mass and stiffness matrices take the forms

$$\mathbf{M} = m\mathbf{1}, \quad \mathbf{K} = k\begin{bmatrix} 1+\alpha & -\alpha \\ -\alpha & 1+\alpha \end{bmatrix}$$

with $\mathbf{1}$ representing the 2×2 identity matrix. Hence, the frequency matrix-squared is readily computed as

$$\boldsymbol{\Omega}^2 = \omega^2\begin{bmatrix} 1+\alpha & -\alpha \\ -\alpha & 1+\alpha \end{bmatrix}, \quad \omega \equiv \sqrt{\frac{k}{m}}$$

its Mohr circle being shown in Fig. 5.13a. From this figure, it is apparent that the radius of the Mohr circle of $\boldsymbol{\Omega}^2$ is quite small when compared with the abscissa of its center.

Furthermore, from the same figure, the eigenvalues of $\boldsymbol{\Omega}^2$ are readily derived as

$$\omega_1^2 = \omega^2, \quad \omega_2^2 = (1+2\alpha)\omega^2$$

and hence, the natural frequencies are

$$\omega_1 = \omega, \quad \omega_2 = \sqrt{1+2\alpha}\,\omega$$

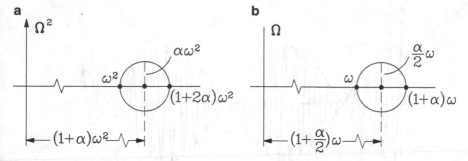

Fig. 5.13 The Mohr circle of: (**a**) the frequency matrix-squared; and (**b**) the frequency matrix, of a slightly coupled mechanical system

Now, if we take into account that $\alpha \ll 1$, the linear approximation of ω_2 becomes

$$\omega_2 \approx (1 + \alpha)\omega$$

and hence, the Mohr circle of the frequency matrix becomes as in Fig. 5.13b. Note that, from the above approximation, we have

$$\frac{\omega_2}{\omega_1} \approx 1 + \alpha$$

i.e., the ratio of the two frequencies is almost unity, the system thus showing an *almost periodic* response *for any initial conditions*, which is a characteristic of beat. From the circle of Fig. 5.13b, it is apparent that the mean frequency and the frequency radius take the approximate values

$$\overline{\omega} \approx \left(1 + \frac{\alpha}{2}\right)\omega, \quad \rho \approx \frac{\alpha}{2}\omega$$

and hence, $\overline{\omega}$ and ρ obey the relation

$$\overline{\omega} \approx \omega + \rho$$

the ratio $\rho/\overline{\omega}$ thus becoming

$$\frac{\rho}{\overline{\omega}} \approx \frac{\alpha}{2 + \alpha} \ll 1$$

which follows because of the assumption on the order of magnitude of α. The frequency matrix, in turn, can be approximated as

$$\Omega \approx \frac{\omega}{2}\begin{bmatrix} 2 + \alpha & -\alpha \\ -\alpha & 2 + \alpha \end{bmatrix}$$

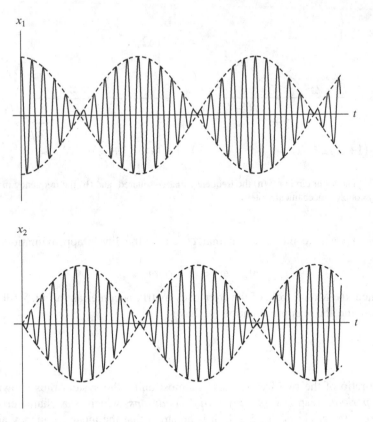

Fig. 5.14 A typical time response of a weakly-coupled two-dof system

Now, in order to determine the time response of this system, we find first the harmonic functions of the frequency matrix times t. This is most simply done by recalling the expressions derived in Sect. A.4 of Appendix A. Upon consideration of the approximations given above, these matrices take the forms

$$\cos(\mathbf{\Omega}t) \approx \begin{bmatrix} \cos\overline{\omega}t\cos\rho t & -\sin\overline{\omega}t\sin\rho t \\ -\sin\overline{\omega}t\sin\rho t & \cos\overline{\omega}t\cos\rho t \end{bmatrix}$$

$$\sin(\mathbf{\Omega}t) \approx \begin{bmatrix} \sin\overline{\omega}t\cos\rho t & \cos\overline{\omega}t\sin\rho t \\ \cos\overline{\omega}t\sin\rho t & \sin\overline{\omega}t\cos\rho t \end{bmatrix}$$

The time response is now obtained as a sum of the two terms below:

$$[\cos(\mathbf{\Omega}t)]\mathbf{x}_0 = \begin{bmatrix} (a_1\cos\rho t)\cos\overline{\omega}t - a_2(\sin\rho t)\sin\overline{\omega}t \\ -a_1(\sin\rho t)\sin\overline{\omega}t + (a_2\cos\rho t)\cos\overline{\omega}t \end{bmatrix}$$

$$\left[\Omega^{-1}\sin(\Omega t)\right]\mathbf{v}_0 = \frac{1}{2\omega(1+\alpha)} \begin{bmatrix} [(2+\alpha)b_2 + \alpha b_1]s\bar{c} + 2[(2+\alpha)b_1 + \alpha b_2]c\bar{s} \\ [\alpha b_2 + (2+\alpha)b_1]s\bar{c} + 2[\alpha b_1 + (2+\alpha)b_2]c\bar{s} \end{bmatrix}$$

where $c \equiv \cos pt$, $s \equiv \sin pt$, $\bar{c} \equiv \cos \bar{\omega}t$, $\bar{s} \equiv \sin \bar{\omega}t$. It is thus apparent that the time response can be regarded as a linear combination of fast varying harmonics, $\cos \bar{\omega}t$ and $\sin \bar{\omega}t$, of slowly-varying amplitude, $A\cos pt$ and $B\sin pt$. A typical response of this system is shown in Fig. 5.14.

5.4   The Classical Modal Method

Regardless of whether the system is semidefinite or definite, we can find its zero-input response by decoupling its two modes by means of a change of variable. Indeed, let us multiply both sides of Eq. 5.1 by $\mathbf{M}^{-1}$:

$$\ddot{\mathbf{x}} + \mathbf{D}\mathbf{x} = \mathbf{0}, \quad t \geq 0, \quad \mathbf{x}(0) = \mathbf{x}_0, \quad \dot{\mathbf{x}}(0) = \mathbf{v}_0 \tag{5.83a}$$

where we recall that the dynamic matrix $\mathbf{D}$ was defined in Eq. 5.23. Furthermore, let us introduce the *vector of modal coordinates* $\boldsymbol{\xi}$ as

$$\mathbf{x} = \mathbf{F}\boldsymbol{\xi} \quad \text{or} \quad \boldsymbol{\xi} = \mathbf{F}^{-1}\mathbf{x} \tag{5.83b}$$

whence,

$$\boldsymbol{\xi}(0) = \mathbf{F}^{-1}\mathbf{x}_0, \quad \dot{\boldsymbol{\xi}}(0) = \mathbf{F}^{-1}\mathbf{v}_0, \quad \ddot{\mathbf{x}} = \mathbf{F}^{-1}\ddot{\boldsymbol{\xi}} \tag{5.83c}$$

Upon substitution of Eqs. 5.83b, c into Eq. 5.83a, we obtain

$$\ddot{\boldsymbol{\xi}} + \Omega_d^2\boldsymbol{\xi} = \mathbf{0}, \quad \boldsymbol{\xi}(0) = \mathbf{F}^{-1}\mathbf{x}_0 \equiv \boldsymbol{\alpha}, \quad \dot{\boldsymbol{\xi}}(0) = \mathbf{F}^{-1}\mathbf{v}_0 \equiv \boldsymbol{\beta} \tag{5.84}$$

where Ω_d is the diagonal form of the frequency matrix. The above equation can now be written in component form as

$$\ddot{\xi}_1 + \omega_1^2\xi_1 = 0, \quad \xi_1(0) = \alpha_1, \quad \dot{\xi}_1(0) = \beta_1 \tag{5.85a}$$
$$\ddot{\xi}_2 + \omega_2^2\xi_2 = 0, \quad \xi_2(0) = \alpha_2, \quad \dot{\xi}_2(0) = \beta_2 \tag{5.85b}$$

We have thus *decoupled* the two modes of the system, the mathematical model now simplifying to two *independent* single-dof systems, whose zero-input response was studied in Sect. 2.3. Therefore, the time response of the system in modal coordinates is

$$\xi_i(t) = \alpha_i \cos \omega_i t + \frac{\beta_i}{\omega_i}\sin \omega_i t, \quad i = 1, 2 \tag{5.86}$$

Now, in order to obtain the time response of the system in the given generalized coordinates, all we need is go back to the original coordinates by multiplying vector ξ by $\mathbf{F}$, as indicated in Eq. 5.83b.

Example 5.4.1 (The Two-dof Test Pad Revisited). Find the time response of the test pad of Example 5.3.3 using modal coordinates.

Solution: The dynamic matrix of the system can be readily computed as

$$\mathbf{D} = \mathbf{M}^{-1}\mathbf{K} = \omega^2 \begin{bmatrix} 2 & -1 \\ -1 & 2 \end{bmatrix}, \quad \omega \equiv \sqrt{\frac{2k}{m}}$$

which happens to coincide, in this particular case, with the frequency matrix-squared—can the reader explain this coincidence?—whose eigenvalues were computed in Example 5.3.3 and found to be

$$\omega_1 = \omega = \sqrt{\frac{2k}{m}}, \quad \omega_2 = \sqrt{3}\omega = \sqrt{\frac{6k}{m}}$$

Furthermore, the eigenvectors of the frequency matrix, $\mathbf{e}_1$ and $\mathbf{e}_2$, are shown in Fig. 5.11, whence,

$$\mathbf{e}_1 = \frac{\sqrt{2}}{2}\begin{bmatrix} 1 \\ 1 \end{bmatrix}, \quad \mathbf{e}_2 = \frac{\sqrt{2}}{2}\begin{bmatrix} 1 \\ -1 \end{bmatrix}$$

Now we need the modal matrix, and hence, the modal vectors. Upon arranging the two foregoing vectors in the eigenmatrix, we have

$$\mathbf{E} = \frac{\sqrt{2}}{2}\begin{bmatrix} 1 & 1 \\ 1 & -1 \end{bmatrix}$$

Then, $\mathbf{F}$ is calculated as in Eq. 5.39, with $\mathbf{N}$, the positive-definite square root of $\mathbf{M}$, given by

$$\mathbf{N} = \frac{1}{2}\sqrt{\frac{m}{6}}\begin{bmatrix} 1+\sqrt{3} & \sqrt{3}-1 \\ \sqrt{3}-1 & 1+\sqrt{3} \end{bmatrix}$$

as the reader is invited to verify. Hence,

$$\mathbf{N}^{-1} = 2\sqrt{\frac{6}{m}}\frac{\sqrt{3}}{12}\begin{bmatrix} 1+\sqrt{3} & 1-\sqrt{3} \\ 1-\sqrt{3} & 1+\sqrt{3} \end{bmatrix}$$

Therefore,

$$\mathbf{F} = \frac{1}{2}\sqrt{\frac{2}{m}}\begin{bmatrix} 1+\sqrt{3} & 1-\sqrt{3} \\ 1-\sqrt{3} & 1+\sqrt{3} \end{bmatrix}\frac{\sqrt{2}}{2}\begin{bmatrix} 1 & 1 \\ 1 & -1 \end{bmatrix} = \frac{1}{2}\sqrt{\frac{1}{m}}\begin{bmatrix} 2 & 2\sqrt{3} \\ 2 & -2\sqrt{3} \end{bmatrix}$$

That is

$$F = \sqrt{\frac{1}{m}} \begin{bmatrix} 1 & \sqrt{3} \\ 1 & -\sqrt{3} \end{bmatrix}$$

The initial conditions of the modal coordinates are thus

$$\alpha = F^{-1}x_0 = 0$$

$$\beta = F^{-1}v_0 = -\frac{1}{2\sqrt{3}}\sqrt{m} \begin{bmatrix} -\sqrt{3} & -\sqrt{3} \\ -1 & 1 \end{bmatrix} \frac{\alpha\sqrt{2gh}}{1+7\alpha/4} \begin{bmatrix} -5 \\ 1 \end{bmatrix}$$

where vector α should not be confused with the scalar α, representing the ratio of the mass of the ball to the mass of the pad. Thus,

$$\alpha = 0$$

$$\beta = \frac{\sqrt{3m}}{6}\frac{\alpha\sqrt{2gh}}{1+7\alpha/4}\begin{bmatrix} -4\sqrt{3} \\ -6 \end{bmatrix} = -\frac{\sqrt{3m}}{3}\frac{\alpha\sqrt{2gh}}{1+7\alpha/4}\begin{bmatrix} 2\sqrt{3} \\ 3 \end{bmatrix} \qquad (5.87)$$

Therefore,

$$\xi_1 = -\frac{\sqrt{3m}}{3}\frac{\alpha\sqrt{2gh}}{1+7\alpha/4}2\sqrt{3}\sqrt{\frac{m}{2k}}\sin\sqrt{\frac{2k}{m}}t$$

$$\xi_2 = -\frac{\sqrt{3m}}{3}\frac{\alpha\sqrt{2gh}}{1+7\alpha/4}3\sqrt{\frac{m}{6k}}\sin\sqrt{\frac{6k}{m}}t$$

which simplify to

$$\xi_1 = -\frac{2m}{\sqrt{k}}\frac{\alpha\sqrt{2gh}}{1+7\alpha/4}\sin\sqrt{\frac{2k}{m}}t$$

$$\xi_2 = -\frac{m}{\sqrt{k}}\frac{\alpha\sqrt{2gh}}{1+7\alpha/4}\sin\sqrt{\frac{6k}{m}}t$$

Note that, up to scaling factors, the modal coordinates look like functions $x(t)$ and $\theta(t)$, which represent the motions of the center of mass of the pad and of the pad about its center of mass, as derived in Example 5.3.3.

Now, transforming back to the given coordinates, we have

$$x = F\xi = -\frac{\alpha}{1+7\alpha/4}\sqrt{\frac{m}{k}}\begin{bmatrix} 2\sin\sqrt{\frac{2k}{m}}t + \sqrt{3}\sin\sqrt{\frac{6k}{m}}t \\ 2\sin\sqrt{\frac{2k}{m}}t - \sqrt{3}\sin\sqrt{\frac{6k}{m}}t \end{bmatrix}$$

which is identical to the expression obtained with the time-response formula.

Example 5.4.2 (The Belt-Pulley Transmission Revisited). Find the time response of the belt-pulley transmission introduced in Example 5.3.1 using modal coordinates.

Solution: We start by calculating the dynamic matrix of the given system:

$$\mathbf{D} = \begin{bmatrix} 1/J_1 & 0 \\ 0 & 1/J_2 \end{bmatrix} 2k \begin{bmatrix} r_1^2 & -r_1 r_2 \\ -r_1 r_2 & r_2^2 \end{bmatrix} = \begin{bmatrix} 2kr_1^2/J_1 & -2kr_1 r_2/J_1 \\ -2kr_1 r_2/J_2 & 2kr_2^2/J_2 \end{bmatrix}$$

whose eigenvalues are calculated from

$$\det(\omega^2 \mathbf{1} - \mathbf{D}) = \det\left(\begin{bmatrix} \omega^2 - 2kr_1^2/J_1 & 2kr_1 r_2/J_1 \\ 2kr_1 r_2/J_2 & \omega^2 - 2kr_2^2/J_2 \end{bmatrix}\right) = 0$$

whence,

$$\omega^2 \left[\omega^2 - 2k\left(\frac{r_1^2}{J_1} + \frac{r_2^2}{J_2}\right)\right] = 0$$

with roots

$$\omega_1 = 0, \quad \omega_2 = \sqrt{\frac{2k}{J_1 J_2}(r_1^2 J_2 + r_2^2 J_1)}$$

Now, the modal vectors $\mathbf{f}_1$ and $\mathbf{f}_2$ are calculated from

$$(\omega_i^2 \mathbf{1} - \mathbf{D})\mathbf{f}_i = \mathbf{0}, \quad i = 1, 2$$

under condition (5.29), repeated below for quick reference:

$$\mathbf{f}_1^T \mathbf{M} \mathbf{f}_1 = \mathbf{f}_2^T \mathbf{M} \mathbf{f}_2 = 1, \quad \mathbf{f}_1^T \mathbf{M} \mathbf{f}_2 = 0$$

the last condition being necessarily satisfied from the algebraic properties of the modal vectors. Thus, for $i = 1$,

$$(\omega_1^2 \mathbf{1} - \mathbf{D})\mathbf{f}_1 \equiv \mathbf{D} \mathbf{f}_1 = \mathbf{0}$$

or, if we let $\mathbf{f}_1 = [x \quad y]^T$, then

$$\begin{bmatrix} r_1^2/J_1 & -r_1 r_2/J_1 \\ -r_1 r_2/J_1 & r_2^2/J_2 \end{bmatrix} \begin{bmatrix} x \\ y \end{bmatrix} = \begin{bmatrix} 0 \\ 0 \end{bmatrix}$$

whose first scalar equation leads to

$$\frac{r_1}{J_1}(r_1 x - r_2 y) = 0$$

the second scalar equation of that vector equation not adding further information on $\mathbf{f}_1$ because it is linearly dependent with the first one. We thus have

$$r_1 x - r_2 y = 0 \quad \text{or} \quad y = \frac{r_1}{r_2} x$$

Furthermore, $\mathbf{f}_1$ must have a unit magnitude with respect to the mass matrix, and hence,

$$[x \ y] \begin{bmatrix} J_1 & 0 \\ 0 & J_2 \end{bmatrix} \begin{bmatrix} x \\ y \end{bmatrix} = 1$$

which leads to

$$J_1 x^2 + J_2 y^2 = 1$$

Solving for x^2 from the two above equations,

$$x^2 = \frac{r_2^2}{r_1^2 J_2 + r_2^2 J_1}$$

and hence, if we choose the positive square root of the above expression,

$$x = r_2 \frac{\sqrt{r_1^2 J_2 + r_2^2 J_1}}{r_1^2 J_2 + r_2^2 J_1}, \quad y = r_1 \frac{\sqrt{r_1^2 J_2 + r_2^2 J_1}}{r_1^2 J_2 + r_2^2 J_1}$$

whence,

$$\mathbf{f}_1 = \frac{1}{\sqrt{r_1^2 J_2 + r_2^2 J_1}} \begin{bmatrix} r_2 \\ r_1 \end{bmatrix}$$

Now, for $i = 2$, we let $\mathbf{f}_2 = [u \ v]^T$ and find $\mathbf{f}_2$ from

$$(\omega_2^2 \mathbf{1} - \mathbf{D})\mathbf{f}_2 = \mathbf{0}$$

or

$$2k \begin{bmatrix} (r_1^2 J_2 + r_2^2 J_1)/J_1 J_2 - r_1^2/J_1 & r_1 r_2/J_1 \\ r_1 r_2/J_1 & (r_1^2 J_2 + r_2^2 J_1)/J_1 J_2 - r_2^2/J_2 \end{bmatrix} \begin{bmatrix} u \\ v \end{bmatrix} = \begin{bmatrix} 0 \\ 0 \end{bmatrix}$$

whose first scalar equation leads to

$$2k \left(\frac{r_2^2}{J_2} u + \frac{r_1 r_2}{J_1} v \right) = 0$$

its second scalar equation adding no further information on $\mathbf{f}_2$ because that equation is linearly dependent with the first one. Moreover,

$$[u \; v]^T \begin{bmatrix} J_1 & 0 \\ 0 & J_2 \end{bmatrix} \begin{bmatrix} u \\ v \end{bmatrix} = 1$$

or

$$\frac{r_1^2 J_2 + r_2^2 J_1}{r_1^2 J_2} u^2 = \frac{1}{J_1}$$

If we take the positive root in the above equation, then

$$u = r_1 \sqrt{\frac{J_2/J_1}{r_1^2 J_2 + r_2^2 J_1}}, \quad v = -r_2 \sqrt{\frac{J_1/J_2}{r_1^2 J_2 + r_2^2 J_1}}$$

whence,

$$\mathbf{f}_2 = \frac{1}{r_1^2 J_2 + r_2^2 J_1} \begin{bmatrix} r_1 \sqrt{J_2/J_1} \\ -r_2 \sqrt{J_1/J_2} \end{bmatrix}$$

Therefore, the modal matrix takes the form

$$\mathbf{F} = \frac{1}{r_1^2 J_2 + r_2^2 J_1} \begin{bmatrix} r_2 & r_1 \sqrt{J_2/J_1} \\ r_1 & -r_2 \sqrt{J_1/J_2} \end{bmatrix}$$

its inverse being

$$\mathbf{F}^{-1} = \begin{bmatrix} r_2 J_1 & r_1 J_2 \\ r_1 \sqrt{J_1 J_2} & -r_2 \sqrt{J_1 J_2} \end{bmatrix}$$

and, if we define the *equivalent moment of inertia* J_{eq} as

$$J_{eq} \equiv \sqrt{J_1 J_2}$$

then,

$$\mathbf{F}^{-1} = J_{eq} \begin{bmatrix} r_2 \sqrt{J_2/J_1} & r_1 \sqrt{J_1/J_2} \\ r_1 & -r_2 \end{bmatrix}$$

from which we obtain the initial values of the modal coordinates and their time derivatives as

$$\boldsymbol{\alpha} = \begin{bmatrix} \xi_1(0) \\ \xi_2(0) \end{bmatrix} J_{eq} \begin{bmatrix} r_2 \sqrt{J_1/J_2} \theta_1(0) + r_1 \sqrt{J_2/J_1} \theta_2(0) \\ r_1 \theta_1(0) - r_2 \theta_2(0) \end{bmatrix}$$

$$\boldsymbol{\beta} = \begin{bmatrix} \dot{\xi}_1(0) \\ \dot{\xi}_2(0) \end{bmatrix} J_{eq} \begin{bmatrix} r_2 \sqrt{J_1/J_2} \dot{\theta}_1(0) + r_1 \sqrt{J_2/J_1} \dot{\theta}_2(0) \\ r_1 \dot{\theta}_1(0) - r_2 \dot{\theta}_2(0) \end{bmatrix}$$

The time response of the system in modal coordinates is

$$\xi_1(t) = \alpha_1 + \beta_1 t$$

$$\xi_2(t) = \alpha_2 \cos \omega_2 t + \frac{\beta_2}{\omega_2} \sin \omega_2 t$$

It is apparent that the first modal coordinate represents the rigid mode of the system, while the second one the flexible mode. Now, going back to the original generalized coordinates,

$$\begin{bmatrix} \theta_1(t) \\ \theta_2(t) \end{bmatrix} = \frac{\sqrt{J_{eq}/J_1 J_2}}{r_1^2 J_2 + r_2^2 J_1} \begin{bmatrix} r_2(\alpha_1 + \beta_1 t) + r_1\sqrt{J_2/J_1}[\alpha_2 \cos \omega_2 t + (\beta_2/\omega_2)\sin \omega_2 t] \\ r_1(\alpha_1 + \beta_1 t) - r_2\sqrt{J_1/J_2}[\alpha_2 \cos \omega_2 t + (\beta_2/\omega_2)\sin \omega_2 t] \end{bmatrix}$$

Upon substitution of the values of α_i and β_i, for $i = 1,2$ in terms of the initial conditions of the original coordinates in the above expression, we obtain

$$\theta_1(t) = r_2(A + Bt) + r_1 J_2(C\cos \omega_n t + D\sin \omega_n t)$$

$$\theta_2(t) = r_1(A + Bt) + r_2 J_1(C\cos \omega_n t + D\sin \omega_n t) \tag{5.88}$$

where coefficients $A, B, C,$ and D are identical to those introduced in Example 5.3.1, thereby confirming the validity of the results in the two examples.

5.5   The Zero-State Response of Two-dof Systems

In this section we will be concerned with the time response of a two-dof system under a nonzero forcing term and zero initial conditions, i.e.,

$$\mathbf{M}\ddot{\mathbf{x}} + \mathbf{K}\mathbf{x} = \boldsymbol{\phi}(t), \quad \mathbf{x}(0) = \mathbf{0}, \quad \dot{\mathbf{x}}(0) = \mathbf{0}, \quad t \geq 0 \tag{5.89}$$

or, in normal form,

$$\ddot{\mathbf{y}} + \boldsymbol{\Omega}^2 \mathbf{y} = \mathbf{g}(t), \quad \mathbf{y}(0) = \mathbf{0}, \quad \dot{\mathbf{y}}(0) = \mathbf{0} \tag{5.90}$$

where $\mathbf{y}$ is defined as in Eq. 5.6, while $\mathbf{g}(t)$ is given as

$$\mathbf{g}(t) \equiv \mathbf{N}^{-1}\boldsymbol{\phi}(t) \tag{5.91}$$

As in Sect. 5.3, we distinguish between systems with a positive-semidefinite frequency matrix and those with a positive-definite frequency matrix.

5.5.1   Semidefinite Systems

In this case, the frequency matrix is positive-semidefinite because the stiffness matrix is so, the derivation of the zero-state response being simplified as in the case of the zero-input response. To illustrate this derivation, we go back to our previous example of Fig. 5.7. We shall assume that the abscissa of the center of mass of the system, x, is measured in a coordinate system located such that $x_0 = 0$, with $v_0 = 0$. As well, a force $f_k(t)$ is assumed to act upon mass m_k, for $k = 1, 2$, the governing equations thus becoming

$$m_1 \ddot{x}_1 + k(x_1 - x_2) = f_1(t) \tag{5.92a}$$

$$m_2 \ddot{x}_2 + k(x_2 - x_1) = f_2(t) \tag{5.92b}$$

with initial conditions all zero, i.e.,

$$x_1(0) = x_2(0) = 0, \quad \dot{x}_1(0) = \dot{x}_2(0) = 0 \tag{5.92c}$$

Upon summation of the two equations (5.92a & b), we obtain

$$m_1 \ddot{x}_1 + m_2 \ddot{x}_2 = f_1(t) + f_2(t) \tag{5.93}$$

whose left-hand side can be readily recognized to be the product of the total mass $m_1 + m_2$ by the acceleration of the center of mass, $\ddot{x}$, i.e.,

$$(m_1 + m_2)\ddot{x} = f_1(t) + f_2(t) \tag{5.94}$$

and hence, upon simple quadrature of the above equation, we have

$$x(t) = \frac{1}{m_1 + m_2} \int_{\theta=0}^{\theta=t} \left[\int_{\tau=0}^{\tau=\theta} [f_1(\tau) + f_2(\tau)] d\tau \right] d\theta \tag{5.95}$$

which is related to the generalized coordinates by

$$m_1 x_1 + m_2 x_2 = (m_1 + m_2)x \tag{5.96}$$

Upon solving for x_2 from Eq. 5.96, we obtain

$$x_2(t) = \frac{m_1 + m_2}{m_2} x(t) - \frac{m_1}{m_2} x_1(t) \tag{5.97}$$

We then substitute Eq. 5.97 into Eq. 5.92a, thus obtaining

$$\ddot{x}_1 + \frac{k}{m_{eq}} x_1 = \frac{k}{m_{eq}} x(t) + \frac{1}{m_1} f_1(t), \quad x_1(0) = 0, \quad \dot{x}_1(0) = 0$$

with the equivalent mass m_{eq} defined as

$$m_{eq} \equiv \frac{m_1 m_2}{m_1 + m_2}$$

The above governing equation can thus be recast in a more familiar form:

$$\ddot{x}_1 + \omega_n^2 x_1 = \omega_n^2 x(t) + \frac{1}{m_1} f_1(t), \quad x_1(0) = 0, \quad \dot{x}_1(0) = 0 \qquad (5.98)$$

the natural frequency ω_n of the above system being

$$\omega_n \equiv \sqrt{\frac{k}{m_{eq}}}$$

Now the time response of the system described by Eq. 5.98 is readily derived using the convolution, namely,

$$x_1(t) = \frac{1}{\omega_n} \int_0^t \left[\omega_n^2 x(t - \tau) + \frac{1}{m_1} f_1(t - \tau) \right] \sin \omega_n \tau \, d\tau \qquad (5.99)$$

where $f_1(t)$ is given and $x(t)$ is computed from Eq. 5.95. Moreover, $x_2(t)$ is determined from Eq. 5.96 in terms of $x(t)$ and $x_1(t)$, thereby completing the analysis.

In summary, we have derived a step-by-step procedure to find the zero-state response of semidefinite two-dof systems of the kind discussed here. Now, the feature distinguishing these systems from others is the *conservation of momentum* in the absence of external forces. Note that the belt-pulley system of Example 5.3.1 does not have this feature, and hence, the procedure cannot be applied to it. In the case of that system, however, the procedure would be, while different from the one described here, more straightforward to establish. The derivation of the corresponding procedure is not discussed here, but is left as an exercise to the reader.

The procedure to find the zero-state response of momentum-conserving systems[4] is now described:

1. Determine $x(t)$, the motion of the center of mass, from quadrature
2. Find the time response of the first mass by application of the convolution integral
3. Find the time response of the second mass as a linear combination of the two foregoing responses

Example 5.5.1 (A Momentum-Preserving Semidefinite System). Given the model of a two-rotor turbine, as depicted in Fig. 5.15, which is composed of two identical rigid disks of moments of inertia J about their c.o.m., coupled by an elastic shaft of stiffness k. When the system is at rest, a constant torque τ_0 is suddenly applied on rotor A. Find the ensuing motion of the system.

---

[4] A system can be momentum-preserving only in the absence of an input. In this section, the time response of this kind of systems under zero initial conditions and non-zero input is studied.

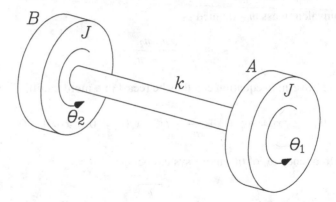

Fig. 5.15 A two-rotor system

Solution: It is a simple matter to show that the mass and stiffness matrices of this system are

$$\mathbf{M} = J \begin{bmatrix} 1 & 0 \\ 0 & 1 \end{bmatrix}, \quad \mathbf{K} = k \begin{bmatrix} 1 & -1 \\ -1 & 1 \end{bmatrix}$$

whence it is apparent that the stiffness matrix is positive-semidefinite. As a matter of fact, the rigid mode of the system is readily identified as one for which $\theta_1 = \theta_2$. The mathematical model of the system in scalar form is then readily derived as

$$J\ddot{\theta}_1 + k(\theta_1 - \theta_2) = \tau_0 u(t)$$
$$J\ddot{\theta}_2 + k(\theta_2 - \theta_1) = 0$$

where $u(t)$ is the unit-step function and the initial conditions are assumed to be all zero. Note that, in the absence of input torque, the sum of the two above equations leads to $J\ddot{\theta}_1 + J\ddot{\theta}_2 = 0$, which in turn leads to $J\dot{\theta}_1 + J\dot{\theta}_2 = $ const, thereby realizing that the system is momentum-preserving. As a consequence, the procedure of Sect. 5.5.1 can be applied to find its time response.

In this case, however, we do not have a 'center of mass' and, hence, neither have we a 'c.o.m. coordinate'. This does not prevent us from defining the variable $\theta(t)$ in exactly the same way as we defined the variable $x(t)$ in Sect. 5.5.1, i.e., as[5]

$$\theta \equiv \frac{\theta_1 + \theta_2}{2}$$

---

[5]In this case, θ is simply the average of θ_1 and θ_2 because the rotors have identical moments of inertia; in general, θ is defined as the *weighted average*.

Hence, upon summation of the two scalar governing equations displayed above, we obtain a second-order ODE for θ, namely,

$$J\ddot{\theta} = \frac{\tau_0}{2}u(t), \quad \theta(0^-) = 0, \quad \dot{\theta}(0^-) = 0$$

and hence, by simple quadrature,

$$\dot{\theta}(t) = \frac{\tau_0}{2J}r(t)$$

where $r(t)$ is the ramp function. One more quadrature leads to

$$\theta(t) = \frac{\tau_0}{4J}t^2 u(t)$$

Now, because of the simple structure of the mass and stiffness matrices of this system, its time response can be determined by introducing some shortcuts in the foregoing general procedure. Indeed, if we subtract the second equation from the first of the governing scalar equations, we obtain

$$2J\ddot{\phi} + 4k\phi = \tau_0 u(t), \quad \phi(0^-) = 0, \quad \dot{\phi}(0^-) = 0$$

with ϕ defined as

$$\phi \equiv \frac{\theta_1 - \theta_2}{2}$$

The above ODE for ϕ can now be rewritten in normal form as

$$\ddot{\phi} + \omega_n^2 \phi = \frac{\tau_0}{2J}u(t), \quad \phi(0^-) = 0, \quad \dot{\phi}(0^-) = 0$$

with the natural frequency ω_n defined as

$$\omega_n \equiv \sqrt{\frac{2k}{J}}$$

Thus, all we need to find the response of the system described by the above equation is the step response of a second-order undamped system. This response was derived in Sect. 2.5.2. Hence, from linearity,

$$\phi(t) = \frac{\tau_0}{2J\omega_n^2}(1 - \cos\omega_n t)u(t)$$

Now, finding the time response of the given system in the given generalized coordinates is a simple matter, for all we do is invert the relations defining θ and ϕ, thereby obtaining

$$\theta_1 = \frac{\tau_0}{2J} \left(\frac{1}{2} t^2 + \frac{1 - \cos \omega_n t}{\omega_n^2} \right) u(t)$$

$$\theta_2 = \frac{\tau_0}{2J} \left(\frac{1}{2} t^2 - \frac{1 - \cos \omega_n t}{\omega_n^2} \right) u(t)$$

Therefore, the motion of the system is the superposition of a uniformly accelerated motion and a harmonic motion, as one should have expected.

5.5.2 Definite Systems

The time response of the system governed by Eq. 5.90, when its frequency matrix is positive-definite, can be derived by mimicking that of undamped scalar systems. This is done by recalling the time response of a single-dof mass-spring system to an excitation $g(t)$, under zero initial conditions, which is reproduced below for quick reference:

$$y(t) = \int_0^t \frac{1}{\omega_n} \sin \omega_n (t - \tau) g(\tau) d\tau \qquad (5.100)$$

For a two-dof system, the zero-state response is derived by simply replacing $y(t)$ and $g(t)$ by their vector counterparts and the natural frequency ω_n by the frequency matrix Ω in the response given in Eq. 5.100, namely,

$$\mathbf{y}(t) = \Omega^{-1} \int_0^t \sin \Omega (t - \tau) \mathbf{g}(\tau) d\tau \qquad (5.101)$$

where Ω^{-1} exists because Ω is positive-definite, and hence, nonsingular. While this form of deriving the time response of the system at hand seems plausible, it is rather informal, and hence, a verification is warranted. In order to test whether a proposed solution to an ODE, or to a system of ODEs, with prescribed initial conditions, is in fact *the solution*, two items must be verified: (1) when the solution is substituted into the ODE, this equation, or this system, must hold, and (2) the solution proposed must verify the initial conditions. From results on the existence and unicity of the solution of linear systems of ODEs,[6] then, the solution proposed is, in fact, *the solution* if it passes these two tests.

Now, in order to verify the first condition above, we will need to recall how to differentiate an integral with variable integration extremes. Specifically, if we have the function of time, $z(t)$, defined as

$$z(t) \equiv \int_0^t f(t, \tau) d\tau \qquad (5.102)$$

---

[6]In the theory of ODE it is known that, given a system of linear ODE with prescribed initial conditions, the system (a) admits a solution and (b) this solution is unique.

then

$$\dot{z}(t) = \int_0^t \left[\frac{\partial}{\partial t} f(t, \tau) \right] d\tau + f(t, \tau) \Big|_{\tau=t} \qquad (5.103)$$

and hence, the time-derivative of $\mathbf{y}(t)$ given in Eq. 5.101 takes the form

$$\dot{\mathbf{y}}(t) = \int_0^t [\cos \Omega(t - \tau)] \mathbf{g}(\tau) d\tau \qquad (5.104)$$

One more time-differentiation leads to

$$\ddot{\mathbf{y}}(t) = -\Omega \int_0^t [\sin \Omega(t - \tau)] \mathbf{g}(\tau) d\tau + [\cos \mathbf{O}] \mathbf{g}(t)$$

where $\mathbf{O}$ stands for the 2×2 zero matrix, and hence,

$$\cos(\mathbf{O}) = \mathbf{1}$$

with $\mathbf{1}$ defined, as before, as the 2×2 identity matrix, thereby ending up with a relation that mimics the scalar case. Therefore, the above expression for $\ddot{\mathbf{y}}(t)$ becomes

$$\ddot{\mathbf{y}}(t) = -\Omega \int_0^t [\sin \Omega(t - \tau)] \mathbf{g}(\tau) d\tau + \mathbf{g}(t) \qquad (5.105)$$

Hence, it is a simple matter to test that the proposed $\mathbf{y}(t)$ verifies indeed the given ODE. To test whether the same expression verifies the initial conditions, we simply evaluate $\mathbf{y}(t)$ and its time-derivative, as given in Eqs. 5.101 and 5.104, at $t = 0$, as we do below:

$$\mathbf{y}(0) = \Omega^{-1} \int_0^0 [\sin \Omega(-\tau)] \mathbf{g}(\tau) d\tau = \mathbf{0}$$

$$\dot{\mathbf{y}}(0) = \int_0^0 [\cos \Omega(-\tau)] \mathbf{g}(\tau) d\tau = \mathbf{0}$$

the initial conditions thus being verified as well. Therefore, the proposed function $\mathbf{y}(t)$ is actually the time response sought.

Now, going back to the generalized coordinate $\mathbf{x}$ and the generalized speed $\dot{\mathbf{x}}$, we have

$$\mathbf{x}(t) = \mathbf{N}^{-1} \Omega^{-1} \int_0^t \sin \Omega(t - \tau) \mathbf{N}^{-1} \mathbf{f}(\tau) d\tau \qquad (5.106a)$$

$$\dot{\mathbf{x}}(t) = \mathbf{N}^{-1} \int_0^t [\cos \Omega(t - \tau)] \mathbf{N}^{-1} \mathbf{f}(\tau) d\tau \qquad (5.106b)$$

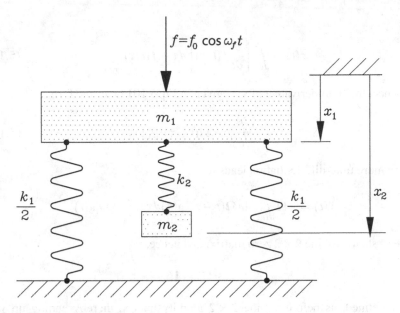

Fig. 5.16 The vibration absorber

Example 5.5.2 (The Vibration Absorber). An interesting application of the zero-state response of two-dof systems is found in the elimination of resonance in single-dof systems. Show that by adding one extra degree of freedom to such a system, the natural frequency of the system can be shifted away from the excitation frequency.

Solution: We are concerned here with a single-dof undamped system of mass m_1 and stiffness k_1 that is acted upon by a harmonic force $f_0 \cos \omega_f t$, where ω_f happens to coincide with the natural frequency of the system, $\sqrt{k_1/m_1}$. In order to suppress the undesired resonance, we add a second mass m_2 to the foregoing system, coupled to m_1 via a spring of stiffness k_2, thereby ending up with the two-dof system of Fig. 5.16.

As we will show presently, the resonant condition of the single-dof system can be suppressed by properly selecting the values of m_2 and k_2. In order to find means of selecting these values, we derive first the mathematical model of the two-dof system of Fig. 5.16, which takes the form

$$\mathbf{M}\ddot{\mathbf{x}} + \mathbf{K}\mathbf{x} = \phi(t)$$

with $\mathbf{M}$ and $\mathbf{K}$ defined as

$$\mathbf{M} = \begin{bmatrix} m_1 & 0 \\ 0 & m_2 \end{bmatrix}, \quad \mathbf{K} = \begin{bmatrix} k_1 + k_2 & -k_2 \\ -k_2 & k_2 \end{bmatrix}$$

whence,

$$N = \begin{bmatrix} \sqrt{m_1} & 0 \\ 0 & \sqrt{m_2} \end{bmatrix}$$

and so

$$\Omega^2 = \frac{1}{m_1 m_2} \begin{bmatrix} m_2(k_1 + k_2) & -\sqrt{m_1 m_2}k_2 \\ -\sqrt{m_1 m_2}k_2 & m_1 k_2 \end{bmatrix}$$

Moreover,

$$\mathbf{x} \equiv \begin{bmatrix} x_1 \\ x_2 \end{bmatrix}, \quad \boldsymbol{\phi}(t) \equiv f_0 \cos \omega_f t \begin{bmatrix} 1 \\ 0 \end{bmatrix}$$

Thus, the system takes the normal form of Eq. 5.90 with $\mathbf{g}(t)$ defined as

$$\mathbf{g}(t) \equiv \mathbf{N}^{-1} \boldsymbol{\phi}(t) = \mathbf{a}(\cos \omega_f t), \quad \mathbf{a} \equiv \frac{f_0 \sqrt{m_1}}{m_1} \begin{bmatrix} 1 \\ 0 \end{bmatrix}$$

Since resonance occurs regardless of the initial conditions, we are interested in the steady-state response of the system at hand. Hence, initial conditions become irrelevant in this analysis. Now, the steady-state response of the foregoing system, as modeled by Eq. 5.90, can be derived by mimicking the corresponding response of its single-dof counterpart, as derived in Sect. 2.7.3, which is reproduced below for quick reference:

$$y(t) = \frac{a}{\omega_n^2 - \omega_f^2} \cos \omega_f t$$

where a is the scalar counterpart of vector $\mathbf{a}$. Now, by properly mimicking the above response, we obtain

$$\mathbf{y}(t) = (\Omega^2 - \omega_f^2 \mathbf{1})^{-1} \mathbf{a}(\cos \omega_f t)$$

Again, this form of producing the time response of the system under study is rather informal, a verification thus being warranted. It is left to the reader the task of verifying that $\mathbf{y}(t)$, as given above, actually verifies the ODE, but the initial conditions need not be verified, for none are given, the reason being that we are interested here in the steady-state response of the system.

The sole item to determine is now the inverse appearing in the above expression for $\mathbf{y}(t)$. Note that this inverse exists as long as the excitation frequency ω_f is not a natural frequency of the two-dof system. Under the assumption that this is not the case, we proceed to calculate that inverse. To this end, we expand first the matrix to be inverted, namely,

$$\Omega^2 - \omega_f^2 \mathbf{1} = \begin{bmatrix} \dfrac{2\bar{k}}{m_1} - \omega_f^2 & -\dfrac{k_2}{\tilde{m}} \\[2ex] -\dfrac{k_2}{\tilde{m}} & \dfrac{k_2}{m_2} - \omega_f^2 \end{bmatrix}$$

with

$$\bar{k} \equiv \frac{1}{2}(k_1 + k_2), \quad \tilde{m} \equiv \sqrt{m_1 m_2}$$

i.e., $\bar{k}$ is the arithmetic mean of k_1 and k_2, $\tilde{m}$ the geometric mean of m_1 and m_2. If we recall the expression derived in Eq. A.18 of Appendix A, the inverse of the foregoing matrix can be readily derived as

$$(\Omega^2 - \omega_f^2 \mathbf{1})^{-1} = \frac{1}{\Delta_\Omega} \begin{bmatrix} \dfrac{k_2}{m_2} - \omega_f^2 & \dfrac{k_2}{\tilde{m}} \\[2ex] \dfrac{k_2}{\tilde{m}} & \dfrac{2\bar{k}}{m_1} - \omega_f^2 \end{bmatrix}$$

with Δ_Ω defined as

$$\Delta_\Omega \equiv \det\left(\Omega^2 - \omega_f^2 \mathbf{1}\right) = \omega_f^4 - \left(\frac{\bar{k}}{m_1} + \frac{k_2}{m_2}\right)\omega_f^2 + \frac{\bar{k}k_2}{\tilde{m}^2}$$

Hence,

$$\mathbf{y}(t) = \frac{\sqrt{m_1} f_0}{m_1 \Delta_\Omega} \begin{bmatrix} \dfrac{k_2}{m_2} - \omega_f^2 \\[2ex] \dfrac{k_2}{\tilde{m}} \end{bmatrix} \cos \omega_f t$$

Therefore,

$$\mathbf{x}(t) = \frac{f_0}{m_1 \Delta_\Omega} \begin{bmatrix} \dfrac{k_2}{m_2} - \omega_f^2 \\[2ex] \dfrac{k_2}{\tilde{m}} \end{bmatrix} \cos \omega_f t$$

It is now apparent that if we choose m_2 and k_2 so that they obey the relation

$$\frac{k_2}{m_2} = \omega_f^2$$

then it is possible to keep the first mass stationary, thereby suppressing the resonance. It is to be noted, however, that this technique of resonance-suppression works only for the excitation frequency ω_f. If this frequency changes, the secondary system will have to be changed correspondingly. Moreover, suppressing the above frequency has been accomplished at the expense of introducing two possible resonant frequencies, namely, the two natural frequencies of the foregoing two-dof system. It is left as an exercise for the reader to determine these two natural frequencies.

An example of time response of a two-dof system to a triangular bump is included in 5-UDampedTriangPulse2dof.mw.

5.6 The Total Response of Two-dof Systems

The total time response of the system under study is simply the sum of the zero-input and the zero-state responses given above. Thus, for our semidefinite example of Sect. 5.3.1, we have

$$x_1(t) = x_0 + v_0 t + (a_1 - x_0)\cos\omega_n t + \frac{b_1 - v_0}{\omega_n}\sin\omega_n t$$

$$+ \frac{1}{\omega_n}\int_0^t \left[\omega_n^2 x(t-\tau) + \frac{1}{m_1}f_1(t-\tau)\right]\sin\omega_n\tau d\tau \tag{5.107}$$

$$x_2(t) = x_0 + v_0 t - \frac{m_1}{m_2}\left[(a_1 - x_0)\cos\omega_n t + \frac{b_1 - v_0}{\omega_n}\sin\omega_n t\right] + \frac{m_1 + m_2}{m_2}x(t)$$

$$- \frac{m_1}{m_2\omega_n}\int_0^t \left[\omega_n^2 x(t-\tau) + \frac{1}{m_1}f_1(t-\tau)\right]\sin\omega_n\tau d\tau \tag{5.108}$$

On the other hand, for definite systems,

$$\mathbf{x}(t) = \mathbf{N}^{-1}(\cos\boldsymbol{\Omega}t)\mathbf{N}\mathbf{x}_0 + \mathbf{N}^{-1}\boldsymbol{\Omega}^{-1}(\sin\boldsymbol{\Omega}t)\mathbf{N}\mathbf{v}_0$$

$$+ \mathbf{N}^{-1}\boldsymbol{\Omega}^{-1}\int_0^t \sin\boldsymbol{\Omega}(t-\tau)\mathbf{N}^{-1}\boldsymbol{\phi}(\tau)d\tau \tag{5.109}$$

For the simulation of the systems under study, we will need an expression for $\dot{\mathbf{x}}$, which can be readily derived by simple superposition of the expressions obtained in Eqs. 5.69b and 5.106b, namely,

$$\dot{\mathbf{x}} = -\mathbf{N}^{-1}\boldsymbol{\Omega}(\sin\boldsymbol{\Omega}t)\mathbf{N}\mathbf{x}_0 + \mathbf{N}^{-1}\cos\boldsymbol{\Omega}t\mathbf{N}\mathbf{v}_0$$

$$+ \mathbf{N}^{-1}\int_0^t \cos\boldsymbol{\Omega}(t-\tau)\mathbf{N}^{-1}\boldsymbol{\phi}(\tau)d\tau \tag{5.110}$$

5.6.1 The Classical Modal Method Applied to the Total Response

An alternative procedure to calculate the total time response of two-dof systems is based on the introduction of modal coordinates, following an approach similar to that introduced for the zero-input response. To illustrate the procedure, we recall the mathematical model of the system at hand under nonzero initial conditions and nonzero input, namely,

$$\mathbf{M\ddot{x}} + \mathbf{Kx} = \phi(t), \quad \mathbf{x}(0) = \mathbf{x}_0, \dot{\mathbf{x}}(0) = \mathbf{v}_0 \qquad (5.111)$$

Upon multiplying both sides of the foregoing equation by $\mathbf{M}^{-1}$, we obtain

$$\ddot{\mathbf{x}} + \mathbf{Dx} = \mathbf{M}^{-1}\phi(t), \quad \mathbf{x}(0) = \mathbf{x}_0, \quad \dot{\mathbf{x}}(0) = \mathbf{v}_0 \qquad (5.112)$$

Now, modal coordinates are introduced:

$$\mathbf{x} = \mathbf{F}\boldsymbol{\xi}, \quad \dot{\mathbf{x}} = \mathbf{F}\dot{\boldsymbol{\xi}}, \quad \ddot{\mathbf{x}} = \mathbf{F}\ddot{\boldsymbol{\xi}}$$

Again,

$$\boldsymbol{\xi}(0) = \boldsymbol{\alpha}, \quad \dot{\boldsymbol{\xi}}(0) = \boldsymbol{\beta}$$

and so,

$$\boldsymbol{\alpha} = \mathbf{F}^{-1}\mathbf{x}_0, \quad \boldsymbol{\beta} = \mathbf{F}^{-1}\mathbf{v}_0$$

The mathematical model of Eq. 5.111 thus becomes

$$\mathbf{F}\ddot{\boldsymbol{\xi}} + \mathbf{DF}\boldsymbol{\xi} = \mathbf{M}^{-1}\phi(t), \quad \boldsymbol{\xi}(0) = \boldsymbol{\alpha}, \quad \dot{\boldsymbol{\xi}}(0) = \boldsymbol{\beta}$$

Next we multiply both sides of the foregoing equation by $\mathbf{F}^{-1}$ to obtain

$$\ddot{\boldsymbol{\xi}} + \boldsymbol{\Delta}_D^2\boldsymbol{\xi} = \phi(t), \quad \boldsymbol{\xi}(0) = \boldsymbol{\alpha}, \quad \dot{\boldsymbol{\xi}}(0) = \boldsymbol{\beta} \qquad (5.113a)$$

where

$$\boldsymbol{\Delta}_D^2 = \mathbf{F}^{-1}\mathbf{DF}, \quad \mathbf{f}(t) = \mathbf{F}^{-1}\mathbf{M}^{-1}\phi(t), \qquad (5.113b)$$

Equation 5.113a can be written in component form as

$$\ddot{\xi}_1 + \omega_1^2\xi = f_1(t), \quad \xi_1(0) = \alpha_1, \quad \dot{\xi}_1(0) = \beta_1$$
$$\ddot{\xi}_2 + \omega_2^2\xi = f_2(t), \quad \xi_2(0) = \alpha_2, \quad \dot{\xi}_2(0) = \beta_2$$

thereby obtaining two single-dof decoupled systems, whose response was obtained in Sect. 2.6 in the form

$$\xi_i(t) = \alpha_i \cos \omega_i t + \frac{\beta_i}{\omega_i} \sin \omega_i t + \frac{1}{\omega_i} \int_0^t f_i(\tau) \sin \omega_i(t - \tau)d\tau, \quad i = 1, 2 \quad (5.114)$$

The total time response in the original given coordinates is obtained by a simple change of variable.

Example 5.6.1 (The Momentum-Preserving System Revisited). Obtain the time response of the system introduced in Example 5.5.1, under the same initial and excitation conditions, using modal coordinates.

Solution: We start by calculating the dynamic matrix:

$$\mathbf{D} = \mathbf{M}^{-1}\mathbf{K} = \omega_0^2 \begin{bmatrix} 1 & -1 \\ -1 & 1 \end{bmatrix}, \quad \omega_0^2 \equiv \frac{k}{m}$$

whose characteristic equation is readily derived as

$$\det \begin{bmatrix} \omega^2 - \omega_0^2 & \omega_0^2 \\ \omega_0^2 & \omega^2 - \omega_0^2 \end{bmatrix} = 0$$

or

$$\omega^2(\omega^2 - 2\omega_0^2) = 0$$

The natural frequencies are thus found as

$$\omega_1 = 0, \quad \omega_2 = \sqrt{2}\omega_0$$

thereby confirming the result obtained in Example 5.5.1. The modal vectors are then found to be

$$\mathbf{f}_1 = \frac{\sqrt{2J}}{2J}\begin{bmatrix} 1 \\ 1 \end{bmatrix}, \quad \mathbf{f}_2 = \frac{\sqrt{2J}}{2J}\begin{bmatrix} 1 \\ -1 \end{bmatrix}$$

Hence, the modal matrix is

$$\mathbf{F} = \frac{\sqrt{2J}}{2J}\begin{bmatrix} 1 & 1 \\ 1 & -1 \end{bmatrix}$$

whence,

$$\mathbf{F}^{-1} = -\frac{\sqrt{2J}}{2}\begin{bmatrix} -1 & -1 \\ -1 & 1 \end{bmatrix} = \frac{\sqrt{2J}}{2}\begin{bmatrix} 1 & 1 \\ 1 & -1 \end{bmatrix}$$

Also,

$$\mathbf{f}(t) = \mathbf{F}^{-1}\mathbf{M}^{-1}\boldsymbol{\phi}(t) = \frac{\sqrt{2J}}{2J}\begin{bmatrix} 1 \\ 1 \end{bmatrix}\tau_0 u(t)$$

the mathematical model in modal coordinates thus becoming

$$\ddot{\xi}_1 = \frac{\sqrt{2J}}{2J}\tau_0 u(t), \quad \xi_1(0) = \alpha_1, \quad \dot{\xi}_1(0) = \beta_1$$

$$\ddot{\xi}_2 + 2\omega_0^2\xi_2 = \frac{\sqrt{2J}}{2J}\tau_0 u(t), \quad \xi_2(0) = \alpha_2, \quad \dot{\xi}_2(0) = \beta_2$$

The time responses of these uncoupled systems are readily found to be

$$\xi_1(t) = \alpha_1 + \beta_1 t + \frac{\sqrt{2J}}{4J} t^2 \tau_0 u(t)$$

$$\xi_2(t) = \alpha_2 \cos \sqrt{2}\omega_0 t + \frac{\beta_2}{\sqrt{2}\omega_0} \sin \sqrt{2}\omega_0 t + \frac{\sqrt{2J}}{2J} \frac{\tau_0}{\sqrt{2}\omega_0} \int_0^t u(\theta) \sin \sqrt{2}\omega_0 (t - \theta) d\theta$$

The convolution integral in the above equation can be evaluated in closed form upon realizing that $u(t) = 1$ in the interval of integration, thereby obtaining

$$\xi_2(t) = \alpha_2 \cos \sqrt{2}\omega_0 t + \frac{\beta_2}{\sqrt{2}\omega_0} \sin \sqrt{2}\omega_0 t + \frac{\sqrt{2J}}{4J} \frac{\tau_0}{\omega_0^2}(1 - \cos \sqrt{2}\omega_0 t)u(t)$$

Since the initial conditions in the given example are all zero, the above time responses reduce to

$$\xi_1(t) = \frac{\sqrt{2J}}{4J} \tau_0 t^2 u(t)$$

$$\xi_2(t) = \frac{\sqrt{2J}}{4J} \frac{\tau_0}{\omega_0^2}(1 - \cos \sqrt{2}\omega_0 t)u(t) \tag{5.115}$$

Upon changing back to the original generalized coordinates, the foregoing responses will reveal that they yield exactly the same result as that of Example 5.5.1.

5.7 Damped Two-dof Systems

Here we will go back to the system of Fig. 5.1, if with added dashpots of coefficients c_1, c_2 and c_3, as displayed in Fig. 5.17.

The governing equations of the system under study take the form

$$\mathbf{M}\ddot{\mathbf{x}} + \mathbf{C}\dot{\mathbf{x}} + \mathbf{K}\mathbf{x} = \boldsymbol{\phi}(t), \quad \mathbf{x}(0) = \mathbf{x}_0, \quad \dot{\mathbf{x}}(0) = \mathbf{v}_0 \tag{5.116}$$

with $\mathbf{M}$ and $\mathbf{K}$ already defined in Eq. 5.2, and reproduced below for quick reference, where $\mathbf{C}$ is included. Here, the latter can be readily derived as the Hessian matrix of the dissipation function of the system, whence,

$$\mathbf{M} = \begin{bmatrix} m_1 & 0 \\ 0 & m_2 \end{bmatrix}, \quad \mathbf{C} = \begin{bmatrix} c_1 + c_2 & -c_2 \\ -c_2 & c_2 + c_3 \end{bmatrix}, \quad \mathbf{K} = \begin{bmatrix} k_1 + k_2 & -k_2 \\ -k_2 & k_2 + k_3 \end{bmatrix} \tag{5.117a}$$

Moreover,

$$\mathbf{x} \equiv \begin{bmatrix} x_1 \\ x_2 \end{bmatrix}, \quad \boldsymbol{\phi}(t) \equiv \begin{bmatrix} \phi_1(t) \\ \phi_2(t) \end{bmatrix} \tag{5.117b}$$

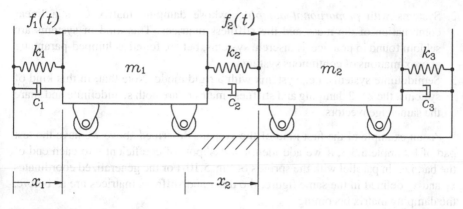

Fig. 5.17 A damped two-dof system

We thus have that $\mathbf{M}$ is, in general, positive-definite, while $\mathbf{C}$ and $\mathbf{K}$ are at least positive-semidefinite. Moreover, in order to derive the time response of this system, we begin by expressing the foregoing model in *normal form*. This is done by factoring $\mathbf{M}$ as in Sect. 5.2 and introducing the same change of variable as before, which leads to the model

$$\ddot{\mathbf{y}} + \mathbf{\Delta}\dot{\mathbf{y}} + \mathbf{\Omega}^2\mathbf{y} = \mathbf{g}(t), \quad \mathbf{y}(0) = \mathbf{y}_0, \quad \dot{\mathbf{y}}(0) = \mathbf{s}_0 \tag{5.118}$$

with $\mathbf{\Omega}$ and $\mathbf{g}(t)$ defined as in the undamped case, while $\mathbf{\Delta}$ is defined as

$$\mathbf{\Delta} \equiv \mathbf{N}^{-1}\mathbf{C}\mathbf{N}^{-1} \tag{5.119}$$

and should not be mistaken by the dynamic matrix introduced in Eq. 5.23. Note that, by virtue of their definitions, $\mathbf{\Omega}$ is either positive-definite or positive-semidefinite, depending on whether $\mathbf{K}$ is positive-definite or semidefinite. By the same token, $\mathbf{\Delta}$ is positive-definite if $\mathbf{C}$ is; if $\mathbf{C}$ is positive-semidefinite, then $\mathbf{\Delta}$ is correspondingly semidefinite.

The question that arises naturally is whether we can perform a modal analysis of damped systems, similar to that of undamped systems. That is, *what are the natural frequencies and the natural modes of damped systems?* The answer to this question, in general, is far from obvious and constitutes, even nowadays, when tremendous progress has been achieved in the modal analysis of mechanical systems, an open question. To be sure, particular instances of two-dof systems exist, and, for that matter, of n-dof systems as well, for which a modal analysis in the form of that pertaining to undamped systems, is possible. Such instances comprise basically three cases, namely,

1. Systems whose damping and stiffness matrices are both proportional to a third matrix.

2. Systems with *proportional damping*, whose damping matrix **C** is a linear combination of the mass and the stiffness matrices. This kind of systems are seldom found in practice as discrete systems, but are found as lumped-parameter approximations of continuous systems.
3. Semidefinite systems, i.e., systems with a rigid mode. Note that, in this kind of systems, the 2×2 damping and stiffness matrices are both semidefinite and share the same eigenvectors.

As an example of the first kind of the systems described above we have the test pad of Example 5.3.3, if we add identical dashpots of coefficient c to each end of the bar, i.e., in parallel with the springs of Fig. 5.10. For the generalized coordinates x_1 and x_2 defined in the same figure, the mass and stiffness matrices are as before, the damping matrix becoming

$$\mathbf{C} = \begin{bmatrix} c & 0 \\ 0 & c \end{bmatrix} \tag{5.120}$$

It is now apparent that both the damping and the stiffness matrices are proportional to the same matrix, namely, the 2×2 identity matrix. The governing equations, in scalar form, become now

$$\frac{1}{6}m(2\ddot{x}_1 + \ddot{x}_2) + c\dot{x}_1 + kx_1 = 0 \tag{5.121a}$$

$$\frac{1}{6}m(\ddot{x}_1 + 2\ddot{x}_2) + c\dot{x}_2 + kx_2 = 0 \tag{5.121b}$$

If we now add the two above equations sidewise and subtract the first equation from the second likewise, we obtain, after simplifications,

$$\ddot{x} + 2\frac{c}{m}\dot{x} + 2\frac{k}{m}x = 0 \tag{5.122a}$$

$$\ddot{\theta} + 6\frac{c}{m}\dot{\theta} + 6\frac{k}{m}\theta = 0 \tag{5.122b}$$

where $x(t)$ denotes the coordinate of the c.o.m., while $\theta(t)$ the rotation of the bar, as indicated in Fig. 5.10. It is now apparent that, as in the undamped case, the motion of the c.o.m. is decoupled from the rotation about the c.o.m. We can thus distinguish two sets of system parameters, namely, the natural frequency and the damping ratio of the translation of the c.o.m., ω_C and ζ_C, respectively, and the same items of the rotation about the c.o.m., namely, ω_θ and ζ_θ, defined below:

$$\omega_C \equiv \sqrt{2\frac{k}{m}}, \quad \zeta_C \equiv \frac{\sqrt{2km}}{km}c, \quad \omega_\theta \equiv \sqrt{\frac{6k}{m}}, \quad \zeta_\theta \equiv \frac{\sqrt{6km}}{2km}c,$$

Hence, it is apparent that $\omega_\theta > \omega_C$ and $\zeta_\theta > \zeta_C$. It is thus possible that the rotational mode be overdamped, while the translational mode be underdamped. The zero-input response of each of these modes can be readily found from the results of Sect. 2.3.2. Under the assumption that the two modes are underdamped, we obtain

$$x(t) = \frac{e^{-\zeta_C \omega_C t}}{\sqrt{1 - \zeta_C^2}} \left[\left(\sqrt{1 - \zeta_C^2} \cos \overline{\omega}_C t + \zeta_C \sin \overline{\omega}_C t \right) x_0 + \frac{1}{\overline{\omega}_C} (\sin \overline{\omega}_C t) v_0 \right]$$

$$\theta(t) = \frac{e^{-\zeta_\theta \omega_\theta t}}{\sqrt{1 - \zeta_\theta^2}} \left[\left(\sqrt{1 - \zeta_\theta^2} \cos \overline{\omega}_\theta t + \zeta_\theta \sin \overline{\omega}_\theta t \right) \theta_0 + \frac{1}{\overline{\omega}_\theta} (\sin \overline{\omega}_\theta t) \omega_0 \right]$$

where $\overline{\omega}_C$ and $\overline{\omega}_\theta$ are the damped frequencies of the translational and rotational modes, respectively. Moreover, x_0, v_0, θ_0 and ω_0 are the initial values of $x(t)$, $\dot{x}(t)$, $\theta(t)$ and $\dot{\theta}(t)$, respectively. Now, the response of the system in the generalized coordinates is obtained from the relations

$$x_1(t) = x(t) - \frac{l}{2}\theta, \quad x_2(t) = x(t) + \frac{l}{2}\theta \tag{5.123}$$

and hence,

$$x_1(t) = \frac{e^{-\zeta_C \omega_C t}}{\sqrt{1 - \zeta_C^2}} \left[\left(\sqrt{1 - \zeta_C^2} \cos \overline{\omega}_C t + \zeta_C \sin \overline{\omega}_C t \right) x_0 + \frac{1}{\overline{\omega}_C} (\sin \overline{\omega}_C t) v_0 \right]$$
$$- \frac{l}{2} \frac{e^{-\zeta_\theta \omega_\theta t}}{\sqrt{1 - \zeta_\theta^2}} \left[\left(\sqrt{1 - \zeta_\theta^2} \cos \overline{\omega}_\theta t + \zeta_\theta \sin \overline{\omega}_\theta t \right) \theta_0 + \frac{1}{\overline{\omega}_\theta} (\sin \overline{\omega}_\theta t) \omega_0 \right]$$

$$x_2(t) = \frac{e^{-\zeta_C \omega_C t}}{\sqrt{1 - \zeta_C^2}} \left[\left(\sqrt{1 - \zeta_C^2} \cos \overline{\omega}_C t + \zeta_C \sin \overline{\omega}_C t \right) x_0 + \frac{1}{\overline{\omega}_C} (\sin \overline{\omega}_C t) v_0 \right]$$
$$+ \frac{l}{2} \frac{e^{-\zeta_\theta \omega_\theta t}}{\sqrt{1 - \zeta_\theta^2}} \left[\left(\sqrt{1 - \zeta_\theta^2} \cos \overline{\omega}_\theta t + \zeta_\theta \sin \overline{\omega}_\theta t \right) \theta_0 + \frac{1}{\overline{\omega}_\theta} (\sin \overline{\omega}_\theta t) \omega_0 \right]$$

whence it is apparent that the time response in the given generalized coordinates is a linear combination of exponentially decaying harmonics of the type of those occurring in the zero-input response of single-dof systems.

In order to exemplify the second kind of the above-mentioned damped systems, let us consider the system shown in Fig. 4.10, which is reproduced here as Fig. 5.18 for quick reference.

We derived in Sect. 4.6 the governing equations of this system in the form

$$m_1 \ddot{x}_1 + c(\dot{x}_1 - \dot{x}_2) + k(x_1 - x_2) = 0$$
$$m_2 \ddot{x}_2 - c(\dot{x}_1 - \dot{x}_2) - k(x_1 - x_2) = 0$$

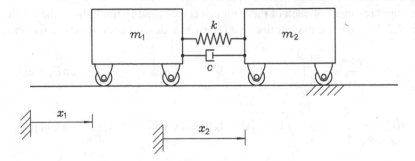

Fig. 5.18 A two-dof semidefinite system

Moreover, in the same section, we decoupled the two modes of the system, thus obtaining

$$\ddot{x}_1 + 2\zeta_{rmeq}\omega_{eq}\dot{x}_1 + \omega_{eq}^2 x_1 = A + Bt$$

Furthermore, we assume that the initial conditions are given in the form

$$\mathbf{x}(0) \equiv \mathbf{x}_0 = \begin{bmatrix} a_1 \\ a_2 \end{bmatrix}, \quad \dot{\mathbf{x}}(0) \equiv \dot{\mathbf{x}}_0 = \begin{bmatrix} b_1 \\ b_2 \end{bmatrix}$$

and hence, constants C_1 and C_2 of Sect. 4.6 become

$$C_1 = m_1 b_1 + m_2 b_2, \quad C_2 = m_1 a_1 + m_2 a_2$$

whence A and B turn out to be

$$A = \frac{c(m_1 b_1 + m_2 b_2) + k(m_1 a_1 + m_2 a_2)}{m_2}, \quad B = \frac{k(m_1 b_1 + m_2 b_2)}{m_2}$$

Now, the response $x_1(t)$ of the above system consists of three parts: (1) the zero-input response; (2) the response to a constant excitation; and (3) the response to a linear excitation, the last two under zero initial conditions. The first part was derived in Sect. 2.3.2, the last two in Sect. 2.7.4, and need not be repeated here.

Finally, the third case of damped systems can be exemplified also with the system of Fig. 5.18 for, in this case, $\mathbf{K}$ and $\mathbf{C}$ are proportional to the same matrix $\mathbf{A}$ in the form

$$\mathbf{K} = k\mathbf{A}, \quad \mathbf{C} = c\mathbf{A}$$

with $\mathbf{A}$ given by

$$\mathbf{A} = \begin{bmatrix} 1 & -1 \\ -1 & 1 \end{bmatrix}$$

We now turn to the *general* case of damped two-dof systems. As a consequence of the foregoing discussion, the zero-input response of damped second-order systems, in general, does not take the form of a single exponentially decaying harmonic. That is, similar to the zero-input response of undamped two-dof systems, the zero-input response of damped two-dof systems is not, in general, an exponentially decaying harmonic. The natural question to ask is, then, whether, under special circumstances, the said response is shaped by a simple exponentially decaying harmonic. By the latter we mean a function $\phi(t)$ of the form

$$\phi(t) = Ae^{\lambda t} \tag{5.124}$$

where λ is, in general, a complex number—in order to account for undamped, underdamped, critically damped and overdamped systems—and A is a constant. In order to answer this question, then, we would have to look for zero-input responses of the form

$$\mathbf{x}(t) = e^{\lambda t}\mathbf{f} \tag{5.125a}$$

where $\mathbf{f}$ would play the role of a modal vector. Note that, if the zero-input response of a damped two-dof system takes the form of Eq. 5.125a, then

$$\dot{\mathbf{x}}(t) = \lambda e^{\lambda t}\mathbf{f} \tag{5.125b}$$

$$\ddot{\mathbf{x}}(t) = \lambda^2 e^{\lambda t}\mathbf{f} \tag{5.125c}$$

Upon substitution of the foregoing expressions in the governing equation, Eq. 5.118, with $\mathbf{g}(t) = \mathbf{0}$, we obtain

$$(\lambda^2 \mathbf{1} + \lambda\mathbf{D} + \mathbf{\Omega}^2)\mathbf{f} = \mathbf{0} \tag{5.126}$$

Therefore, simple exponentially decaying motions are possible with complex exponent λ, provided that λ verifies Eq. 5.126, which represents a *generalized eigenvalue problem*. Contrary to the simple eigenvalue problem derived in connection with undamped two-dof systems, solutions to the more general problem of Eq. 5.126 are less known. Note that λ in the above problem plays the role of an eigenvalue, while $\mathbf{f}$ that of an eigenvector. A trivial solution $\mathbf{f}$ of Eq. 5.126, $\mathbf{f} = \mathbf{0}$, is obviously, of no interest to us, and will henceforth be discarded. We are thus interested in nontrivial solutions $\mathbf{f} \neq \mathbf{0}$, which calls for the matrix in parenthesis in Eq. 5.126 to be singular, i.e.,

$$\det(\lambda^2 \mathbf{1} + \lambda\mathbf{D} + \mathbf{\Omega}^2) = 0 \tag{5.127}$$

A close look at Eq. 5.127 reveals that each entry of the matrix in parenthesis is *quadratic* in λ. Moreover, since that matrix is of 2×2, its determinant is *quartic* in λ. As a consequence, the characteristic equation of the problem under study is a fourth-order polynomial, of the form

$$P(\lambda) \equiv a_0 + a_1\lambda + a_2\lambda^2 + a_3\lambda^3 + \lambda^4 = 0 \tag{5.128}$$

which thus admits up to four complex solutions. Moreover, since the entries of the $\mathbf{D}$ and $\mathbf{\Omega}^2$ matrices are real, the $\{a_k\}_0^3$ coefficients of the characteristic polynomial $P(\lambda)$ are all real, and hence, its complex roots are bound to appear in complex-conjugate pairs. Computing the four roots of the foregoing equation is routine work with currently available software. Indeed, to compute the roots of a quadratic polynomial one can resort to either numerical or symbolic-computations software. Of the former, one can name IMSL or Matlab, while Macsyma, Maple and Mathematica are meant for symbolic computations.

Once the four complex eigenvalues $\{\lambda_k\}_1^4$ of the problem at hand are available, the corresponding normal modes, $\{\mathbf{f}_k\}_1^4$ are calculated from Eq. 5.126. Note that, since the eigenvalues are, in general, complex, the modal vectors are expected to be complex as well. Hence, the computations must be performed with complex arithmetic. However, the solution of the eigenvalue problem can be readily formulated in real arithmetic, which, additionally, allows for a graphical solution of the foregoing eigenvalue problem. Indeed, all it takes to transform the problem into the real field is represent the eigenvalue λ as

$$\lambda \equiv x + jy \tag{5.129}$$

where, x and y are two real numbers, representing the real and the complex parts of λ. Under these conditions, then, Eq. 5.126 leads to two real equations, one for its real part and one for its imaginary part, namely,

$$f_1(x,y) = 0, \quad f_2(x,y) = 0 \tag{5.130}$$

Now, the solution of the foregoing system of two equations in two unknowns can be readily computed using either a purely numerical approach, e.g., via a Newton-type method [1], or a semigraphical approach using the plotting capabilities of commercial software. We outline the latter with a numerical example in the subsection below.

Example 5.7.1 (Modal Analysis of a Damped Test Pad). Shown in Fig. 5.19 is an *asymmetric* test pad, similar to the one of Example 5.3.3. The difference with the latter is that the pad of this example has a spring-dashpot suspension on each of the ends of the the rod and, moreover, the c.o.m. of the pad is offset with respect to the centerline. The model at hand consists of a body with mass m and centroidal moment of inertia J, supported by a damped suspension. Moreover, the c.o.m. C of the pad is located a distance d from the centerline. Find the eigenvalues of the system and decide whether each mode is underdamped, critically damped or overdamped.

Solution: The dynamics model takes the form of Eq. 5.116, with

$$\mathbf{x} \equiv \begin{bmatrix} x \\ \theta \end{bmatrix}, \quad \dot{\mathbf{x}} \equiv \begin{bmatrix} \dot{x} \\ \dot{\theta} \end{bmatrix}, \quad \ddot{\mathbf{x}} \equiv \begin{bmatrix} \ddot{x} \\ \ddot{\theta} \end{bmatrix}, \quad \phi \equiv \begin{bmatrix} 0 \\ 0 \end{bmatrix}$$

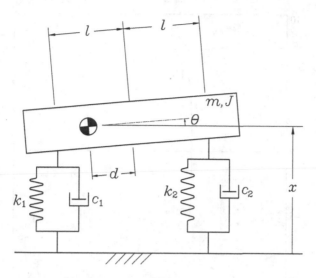

Fig. 5.19 Asymmetric test pad with damped suspension

and the coefficient matrices are

$$\mathbf{M} \equiv \begin{bmatrix} m & 0 \\ 0 & J \end{bmatrix} \tag{5.131}$$

$$\mathbf{C} \equiv \begin{bmatrix} c_1 + c_2 & c_2(l+d) - c_1(l-d) \\ c_2(l+d) - c_1(l-d) & c_1(l-d)^2 + c_2(l+d)^2 \end{bmatrix} \tag{5.132}$$

$$\mathbf{K} \equiv \begin{bmatrix} k_1 + k_2 & k_2(l+d) - k_1(l-d) \\ k_2(l+d) - k_1(l-d) & k_1(l-d)^2 + k_2(l+d)^2 \end{bmatrix} \tag{5.133}$$

Moreover,

$$\mathbf{N} = \begin{bmatrix} \sqrt{m} & 0 \\ 0 & \sqrt{J} \end{bmatrix}$$

Now, using matrix $\mathbf{N}$ we can obtain the normal equation of the system at hand, and hence, its characteristic equation, from which the two nonlinear equations (5.130) can be derived. We assume the numerical data shown below:

$m = 1459$ kg, $\quad J = 2168$ kg m^2, $\quad l = 1.524$ m, $\quad d = 0.1524$ m,

$k_1 = 35277$ N/m, $\quad k_2 = 38217$ N/m, $\quad c_1 = 3500$ Ns/m, $\quad c_2 = 3800$ Ns/m

Using the foregoing data, the two matrices $\mathbf{\Delta}$ and $\mathbf{\Omega}^2$ are found to be

$$\mathbf{\Delta} = \begin{bmatrix} 5.0034 & 0.8826 \\ 0.8826 & 7.9630 \end{bmatrix}, \quad \mathbf{\Omega}^2 = \begin{bmatrix} 50.3729 & 8.8169 \\ 8.8169 & 80.1513 \end{bmatrix}$$

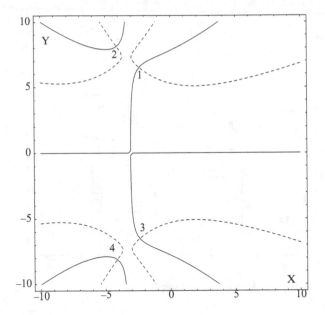

Fig. 5.20 $\mathscr{C}_1$ (*dashed*) and $\mathscr{C}_2$ (*solid*) contours of Eq. 5.134

Table 5.1 Numerical values of each eigenvalue

Intersection point	1	2	3	4
Real part (x)	−2.380	−4.103	−2.380	−4.103
Imag part (y)	6.503	8.107	−6.503	−8.107

The two scalar equations derived from the associated characteristic equation are then

$$f_1(x,y) \equiv 3959.71 + 786.58x + 169.59x^2 + 12.97x^3 + x^4 - 169.59y^2$$
$$- 38.90xy^2 - 6x^2y^2 + y^4 = 0$$

$$f_2(x,y) \equiv y(786.58 + 339.17x + 38.90x^2 + 4x^3 - 12.97y^2 - 4xy^2) = 0 \quad (5.134)$$

Each of the two foregoing equations, then, defines a contour in the x-y plane. Furthermore, we denote by $\mathscr{C}_k$ the contour defined by the kth equation, and obtain their plots as shown in Fig. 5.20, in which $\mathscr{C}_2$ is shown with solid line.

It should be noted that the x and the y axes represent the real and imaginary parts of the system eigenvalues, respectively. Moreover, the two contours intersect at four points, corresponding to the four complex eigenvalues of the system. The coordinates of the intersection points are given in Table 5.1.

Furthermore, the four eigenvalues occur as two complex conjugate pairs, as expected. Now, we recall here that a single-dof system can be underdamped, critically damped or overdamped, depending, on the value of ζ. In two-dof systems, one mode may be underdamped, while the other overdamped. For that matter, one mode may even be undamped. In the case at hand, given that each mode admits complex conjugate eigenvalues, both are underdamped.

The reader is invited to find the damping ratio and the natural frequency associated with each of the two pairs of eigenvalues of Table 5.1, which can be done using the results of Example A.3.4.

5.7.1 Total Response of Damped Two-dof Systems

In order to find the total response of damped two-dof systems, all we need is their zero-state response, and then, superimpose it with the foregoing zero-input response. To derive the total response, we define $s(t) \equiv \dot{y}(t)$ and solve for $\dot{s} \equiv \ddot{y}$ from Eq. 5.118, thereby obtaining a system of four *first-order* linear ODEs, namely,

$$\dot{y} = s \tag{5.135a}$$

$$\dot{s} = -\Omega^2 y - \Delta s + g(t) \tag{5.135b}$$

with the initial conditions $y(0) = y_0$ and $\dot{y}(0) = s_0$. We can now write Eqs. 5.135a and b in *state-variable form* as a four-dimensional system of linear ODEs, namely,

$$\dot{\zeta} = A\zeta + Bg(t), \quad \zeta(0) = \zeta_0 \tag{5.136a}$$

where

$$A \equiv \begin{bmatrix} O & 1 \\ -\Omega^2 & -\Delta \end{bmatrix}, \quad B \equiv \begin{bmatrix} O \\ 1 \end{bmatrix}, \quad \zeta \equiv \begin{bmatrix} y \\ s \end{bmatrix} \tag{5.136b}$$

In the above definitions, A is a 4×4 matrix, while B is a 4×2 matrix, ζ is a four-dimensional state-variable vector, and $g(t)$ was defined in Eq. 5.91 as $N^{-1}f(t)$.

The total response of damped one-dof systems in state-variable form was obtained in Chap. 2. In that case, the system was represented by a first-order equation formally identical to Eq. 5.136a, with matrix A of 2×2 and vector z of dimension 2. Moreover, rather than a 4×2 matrix B, we had in this model a two-dimensional vector b and the input to the system, $f(t)$, was a scalar. Here, the input is a two-dimensional vector. By introducing the proper substitutions, then, the total response of the system takes the form

$$\zeta(t) = e^{At}\zeta_0 + \int_0^t e^{A(t-\tau)}Bg(\tau)\,d\tau \tag{5.137}$$

Of course, the foregoing time response is given in terms of the transformed variables $y = Nx(t)$, $\dot{y}(t) \equiv s(t) = N\dot{x}(t) = Nv(t)$. In order to express the same response in the original state variables $x(t)$ and $v(t)$, a back-transformation is in order:

$$\zeta(t) = Zz(t), \quad Z = \begin{bmatrix} N & O \\ O & N \end{bmatrix} \tag{5.138}$$

The derivations of this expression is left as as exercise.

It is now apparent that the calculation of the time response of the system at hand involves the calculation of the exponential of a 4×4 matrix and of its convolution with a given function of time. Here, unfortunately, we cannot mimic the response of the system under study as we did in the case of undamped systems. The fundamental reason why this mimicking is not possible here lies in that one 'mode' of the system may be underdamped, while the other is overdamped, and so, the general time response cannot be derived from any particular time response of single-dof systems. Note, moreover, that the calculation of the aforementioned matrix exponential involves the four eigenvalues of matrix $\mathbf{A}$, which can be found using a numerical procedure for general square matrices, for the said matrix is apparently non-symmetric. Hence, complex eigenvalues are likely to occur, but, since the matrix entries are all real, these eigenvalues should appear in complex-conjugate pairs, such as in the case of the general eigenvalue problem of Eq. 5.126. As a matter of fact, the two characteristic equations, that were obtained in the foregoing equation, and that associated with the 4×4 matrix $\mathbf{A}$, yield the same eigenvalues.

Example 5.7.2 (The Whirling of Shafts, a Simple Case). Obtain the trajectory of point C of the rotor fixedly attached to the whirling shaft introduced in Example 4.5.1 when the shaft is turning at a constant angular velocity ω.

Solution: Since this system is *decoupled*, i.e., its two generalized coordinates do not appear simultaneously in the same governing equation, its response reduces to that of two independent single-dof systems. For quick reference, we reproduce below the governing equations of the system under study:

$$\ddot{x} + 2\zeta_1\omega_1\dot{x} + \omega_1^2 x = \omega^2 e \cos\omega t$$
$$\ddot{y} + 2\zeta_2\omega_2\dot{y} + \omega_2^2 y = \omega^2 e \sin\omega t$$

with ω_k and ζ_k defined, for $k = 1, 2$, as

$$\omega_k \equiv \sqrt{\frac{k_k}{m}}, \quad \zeta_k \equiv \frac{c_k}{2m\omega_k}$$

We are interested in the steady-state response of the system. Moreover, since the two foregoing equations are already decoupled, their steady-state responses are as derived in Sect. 2.7.3, i.e., as

$$x(t) = M_x \cos(\omega t + \phi_x)$$
$$y(t) = M_y \sin(\omega t + \phi_y)$$

where M_x and M_y denote the magnitude of the response, while ϕ_x and ϕ_y the corresponding phase, namely,

$$M_x = \frac{e r_1}{\sqrt{(1 - r_1^2)^2 + (2\zeta_1 r_1)^2}}, \quad \phi_x = -\tan^{-1}\left(\frac{2\zeta_1 r_1}{1 - r_1^2}\right)$$

$$M_y = \frac{er_2}{\sqrt{(1-r_2^2)^2 + (2\zeta_2 r_2)^2}}, \quad \phi_y = -\tan^{-1}\left(\frac{2\zeta_2 r_2}{1-r_2^2}\right)$$

with r_k defined as the frequency ratio in the x and y directions, for $k = 1, 2$, namely,

$$r_k \equiv \frac{\omega}{\omega_k}$$

What we want now is to find the trajectory described by point C of the rotor of Fig. 4.6. To this end, we have to eliminate the parameter t from the two above expressions for $x(t)$ and $y(t)$. We thus expand the foregoing expressions in the forms

$$x(t) = A\cos\omega t - B\sin\omega t$$

$$y(t) = C\cos\omega t + D\sin\omega t$$

with coefficients A, B, C and D defined as

$$A \equiv M_x\cos\phi_x, \quad B \equiv M_x\sin\phi_x$$
$$C \equiv M_y\cos\phi_y, \quad D \equiv M_y\sin\phi_y$$

If we want to eliminate t from the above relations, then, we must solve for the trigonometric functions involved in the above equations, and then equate the sum of the squares of the expressions thus resulting to unity, thereby obtaining an equation in x and y, free of t, which describes the trajectory of point C. We thus write the above-mentioned equations in vector form as

$$\begin{bmatrix} A & -B \\ C & D \end{bmatrix} \begin{bmatrix} \cos\omega t \\ \sin\omega t \end{bmatrix} = \begin{bmatrix} x \\ y \end{bmatrix}$$

Now, if we denote by Δ the determinant of the above matrix coefficient, we obtain

$$\begin{bmatrix} \cos\omega t \\ \sin\omega t \end{bmatrix} = \frac{1}{\Delta} \begin{bmatrix} D & B \\ -C & A \end{bmatrix} \begin{bmatrix} x \\ y \end{bmatrix}$$

with Δ reducing to

$$\Delta = \det\begin{bmatrix} A & -B \\ C & D \end{bmatrix} = AD - BC = M_x M_y \cos(\phi_x + \phi_y)$$

Moreover, let

$$\mathbf{e} \equiv \begin{bmatrix} \cos\omega t \\ \sin\omega t \end{bmatrix}, \quad \mathbf{r} \equiv \begin{bmatrix} x \\ y \end{bmatrix}, \quad \mathbf{P} \equiv \begin{bmatrix} D & B \\ C & A \end{bmatrix}$$

and hence, the above expression for the vector of harmonic functions of ωt can be rewritten as

$$\mathbf{e} = \frac{1}{\Delta}\mathbf{P}\mathbf{r}$$

Now, $\mathbf{e}$ is a unit vector, and hence, upon making its magnitude equal to unity in the above equation, a quadratic equation in $\mathbf{r}$ is obtained, namely,

$$\mathbf{r}^T\mathbf{P}^T\mathbf{P}\mathbf{r} = \Delta^2, \quad \mathbf{P}^T\mathbf{P} = \begin{bmatrix} M_y^2 & M_xM_y\sin(\phi_x - \phi_y) \\ M_xM_y\sin(\phi_x - \phi_y) & M_x^2 \end{bmatrix}$$

whence the desired trajectory, $f(x,y) = 0$, can be readily derived, namely,

$$f(x,y) \equiv \mathbf{r}^T\mathbf{P}^T\mathbf{P}\mathbf{r} - \Delta^2 = 0$$

which is, apparently, a quadratic equation, and hence, the trajectory sought is a conic section. From elementary analytic geometry, the conic is (a) an ellipse if $\mathbf{P}^T\mathbf{P}$ is positive-definite; (b) a parabola if the same matrix is semidefinite; and (c) a hyperbola or two lines passing through the origin if the matrix is sign-indefinite. Because of the form of the matrix involved, which can only be positive-definite or semidefinite, the third case is ruled out. All we are left with is to investigate the first two cases. From the nature of the problem it should be expected that the trajectory be an ellipse, which is the case, because $\mathbf{P}^T\mathbf{P}$ is, indeed, positive-definite. To show this, all we have to do is verify that its determinant is in fact positive, but this is the case, for

$$\det(\mathbf{P}^T\mathbf{P}) = \det\begin{bmatrix} C^2 + D^2 & BD - AC \\ BD - AC & A^2 + B^2 \end{bmatrix}$$

i.e.,

$$\det(\mathbf{P}^T\mathbf{P}) = \det\begin{bmatrix} M_y^2 & M_xM_y\sin(\phi_x - \phi_y) \\ M_xM_y\sin(\phi_x - \phi_y) & M_x^2 \end{bmatrix}$$

$$= M_x^2 M_y^2 \cos^2(\phi_x - \phi_y) \geq 0$$

the said matrix becoming singular only if the difference between the two phase angles is a multiple of $90°$. We will assume that this is not the case and, moreover, that $M_x^2 > M_y^2$ and $\sin(\phi_x - \phi_y) > 0$, for purposes of illustration. Note that, if the above assumptions do not hold, the ensuing analysis is still possible, if with the necessary adaptations to the new conditions. We thus determine below the parameters of the elliptical trajectory, namely, the magnitude and the orientation of its semiaxes, the former being given by the eigenvalues, the latter by the eigenvectors of $\mathbf{P}^T\mathbf{P}$. Thus, all we need is an eigenvalue analysis of the same matrix, which can be readily accomplished with the aid of the Mohr circle. We show in Fig. 5.21 this circle, along with its parameters, from which the ellipse of Fig. 5.22 is derived.

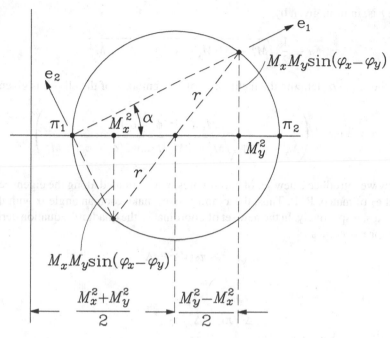

Fig. 5.21 The Mohr circle of $\mathbf{P}^T\mathbf{P}$

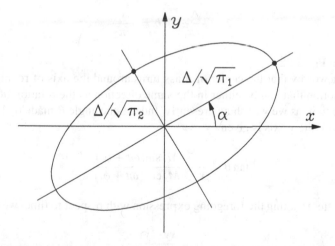

Fig. 5.22 The elliptical trajectory of point C of the rotor of mounted on a shaft under whirling

From the Mohr circle of Fig. 5.21 it is apparent that the two eigenvalues of $\mathbf{P}^T\mathbf{P}$, π_1 and π_2, are given by

$$\pi_1 = \frac{M_x^2 + M_y^2}{2} - r, \quad \pi_2 = \frac{M_x^2 + M_y^2}{2} + r$$

where r is, in turn, given by

$$r \equiv \frac{1}{2}\sqrt{M_x^4 - 2M_x^2 M_y^2 \cos 2(\phi_x - \phi_y) + M_y^4}$$

while the angle α defining the inclination of the semiaxes of the ellipse is given by

$$\alpha = \tan^{-1}\left(\frac{2M_x M_y \sin(\phi_x - \phi_y)}{M_y^2 - M_x^2 + \sqrt{M_x^4 - 2M_x^2 M_y^2 \cos 2(\phi_x - \phi_y) + M_y^4}}\right)$$

Now we introduce a new set of coordinates, x'-y', oriented along the eigenvectors $\mathbf{e}_1$ and $\mathbf{e}_2$ of matrix $\mathbf{P}^T\mathbf{P}$. Thus, the x' and y' axes make each an angle α with the x and y axes, respectively. In the new set of coordinates, the quadratic equation derived above for $\mathbf{r}$ becomes

$$\pi_1(x')^2 + \pi_2(y')^2 = \Delta^2$$

or

$$\frac{(x')^2}{\Delta^2/\pi_1} + \frac{(y')^2}{\Delta^2/\pi_2} = 1$$

which is the equation of an ellipse of semiaxes

$$a = \frac{\Delta}{\sqrt{\pi_1}}, \quad b = \frac{\Delta}{\sqrt{\pi_2}}$$

as shown in Fig. 5.22.

It is noteworthy that the center of mass turns around the axis of rotation with a periodic motion that can be either in the same direction as the rotation of the rotor or opposite to it, as we will show presently. Indeed, the angle θ made by line OC of Fig. 4.6d with the x axis is given by

$$\tan\theta = \frac{y}{x} = \frac{M_y \sin(\omega t + \phi_y)}{M_x \cos(\omega t + \phi_x)}$$

Upon differentiating the foregoing expression with respect to time, we obtain

$$\dot{\theta} = \frac{x\dot{y} - y\dot{x}}{x^2 + y^2}$$

and hence,

$$\text{sgn}(\dot{\theta}) = \text{sgn}(x\dot{y} - y\dot{x})$$

with

$$x\dot{y} - y\dot{x} = \omega M_x M_y \cos(\phi_x - \phi_y)$$

and hence,

$$\text{sgn}(\dot{\theta}) = \text{sgn}[\cos(\phi_x - \phi_y)]$$

From the expressions for $\tan\phi_x$ and $\tan\phi_y$ given above, the harmonic functions of the phase angles are readily derived as

$$\cos\phi_x = \frac{r_1^2 - 1}{\sqrt{(r_1^2 - 1)^2 + (2\zeta_1 r_1)^2}}, \quad \cos\phi_y = \frac{r_2^2 - 1}{\sqrt{(r_2^2 - 1)^2 + (2\zeta_2 r_2)^2}}$$

$$\sin\phi_x = \frac{2\zeta_1 r_1}{\sqrt{(r_1^2 - 1)^2 + (2\zeta_1 r_1)^2}}, \quad \sin\phi_y = \frac{2\zeta_1 r_2}{\sqrt{(r_2^2 - 1)^2 + (2\zeta_2 r_2)^2}}$$

whence,

$$\cos(\phi_x - \phi_y) = \frac{(r_1^2 - 1)(r_2^2 - 1) - 4\zeta_1\zeta_2 r_1 r_2}{\sqrt{(r_1^2 - 1)^2 + (2\zeta_1 r_1)^2}\sqrt{(r_2^2 - 1)^2 + (2\zeta_2 r_2)^2}}$$

Therefore,

$$\text{sgn}(\dot{\theta}) = \text{sgn}[(r_1^2 - 1)(r_2^2 - 1) - 4\zeta_1\zeta_2 r_1 r_2]$$

thus concluding that if the foregoing expression inside the brackets is positive, the ellipse is traced in the same direction as the rotation of the rotor; if negative, then the ellipse is traced in the opposite direction. The reader is invited to analyze under which conditions the term inside the brackets of the above equation vanishes.

In summary, then, the trajectory of the center C of the rotor, described upon whirling of the shaft, is an ellipse of semiaxes given by the eigenvalues and the eigenvectors of matrix $\mathbf{P}^T\mathbf{P}$. Note that the entries of this matrix depend on the mechanical parameters of the system, and hence, if the ellipse is determined experimentally, the said parameters can be estimated from these measurements. Knowing these parameters is essential for a proper operation, i.e., away from resonant conditions.

5.8 Exercises

5.1. (To be assigned only if Exercise 4.7 was previously assigned) We refer to the system of Fig. 4.12, representing the two-dof iconic model of a terrestrial vehicle, under the conditions described in Exercise 4.7. If we assume that the bump does not affect the horizontal, uniform motion of the vehicle, determine its time response, namely $x_1(t)$ and $x_2(t)$, for $t \geq 0$, if time is set equal to zero at the instant in which the vehicle hits the bump.

Hint: It will be helpful to recall the total time response of a single-dof undamped system to the function $\sin(\omega t)u(t)$, as given in Eq. 2.143, which is reproduced below in a form suitable for the problem at hand:

$$x_S(t) = (\omega_n^2 - \omega^2)^{-1}(\sin \omega t - \omega \omega_n^{-1} \sin \omega_n t)u(t)$$

Also note that

$$\sin[(\omega t)\mathbf{1}] = \begin{bmatrix} \sin \omega t & 0 \\ 0 & \sin \omega t \end{bmatrix} = (\sin \omega t)\mathbf{1}$$

5.2. (To be assigned only if Exercise 4.4 was previously assigned) With reference to the system of Fig. 4.15, a horizontal force $f(t) = F_0 \cos \omega_f t$ is applied to the frame, that causes the system to oscillate with a "small-amplitude" motion. Obtain an expression for the magnitude and the phase angle of the rod angle θ vs. ω_f, and sketch the corresponding Bode plots for light damping. Discuss the cases (1) $c \rightarrow 0$ and (2) $k \rightarrow 0$.

5.3. If we assume that the tires of the subway car of Fig. 4.9 are rigid, the system reduces to one with two dof. Derive expressions for the associated 2×2 mass and stiffness matrices, and find the natural frequencies and the natural modes of the system.

5.4. A more realistic model of the belt-pulley transmission of Fig. 5.8 should include a dashpot of damping coefficient c in parallel with each of the springs shown in that figure. For the damped model, assume that a torque $\tau = \tau_0 \cos \omega t$ acts on the pulley of moment of inertia J_1. Find the Bode plots of the steady-state response of the *flexible mode* $\psi_2(t)$ defined as in Example 5.3.1, i.e., as

$$\psi_2(t) \equiv r_1 \theta_1(t) - r_2 \theta_2(t)$$

5.5. Let us assume that the parameters of the system of Fig. 4.16 have numerical values yielding the numerical mass and stiffness matrices

$$\mathbf{M} = \begin{bmatrix} 5 & 0 \\ 0 & 5 \end{bmatrix} \mathrm{Kg} \cdot \mathrm{m}^2, \quad \mathbf{K} = \begin{bmatrix} 10 & -10 \\ -10 & 10 \end{bmatrix} \mathrm{Nm}$$

For the foregoing matrices, find the time response of the system under the conditions described below: As the system is at rest, an impulsive moment $\tau_0 \delta(t)$ is applied on each of the two large disks, with τ_0 being a constant with units of Nms and $\delta(t)$ being the unit-impulse function. Moreover, the two foregoing moments balance each other.

5.6. The iconic model of Fig. 4.10 represents a locomotive of mass m_2 pulling a car of mass m_1. Consider that the coupling between locomotive and car is viscoelastic, with an equivalent natural frequency of $10\,\mathrm{Hz}$ and a damping ratio of 0.7071. Moreover, the locomotive is being driven by an electric motor that delivers a force $F(t) = F_0(10 + \cos \omega t)$ pulling to the right. Find the steady-state response of the system.

5.7. Find the natural frequencies and the natural modes of the overhead crane of Fig. 1.19 for "small-amplitude" oscillations about a stable equilibrium state, in the absence of damping. As well, assume that the slider travels at a uniform speed to the right, with the rod in a vertical, stable position, and subject to no external force. Under these conditions, the slider encounters a stationary obstacle on its way, with which it undergoes a perfectly elastic collision. Find the time response of the system if the speed of the slider just prior to the collision is v_0.

5.8. Here, we consider the iconic model shown in Fig. 5.10, but disregarding the ball. This model represents now an aircraft engine undergoing tests on an elastic foundation. Find the time response of the pad when the engine exerts on the pad the excitation described below:

(a) A moment $\tau(t)$ and a vertical force $f(t)$ applied at its center of mass
(b) A vertical force $f(t)$ applied at point R

For the two above items, assume that $\tau(t)$ and $f(t)$ are given as

$$\tau(t) = \frac{1}{2}\tau_0\left[1+\text{sat}\left(\frac{2t-T}{T}\right)\right], \quad f(t) = \frac{1}{2}f_0\left[1+\text{sat}\left(\frac{2t-T}{T}\right)\right]$$

where the *saturation function* sat(x) is defined in Eq. 1.37 and sketched in Fig. 1.24b.

5.9. Two 2×2 matrices with real entries are given below:

$$\mathbf{M} = \begin{bmatrix} a & b \\ c & d \end{bmatrix}, \quad \mathbf{K} = \begin{bmatrix} e & f \\ g & h \end{bmatrix}$$

(a) State the conditions on the entries of $\mathbf{M}$ under which it can represent the mass matrix of a two-degree-of-freedom mechanical system.
(b) State the conditions on the entries of $\mathbf{K}$ under which it can represent the stiffness matrix of a two-degree-of-freedom mechanical system. Here, allow for the possibility of rigid modes.
(c) We now assume that

$$\mathbf{M} = \begin{bmatrix} a & b \\ b & a \end{bmatrix}, \quad \mathbf{K} = \begin{bmatrix} e & -e \\ -e & e \end{bmatrix}, \quad a>0, \quad a^2>b^2, \quad e>0$$

are the mass and the stiffness matrices of a certain two-degree-of-freedom mechanical system. Find its natural frequencies and its natural modes.

5.10. A robotic joint is modeled as a mechanical system with rotors of moments of inertia J and $2J$, connected by a viscoelastic coupling (a parallel array of a torsional spring and a torsional dashpot) with torsional stiffness k and torsional damping coefficient c, as indicated in Fig. 5.23.

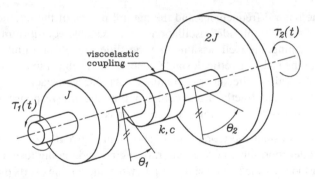

Fig. 5.23 Robotic joint

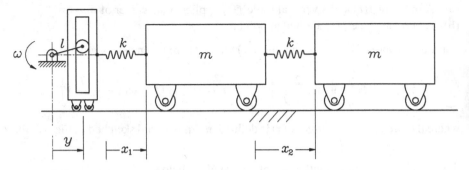

Fig. 5.24 An undamped model of an unloaded press

(a) Derive the mathematical model of the system in terms of the angles θ_1 and θ_2 shown in Fig. 5.23 if we know that the spring in the coupling is unloaded when these two coordinates are identical.

(b) Under the assumption that $\tau_2(t) = -\tau_1(t)$ and that all initial conditions are zero, the torque $\tau(t)$ experienced by the coupling at its ends can be expressed in the form

$$\tau(t) = -(k_e\theta_1 + c_e\dot{\theta}_1)$$

Find expressions for k_e and c_e in terms of the system parameters.

(c) Under the same conditions as in item (b) above, $\tau_1(t)$ is modeled as a harmonic torque of amplitude τ_0 and frequency ω. Thus, the steady-state torque transmitted to the coupling is also harmonic with an amplitude τ_T and a phase angle ψ. Using Bode plots, rather than lengthy calculations, find the ratio τ_T/τ_0, if $\zeta_e = 0.5$ and $\omega_e = \omega/5$, where $c_e = 2\zeta_e\omega_e J$ and $k_e = \omega_e^2 J$.

5.11. A more elaborate, if undamped, model of the press described in Exercise 2.29 is shown in Fig. 5.24. For the case in which $\omega = 10\sqrt{k/m}$, find the amplitude of the oscillations undergone by the mass at the right.

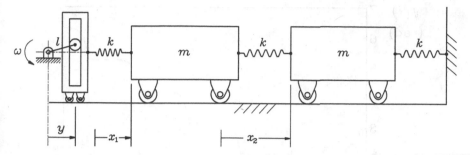

Fig. 5.25 An undamped model of a loaded press

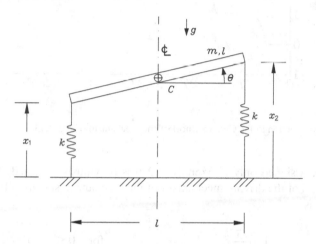

Fig. 5.26 A test pad mounted on an elastic suspension

5.12. When the press of Fig. 5.24 is loaded, the press-load system can be represented by the iconic model of Fig. 5.25. Find the time response of the system, still under the assumption that $\omega = 10\sqrt{k/m}$.

5.13. Here, we consider the iconic model of Fig. 5.26 that represents an aircraft engine undergoing tests on an elastic foundation. Find the amplitudes of the oscillations exhibited by each end of the pad when the engine exerts a moment $\tau(t) = \tau_0 \cos \omega_f t$ on the pad, where $\omega_f = (3/2)\sqrt{k/m}$. To do this, note that the mathematical model of the system can be cast in the form

$$\mathbf{M}\ddot{\mathbf{x}} + \mathbf{K}\mathbf{x} = \frac{\tau_0}{l}(\cos \omega_f t)\mathbf{b}$$

It will be helpful to recall the steady-state time response of a single-dof undamped system to the function $A\cos(\omega_f t)$, as derived in Sect. 2.7.3, which is reproduced below in a form suitable for the problem at hand:

$$y(t) = (\omega_n^2 - \omega_f^2)^{-1}A\cos \omega_f t$$

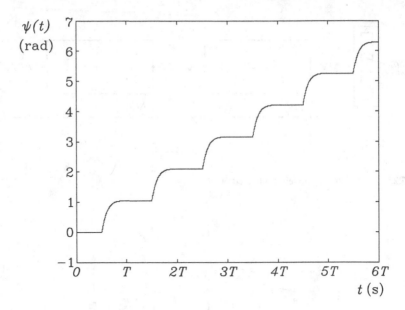

Fig. 5.27 Drive function of the Geneva wheel of the mechanism of Fig. 4.13

5.14. (To be assigned only if Exercise 4.2 was previously assigned) We analyze here the model of the driving mechanism of the mechanism shown in Fig. 4.13, for $\psi_p(t)$ given as

$$
\psi_p(t) = \begin{cases} -\dfrac{\pi}{3T}t, & \text{for} \quad 0 \le t \le T/2; \\[2mm] \dfrac{\pi}{3T}(t-T) + \arctan[f(t)], & \text{for} \quad T/2 \le t \le T \end{cases}
$$

and displayed in Fig. 5.27, while $f(t)$ is given, in turn, as

$$
f(t) = -\frac{\sin \omega_0 t}{2 + \cos \omega_0 t}
$$

(a) Compute the coefficients of the first N_h harmonic components of the foregoing periodic function, by dividing the period in $2N_h$ intervals of equal length. Produce a table of error e in the Fourier approximation of $\psi_p(t)$ vs. N_h, for $N_h = 1, 2, 3, 4, 5, 10$, and 20. Comment on your tabulated results.

(b) Find the steady-state response of the system, i.e., $\theta_1(t)$ and $\theta_2(t)$, for N_h harmonics of the Fourier expansion of $\psi_p(t)$. Choose the smallest value of N_h that will give you an error smaller than 0.05 in the approximation of $\psi_p(t)$. Plot $\theta_1(t)$ and $\theta_2(t)$ vs. time in the $[0, 4T]$ interval. Note that these plots are composed of a periodic and a nonperiodic part.

Fig. 5.28 A gearbox with
elastic shafts

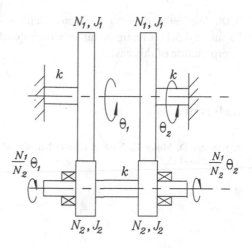

For the calculation, use the numerical values

$$\omega_n = 10\omega_0 = \sqrt{\frac{k}{J}}, \quad \omega_0 = 0.3 \text{ s}^{-1}$$

(c) In order to design the two shafts of the mechanism, we need the *root-mean-square* value of the torque experienced by each shaft, which you are asked to provide. Find also the maximum absolute values of these torques.

Hint: Example 5.5.1 may be helpful in solving this problem.

5.15. A gearbox is modeled as shown in Fig. 5.28, which comprises two identical gears of moments of inertia J_1 and N_1 teeth, meshing with two identical pinions of moments of inertia J_2 and N_2 teeth. All three shafts have stiffness k. For the relations given below:

$$J_1 = J, \quad J_2 = \frac{9}{16}J, \quad \frac{N_1}{N_2} = \frac{4}{3},$$

find the natural frequencies and natural modes of the system. Sketch the latter.

5.16. Derive a procedure to obtain the zero-state response of a two-dof semidefinite system that *does not preserve the generalized—translational or angular—momentum*. To this end, use as an example the belt-pulley system of Fig. 5.8.

5.17. We revisit here the vibration absorber introduced in Example 5.5.2. Show that the solution $\mathbf{x}(t)$ derived from $\mathbf{y}(t) = (\Omega^2 - \omega_f^2 \mathbf{1})^{-1}\mathbf{a}(\cos \omega_f t)$ indeed verifies the mathematical model of the two-dof system at hand. Furthermore, find the natural frequencies and the modal vectors of the same two-dof system.

5.18. We refer here to the rotor-shaft system of Fig. 4.6a, as introduced in Example 4.5.1. Investigate under which conditions $\dot{\theta}$ vanishes and give a physical interpretation of this case.

Reference

1. Kahaner D, Moler C, Nash S (1989) Numerical methods and software. Prentice-Hall, Inc., Englewood Cliffs, NJ

Chapter 6
Vibration Analysis of n-dof Systems

Time is the measure of change.

Aristotle's *Physics,* Chapter 12

6.1 Introduction

The vibration analysis of n-dof systems is the subject of this chapter, the focus being undamped systems, while only outlining the analysis of damped systems in Sect. 6.6. The analysis of n-dof undamped systems parallels that of their two-dof counterparts, the only difference being that the calculations that were possible in graphical form for the latter are not possible for the former. Hence, the analyst has to resort to a numerical procedure to implement these calculations. The widespread availability of pertinent mathematical software, however, eases the analysis tremendously. Therefore, frequent mention will be made to suitable software, as the need arises.

The methods applicable to undamped systems, however, cannot be readily ported into their damped counterparts. Indeed, a closed-form expression for the time response of the latter in a form that would mimic the total response of single-dof systems à la Sect. 2.6.2 is not possible. The reason is that we would need one single formula, but this formula is evasive, given that in Sect. 2.6.2 we do not have one, but rather three distinct formulas, depending on the nature of the system at hand, which can be underdamped, critically damped or overdamped. Most frequently, the n-dof damped system would have some modes—if one could speak of *modes* in the same way that one does with reference to undamped systems—that are underdamped, some that are overdamped, and possibly some critically damped modes. For this reason, the time response of damped systems is only outlined in Sect. 6.6. A more detailed discussion of damped systems, in the context of simulation, is included in Chap. 7.

J. Angeles, *Dynamic Response of Linear Mechanical Systems: Modeling, Analysis and Simulation*, Mechanical Engineering Series, DOI 10.1007/978-1-4419-1027-1_6, © Springer Science+Business Media, LLC 2011

6.2   The Natural Frequencies and the Natural Modes of n-dof Undamped Systems

In this case, the mathematical model of the system at hand is formally identical to that of a two-dof system, namely,

$$\mathbf{M\ddot{x}} + \mathbf{Kx} = \mathbf{0}, \quad \mathbf{x}(0) = \mathbf{x}_0, \quad \dot{\mathbf{x}}(0) = \mathbf{v}_0, \quad t \geq 0 \tag{6.1}$$

where, now, $\mathbf{M}$ and $\mathbf{K}$ are symmetric $n \times n$ matrices, the former being, additionally, positive-definite, while the latter is either positive-semidefinite or positive-definite. Moreover, the vector of generalized coordinates $\mathbf{x}$ is n-dimensional. As shown in Sect. 5.2, the foregoing model leads to a simpler model in normal form, namely,

$$\ddot{\mathbf{y}} + \mathbf{\Omega}^2 \mathbf{y} = \mathbf{0}, \quad \mathbf{y}(0) = \mathbf{y}_0, \quad \dot{\mathbf{y}}(0) = \mathbf{s}_0, \quad t \geq 0 \tag{6.2}$$

with $\mathbf{\Omega}^2$, the frequency matrix, defined as in Chap. 4, namely,

$$\mathbf{\Omega}^2 \equiv \mathbf{N}^{-1} \mathbf{K} \mathbf{N}^{-1} \tag{6.3}$$

Likewise, $\mathbf{N}$ is defined, as in Chap. 4 as well, as the positive-definite square root of $\mathbf{M}$:

$$\mathbf{N} \equiv \sqrt{\mathbf{M}} \tag{6.4}$$

and $\mathbf{y}$ is given by the transformation below:

$$\mathbf{y} = \mathbf{N}\mathbf{x} \quad \Rightarrow \quad \ddot{\mathbf{y}} = \mathbf{N}\ddot{\mathbf{x}} \tag{6.5}$$

Now, the frequency matrix of the system at hand is a $n \times n$ symmetric positive-semidefinite matrix. If $\mathbf{K}$ is positive-definite, then $\mathbf{\Omega}$ is, correspondingly, positive-definite as well. Moreover, $\mathbf{y}$, like $\mathbf{x}$, is a n-dimensional vector.

Again, such as in the two-dof case, the mathematical model of a n-dof undamped system is formally identical to that of a single-dof mass-spring system, where, now, the natural frequency ω_n is replaced by the $n \times n$ frequency matrix $\mathbf{\Omega}$. Furthermore, the n eigenvalues of the frequency matrix are non-negative and its n eigenvectors are mutually orthogonal.

Let $\{\omega_i\}_1^n$ be the n—non-negative—eigenvalues of $\mathbf{\Omega}$. These are the *natural frequencies* of the system under study. Moreover, let $\{\mathbf{e}_i\}_1^n$ denote the *unit* eigenvectors of $\mathbf{\Omega}$, which are necessarily mutually orthogonal, i.e.,

$$\|\mathbf{e}_1\| = \|\mathbf{e}_2\| = \cdots = \|\mathbf{e}_n\| = 1, \quad \mathbf{e}_i^T \mathbf{e}_j = 0, \quad \text{for} \quad i \neq j \tag{6.6}$$

That is, the sets $\{\omega_i\}_1^n$ and $\{\mathbf{e}_i\}_1^n$ verify the relations

$$\mathbf{\Omega}\mathbf{e}_i = \omega_i \mathbf{e}_i, \quad i = 1, \ldots, n \tag{6.7}$$

The calculation of $\{\omega_i\}_1^n$ and $\{e_i\}_1^n$ cannot be accomplished, in general, using the Mohr circle, which is applicable only to 2×2 symmetric matrices. Likewise, matrix $\sqrt{M}$ cannot be calculated graphically, for M is also a $n \times n$ matrix. Moreover, what we will need to compute the frequency matrix is not $\sqrt{M}$ itself, but its inverse. We explain below how to compute this inverse. To this end, we notice that M being symmetric, it has n real eigenvalues and n mutually orthogonal eigenvectors, its eigenvalues being, additionally, positive, for we have assumed at the outset that M is positive-definite. We start by defining

$$\mathbf{E} \equiv \begin{bmatrix} \mathbf{e}_1 & \mathbf{e}_2 & \cdots & \mathbf{e}_n \end{bmatrix}, \quad \Omega_d \equiv \begin{bmatrix} \omega_1 & 0 & \cdots & 0 \\ 0 & \omega_2 & \cdots & 0 \\ \vdots & \vdots & \ddots & \vdots \\ 0 & 0 & \cdots & \omega_n \end{bmatrix} \tag{6.8}$$

Such as in the two-dof case, matrix $\mathbf{E}$ diagonalizes Ω in the sense

$$\Omega_d = \mathbf{E}^T \Omega \mathbf{E} \tag{6.9}$$

whence,

$$\Omega = \mathbf{E}\Omega_d \mathbf{E}^T \tag{6.10}$$

Moreover, let $\{\mu_i\}_1^n$ and $\{\mathbf{m}_i\}_1^n$ be the sets of eigenvalues and eigenvectors of M, respectively. We now define

$$\overline{\mathbf{M}} \equiv \begin{bmatrix} \mathbf{m}_1 & \mathbf{m}_2 & \cdots & \mathbf{m}_n \end{bmatrix}, \quad \mathbf{M}_d \equiv \begin{bmatrix} \mu_1 & 0 & \cdots & 0 \\ 0 & \mu_2 & \cdots & 0 \\ \vdots & \vdots & \ddots & \vdots \\ 0 & 0 & \cdots & \mu_n \end{bmatrix} \tag{6.11}$$

Since any symmetric matrix is diagonalizable in the same way as Ω is, as indicated in Eq. 6.9, $\mathbf{M}$ is diagonalizable with the aid of $\overline{\mathbf{M}}$, namely,

$$\overline{\mathbf{M}}^T \mathbf{M} \overline{\mathbf{M}} = \mathbf{M}_d \tag{6.12}$$

Furthermore, if we recall Fact 3 of Appendix A, $\sqrt{M}$ and M share the same eigenvectors. Moreover, $\sqrt{M}$, like M, is a symmetric matrix and hence, it is diagonalizable with matrix $\overline{\mathbf{M}}$, i.e.,

$$\overline{\mathbf{M}}^T \sqrt{\mathbf{M}} \overline{\mathbf{M}} = \sqrt{\mathbf{M}_d} \tag{6.13}$$

Now, taking the *positive-definite* square root of $\mathbf{M}_d$ is a simple matter, for this is a diagonal matrix, and hence,

$$\sqrt{\mathbf{M}_d} = \begin{bmatrix} \sqrt{\mu_1} & 0 & \cdots & 0 \\ 0 & \sqrt{\mu_2} & \cdots & 0 \\ \vdots & \vdots & \ddots & \vdots \\ 0 & 0 & \cdots & \sqrt{\mu_n} \end{bmatrix} \tag{6.14}$$

Once we have $\sqrt{\mathbf{M}_d}$, calculating $\sqrt{\mathbf{M}}$ is trivial, for all it takes is the reverse transformation of Eq. 6.13, namely,

$$\sqrt{\mathbf{M}} = \overline{\mathbf{M}}\sqrt{\mathbf{M}_d}\overline{\mathbf{M}}^T \tag{6.15}$$

However, to calculate $\boldsymbol{\Omega}$ we need, in fact, $\sqrt{\mathbf{M}^{-1}}$, rather than $\sqrt{\mathbf{M}}$ itself. Obviously, we can calculate $\sqrt{\mathbf{M}^{-1}}$ once we have $\sqrt{\mathbf{M}}$ because of the relation

$$\sqrt{\mathbf{M}^{-1}} = (\sqrt{\mathbf{M}})^{-1} \tag{6.16}$$

which the reader is invited to verify. However, notice that, while $\sqrt{\mathbf{M}_d}$ is diagonal, $\sqrt{\mathbf{M}}$ is, in general, full, and hence, more time-consuming to invert than $\sqrt{\mathbf{M}_d}$. The way to calculate the foregoing matrix is, then, in the form

$$\sqrt{\mathbf{M}^{-1}} \equiv (\sqrt{\mathbf{M}})^{-1} = \overline{\mathbf{M}}\sqrt{\mathbf{M}_d^{-1}}\overline{\mathbf{M}}^T \tag{6.17}$$

where $\sqrt{\mathbf{M}_d^{-1}}$ is readily calculated as

$$\sqrt{\mathbf{M}_d^{-1}} = \begin{bmatrix} 1/\sqrt{\mu_1} & 0 & \cdots & 0 \\ 0 & 1/\sqrt{\mu_2} & \cdots & 0 \\ \vdots & \vdots & \ddots & \vdots \\ 0 & 0 & \cdots & 1/\sqrt{\mu_n} \end{bmatrix} \tag{6.18}$$

Additionally, in order to calculate the foregoing eigenvalues and eigenvectors, we have to resort to a numerical method. In practice, the eigenvalues and eigenvectors of $n \times n$ matrices are best computed using scientific software—Matlab, Maple or Mathematica—where a transformation method is applicable that obviates the process of polynomial root-finding, which is prone to ill-conditioning—numerical instability—for large values of n. For pedagogical reasons, eigenvalues are computed in this chapter via the characteristic polynomial, the eigenvectors following from linear-equation solving. We will illustrate the foregoing ideas with examples. Prior to describing the steps in the underlying computations, however, we introduce some properties of the mass and stiffness matrices, that parallel those of two-dof systems. By virtue of Fact 1 of Appendix A, ω_i and $\mathbf{e}_i$ verify

$$\Omega^2 \mathbf{e}_i = \omega_i^2 \mathbf{e}_i, \quad i = 1, \ldots, n \tag{6.19}$$

and hence, if we express Ω^2 in terms of $\mathbf{K}$ and $\mathbf{N}$, as given by Eq. 6.3, we have

$$\mathbf{N}^{-1} \mathbf{K} \mathbf{N}^{-1} \mathbf{e}_i = \omega_i^2 \mathbf{e}_i$$

or

$$\mathbf{K} \mathbf{N}^{-1} \mathbf{e}_i = \omega_i^2 \mathbf{N} \mathbf{e}_i \tag{6.20}$$

Vector $\mathbf{N}^{-1} \mathbf{e}_i$ appearing in Eq. 6.20 is called, such as in Chap. 4, the ith *modal vector* of the given system. If we let the ith modal vector be represented by $\mathbf{f}_i$, then,

$$\mathbf{f}_i \equiv \mathbf{N}^{-1} \mathbf{e}_i, \quad i = 1, \ldots, n \tag{6.21a}$$

and hence,

$$\mathbf{e}_i = \mathbf{N} \mathbf{f}_i, \quad i = 1, \ldots, n \tag{6.21b}$$

Upon substitution of Eq. 6.21a into the left-hand side of Eq. 6.20, and of Eq. 6.21b into the right-hand side of the same equation, we have

$$\mathbf{K} \mathbf{f}_i = \omega_i^2 \mathbf{N}^2 \mathbf{f}_i \tag{6.22a}$$

i.e.,

$$\omega_i^2 \mathbf{M} \mathbf{f}_i = \mathbf{K} \mathbf{f}_i \tag{6.22b}$$

or

$$(\omega_i^2 \mathbf{M} - \mathbf{K}) \mathbf{f}_i = \mathbf{0}, \quad i = 1, \ldots, n \tag{6.22c}$$

Note that the eigenvectors of Ω are, by definition, of unit magnitude. However, the modal vectors, such as in the two-dof case, turn out to be of unit magnitude, but not in the usual sense; these vectors are, in fact, of unit magnitude *with respect to the mass matrix*, i.e.,

$$\mathbf{f}_i^T \mathbf{M} \mathbf{f}_i = 1, \quad i = 1, \ldots, n \tag{6.23a}$$

or

$$\|\mathbf{f}_i\|_\mathbf{M} = 1, \quad i = 1, \ldots, n \tag{6.23b}$$

Furthermore, when the magnitude of the modal vectors is computed *with respect to the stiffness matrix*, it turns out that the magnitude-squared of the ith modal vector is, such as in the two-dof case, identical to the square of the ith natural frequency, i.e.,

$$\mathbf{f}_i^T \mathbf{K} \mathbf{f}_i = \omega_i^2, \quad i = 1, \ldots, n \tag{6.24a}$$

or

$$\|\mathbf{f}_i\|_\mathbf{K} = \omega_i, \quad i = 1, \ldots, n \tag{6.24b}$$

One more property of the modal vectors follows from Eq. 6.22b, if we multiply its two sides by $\mathbf{M}^{-1}$, namely,

$$\mathbf{M}^{-1}\mathbf{K}\mathbf{f}_i = \omega_i^2 \mathbf{f}_i \qquad (6.25)$$

where the product $\mathbf{M}^{-1}\mathbf{K}$ is readily identified as the *dynamic matrix* of the system at hand, such as in the case of two-dof systems, i.e.,

$$\mathbf{D} \equiv \mathbf{M}^{-1}\mathbf{K} \qquad (6.26)$$

Therefore, the eigenvalues of the dynamic matrix are the natural frequencies-squared, its eigenvectors being the modal vectors. Note, however, that these eigenvectors are not of unit magnitude.

As we studied in Sect. 4.3, the zero-input response of two-dof systems is, in general, not harmonic, the same holding for n-dof systems. Therefore, harmonic motions of n-dof systems occur only if they are shaped by the modal vectors. That is, if $\mathbf{x}(t)$ has the shape

$$\mathbf{x}(t) = A_i(\cos \omega_i t)\mathbf{f}_i \qquad (6.27a)$$

then

$$\ddot{\mathbf{x}}(t) = -\omega_i^2 A_i(\cos \omega_i t)\mathbf{f}_i \qquad (6.27b)$$

Upon substitution of $\mathbf{x}(t)$ and $\ddot{\mathbf{x}}(t)$, as given above, in the governing Eq. 6.1, we have

$$\mathbf{M}\ddot{\mathbf{x}} + \mathbf{K}\mathbf{x} = -A_i(\cos \omega_i t)(\omega_i^2 \mathbf{M} - \mathbf{K})\mathbf{f}_i, \quad i = 1,\ldots,n$$

By virtue of Eq. 6.22c, the right-hand side of the foregoing equation vanishes, and hence, $\mathbf{x}(t)$, as given by Eq. 6.27a, verifies the governing equation. Since we have now n modal vectors, harmonic motions of n distinct shapes are possible. Moreover, each of these motions oscillates at one of the natural frequencies of the system. Each of the n shapes is called a *mode*, hence the name given to vectors $\mathbf{f}_i$. Furthermore, any of the three Eqs. 6.22a–c above is termed the *modal equation* of the system at hand. Now it is a simple matter to realize that the modal vectors $\{\mathbf{f}_i\}_1^n$ are *orthogonal with respect to the mass matrix*. Indeed, if we express the orthogonality of the eigenvectors $\mathbf{e}_i$ and $\mathbf{e}_j$, for $i \neq j$, in terms of Eq. 6.21b, we obtain

$$\mathbf{f}_i^T \mathbf{N}^2 \mathbf{f}_j = 0, \quad i, j = 1,\ldots,n; \quad i \neq j$$

and, by virtue of definition (6.4), the foregoing equation leads to

$$\mathbf{f}_i^T \mathbf{M} \mathbf{f}_j = 0, \quad i, j = 1,\ldots,n; \quad i \neq j \qquad (6.28)$$

thereby showing the orthogonality of the modal vectors with respect to the mass matrix. Furthermore, if both sides of Eq. 6.22b, for $j \neq i$, are multiplied from the left by $\mathbf{f}_i^T$, then, by virtue of Eq. 6.28, one can readily verify that the two foregoing vectors are orthogonal with respect to the stiffness matrix as well, i.e.,

$$\mathbf{f}_i^T \mathbf{K} \mathbf{f}_j = 0, \quad i, j = 1,\ldots,n; \quad i \neq j \qquad (6.29)$$

If, as in the two-dof case, the modal vectors are arrayed columnwise in the $n \times n$ matrix $\mathbf{F}$, then $\mathbf{F}$ obeys the relations

$$\mathbf{F}^T \mathbf{M} \mathbf{F} = 1, \quad \mathbf{F}^T \mathbf{K} \mathbf{F} = \Omega_d^2, \quad \mathbf{F}^{-1} \mathbf{D} \mathbf{F} = \Omega_d^2 \qquad (6.30)$$

similar to those derived in Sect. 5.2.1, where Ω_d was introduced to represent the diagonal form of the frequency matrix.

Example 6.2.1 (Torsional Vibrations of an Aircraft Wing). The system of Fig. 4.7 is revisited here, with matrices $\mathbf{M}$ and $\mathbf{K}$ given as

$$\mathbf{M} = J_0 \begin{bmatrix} 1 & 0 & 0 & 0 & 0 \\ 0 & 4 & 0 & 0 & 0 \\ 0 & 0 & 9 & 0 & 0 \\ 0 & 0 & 0 & 4 & 0 \\ 0 & 0 & 0 & 0 & 1 \end{bmatrix}, \quad \mathbf{K} = k \begin{bmatrix} 1 & -1 & 0 & 0 & 0 \\ -1 & 3 & -2 & 0 & 0 \\ 0 & -2 & 4 & -2 & 0 \\ 0 & 0 & -2 & 3 & -1 \\ 0 & 0 & 0 & -1 & 1 \end{bmatrix}$$

Find the natural frequencies and the natural modes of the system.

Solution: We begin by computing $\mathbf{N}$ as the positive-definite square root of $\mathbf{M}$, which in this case is straightforward, since $\mathbf{M}$ is diagonal. Therefore,

$$\mathbf{N} = \sqrt{J_0} \begin{bmatrix} 1 & 0 & 0 & 0 & 0 \\ 0 & 2 & 0 & 0 & 0 \\ 0 & 0 & 3 & 0 & 0 \\ 0 & 0 & 0 & 2 & 0 \\ 0 & 0 & 0 & 0 & 1 \end{bmatrix}, \quad \mathbf{N}^{-1} = \frac{1}{\sqrt{J_0}} \begin{bmatrix} 1 & 0 & 0 & 0 & 0 \\ 0 & 1/2 & 0 & 0 & 0 \\ 0 & 0 & 1/3 & 0 & 0 \\ 0 & 0 & 0 & 1/2 & 0 \\ 0 & 0 & 0 & 0 & 1 \end{bmatrix}$$

and hence,

$$\Omega^2 = \frac{k}{\sqrt{J_0}} \mathbf{N}^{-1} \begin{bmatrix} 1 & -1 & 0 & 0 & 0 \\ -1 & 3 & -2 & 0 & 0 \\ 0 & -2 & 4 & -2 & 0 \\ 0 & 0 & -2 & 3 & -1 \\ 0 & 0 & 0 & -1 & 1 \end{bmatrix} \begin{bmatrix} 1 & 0 & 0 & 0 & 0 \\ 0 & 1/2 & 0 & 0 & 0 \\ 0 & 0 & 1/3 & 0 & 0 \\ 0 & 0 & 0 & 1/2 & 0 \\ 0 & 0 & 0 & 0 & 1 \end{bmatrix}$$

i.e.,

$$\Omega^2 = \frac{k}{J_0} \begin{bmatrix} 1 & -1/2 & 0 & 0 & 0 \\ -1/2 & 3/4 & -1/3 & 0 & 0 \\ 0 & -1/3 & 4/9 & -1/3 & 0 \\ 0 & 0 & -1/3 & 3/4 & -1/2 \\ 0 & 0 & 0 & -1/2 & 1 \end{bmatrix} \equiv \omega_0^2 \mathbf{A}$$

where ω_0 is defined as

$$\omega_0 \equiv \sqrt{\frac{k}{J_0}}$$

As expected, Ω^2 is a symmetric matrix. Moreover, the system admits apparently one rigid mode, under which all rotors turn through the same angle θ, which thus leads to the displacement vector $\mathbf{x} = \theta [1, 1, 1, 1, 1]^T$. It is now obvious that the product $\mathbf{Kx}$ vanishes, and the potential energy of the system under this motion vanishes as well. Now, since $\mathbf{K}$ is positive-semidefinite, Ω^2 and, hence, Ω itself, are positive-semidefinite as well, which means that at least one of the natural frequencies of the system is zero. Further, we compute the natural frequencies of the system as the square roots of the eigenvalues of Ω^2. To do this, we first derive the characteristic polynomial of $\mathbf{A}$. From Fact 2 of Appendix A, the eigenvalues of Ω^2 are those of $\mathbf{A}$ multiplied by ω_0^2. We thus have, if we let $P(\lambda)$ denote the characteristic polynomial of $\mathbf{A}$,

$$P(\lambda) = \det(\lambda \mathbf{1} - \mathbf{A}) = \det \begin{bmatrix} \lambda - 1 & 1/2 & 0 & 0 & 0 \\ 1/2 & \lambda - 3/4 & 1/3 & 0 & 0 \\ 0 & 1/3 & \lambda - 4/9 & 1/3 & 0 \\ 0 & 0 & 1/3 & \lambda - 3/4 & 1/2 \\ 0 & 0 & 0 & 1/2 & \lambda - 1 \end{bmatrix}$$

Upon expanding the foregoing determinant by the cofactors of its first row, we have

$$\det(\lambda \mathbf{1} - \mathbf{A}) = (\lambda - 1)\Delta_{11} - \frac{1}{2}\Delta_{12}$$

where

$$\Delta_{11} = \det \begin{bmatrix} \lambda - 3/4 & 1/3 & 0 & 0 \\ 1/3 & \lambda - 4/9 & 1/3 & 0 \\ 0 & 1/3 & \lambda - 3/4 & 1/2 \\ 0 & 0 & 1/2 & \lambda - 1 \end{bmatrix}$$

and

$$\Delta_{12} = \det \begin{bmatrix} 1/2 & 1/3 & 0 & 0 \\ 0 & \lambda - 4/9 & 1/3 & 0 \\ 0 & 1/3 & \lambda - 3/4 & 1/2 \\ 0 & 0 & 1/2 & \lambda - 1 \end{bmatrix}$$

Now, we expand below each of the two foregoing 4×4 subdeterminants. We do this by cofactors of their first column, namely,

$$\Delta_{11} = \left(\lambda - \frac{3}{4}\right)\Delta_{1111} - \frac{1}{3}\Delta_{1121}$$

$$\Delta_{12} = \frac{1}{2}\Delta_{1111} \tag{6.31}$$

where Δ_{1111}, appearing in the two foregoing subdeterminants, is given below:

$$\Delta_{1111} = \det \begin{bmatrix} \lambda - 4/9 & 1/3 & 0 \\ 1/3 & \lambda - 3/4 & 1/2 \\ 0 & 1/2 & \lambda - 1 \end{bmatrix}$$

and

$$\Delta_{1121} = \det \begin{bmatrix} 1/3 & 0 & 0 \\ 1/3 & \lambda - 3/4 & 1/2 \\ 0 & 1/2 & \lambda - 1 \end{bmatrix}$$

All we need now is expand the foregoing 3×3 subdeterminants. We do this by cofactors of their first row, namely,

$$\Delta_{1111} = \left(\lambda - \frac{4}{9} \right) \left[\left(\lambda - \frac{3}{4} \right) (\lambda - 1) - \frac{1}{4} \right] - \frac{1}{3} \left(\frac{1}{3} (\lambda - 1) \right)$$

$$\Delta_{1121} = \frac{1}{3} \left[\left(\lambda - \frac{3}{4} \right) (\lambda - 1) - \frac{1}{4} \right]$$

Finally, upon back-substituting the foregoing subdeterminants in the above expression for $P(\lambda)$, we obtain, after several reduction steps that are left to the reader to verify,

$$P(\lambda) = \left(\lambda^2 - \frac{7}{4}\lambda + \frac{1}{2} \right) \left(\lambda^2 - \frac{79}{36}\lambda + \frac{19}{18} \right) \lambda$$

It is apparent that the above polynomial admits the root $\lambda = 0$, which is just a consequence of matrix $\mathbf{K}$ being semidefinite, and hence, singular. Furthermore, the roots of the first quadratic factor of $P(\lambda)$ are readily computed as

$$\lambda = \frac{7 \pm \sqrt{17}}{8}$$

Those of the second quadratic factor of $P(\lambda)$ being

$$\lambda = \frac{79 \pm \sqrt{769}}{72}$$

Thus, the five eigenvalues of $\mathbf{A}$ are, in ascending order,

$$\lambda_1 = 0, \quad \lambda_2 = 0.35961, \quad \lambda_3 = 0.71207, \quad \lambda_4 = 1.39039, \quad \lambda_5 = 1.48237$$

Therefore, the natural frequencies sought are

$$\omega_1 = 0, \quad \omega_2 = 0.59967\omega_0, \quad \omega_3 = 0.84384\omega_0, \quad \omega_4 = 1.17915\omega_0, \quad \omega_5 = 1.21753\omega_0$$

Now we calculate the eigenvectors of $\mathbf{\Omega}^2$, and hence, the normal modes of the system under study. Since $\mathbf{A}$ and $\mathbf{\Omega}$ share the same eigenvectors, we calculate the eigenvectors of the former, which we do for each eigenvalue of $\mathbf{A}$. Let, in every case,

$$\mathbf{e}_i \equiv \begin{bmatrix} v & w & x & y & z \end{bmatrix}^T, \quad i = 1,\dots,5$$

subject to the condition that each vector $\mathbf{e}_i$ be of unit magnitude, i.e.,

$$v^2 + w^2 + x^2 + y^2 + z^2 = 1$$

Then, the components of each eigenvector $\mathbf{e}_i$ are calculated from the linear homogeneous system

$$(\mathbf{A} - \lambda_i \mathbf{1})\mathbf{e}_i = \mathbf{0}$$

We thus have

For $\lambda_1 = 0$:

$$v - 0.5w = 0$$
$$-0.5v + 0.75w - 0.\overline{3}x = 0$$
$$-0.\overline{3}w + 0.\overline{4}x - 0.\overline{3}y = 0$$
$$-0.\overline{3}x + 0.75y - 0.5z = 0$$
$$-0.5y + z = 0$$

From the first equation we have

$$w = 2v$$

Upon substituting the foregoing value of w into the second equation,

$$x = 3v$$

The remaining unknowns are found following the foregoing pattern, which yields

$$y = w = 2v, \quad z = v$$

and hence,

$$\mathbf{e}_1 = \begin{bmatrix} 1 & 2 & 3 & 2 & 1 \end{bmatrix}^T v$$

where the value of v will be chosen so as to render $\mathbf{e}_1$ of unit magnitude.
For $\lambda_2 = 0.35961$:

$$0.64039v - 0.5w = 0$$
$$-0.5v + 0.39039w - 0.\overline{3}x = 0$$
$$-0.\overline{3}w + 0.084830x - 0.\overline{3}y = 0$$

$$-0.\overline{3}x + 0.39039y - 0.5z = 0$$

$$-0.5y + 0.64039z = 0$$

Proceeding as in the first case, we obtain, from the first equation,

$$w = 1.28078v$$

Upon substituting the foregoing value into the second equation, we obtain

$$x = 0$$

Likewise, we obtain, successively,

$$y = -w = -1.28078v, \quad z = -v$$

Hence,

$$\mathbf{e}_2 = \begin{bmatrix} 1 & 1.28078 & 0 & -1.28078 & -1 \end{bmatrix}^T v$$

For $\lambda_3 = 0.71207$:

$$0.28793v - 0.5w = 0$$

$$-0.5v + 0.037930w - 0.\overline{3}x = 0$$

$$-0.\overline{3}w - 0.26763x - 0.\overline{3}y = 0$$

$$-0.\overline{3}x + 0.037930y - 0.5z = 0$$

$$-0.5y + 0.28793z = 0$$

Proceeding exactly as in the two previous cases, we have

$$\mathbf{e}_3 = \begin{bmatrix} 1 & 0.57586 & -1.43449 & 0.57586 & 1 \end{bmatrix}^T v$$

Furthermore,
For $\lambda_4 = 1.39039$:

$$-0.39039v - 0.5w = 0$$

$$-0.5v - 0.64039w - 0.\overline{3}x = 0$$

$$-0.\overline{3}w - 0.94595x - 0.\overline{3}y = 0$$

$$-0.\overline{3}x - 0.64039y - 0.5z = 0$$

$$-0.y + 0.39039z = 0$$

whence,

$$\mathbf{e}_4 = \begin{bmatrix} 1 & -0.78078 & 0 & 0.78078 & -1 \end{bmatrix}^T v$$

Finally,
For $\lambda_5 = 1.48237$:

$$-0.48237v - 0.5w = 0$$

$$-0.5v - 0.73237w - 0.\overline{3}x = 0$$

$$-0.\overline{3}w - 1.03793x - 0.\overline{3}y = 0$$

$$-0.\overline{3}x - 0.73237y - 0.5z = 0$$

$$-0.5y - 0.48237z = 0$$

which leads to

$$\mathbf{e}_5 = \begin{bmatrix} 1 & -0.96474 & 0.61966 & -0.96474 & 1 \end{bmatrix}^T v$$

thereby completing the calculation of all the eigenvectors of $\mathbf{A}$. It is left to the reader to verify that these eigenvectors are mutually orthogonal.

Now, in order to calculate the normal modes, we proceed as in Chap. 4, which requires matrices $\mathbf{E}$ and $\mathbf{F}$, given as

$$\mathbf{E} \equiv \begin{bmatrix} \mathbf{e}_1 & \mathbf{e}_2 & \mathbf{e}_3 & \mathbf{e}_4 & \mathbf{e}_5 \end{bmatrix}, \quad \mathbf{F} \equiv \begin{bmatrix} \mathbf{f}_1 & \mathbf{f}_2 & \mathbf{f}_3 & \mathbf{f}_4 & \mathbf{f}_5 \end{bmatrix}$$

We thus have

$$\mathbf{E} = \begin{bmatrix} 0.22942 & 0.43516 & 0.46024 & 0.55734 & 0.48533 \\ 0.45883 & 0.55735 & 0.26503 & -0.43516 & -0.46822 \\ 0.68825 & 0.0 & -0.66021 & 0.0 & 0.30074 \\ 0.45883 & -0.55735 & 0.26503 & 0.43516 & -0.46822 \\ 0.22942 & -0.43516 & 0.46024 & -0.55734 & 0.48533 \end{bmatrix}$$

Then, $\mathbf{F}$ is calculated as in Chap. 4 from

$$\mathbf{F} = \mathbf{N}^{-1}\mathbf{E}$$

where $\mathbf{N}$, it is recalled, is $\sqrt{\mathbf{M}}$, i.e.,

$$\mathbf{N}^{-1} = \frac{\sqrt{J_0}}{J_0} \begin{bmatrix} 1 & 0 & 0 & 0 & 0 \\ 0 & 0.5 & 0 & 0 & 0 \\ 0 & 0 & 0.\overline{3} & 0 & 0 \\ 0 & 0 & 0 & 0.5 & 0 \\ 0 & 0 & 0 & 0 & 1 \end{bmatrix}$$

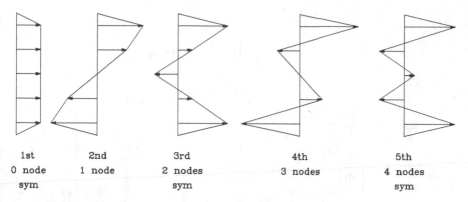

1st	2nd	3rd	4th	5th
0 node	1 node	2 nodes	3 nodes	4 nodes
sym		sym		sym

Fig. 6.1 Representation of the natural modes of vibration of an aircraft wing

Upon calculating the foregoing product, we obtain

$$
\mathbf{F} = \frac{\sqrt{J_0}}{J_0}
\begin{bmatrix}
0.22942 & 0.43516 & 0.46024 & 0.55734 & 0.48533 \\
0.22942 & 0.27868 & 0.13252 & -0.21758 & -0.23411 \\
0.22942 & 0.0 & -0.22007 & 0.0 & 0.10025 \\
0.22942 & -0.27868 & 0.13252 & 0.21758 & -0.23411 \\
0.22942 & -0.43516 & 0.46024 & -0.55734 & 0.48533
\end{bmatrix}
$$

The modes thus calculated are displayed in Fig. 6.1. From that figure, it is apparent that the first, third and fifth modes are symmetric, while the other two modes are antisymmetric. Note also that the number of *nodes*, i.e., points at which the mode diagram traverses the vertical axis, increases by one as the natural frequency increases to the next higher value. That is, the rigid mode has no node, the second mode has one node, while the third mode has two nodes, the fourth mode has three, and the fifth mode four.

Example 6.2.2 (Model for the Roll Vibrations of a Terrestrial Vehicle). An iconic model for the study of the *roll* vibrations of a terrestrial vehicle is shown in Fig. 6.2. In this model, the two identical masses with the two identical springs of stiffness k_1 represent the wheels, while the block of length $2l$ represents the body and the two identical springs of stiffness k_2 represent the suspension. Note that, under this type of motion, we can safely assume that the center of mass of the vehicle lies at midspan between the two wheels, i.e., along the centerline of the vehicle. Find the natural frequencies and the natural modes of the system.

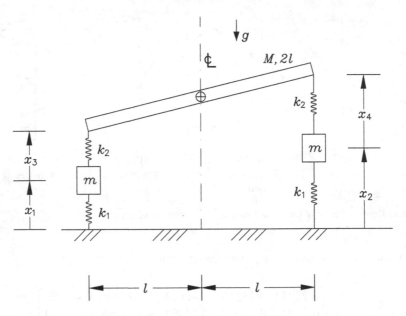

Fig. 6.2 Iconic model for the *roll* vibrations of a terrestrial vehicle

Solution: As the reader is invited to verify, the mass and stiffness matrices of the system under study take the forms

$$\mathbf{M} = \begin{bmatrix} (1/4)M\rho_+^2 + m & (1/4)M\rho_-^2 & (1/4)M\rho_+^2 & (1/4)M\rho_-^2 \\ (1/4)M\rho_-^2 & (1/4)M\rho_+^2 + m & (1/4)M\rho_-^2 & (1/4)M\rho_+^2 \\ (1/4)M\rho_+^2 & (1/4)M\rho_-^2 & (1/4)M\rho_+^2 & (1/4)M\rho_-^2 \\ (1/4)M\rho_-^2 & (1/4)M\rho_+^2 & (1/4)M\rho_-^2 & (1/4)M\rho_+^2 \end{bmatrix},$$

$$\mathbf{K} = \begin{bmatrix} k_1 & 0 & 0 & 0 \\ 0 & k_1 & 0 & 0 \\ 0 & 0 & k_2 & 0 \\ 0 & 0 & 0 & k_2 \end{bmatrix}, \quad \rho_+^2 = 1 + \rho^2, \quad \rho_-^2 = 1 - \rho^2$$

where $\rho \equiv r/l$, with r defined as the radius of gyration of the vehicle body. Now, let us assume the numerical values given below:

$$M = 16m, \quad \rho = \sqrt{2}/2, \quad k_1 = k, \quad k_2 = 9k$$

Under these conditions, the mass and stiffness matrices become

$$\mathbf{M} = m \begin{bmatrix} 7 & 2 & 6 & 2 \\ 2 & 7 & 2 & 6 \\ 6 & 2 & 6 & 2 \\ 2 & 6 & 2 & 6 \end{bmatrix}, \quad \mathbf{K} = k \begin{bmatrix} 1 & 0 & 0 & 0 \\ 0 & 1 & 0 & 0 \\ 0 & 0 & 9 & 0 \\ 0 & 0 & 0 & 9 \end{bmatrix}$$

What we observe now is that the mass matrix is full, its square root being now more elaborate to compute. A straightforward way of computing the square root of **M** is to first diagonalize it, using its eigenvectors. Once it is in diagonal form, its square root is simply the matrix of the square roots of its diagonal entries. Moreover, what will be needed is not just the square root of **M**, but the inverse of this. As explained above, this inverse is most easily computed via the inverse of **M** in diagonal form. Alternatively, the said inverse can be directly obtained using the Cayley-Hamilton Theorem, an approach that is not recommended in this case because it requires, besides the calculation of the eigenvalues of **M**, the solution of a system of four linear equations in four unknowns. This solution is straightforward, but, if done by hand, may quickly lead to arithmetic errors. The most straightforward way of calculating the said inverse is by using the sqrtm function of MATLAB, followed by the inv function of the same software package. For matrices of moderate numerical complexity, like the one at hand, we can also use software for symbolic computations. We reproduce below the steps for the calculation of the frequency matrix:

Let

$$\mathbf{M} = m\mathbf{A}, \quad \mathbf{K} = k\mathbf{B}$$

with **A** and **B** defined, obviously, as

$$\mathbf{A} = \begin{bmatrix} 7 & 2 & 6 & 2 \\ 2 & 7 & 2 & 6 \\ 6 & 2 & 6 & 2 \\ 2 & 6 & 2 & 6 \end{bmatrix}, \quad \mathbf{B} = \begin{bmatrix} 1 & 0 & 0 & 0 \\ 0 & 1 & 0 & 0 \\ 0 & 0 & 9 & 0 \\ 0 & 0 & 0 & 9 \end{bmatrix}$$

Hence, the frequency matrix-squared takes now the form

$$\mathbf{\Omega}^2 = \omega^2 \sqrt{\mathbf{A}^{-1}} \mathbf{B} \sqrt{\mathbf{A}^{-1}}$$

with ω^2 defined, in turn, as

$$\omega^2 \equiv \frac{k}{m}$$

The steps to calculate the frequency matrix-squared are summarized below:

1. Square root of **A**. This is done using computer-algebra software, which produces

$$\sqrt{\mathbf{A}} = \begin{bmatrix} 2.2132 & 0.2718 & 1.3951 & 0.2856 \\ 0.2718 & 2.2132 & 0.2856 & 1.3951 \\ 1.3951 & 0.2856 & 1.9695 & 0.3054 \\ 0.2856 & 1.3951 & 0.3054 & 1.9695 \end{bmatrix}$$

2. Square root of $\mathbf{A}^{-1}$. This is calculated as

$$\sqrt{\mathbf{A}^{-1}} \equiv (\sqrt{\mathbf{A}})^{-1} = \begin{bmatrix} 0.8182 & -0.0139 & -0.5745 & -0.0198 \\ -0.0139 & 0.8182 & -0.0198 & -0.5745 \\ -0.5745 & -0.0198 & 0.9247 & -0.0461 \\ -0.0198 & -0.5745 & -0.0461 & 0.9247 \end{bmatrix}$$

which is done also using computer-algebra software.

3. Calculation of the frequency matrix-squared. We do this via $\sqrt{\mathbf{A}^{-1}}\mathbf{B}\sqrt{\mathbf{A}^{-1}}$, namely,

$$\sqrt{\mathbf{A}^{-1}}\mathbf{B}\sqrt{\mathbf{A}^{-1}} = \begin{bmatrix} 3.6431 & 0.1816 & -5.2421 & 0.0656 \\ 0.1816 & 3.6431 & 0.0656 & -5.2421 \\ -5.2421 & 0.0656 & 8.0444 & -0.7441 \\ 0.0656 & -5.2421 & -0.7441 & 8.0444 \end{bmatrix} \equiv \mathbf{W}$$

and hence,

$$\Omega^2 = \omega^2 \mathbf{W}$$

Now, the calculation of the natural frequencies and natural modes is done via the eigenvalues and eigenvectors of matrix $\mathbf{W}$. Let the eigenvalues of this matrix be $\{\lambda_i\}_1^4$, its eigenvectors being $\{\mathbf{e}_i\}_1^4$, which are identical to those of the frequency matrix. We then have, by resorting to MATLAB's routine eig,[1]

$$\lambda_1 = 0.1021, \quad \lambda_2 = 0.1865, \quad \lambda_3 = 11.0229, \quad \lambda_4 = 12.0635$$

and hence,

$$\omega_1 = 0.3195\omega, \quad \omega_2 = 0.4319\omega, \quad \omega_3 = 3.3201\omega, \quad \omega_4 = 3.4733\omega$$

The eigenvectors are displayed below, as the columns of the 4×4 matrix $\mathbf{E}$:

$$\mathbf{E} = \begin{bmatrix} 0.5741 & 0.6018 & 0.4128 & 0.3713 \\ 0.5741 & -0.6018 & 0.4128 & -0.3713 \\ 0.4128 & 0.3713 & -0.5741 & -0.6018 \\ 0.4128 & -0.3713 & -0.5741 & 0.6018 \end{bmatrix}$$

Hence,

$$\mathbf{F} = \mathbf{N}^{-1}\mathbf{E} = \frac{\sqrt{m}}{m}\sqrt{\mathbf{A}^{-1}}\mathbf{E}$$

---

[1] Computer-algebra software can also be used here.

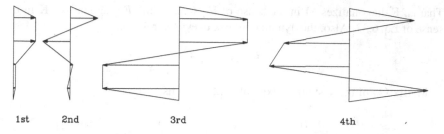

Fig. 6.3 Representation of the natural modes of a terrestrial vehicle under roll vibrations

i.e.,

$$
\mathbf{F} = \frac{\sqrt{m}}{m}
\begin{bmatrix}
0.2164 & 0.2947 & 0.6732 & 0.6428 \\
0.2164 & -0.2947 & 0.6732 & -0.6428 \\
0.0216 & 0.0266 & -0.7497 & -0.7901 \\
0.0216 & -0.0266 & -0.7497 & 0.7901
\end{bmatrix}
$$

The natural modes are displayed in Fig. 6.3, where one can notice that the second mode exhibits more modes than its third and fourth counterparts.

6.2.1 Algebraic Properties of the Normal Modes

We summarize below the properties that were discussed at the beginning of this section. By virtue of the orthogonality and the normality of the n eigenvectors of the frequency matrix, the eigenmatrix is orthogonal, i.e.,

$$
\mathbf{E}^T\mathbf{E} =
\begin{bmatrix}
\mathbf{e}_1^T\mathbf{e}_1 & \mathbf{e}_1^T\mathbf{e}_2 & \cdots & \mathbf{e}_1^T\mathbf{e}_n \\
\mathbf{e}_2^T\mathbf{e}_1 & \mathbf{e}_2^T\mathbf{e}_2 & \cdots & \mathbf{e}_2^T\mathbf{e}_n \\
\vdots & \vdots & \ddots & \vdots \\
\mathbf{e}_n^T\mathbf{e}_1 & \mathbf{e}_n^T\mathbf{e}_2 & \cdots & \mathbf{e}_n^T\mathbf{e}_n
\end{bmatrix}
=
\begin{bmatrix}
1 & 0 & \cdots & 0 \\
0 & 1 & \cdots & 0 \\
\vdots & \vdots & \ddots & \vdots \\
0 & 0 & \cdots & 1
\end{bmatrix}
= \mathbf{1}
$$

where $\mathbf{1}$ denotes the $n \times n$ identity matrix.

Furthermore, it is apparent from Eqs. 6.23a and 6.28 that

$$
\mathbf{F}^T\mathbf{M}\mathbf{F} = \mathbf{1} \tag{6.32}
$$

Moreover, from Eqs. 6.24a and 6.29,

$$
\mathbf{F}^T\mathbf{K}\mathbf{F} = \mathbf{\Omega}_d^2 \equiv
\begin{bmatrix}
\omega_1^2 & 0 & \cdots & 0 \\
0 & \omega_2^2 & \cdots & 0 \\
\vdots & \vdots & \ddots & \vdots \\
0 & 0 & \cdots & \omega_n^2
\end{bmatrix}
\tag{6.33}
$$

That is, $\mathbf{F}$ diagonalizes $\mathbf{M}$ in the sense of Eq. 6.32, while $\mathbf{F}$ diagonalizes $\mathbf{K}$ in the sense of Eq. 6.33. Also, the dynamic matrix obeys the relation

$$\mathbf{F}^{-1}\mathbf{D}\mathbf{F} = \Omega_d^2$$

Matrix Ω_d of the system of Example 6.2.1 is, thus,

$$\Omega_d = \begin{bmatrix} 0 & 0 & 0 & 0 & 0 \\ 0 & 0.59967\omega_0 & 0 & 0 & 0 \\ 0 & 0 & 0.84384\omega_0 & 0 & 0 \\ 0 & 0 & 0 & 1.17915\omega_0 & 0 \\ 0 & 0 & 0 & 0 & 1.21753\omega_0 \end{bmatrix}$$

and the dynamic matrix of the same example is

$$\mathbf{D} = \omega_0^2 \begin{bmatrix} 1 & -1 & 0 & 0 & 0 \\ -1/4 & 3/4 & -1/2 & 0 & 0 \\ 0 & -2/9 & 4/9 & -2/9 & 0 \\ 0 & 0 & -1/2 & 3/4 & -1/4 \\ 0 & 0 & 0 & -1 & 1 \end{bmatrix}$$

6.3 The Zero-input Response of Undamped n-dof Systems

In this section we derive the zero-input response of n-dof undamped systems by mimicking the corresponding response of single-dof systems. However, as we saw in Chap. 4, this approach requires that the frequency matrix, and hence, the stiffness matrix, be positive-definite. Semidefinite systems will be studied in Sect. 6.3.1 using an alternative approach. The systems we will consider here are, thus, assumed to have positive-definite mass and stiffness matrices.

Under the assumption that the stiffness matrix is positive-definite, the frequency matrix, defined in Eq. 6.3, is positive-definite as well. In this case, the response of the system at hand can be derived from the response of a single-dof system, as we did in the two-dof case. The time response of n-dof systems, then, is formally identical to that of two-dof systems, namely,

$$\mathbf{y}(t) = (\cos \Omega t)\mathbf{y}_0 + \Omega^{-1}(\sin \Omega t)\mathbf{v}_0 \tag{6.34}$$

$$\dot{\mathbf{y}}(t) = -\Omega(\sin \Omega t)\mathbf{y}_0 + (\cos \Omega t)\mathbf{v}_0, \quad t \geq 0 \tag{6.35}$$

where, now, $\mathbf{y}$, $\mathbf{y}_0$, $\dot{\mathbf{y}}$, and $\mathbf{v}_0$ are all n-dimensional vectors, while Ω is a $n \times n$ symmetric and positive-definite matrix; Ω is, hence, nonsingular as well. Moreover, $\cos \Omega t$ and $\sin \Omega t$ are the analytic functions of Ωt derived from the corresponding scalar functions $\{\cos \omega_i t\}_1^n$ and $\{\sin \omega_i t\}_1^n$, with ω_i denoting the ith eigenvalue of the frequency matrix. Note from Eq. 6.34 that, if both $\mathbf{y}_0$ and $\mathbf{v}_0$ are proportional to the eigenvector $\mathbf{e}_i$, then $\mathbf{y}(t)$ is harmonic, with frequency ω_i. Otherwise, in general, $\mathbf{y}(t)$ *is not harmonic*.

Now, since the vector of generalized coordinates is $\mathbf{x}$, rather than $\mathbf{y}$, a transformation back to the original coordinates is in order. From Eq. 6.5, we have

$$\mathbf{x}(t) = \mathbf{N}^{-1}(\cos \mathbf{\Omega} t)\mathbf{N}\mathbf{x}_0 + \mathbf{N}^{-1}\mathbf{\Omega}^{-1}(\sin \mathbf{\Omega} t)\mathbf{N}\mathbf{v}_0, \quad t \geq 0 \qquad (6.36)$$

$$\dot{\mathbf{x}}(t) = -\mathbf{N}^{-1}\mathbf{\Omega}(\sin \mathbf{\Omega} t)\mathbf{N}\mathbf{x}_0 + \mathbf{N}^{-1}(\cos \mathbf{\Omega} t)\mathbf{N}\mathbf{v}_0, \quad t \geq 0 \qquad (6.37)$$

where $\dot{\mathbf{x}}(0) \equiv \mathbf{v}_0$ and $\mathbf{w}_0 \equiv \mathbf{N}\mathbf{v}_0$. Note from Eq. 6.36 that, if both $\mathbf{x}_0$ and $\mathbf{v}_0$ are proportional to the modal vector $\mathbf{f}_i$, then both $\mathbf{N}\mathbf{x}_0$ and $\mathbf{N}\mathbf{v}_0$ are proportional to the eigenvector $\mathbf{e}_i$, say of the forms $x_0\mathbf{e}_i$ and $v_0\mathbf{e}_i$, with x_0 and v_0 defined as the proportionality factors. As a consequence, then, and by virtue of Fact 3 of Appendix A, we have that $(\cos \mathbf{\Omega} t)\mathbf{N}\mathbf{x}_0 = (x_0\cos \omega_i t)\mathbf{e}_i$ and $\mathbf{\Omega}^{-1}(\sin \mathbf{\Omega} t)\mathbf{N}\mathbf{v}_0 = (v_0/\omega_i)(\sin \omega_i t)\mathbf{e}_i$. Hence, $\mathbf{x}(t)$ is proportional to $\mathbf{N}^{-1}\mathbf{e}_i$, i.e., to $\mathbf{f}_i$. Therefore, under the given conditions, $\mathbf{x}(t)$ is harmonic, of frequency ω_i.

In computing the time response of the systems at hand, we cannot proceed as we did in Chap. 4, i.e., graphically, with the aid of the Mohr circle. The foregoing calculations will have to be done numerically. Below we outline the computational procedure: under the assumption that we have $\mathbf{\Omega}$ and its eigenvalues and eigenvectors, we can now calculate

$$\cos(\mathbf{\Omega}_d t) = \begin{bmatrix} \cos \omega_1 t & 0 & \cdots & 0 \\ 0 & \cos \omega_2 t & \cdots & 0 \\ \vdots & \vdots & \ddots & \vdots \\ 0 & 0 & \cdots & \cos \omega_n t \end{bmatrix}$$

$$\sin(\mathbf{\Omega}_d t) = \begin{bmatrix} \sin \omega_1 t & 0 & \cdots & 0 \\ 0 & \sin \omega_2 t & \cdots & 0 \\ \vdots & \vdots & \ddots & \vdots \\ 0 & 0 & \cdots & \sin \omega_n t \end{bmatrix}$$

$$\mathbf{\Omega}_d^{-1}\sin(\mathbf{\Omega}_d t) = \begin{bmatrix} \sin \omega_1 t/\omega_1 & 0 & \cdots & 0 \\ 0 & \sin \omega_2 t/\omega_2 & \cdots & 0 \\ \vdots & \vdots & \ddots & \vdots \\ 0 & 0 & \cdots & \sin \omega_n t/\omega_n \end{bmatrix}$$

Now, in order to calculate all matrices involved in the time response given in Eq. 6.36, all we need is a suitable transformation, namely,

$$\cos(\mathbf{\Omega} t) = \mathbf{E}\cos(\mathbf{\Omega}_d t)\mathbf{E}^T \qquad (6.38a)$$

$$\sin(\boldsymbol{\Omega} t) = \mathbf{E}\sin(\boldsymbol{\Omega}_d t)\mathbf{E}^T \tag{6.38b}$$

$$\boldsymbol{\Omega}^{-1}\sin(\boldsymbol{\Omega} t) = \mathbf{E}[\boldsymbol{\Omega}_d^{-1}\sin(\boldsymbol{\Omega}_d t)]\mathbf{E}^T \tag{6.38c}$$

Example 6.3.1 (Time Response of a Vehicle-suspension System). For the system depicted in Fig. 6.2, find the time response upon transmitting to the left wheels an initial speed $\dot{x}_1(0) = v_0$, while leaving all other initial conditions equal to zero.

Solution: We computed already the eigenvalues and eigenvectors of the frequency matrix of this system in Example 6.2.2. We thus have

$$\boldsymbol{\Omega}_d = \begin{bmatrix} 0.3195\omega & 0 & 0 & 0 \\ 0 & 0.4319\omega & 0 & 0 \\ 0 & 0 & 3.3201\omega & 0 \\ 0 & 0 & 0 & 3.4733\omega \end{bmatrix}$$

Now, since $\mathbf{x}(0) = \mathbf{0}$, we will not need $\cos(\boldsymbol{\Omega}_d t)$. All we need is $\boldsymbol{\Omega}_d^{-1}\sin(\boldsymbol{\Omega}_d t)$:

$$\boldsymbol{\Omega}_d^{-1}\sin(\boldsymbol{\Omega}_d t) = \frac{1}{\omega}\mathrm{diag}(s_1, s_2, s_3, s_4)$$

with

$$s_1 = 3.1299\sin(0.3195\omega t), \quad s_2 = 2.3154\sin(0.4319\omega t)$$

$$s_3 = 0.3012\sin(3.3201\omega t), \quad s_4 = 0.2879\sin(3.4733\omega t)$$

Now, we have

$$\boldsymbol{\Omega}^{-1}\sin(\boldsymbol{\Omega} t) = \mathbf{E}[\boldsymbol{\Omega}_d^{-1}\sin(\boldsymbol{\Omega}_d t)]\mathbf{E}^T$$

Hence,

$$\mathbf{y}(t) = \frac{v_0\sqrt{m}}{\omega} \begin{bmatrix} 3.8101\sin(.3195\omega t) + 2.2019\sin(.4319\omega t) \\ +0.0076\sin(3.3201\omega t) + 0.0057\sin(3.4733\omega t) \\ 3.8101\sin(.3195\omega t) - 2.2019\sin(.4319\omega t) \\ +0.0076\sin(3.3201\omega t) - 0.0057\sin(3.4733\omega t) \\ 2.7396\sin(.3195\omega t) + 1.3586\sin(.4319\omega t) \\ -0.0105\sin(3.3201\omega t) - 0.0092\sin(3.4733\omega t) \\ 2.7396\sin(.3195\omega t) - 1.3586\sin(.4319\omega t) \\ -0.0105\sin(3.3201\omega t) + 0.0092\sin(3.4733\omega t) \end{bmatrix}$$

and, finally,

$$\mathbf{x}(t) = \mathbf{N}^{-1}\mathbf{y}(t) = \frac{v_0}{\omega} \begin{bmatrix} 1.4363\sin(.3195\omega t) + 1.0786\sin(.4319\omega t) \\ +0.0124\sin(3.3201\omega t) + 0.0098\sin(3.4733\omega t) \\ 1.4363\sin(.3195\omega t) - 1.0786\sin(.4319\omega t) \\ +0.0124\sin(3.3201\omega t) - 0.0098\sin(3.4733\omega t) \\ 0.1427\sin(.3195\omega t) + 0.0975\sin(.4319\omega t) \\ -0.0138\sin(3.3201\omega t) - 0.0121\sin(3.4733\omega t) \\ 0.1427\sin(.3195\omega t) - 0.0975\sin(.4319\omega t) \\ -0.0138\sin(3.3201\omega t) + 0.0121\sin(3.4733\omega t) \end{bmatrix}$$

6.3.1   The Calculation of the Zero-input Response of n-dof Systems Using the Classical Modal Method

Alternatively, the time response of n-dof systems can be obtained by resorting to *modal coordinates*, also known as *normal coordinates*. These are those in which the system equations are *decoupled*. The underlying ideas parallel those of Sect. 5.3.

We begin by recalling the system mathematical model in normal form, Eq. 6.2. Upon multiplying both sides of that equation by $\mathbf{E}^T$, we have

$$\mathbf{E}^T\ddot{\mathbf{y}} + \mathbf{E}^T \mathbf{\Omega}^2\mathbf{y} = \mathbf{0}, \quad \mathbf{y}(0) = \mathbf{y}_0, \quad \dot{\mathbf{y}}(0) = \mathbf{w}_0, \quad t \geq 0$$

Now we introduce the *coordinate transformation*

$$\boldsymbol{\xi} \equiv \begin{bmatrix} \xi_1 \\ \xi_2 \\ \vdots \\ \xi_n \end{bmatrix} = \mathbf{E}^T\mathbf{y}, \quad \mathbf{y} = \mathbf{E}\boldsymbol{\xi} \tag{6.39a}$$

$$\ddot{\boldsymbol{\xi}} = \mathbf{E}^T\ddot{\mathbf{y}}, \quad \ddot{\mathbf{y}} = \mathbf{E}\ddot{\boldsymbol{\xi}} \tag{6.39b}$$

the above mathematical model thus becoming

$$\ddot{\boldsymbol{\xi}} + \mathbf{E}^T \mathbf{\Omega}^2\mathbf{E}\boldsymbol{\xi} = \mathbf{0}, \quad \boldsymbol{\xi}(0) = \mathbf{a}, \quad \dot{\boldsymbol{\xi}}(0) = \mathbf{b}, \quad t \geq 0 \tag{6.40}$$

In the above equation, we can readily identify the matrix coefficient of $\boldsymbol{\xi}$ as $\mathbf{\Omega}_d^2$, and hence, the model takes the form

$$\ddot{\boldsymbol{\xi}} + \mathbf{\Omega}_d^2\boldsymbol{\xi} = \mathbf{0}, \quad \boldsymbol{\xi}(0) = \mathbf{E}^T\mathbf{y}_0 = \mathbf{a}, \quad \dot{\boldsymbol{\xi}}(0) = \mathbf{E}^T\mathbf{w}_0 = \mathbf{b}, \quad t \geq 0 \tag{6.41a}$$

which can be rewritten in component form as

$$\ddot{\xi}_1 + \omega_1^2 \xi_1 = 0 \tag{6.41b}$$

$$\ddot{\xi}_2 + \omega_2^2 \xi_2 = 0 \tag{6.41c}$$

$$\vdots$$

$$\ddot{\xi}_n + \omega_n^2 \xi_n = 0 \tag{6.41d}$$

with the initial conditions

$$\xi_i(0) = a_i, \quad \dot{\xi}_i(0) = b_i, \quad i = 1, \dots, n \tag{6.41e}$$

That is, the normal coordinates allow us to express the mathematical model of a n-dof system as a set of n uncoupled single-dof systems. Hence, obtaining the response of the system at hand has been reduced to obtaining that of n single-dof systems, which is a straightforward task. Note that the time response of the ith single-dof system of the above set can now be written in the form

$$\xi_i(t) = (\cos \omega_i t)a_i + \frac{1}{\omega_i}(\sin \omega_i t)b_i, \quad \omega_i \neq 0, \quad i = 1, \dots, n \tag{6.42}$$

which is valid as long as ω_i is not zero. In the special case in which $\omega_i = 0$, the treatment follows that discussed in Sect. 5.3. For concreteness, let us assume that $\omega_1 = 0$, the remaining natural frequencies being all greater than zero. The model for the first mode then takes the form of a unit-mass particle under free motion, i.e., under zero external force:

$$\ddot{\xi}_1 = 0, \quad \xi_1(0) = a_1, \quad \dot{\xi}_1(0) = b_1 \tag{6.43}$$

The time response of this system is simply that of uniform motion, i.e.,

$$\xi_1(t) = a_1 + b_1 t \tag{6.44}$$

If the system has more than one rigid mode, the other rigid modes are treated likewise. We now have the time response for all modes, including the rigid ones. Therefore, all we need is this time response in the given coordinates, $\mathbf{x}(t)$. To go back to these coordinates, we first return to $\mathbf{y}(t)$, which is readily done by resorting to Eq. 6.39a, i.e.,

$$\mathbf{y}(t) = \mathbf{E}\boldsymbol{\xi}(t)$$

Finally, the time response in the original coordinates is expressed as

$$\mathbf{x}(t) = \mathbf{N}^{-1}\mathbf{y}(t) \tag{6.45}$$

thereby completing the time response of the system at hand.

Example 6.3.2 (Time Response of the Wing Model to a Gust-wind Disturbance).
For the wing model of Fig. 4.7, that was analyzed in Example 6.2.1, we want to
know how the system will respond to a gust-wind perturbation transmitting a sudden
pitch rate $\dot{\theta}_3(0) = p$ to the aircraft fuselage. To this end, find the time response of
the system to the initial conditions

$$\theta(0) = \mathbf{0}, \quad \dot{\theta}(0) = \begin{bmatrix} 0 \\ 0 \\ p \\ 0 \\ 0 \end{bmatrix}$$

Solution: We adopt the numbering of frequencies in ascending order, as introduced
in Example 6.2.1. The first item to determine is the initial conditions for the normal
coordinates. We begin by calculating those for $\mathbf{y}(t)$, namely,

$$\mathbf{y}(0) = \mathbf{N}\theta_0, \quad \dot{\mathbf{y}}(0) = \mathbf{N}\dot{\theta}_0$$

Therefore,

$$\mathbf{y}(0) = \mathbf{0}, \quad \dot{\mathbf{y}}(0) = 3\sqrt{J_0}\, p \begin{bmatrix} 0 \\ 0 \\ 1 \\ 0 \\ 0 \end{bmatrix}$$

Thus, the initial conditions for the normal coordinates are now

$$\xi(0) = \mathbf{0}, \quad \dot{\xi}(0) = \mathbf{E}^T \dot{\mathbf{y}}(0) = 3\sqrt{J_0}\, p \begin{bmatrix} 0.68825 \\ 0.0 \\ -0.66021 \\ 0.0 \\ 0.30074 \end{bmatrix}$$

whence it is apparent that only the symmetric modes are excited, which is natural,
for the initial conditions are symmetric.
   We thus have, in normal coordinates,

$$\ddot{\xi}_1 = 0, \quad \xi_1(0) = 0, \quad \dot{\xi}_1(0) = 3\sqrt{J_0}\,0.68825 p$$

Hence, the time response of the rigid mode is

$$\xi_1(t) = 3\sqrt{J_0}\,0.68825\, pt$$

Likewise, the time response of the remaining modes, which are all flexible, is

$$\xi_2(t) = 0$$

$$\xi_3(t) = \frac{-0.66021}{0.84384\,\omega_0}(3\sqrt{J_0}\,p)\sin(0.84384\,\omega_0 t)$$

$$= -0.78239\left(3\sqrt{J_0}\,r\right)\sin(0.84384\,\omega_0 t)$$

$$\xi_4(t) = 0$$

$$\xi_5(t) = \frac{0.30074}{1.21753\,\omega_0}(3\sqrt{J_0}\,p)\sin(1.21753\,\omega_0 t)$$

$$= 0.24701\left(3\sqrt{J_0}\,r\right)\sin(1.21753\,\omega_0 t)$$

where r is the ratio $r \equiv p/\omega_0$. Therefore,

$$\mathbf{y}(t)=3\sqrt{J_0}\begin{bmatrix}0.15790pt - 0.36009r\sin(0.84384\,\omega_0 t) + 0.11988r\sin(1.21753\,\omega_0 t)\\0.31579pt - 0.20736r\sin(0.84384\,\omega_0 t) - 0.11566r\sin(1.21753\,\omega_0 t)\\0.47369pt + 0.51654r\sin(0.84384\,\omega_0 t) + 0.07429r\sin(1.21753\,\omega_0 t)\\0.31579pt - 0.20736r\sin(0.84384\,\omega_0 t) - 0.11566r\sin(1.21753\,\omega_0 t)\\0.15790pt - 0.36009r\sin(0.84384\,\omega_0 t) + 0.11988r\sin(1.21753\,\omega_0 t)\end{bmatrix}$$

Finally, in the original coordinates, to four decimals, we have

$$\boldsymbol{\theta}(t)=\begin{bmatrix}3[0.1579pt - 0.3601r\sin(0.8438\,\omega_0 t) + 0.1199r\sin(1.2175\,\omega_0 t)]\\(3/2)[0.3158pt - 0.2074r\sin(0.8438\,\omega_0 t) - 0.1157r\sin(1.2175\,\omega_0 t)]\\0.4737pt + 0.5165r\sin(0.8438\,\omega_0 t) + 0.0743r\sin(1.2175\,\omega_0 t)\\(3/2)[0.3158pt - 0.2074r\sin(0.8438\,\omega_0 t) - 0.1157r\sin(1.2175\,\omega_0 t)]\\3[0.1579pt - 0.3601r\sin(0.8438\,\omega_0 t) + 0.1199r\sin(1.2175\,\omega_0 t)]\end{bmatrix}$$

Therefore, the time response to the given initial conditions is symmetric, which is to be expected, because of the symmetry of (a) the initial conditions and (b) the model. The presence of the linear terms pt in the above response, that arise by virtue of the rigid mode, is to be highlighted.

6.4 The Zero-state Response of n-dof Systems

The time response of a n-dof system under a nonzero forcing term and zero initial conditions is the subject of this section. We thus have

$$\mathbf{M}\ddot{\mathbf{x}} + \mathbf{K}\mathbf{x} = \boldsymbol{\phi}(t), \quad \mathbf{x}(0) = \mathbf{0}, \quad \dot{\mathbf{x}}(0) = \mathbf{0}, \quad t \geq 0 \qquad (6.46)$$

or, in normal form,

$$\ddot{\mathbf{y}} + \mathbf{\Omega}^2 \mathbf{y} = \mathbf{g}(t), \quad \mathbf{y}(0) = \mathbf{0}, \quad \dot{\mathbf{y}}(0) = \mathbf{0} \tag{6.47}$$

where $\mathbf{y}$ is defined as in Eq. 6.5, while $\mathbf{g}(t)$ is given as

$$\mathbf{g}(t) \equiv \mathbf{N}^{-1}\boldsymbol{\phi}(t) \tag{6.48}$$

As in Sect. 6.3, we distinguish between systems with a positive-semidefinite and those with a positive-definite frequency matrix. The former will be discussed in Sect. 6.4.1.

The time response of the system governed by Eq. 6.47, when its frequency matrix is positive-definite, can be derived by mimicking that of undamped scalar systems. This is done by recalling the time response of a single-dof mass-spring system to an excitation $g(t)$, under zero initial conditions, which is reproduced below for quick reference:

$$y(t) = \frac{1}{\omega_n} \int_0^t \sin \omega_n (t - \tau) g(\tau) d\tau \tag{6.49}$$

For a n-dof system, the zero-state response is derived by simply replacing $y(t)$ and $g(t)$ by their vector counterparts and the natural frequency ω_n by the frequency matrix $\mathbf{\Omega}$ in the response given in Eq. 6.49, namely,

$$\mathbf{y}(t) = \mathbf{\Omega}^{-1} \int_0^t \sin \mathbf{\Omega}(t - \tau) \mathbf{g}(\tau) d\tau \tag{6.50}$$

where $\mathbf{\Omega}^{-1}$ exists because $\mathbf{\Omega}$ is positive-definite, and hence, nonsingular. Such as in the two-dof case, the above derivation of the time response is rather informal, but its validity can be readily verified as in Chap. 4, namely, by substituting the foregoing expression in Eq. 6.50, and noticing that this expression satisfies both the ODE and the initial conditions.

Now, going back to the generalized coordinate vector $\mathbf{x}$ and the generalized speed vector $\dot{\mathbf{x}}$, we have

$$\mathbf{x}(t) = \mathbf{N}^{-1} \mathbf{\Omega}^{-1} \int_0^t \sin \mathbf{\Omega}(t - \tau) \mathbf{N}^{-1} \mathbf{f}(\tau) d\tau \tag{6.51}$$

$$\dot{\mathbf{x}}(t) = \mathbf{N}^{-1} \int_0^t [\cos \mathbf{\Omega}(t - \tau)] \mathbf{N}^{-1} \mathbf{f}(\tau) d\tau \tag{6.52}$$

6.4.1   The Calculation of the Zero-state Response of n-dof Systems Using the Classical Modal Method

The modal method is advantageous in that it does not require that the system be definite. In order to apply it, we resort to modal, a.k.a. normal, coordinates, which we do by multiplying both sides of Eq. 6.47 by $\mathbf{E}^T$, thereby obtaining

$$\mathbf{E}^T \ddot{\mathbf{y}} + \mathbf{E}^T \Omega^2 \mathbf{y} = \mathbf{E}^T \mathbf{g}(t), \quad \mathbf{y}(0) = \mathbf{0}, \quad \dot{\mathbf{y}}(0) = \mathbf{0}$$

Again, we introduce the change of variable of Eqs. 6.39a and b, with the additional definition

$$\gamma(t) \equiv \mathbf{E}^T \mathbf{g}(t) \tag{6.53}$$

thereby obtaining

$$\ddot{\xi} + \Omega_d^2 \xi = \gamma(t), \quad \xi(0) = \mathbf{0}, \quad \dot{\xi}(0) = \mathbf{0} \tag{6.54a}$$

or, in component form,

$$\ddot{\xi}_i + \omega_i^2 \xi_i = \gamma_i(t), \quad \xi_i(0) = 0, \quad \dot{\xi}_i(0) = 0 \tag{6.54b}$$

where $\gamma_i(t)$ is the ith component of vector $\gamma(t)$. Thus, we derive, for the ith normal mode

$$\xi_i(t) = \frac{1}{\omega_i} \int_0^t [\sin \omega_i(t - \tau)] \gamma_i(\tau) d\tau, \quad \omega_i \neq 0, \quad i = 1, \ldots, n \tag{6.55}$$

Now, if any of the natural frequencies is zero, we cannot apply, at least directly, the foregoing expression. For concreteness, let us assume that $\omega_1 = 0$, all other natural frequencies being greater than zero. Thus, for $i = 1$, Eq. 6.54b takes the form

$$\ddot{\xi}_1 = \gamma_1(t), \quad \xi_1(0) = 0, \quad \dot{\xi}_1(0) = 0 \tag{6.56}$$

The time response of the foregoing system being derived by simple quadrature, i.e.,

$$\dot{\xi}_1(t) = \int_0^t \gamma_1(\tau) d\tau \tag{6.57}$$

and

$$\xi_1(t) = \int_0^t \left(\int_0^\theta \gamma_1(\tau) d\tau \right) d\theta \tag{6.58}$$

Once we have all time responses in normal coordinates, we can go back to the original coordinates, as in Sect. 6.3. We thus have

$$\mathbf{x}(t) = \mathbf{N}^{-1} \mathbf{E} \xi(t) \tag{6.59}$$

6.5 The Total Response of n-dof Undamped Systems

The total time response of the system under study is simply the sum of the zero-input and the zero-state responses given above. Let $\mathbf{x}_I(t)$ and $\mathbf{x}_S(t)$ denote the zero-input and the zero-state responses of the systems under study. We thus have

$$\mathbf{x}(t) = \mathbf{x}_I(t) + \mathbf{x}_S(t) \tag{6.60}$$

That is, for definite systems,

$$\mathbf{x}(t) = \mathbf{N}^{-1}(\cos \boldsymbol{\Omega} t)\mathbf{N}\mathbf{x}_0 + \mathbf{N}^{-1}\boldsymbol{\Omega}^{-1}(\sin \boldsymbol{\Omega} t)\mathbf{N}\mathbf{c}$$
$$+ \mathbf{N}^{-1}\boldsymbol{\Omega}^{-1} \int_0^t [\sin \boldsymbol{\Omega}(t - \tau)]\mathbf{N}^{-1}\boldsymbol{\phi}(\tau)d\tau, \quad t \geq 0 \tag{6.61a}$$

$$\dot{\mathbf{x}}(t) = -\mathbf{N}^{-1}\boldsymbol{\Omega}(\sin \boldsymbol{\Omega} t)\mathbf{N}\mathbf{x}_0 + \mathbf{N}^{-1}(\cos \boldsymbol{\Omega} t)\mathbf{N}\mathbf{c}$$
$$+ \mathbf{N}^{-1} \int_0^t [\cos \boldsymbol{\Omega}(t - \tau)]\mathbf{N}^{-1}\boldsymbol{\phi}(\tau)d\tau, \quad t \geq 0 \tag{6.61b}$$

The total response in modal coordinates, then, takes the form

$$\xi_i(t) = (\cos \omega_i t)a_i + \frac{1}{\omega_i}(\sin \omega_i t)b_i$$
$$+ \frac{1}{\omega_i} \int_0^t [\sin \omega_i(t - \tau)]\gamma_i(\tau)d\tau, \quad \omega_i \neq 0, \quad i = 1,\ldots,n \tag{6.62a}$$

$$\dot{\xi}_i(t) = -\omega_i(\sin \omega_i t)a_i + (\cos \omega_i t)b_i$$
$$+ \int_0^t [\cos \omega_i(t - \tau)]\gamma_i(\tau)d\tau, \quad \omega_i \neq 0, \quad i = 1,\ldots,n \tag{6.62b}$$

where the initial conditions of the Eq. 6.40 have been recalled. If, say $\omega_1 = 0$, then the total response of this mode takes the form

$$\xi_1(t) = a_1 + b_1 t + \int_0^t \left(\int_0^\theta \gamma_1(\tau)d\tau \right) d\theta \tag{6.63a}$$

$$\dot{\xi}_1(t) = b_1 + \int_0^t \gamma_1(\tau)d\tau \tag{6.63b}$$

6.6 Analysis of n-dof Damped Systems

The analysis of n-dof damped systems deserves special attention, as one formula mimicking the single-dof case, which is possible for undamped systems, is not possible here. We thus take the same approach as in Sect. 5.7.1. To this end, we start by converting the system of n second-order ODEs of Eq. 4.22 into a system of $2n$ first-order ODEs. We thus express the foregoing equations in normal form, by recalling the usual transformations:

$$\Omega^2 = \sqrt{M^{-1}}K\sqrt{M^{-1}}, \quad \Delta = \sqrt{M^{-1}}C\sqrt{M^{-1}}, \quad y = \sqrt{M}x, \quad g(t) = \sqrt{M^{-1}}\phi(t) \tag{6.64}$$

Further, we define

$$\dot{y} = w \tag{6.65a}$$

$$\dot{w} = -\Omega^2 y - \Delta w + g(t) \tag{6.65b}$$

with the initial conditions $y(0) = y_0$ and $v(0) = v_0$. We can now write Eqs. 6.65a and b in *state-variable form* as a $2n$-dimensional system of linear ODEs, namely,

$$\dot{\eta} = A\eta + Bg(t), \quad \eta(0) = \eta_0 \tag{6.66a}$$

where

$$A \equiv \begin{bmatrix} O & 1 \\ -\Omega^2 & -\Delta \end{bmatrix}, \quad B \equiv \begin{bmatrix} O \\ 1 \end{bmatrix}, \quad \eta \equiv \begin{bmatrix} y \\ w \end{bmatrix} \tag{6.66b}$$

In the above definitions, A and B are $2n \times 2n$ and $2n \times n$ matrices, respectively, while η is a $2n$-dimensional vector. Furthermore, O and 1 are the $2n \times 2n$ zero and identity matrices. Apparently, A is not symmetric, and hence, its eigenvalues and eigenvectors are bound to be complex.

The total response of damped two-dof systems in state-variable form was obtained in Eq. 5.137, with matrix A of 4×4, matrix B of 4×2 and vector z of dimension four.[2] The total response of the damped n-dof system at hand takes a form similar to that of Eq. 5.137, if with the foregoing differences, namely,

$$\eta(t) = e^{At}\eta_0 + \int_0^t e^{A(t-\tau)}Bg(\tau)\,d\tau \tag{6.67}$$

which gives the time response of interest in the transformed variables $y = \sqrt{M}x$ and $\dot{y} = w = \sqrt{M}\dot{x}$. In order to obtain the time response in the original coordinates, via the state vector $z = [x^t, \dot{x}^T]^T$, a change of coordinates is in order, which is left as an exercise.

Closed-form expressions of the exponential and the integral of Eq. 6.67 in terms of matrices Ω and Δ will not be pursued, as these expressions are elusive. However, the time response of damped systems will be obtained in Chap. 7 in terms of the impulsive response.

---

[2]Notice that $z = [x^T, \dot{x}^T]^T$ in Eq. 5.137, while in Eqs. 6.66a and b, $\eta = [y^T, w^T]^T$.

6.7 Exercises

6.1. A certain mechanical system has the mass and stiffness matrices shown below

$$\mathbf{M} = m\mathbf{1}, \qquad \mathbf{K} = k \begin{bmatrix} 1 & 1 & 1 & 1 \\ 1 & 2 & 3 & 4 \\ 1 & 3 & 6 & 10 \\ 1 & 4 & 10 & 20 \end{bmatrix}$$

where $\mathbf{1}$ is the 4×4 identity matrix. Find the natural frequencies and the normal modes of the system.

6.2. For the undamped model of the subway car suspension system of Fig. 4.9,

(a) Find the natural frequencies and the modal vectors.
(b) Suppose that the subway is travelling at a constant speed v when it encounters a bump of height $B = 5\,\text{mm}$ and wavelength $\lambda = \pi$ m at time $t = 0$. Write a computer program that will determine the responses of the generalized coordinates $x_1(t)$, $x_2(t)$, and $x_3(t)$ for $0 \le t \le 10$ s under the velocities (1) $v = 20.6\,\text{m/s} = 74.16\,\text{km/h}$, and (2) $v = 23.4\,\text{m/s} = 84.2$ km/h.

Make sure to include a well-documented listing of the source code of your program, as well as plots of the time responses.

6.3. Verify that the eigenmodes of the system of Example 6.2.1 satisfy Eq. 6.30, with Ω_d defined in Eq. 6.9.

6.4. Repeat Exercise 6.3 for the system of Example 6.2.2.

6.5. Plot the time response $(\omega/v_0)\mathbf{x}(t)$ of Example 6.3.1 for one (normalized) longest natural period[3] $T_1 = 1/0.3195$ (dimensionless).

6.6. Repeat Exercise 6.5 for the time response $\theta(t)$ of Example 6.3.2.

6.7. Show that the time response of n-dof undamped systems of Eqs. 6.62a and b are equally valid for semi-definite systems. *Hint: assume, e.g., that $\omega_1 = 0$, then recall that $\lim_{\omega_1 t \to 0}(\omega_1 t / \sin(\omega_1 t)) = 1$.*

6.8. Once the time response (6.67) has been obtained, a simple change of variable can be applied to obtain the response in terms of the original state vector $\mathbf{z} = [\mathbf{x}^T, \dot{\mathbf{x}}^T]^T$, as η and $\mathbf{z}$ are related by

$$\eta(t) = \mathbf{Z}\mathbf{z}(t), \qquad \mathbf{Z} = \begin{bmatrix} \mathbf{N} & \mathbf{O} \\ \mathbf{O} & \mathbf{N} \end{bmatrix}$$

---

[3]The natural periods of a linear mechanical system are the reciprocals of its natural frequencies.

with $\mathbf{N} = \sqrt{\mathbf{M}}$ and $\mathbf{O}$ denoting the $n \times n$ zero matrix. Alternatively, $\mathbf{z}(t)$ can be obtained from the original state equations if the model (5.116) is cast in state-variable form:

$$\dot{\mathbf{z}} = \mathbf{A}^*\mathbf{z} + \mathbf{b}^*\phi(t)$$

Find expressions for $\mathbf{A}^*$ and $\mathbf{B}^*$ in terms of $\mathbf{A}$ and $\mathbf{B}$, respectively. Then, reconcile the response obtained from the above model with that obtained by means of a change of variable in Eq. 6.67. *Hint: Recall Fact 5, which allows you to write*

$$\mathbf{Z}^{-1}e^{\mathbf{A}}\mathbf{Z} = e^{\mathbf{Z}^{-1}\mathbf{A}\mathbf{Z}}$$

Chapter 7
Simulation of n-dof Systems

Computers are good at following instructions,
but not at reading your mind.

Knuth, D.E., 1984, *The TEXbook*, Addison-Wesley, Boston, MA.

7.1 Introduction

The principles introduced in Chap. 6 allow the determination of the time response of n-dof systems when n is either small enough or the system possesses symmetries that render its time response analysis handleable in closed form. More general n-dof systems call for a numerical procedure. This is done here upon extension of the techniques introduced in Chap. 3. The aim of this chapter is thus to derive simulation schemes for the total time response of n-dof systems, for an arbitrary integer n. The principles laid down in Chap. 3 will be applied.

Given that the emphasis is on the fundamentals, at an intermediate level, specialized numerical integration schemes available in the literature are left aside. Instead, algorithms are developed that are robust to roundoff and truncation errors—the latter arise when approximating integrals by sums, and derivatives by finite differences, for example—as they preserve the energy of undamped systems. The algorithms are based on the material introduced in Chaps. 3–6. In engineering practice, continuous structures, e.g., aircraft fuselages, are discretized by a finite number N of linearly elastic elements from which the $n \times n$ mass and stiffness matrices are derived, where $n > N$, and n depends on the type of element used. In this context, these matrices are used to conduct the *modal analysis* of the structure, in the absence of damping.

The simulation algorithms proposed here are based on the zero-order hold, introduced in Chap. 3, and, for undamped systems, on a closed-form time response, as derived in Chap. 5 for two-dof systems of this kind—i.e., undamped. It is shown that, for a n-dof undamped system, the $2n \times 2n$ system matrix, mapping a state at

J. Angeles, *Dynamic Response of Linear Mechanical Systems: Modeling, Analysis and Simulation*, Mechanical Engineering Series, DOI 10.1007/978-1-4419-1027-1_7, © Springer Science+Business Media, LLC 2011

instant t_k into a state at instant t_{k+1}, is proper orthogonal. Hence, the system matrix in question represents a *rotation* in the $2n$-dimensional space of state variables. The outcome is that the state-variable vector, seen as a vector in $2n$-dimensional space, only rotates in this space during simulation, without changing its magnitude—i.e., its *norm*, in a more general sense. The magnitude of the state-variable vector, in fact, equals the total energy of the system, and hence, energy is inherently preserved by the simulation algorithm, as made apparent in Sect. 7.3.1.

Regarding damped systems, these were handled in Chap. 3 using the closed-form expression for their general time response, under non-zero initial conditions and non-zero input. A crucial step in this analysis is the computation of the zero-state response, which calls for the symbolic computation of the integral of the exponential of the system matrix $\mathbf{A}$, as occurring in Eq. 3.34. The integral is displayed in Eq. 3.37 in terms of $e^{\mathbf{A}h}$, labeled $\mathbf{F}$, which is displayed in turn in Eq. 3.40. For n-dof damped systems, matrix $\mathbf{A}$ is of $2n \times 2n$, a symbolic expression for its exponential being quite challenging. For this reason, a numerical evaluation is usually pursued, which is available in scientific software.

Two simulation algorithms for damped systems are developed in this chapter: one is an extension of the algorithm introduced in Chap. 3 for single-dof damped systems, without attempting to derive the matrix exponential and its time integral in terms of the mass, damping and stiffness matrices. The second algorithm does this based on the Laplace transform, outlined in Appendix B, and the impulse response of the system.

7.2   Undamped Systems

The model of a n-dof undamped linear mechanical system was formulated in Sect. 4.5 as

$$\mathbf{M}\ddot{\mathbf{x}} + \mathbf{K}\mathbf{x} = \boldsymbol{\phi}(t), \quad \mathbf{x}(0) = \mathbf{x}_0, \quad \dot{\mathbf{x}}(0) = \mathbf{v}_0, \quad t \geq 0 \tag{7.1}$$

where $\mathbf{M}$ and $\mathbf{K}$ are symmetric $n \times n$ matrices, the former being, additionally, positive-definite, while the latter is at least positive-semidefinite. A semidefinite system entails rigid modes, which can be extracted from the original system by suitably reducing the number of generalized coordinates $\mathbf{x}$; this renders the reduced form of both $\mathbf{M}$ and $\mathbf{K}$ positive-definite. The issue of rigid-mode extraction being rather special, will not be included in this book. It is thus assumed that the two foregoing matrices are positive-definite. Moreover, the vector of generalized coordinates $\mathbf{x}$ is n-dimensional, and so is $\boldsymbol{\phi}(t)$, which represents an array of n input generalized forces.

The foregoing model is now cast in what was introduced in Chap. 5 as the *normal form*, namely,

$$\ddot{\mathbf{y}} + \boldsymbol{\Omega}^2\mathbf{y} = \boldsymbol{\phi}(t), \quad \mathbf{y}(0) = \mathbf{y}_0, \quad \dot{\mathbf{y}}(0) = \mathbf{s}_0, \quad t \geq 0 \tag{7.2}$$

with $\mathbf{\Omega}$, the frequency matrix, defined in Eq. 6.3 as the *positive-definite* square root[1] of

$$\mathbf{\Omega}^2 \equiv \mathbf{N}^{-1}\mathbf{K}\mathbf{N}^{-1} \qquad (7.3)$$

and $\mathbf{N}$ as the *positive-definite* square root of $\mathbf{M}$, i.e.,

$$\mathbf{N} \equiv \sqrt{\mathbf{M}} \qquad (7.4)$$

while $\mathbf{y}$ and $\phi(t)$ are given by the transformations below:

$$\mathbf{y} = \mathbf{N}\mathbf{x}, \quad \mathbf{s} = \dot{\mathbf{y}} = \mathbf{N}\dot{\mathbf{x}}, \quad \mathbf{g}(t) = \mathbf{N}^{-1}\phi(t) \qquad (7.5)$$

7.3 The Discrete-Time Response of Undamped Systems

The simulation algorithm relies on the concept of *discrete-time (DT) system*, introduced in Chap. 3, as derived from its given *continuous-time (CT)* counterpart, Eq. 6.1, or from its normal form, Eq. 6.2, for that matter.[2] The continuous time response of the former was derived in Eq. 6.61a and b, as reproduced below for quick reference:

$$\mathbf{x}(t) = \mathbf{N}^{-1}(\cos \mathbf{\Omega} t)\mathbf{N}\mathbf{x}_0 + \mathbf{N}^{-1}\mathbf{\Omega}^{-1}(\sin \mathbf{\Omega} t)\mathbf{N}\mathbf{v}_0$$
$$| \mathbf{N}^{-1}\mathbf{\Omega}^{-1}\int_0^t \sin \mathbf{\Omega}(t - \tau)\mathbf{N}^{-1}\phi(\tau)d\tau, \quad t \geq 0 \qquad (7.6)$$

$$\dot{\mathbf{x}}(t) = -\mathbf{N}^{-1}\mathbf{\Omega}(\sin \mathbf{\Omega} t)\mathbf{N}\mathbf{x}_0 + \mathbf{N}^{-1}(\cos \mathbf{\Omega} t)\mathbf{N}\mathbf{v}_0$$
$$+ \mathbf{N}^{-1}\int_0^t [\cos \mathbf{\Omega}(t - \tau)]\mathbf{N}^{-1}\phi(\tau)d\tau, \quad t \geq 0 \qquad (7.7)$$

Similar to two-dof systems, the state variable vector of n-dof will be labeled $\mathbf{z}(t)$, i.e.,

$$\mathbf{z} = \begin{bmatrix} \mathbf{x} \\ \dot{\mathbf{x}} \end{bmatrix} \qquad (7.8)$$

which is a $2n$-dimensional array.

By virtue of the relation between the vector of generalized coordinates $\mathbf{x}(t)$ and its transformed counterpart $\mathbf{y}(t)$, as well as between the vector of generalized forces

---

[1] A $n \times n$ positive-definite matrix has 2^n square roots, of which one is positive-definite, one negative-definite, and all others sign-indefinite.

[2] This section is largely based on material that appeared in Al-Widyan et al. [1], © 2003, with permission from Begell House, Inc.

$\mathbf{f}(t)$ and its transformed counterpart $\phi(t)$, as appearing in Eq. 7.5, the time response of the normal-form model reduces to:

$$\mathbf{y}(t) = \cos(\Omega t)\mathbf{y}_0 + \Omega^{-1}\sin(\Omega t)\mathbf{s}_0 + \int_0^t \Omega^{-1}\sin[\Omega(t-\tau)]\mathbf{g}(\tau)d\tau \quad (7.9a)$$

$$\dot{\mathbf{y}}(t) \equiv \mathbf{v}(t) = -\Omega\sin(\Omega t)\mathbf{y}_0 + \cos(\Omega t)\mathbf{s}_0 + \int_0^t \cos[\Omega(t-\tau)]\mathbf{g}(\tau)d\tau \quad (7.9b)$$

Similar to Sect. 3.4, what we need now is an expression for $\mathbf{y}(t_{k+1}) \equiv \mathbf{y}_{k+1}$ in terms of $\mathbf{y}_k \equiv \mathbf{y}(t_k)$, $\dot{\mathbf{y}}_k \equiv \mathbf{v}(t_k)$ and $\mathbf{g}_k \equiv \mathbf{g}(t_k)$. Thus, we regard instant t_k as the initial time and compute $\mathbf{y}_{k+1}$ from instant t_k using Eq. 7.9a, i.e.,

$$\mathbf{y}_{k+1} = \cos(\Omega h)\mathbf{y}_k + \Omega^{-1}\sin(\Omega h)\mathbf{s}_k + \Omega^{-1}\int_{t_k}^{t_k+h}\sin[\Omega(t_k+h-\tau)]\,\mathbf{g}(\tau)d\tau$$

$$= \cos(\Omega h)\mathbf{y}_k + \Omega^{-1}\sin(\Omega h)\mathbf{s}_k + \Omega^{-1}\left\{\int_{t_k}^{t_k+h}\sin[\Omega(t_k+h-\tau)]\,d\tau\right\}\mathbf{g}_k$$

where $\phi_k \equiv \phi(t_k)$ is constant in $t_k \leq t \leq t_{k+1}$, as per the assumption behind the zero-order hold, introduced in Sect. 3.2, while keeping $t_{k+1} - t_k = h$, for all k.

To calculate the integral in curly brackets in the foregoing expression, we let $u \equiv t_k + h - \tau$, which allows us to write,

$$\int_{t_k}^{t_k+h}\sin[\Omega(t_k+h-\tau)]\,d\tau = -\int_h^0 \sin(\Omega u)\,du \equiv \int_0^h \sin(\Omega u)\,du \quad (7.10)$$

and hence, upon mimicking, in matrix form, the integral of the scalar sine function,

$$\int_{t_k}^{t_k+h}\sin[\Omega(t_k+h-\tau)]\,d\tau = \Omega^{-1}[1-\cos(\Omega h)] \equiv [1-\cos(\Omega h)]\Omega^{-1} \quad (7.11)$$

thereby showing[3] that the integral at hand is a constant as long as the sampling takes place at equal time intervals of length h. Thus, $\mathbf{y}_{k+1}$ takes the form

$$\mathbf{y}_{k+1} = \cos(\Omega h)\mathbf{y}_k + \Omega^{-1}\sin(\Omega h)\mathbf{v}_k + \Omega^{-2}[1-\cos(\Omega h)]\,\mathbf{g}_k \quad (7.12)$$

which is the discrete-time response sought. However, the foregoing response requires the updating of $\mathbf{v}_k$, which means that an expression for $\mathbf{v}_{k+1}$ similar to that for $\mathbf{y}_{k+1}$ must be derived; this is done with the aid of Eq. 7.9b. To obtain the

---

[3]The identity in Eq. 7.11 follows from Fact 4 of Appendix A.

desired expression, we have to evaluate, between t_k and t_{k+1}, the integral appearing in Eq. 7.9b labeled $\mathbf{c}$:

$$\mathbf{c} \equiv \int_{t_k}^{t_k+h} \cos[\mathbf{\Omega}(t_k+h-\tau)]\boldsymbol{\phi}_k d\tau \equiv \left\{ \int_{t_k}^{t_k+h} \cos[\mathbf{\Omega}(t_k+h-\tau)] d\tau \right\} \mathbf{g}_k$$

The above integral is evaluated by resorting to the same substitution used in Eq. 7.10, the term $\mathbf{c}$ thus reducing to

$$\mathbf{c} = \mathbf{\Omega}^{-1} \sin(\mathbf{\Omega}h)\mathbf{g}_k \tag{7.13}$$

Then, the final expression for $\mathbf{s}_{k+1}$ becomes

$$\mathbf{s}_{k+1} = -\mathbf{\Omega}\sin(\mathbf{\Omega}h)\mathbf{y}_k + \cos(\mathbf{\Omega}h)\mathbf{v}_k + \mathbf{\Omega}^{-1}\sin(\mathbf{\Omega}h)\mathbf{g}_k \tag{7.14}$$

If we let $\boldsymbol{\eta}_k \equiv [\mathbf{y}_k^T, \mathbf{s}_k^T]^T$ denote the vector of state variables at instant $t = t_k$, then

$$\boldsymbol{\eta}_{k+1} = \mathbf{F}\boldsymbol{\eta}_k + \mathbf{\Gamma}\mathbf{g}_k \tag{7.15a}$$

with $\mathbf{F}$ and $\mathbf{\Gamma}$ defined now as $2n \times 2n$ matrices, namely,

$$\mathbf{F} \equiv \begin{bmatrix} \cos(\mathbf{\Omega}h) & \mathbf{\Omega}^{-1}\sin(\mathbf{\Omega}h) \\ -\mathbf{\Omega}\sin(\mathbf{\Omega}h) & \cos(\mathbf{\Omega}h) \end{bmatrix}, \quad \mathbf{\Gamma} \equiv \begin{bmatrix} \mathbf{\Omega}^{-2}[\mathbf{1}-\cos(\mathbf{\Omega}h)] \\ \mathbf{\Omega}^{-1}\sin(\mathbf{\Omega}h) \end{bmatrix} \tag{7.15b}$$

and hence, the expression for $\boldsymbol{\eta}_N$ in terms of $\boldsymbol{\eta}_0$ is readily derived by mimicking the expression derived for undamped second-order, single-dof systems in Eq. 3.18, which leads to

$$\boldsymbol{\eta}_N = \mathbf{F}^N\boldsymbol{\eta}_0 + \sum_{k=0}^{N-1} \mathbf{F}^k \mathbf{\Gamma}\mathbf{g}_{N-1-k}, \quad N = 1, 2, \cdots \tag{7.16}$$

thereby obtaining the *discrete-time response* of n-dof undamped systems. The output of the system is $\boldsymbol{\eta}_N$ itself. In practice, this response is *not* to be computed with the above formula; it is rather calculated at every sample instant t_k by means of Eq. 7.15a, which only requires two matrix-times-vector multiplications. The time response of Eq. 7.16 makes it apparent that $\boldsymbol{\eta}_N$ involves the first N powers of $\mathbf{F}$, and hence, the first N powers of both $\mathbf{\Omega}$ and $\mathbf{\Omega}^{-1}$. Furthermore, the relation between $\boldsymbol{\eta}_k$ and $\mathbf{z}_k$, the discrete-time counterpart of the state-variable vector given by Eq. 7.8, is

$$\boldsymbol{\eta}_k = \mathbf{Y}\mathbf{z}_k, \quad \mathbf{Y} = \begin{bmatrix} \mathbf{N} & \mathbf{0} \\ \mathbf{0} & \mathbf{N} \end{bmatrix} \tag{7.17}$$

If we realize that the eigenvalues of $\mathbf{\Omega}^N$ are $\{\omega_i^N\}_1^n$ and those of $(\mathbf{\Omega}^{-1})^N$ are $\{1/\omega_i^N\}_1^n$, then it is apparent that the "small" eigenvalues become attenuated in time because of $\mathbf{\Omega}^N$, while their "large" counterparts become amplified likewise. A similar phenomenon, but opposite to this one, occurs because of $(\mathbf{\Omega}^{-1})^N$. This problem arises, apparently, because the entries of $\mathbf{F}$ and $\mathbf{\Gamma}$ have different physical units. A similar remark was made in connection with the discrete-time response of undamped single-dof systems regarding matrix $\mathbf{F}$ and vector $\mathbf{g}$, as appearing in Eq. 3.17b. The way to cope with this problem, introduced in Chap. 3, was a change of variable, which is done here as well.

By looking closely at matrix $\mathbf{F}$, as given by Eq. 7.15b, it is apparent that the absolute values of the entries of its diagonal blocks lie between zero and unity, while the entries of its off-diagonal blocks are unbounded, and can take any real value, depending on the eigenvalues of $\mathbf{\Omega}$. Moreover, roundoff errors will be unavoidably magnified as the simulation proceeds, for exponent N in Eq. 7.16 takes larger and larger values. As a means to alleviate roundoff-error buildup, that can lead to catastrophic results, we rewrite Eqs. 7.12 and 7.14 in a more suitable form. To this end, by virtue of the commutativity of $\mathbf{\Omega}$ and $\mathbf{\Omega}^{-1}$ with any analytic function of $\mathbf{\Omega}$, in particular with $\cos(\mathbf{\Omega}h)$ and $\sin(\mathbf{\Omega}h)$, Eqs. 7.12 and 7.14 are expressed alternatively as:

$$\mathbf{y}_{k+1} = \cos(\mathbf{\Omega}h)\mathbf{y}_k + \sin(\mathbf{\Omega}h)\mathbf{\Omega}^{-1}\mathbf{s}_k + [\mathbf{1} - \cos(\mathbf{\Omega}h)]\,\mathbf{\Omega}^{-2}\mathbf{g}_k \quad (7.18a)$$

$$\mathbf{\Omega}^{-1}\mathbf{s}_{k+1} = -\sin(\mathbf{\Omega}h)\mathbf{y}_k + \cos(\mathbf{\Omega}h)\mathbf{\Omega}^{-1}\mathbf{s}_k + \sin(\mathbf{\Omega}h)\mathbf{\Omega}^{-2}\mathbf{g}_k \quad (7.18b)$$

Now we introduce the change of variable

$$\mathbf{w}_k \equiv \mathbf{\Omega}^{-1}\mathbf{s}_k \quad \text{and} \quad \mathbf{u}_k \equiv \mathbf{\Omega}^{-2}\mathbf{g}_k \equiv (\mathbf{\Omega}^2)^{-1}\mathbf{g}_k \quad (7.19)$$

thereby obtaining the simpler scheme

$$\mathbf{y}_{k+1} = \cos(\mathbf{\Omega}h)\mathbf{y}_k + \sin(\mathbf{\Omega}h)\mathbf{w}_k + [\mathbf{1} - \cos(\mathbf{\Omega}h)]\,\mathbf{u}_k \quad (7.20a)$$

$$\mathbf{w}_{k+1} = -\sin(\mathbf{\Omega}h)\mathbf{y}_k + \cos(\mathbf{\Omega}h)\mathbf{w}_k + \sin(\mathbf{\Omega}h)\mathbf{u}_k \quad (7.20b)$$

One simple way of computing $\mathbf{\Omega}^{-1}$ and $(\mathbf{\Omega}^2)^{-1}$ is via $\mathbf{\Omega}_d$, the diagonal form of the frequency matrix, introduced in Eq. 6.8, its relation with $\mathbf{\Omega}$ appearing in Eqs. 6.9 and 6.10. The reader should be able to prove that

$$\mathbf{\Omega}^{-1} = \mathbf{E}\mathbf{\Omega}_d^{-1}\mathbf{E}^T, \quad (\mathbf{\Omega}^2)^{-1} = \mathbf{E}(\mathbf{\Omega}_d^2)^{-1}\mathbf{E}^T \quad (7.21)$$

The simulation scheme of Eq. 7.20a and b now takes the form

$$\zeta_{k+1} = \mathbf{H}\zeta_k + \mathbf{J}\mathbf{u}_k \qquad (7.22a)$$

$$\mathbf{z}_k = \Xi\zeta_k \qquad (7.22b)$$

with vector $\mathbf{z}_k$ defined as the discrete-time counterpart of $\mathbf{z}(\mathbf{t})$, introduced in Eq. 7.8, while vector ζ_k is defined as the state variable vector of the simulation scheme of Eq. 7.20, i.e.,[4]

$$\mathbf{z}_k \equiv \begin{bmatrix} \mathbf{x}_k \\ \mathbf{v}_k \end{bmatrix} = \begin{bmatrix} \mathbf{x}_k \\ \dot{\mathbf{x}}(t_k) \end{bmatrix}, \qquad \zeta_k \equiv \begin{bmatrix} \mathbf{y}_k \\ \mathbf{w}_k \end{bmatrix} \qquad (7.22c)$$

Matrices $\mathbf{H}$, $\mathbf{J}$ and Ξ are defined, in turn, as

$$\mathbf{H} \equiv \begin{bmatrix} \cos(\Omega h) & \sin(\Omega h) \\ -\sin(\Omega h) & \cos(\Omega h) \end{bmatrix}, \quad \mathbf{J} \equiv \begin{bmatrix} 1 - \cos(\Omega h) \\ \sin(\Omega h) \end{bmatrix}, \quad \Xi \equiv \begin{bmatrix} \mathbf{N}^{-1} & \mathbf{O} \\ \mathbf{O} & \mathbf{N}^{-1}\Omega \end{bmatrix}$$
$$(7.22d)$$

where $\mathbf{O}$ is the $n \times n$ zero matrix.

The output of the foregoing system is the original discrete-time state vector $\mathbf{z}_k$, containing the n components of $\mathbf{x}_k$ and the n components of $\dot{\mathbf{x}}(t_k)$.

Now, matrices $\mathbf{H}$ and $\mathbf{J}$ are better behaved numerically. Note that, in the foregoing scheme, $\mathbf{H}$ is orthogonal, i.e.,

$$\mathbf{H}\mathbf{H}^T = \mathbf{1} \qquad (7.23)$$

with $\mathbf{1}$ denoting the $2n \times 2n$ identity matrix.

The discrete-time response of the system at hand now takes the form

$$\zeta_N = \mathbf{H}^N \zeta_0 + \sum_{k=0}^{N-1} \mathbf{H}^k \mathbf{J} \mathbf{u}_{N-1-k} \qquad (7.24)$$

where, by virtue of the special structure of $\mathbf{H}$,

$$\mathbf{H}^k = \begin{bmatrix} \cos(k\Omega h) & \sin(k\Omega h) \\ -\sin(k\Omega h) & \cos(k\Omega h) \end{bmatrix}, \quad k = 1, 2, \ldots \qquad (7.25)$$

which the reader is invited to verify. This matrix represents a rotation through angles of $k\omega_i h$, $i = 1, \ldots, n$, where ω_i is the ith eigenvalue of Ω, in $2n$-dimensional space, in the same way that, in 2D space, $\mathbf{H}^n$ of Eq. 3.23 represents a rotation through an angle $n\omega_n h$—with ω_n representing the *natural frequency* of the single-dof system in question.

---

[4]Properly speaking, the lower block of $\mathbf{z}_k$, as given by Eq. 7.22b, is different from $\dot{\mathbf{x}}(t_k)$ because of the approximation involved when introducing the ZOH.

The simulation algorithm, for $k = 1, \cdots, N$, is summarized below:

Algorithm UDnDOF

1. calculate the eigenvalues and eigenvectors of $\mathbf{M}$, $\{\mu_i\}_1^n$ and $\{\mathbf{m}_i\}_1^n$, respectively; $k \leftarrow 0$
2. $\overline{\mathbf{M}} \leftarrow [\mathbf{m}_1 \quad \mathbf{m}_2 \quad \cdots \quad \mathbf{m}_n]$
3. $\mathbf{M}_d \leftarrow \operatorname{diag}(\mu_1, \mu_2, \cdots, \mu_n)$;
4. $\mathbf{N}_d^{-1} \leftarrow \operatorname{diag}(1/\sqrt{\mu_1}, 1/\sqrt{\mu_2}, \cdots, 1/\sqrt{\mu_n})$;
5. $\mathbf{N}^{-1} \leftarrow \overline{\mathbf{M}} \mathbf{N}_d^{-1} \overline{\mathbf{M}}^T$
6. $\Omega^2 \leftarrow \mathbf{N}^{-1} \mathbf{K} \mathbf{N}^{-1}$;
7. calculate the eigenvalues and eigenvectors of Ω^2, $\{\omega_i^2\}_1^n$ and $\{\mathbf{e}_i\}_1^n$, respectively;
8. $\mathbf{E} \leftarrow [\mathbf{e}_1 \quad \mathbf{e}_2 \quad \cdots \quad \mathbf{e}_n]$;
9. $\Omega_d \leftarrow \operatorname{diag}(\omega_1, \omega_2, \cdots, \omega_n)$, $\mathbf{C} \leftarrow \operatorname{diag}(\cos \omega_1 h, \cos \omega_2 h, \cdots, \cos \omega_n h)$, $\mathbf{S} \leftarrow \operatorname{diag}(\sin \omega_1 h, \sin \omega_2 h, \cdots, \sin \omega_n h)$
10. $\Omega \leftarrow \mathbf{E} \Omega_d \mathbf{E}^T, \quad \cos(\Omega h) \leftarrow \mathbf{E} \mathbf{C} \mathbf{E}^T, \quad \sin(\Omega h) \leftarrow \mathbf{E} \mathbf{S} \mathbf{E}^T$
11. $\mathbf{H} \leftarrow \begin{bmatrix} \cos(\Omega h) & \sin(\Omega h) \\ -\sin(\Omega h) & \cos(\Omega h) \end{bmatrix}$;
12. $\mathbf{J} \leftarrow \begin{bmatrix} 1 - \cos(\Omega h) \\ \sin(\Omega h) \end{bmatrix}$;
13. $\mathbf{u}_k \leftarrow (\Omega^2)^{-1} \mathbf{g}_k$;
14. $\zeta_{k+1} \leftarrow \mathbf{H} \zeta_k + \mathbf{J} \mathbf{u}_k$;
15. $\mathbf{x}_{k+1} \leftarrow \mathbf{N}^{-1} \mathbf{y}_{k+1}, \quad \dot{\mathbf{x}}_{k+1} \leftarrow \mathbf{N}^{-1} \Omega \mathbf{w}_{k+1}$;
16. if $t_k < T$ then $k \leftarrow k+1$ go to 13; else stop

7.3.1   The Numerical Stability of the Simulation Algorithm of Undamped Systems

As stated in Chap. 3, the powers of orthogonal matrices are orthogonal as well. This means that the powers of $\mathbf{H}$ in Eq. 7.24 are orthogonal, which means, in turn, that the magnitude of the discrete zero-input response $\mathbf{H}^N \zeta_0$ remains equal to that of ζ_0. That is, $\mathbf{H}^N \zeta_0$ is nothing but vector ζ_0 rotated N times by matirx $\mathbf{H}$ in $2n$-dimensional space. Therefore,

$$\|\zeta_k\| = \|\zeta_0\|, \quad k = 1, \ldots, N \tag{7.26}$$

which leads to the conservation of energy of the undamped, zero-input response of the system at hand, as shown presently. For starters, a $2n \times 2n$ nonsingular matrix $\mathbf{Z}$ is introduced[5]:

$$\mathbf{Z} = \begin{bmatrix} \boldsymbol{\Omega} & \mathbf{O} \\ \mathbf{O} & \boldsymbol{\Omega} \end{bmatrix} \tag{7.27}$$

with $\mathbf{O}$ introduced in Eq. 7.22d as the $n \times n$ zero matrix. Now a property of $\mathbf{Z}$ and $\mathbf{H}$ is pointed out:

$$\mathbf{Z}\mathbf{H}\mathbf{Z}^{-1} = \mathbf{H} \tag{7.28}$$

The proof of the above relation is straightforward: For brevity, let $\mathbf{C} \equiv \cos \boldsymbol{\Omega} h$ and $\mathbf{S} \equiv \sin \boldsymbol{\Omega} h$. Then,

$$\begin{aligned} \mathbf{Z}\mathbf{H}\mathbf{Z}^{-1} &= \begin{bmatrix} \boldsymbol{\Omega} & \mathbf{O} \\ \mathbf{O} & \boldsymbol{\Omega} \end{bmatrix} \begin{bmatrix} \mathbf{C} & \mathbf{S} \\ -\mathbf{S} & \mathbf{C} \end{bmatrix} \begin{bmatrix} \boldsymbol{\Omega}^{-1} & \mathbf{O} \\ \mathbf{O} & \boldsymbol{\Omega}^{-1} \end{bmatrix} \\ &= \begin{bmatrix} \boldsymbol{\Omega}\mathbf{C}\boldsymbol{\Omega}^{-1} & \boldsymbol{\Omega}\mathbf{S}\boldsymbol{\Omega}^{-1} \\ -\boldsymbol{\Omega}\mathbf{S}\boldsymbol{\Omega}^{-1} & \boldsymbol{\Omega}\mathbf{C}\boldsymbol{\Omega}^{-1} \end{bmatrix} \end{aligned} \tag{7.29}$$

However, Fact 4 of Appendix A states that any square matrix commutes with its analytic functions. This means that

$$\boldsymbol{\Omega}\mathbf{C}\boldsymbol{\Omega}^{-1} = \boldsymbol{\Omega}\boldsymbol{\Omega}^{-1}\mathbf{C} = \mathbf{C}, \quad \boldsymbol{\Omega}\mathbf{S}\boldsymbol{\Omega}^{-1} = \boldsymbol{\Omega}\boldsymbol{\Omega}^{-1}\mathbf{S} = \mathbf{S}$$

thereby proving relation (7.28). The reader should be able to prove that

$$\mathbf{Z}\mathbf{H}^k\mathbf{Z}^{-1} = \mathbf{H}^k, \quad k = 1, 2, \dots$$

Further, multiplying by $\mathbf{Z}$ the two sides of Eq. 7.24, with $\mathbf{u}_{N-1-k} = \mathbf{0}$, for $k = 0, 1, \dots, N - 1$ and $\mathbf{0}$ denoting the n-dimensional zero vector, while the $2n \times 2n$ identity matrix $\mathbf{1}$ is inserted bewteen $\mathbf{H}^N$ and $\boldsymbol{\zeta}_0$, disguised as the product $\mathbf{Z}^{-1}\mathbf{Z}$, thereby obtaining

$$\mathbf{Z}\boldsymbol{\zeta}_N = \mathbf{Z}\mathbf{H}^N\mathbf{Z}^{-1}\mathbf{Z}\boldsymbol{\zeta}_0 \tag{7.30}$$

Now, by virtue of Eq. 7.28, the foregoing equation leads to

$$\mathbf{Z}\boldsymbol{\zeta}_N = \mathbf{H}^N\mathbf{Z}\boldsymbol{\zeta}_0 \tag{7.31}$$

Moreover,

$$\|\mathbf{Z}\boldsymbol{\zeta}_N\|^2 = \left\| \begin{bmatrix} \boldsymbol{\Omega}\mathbf{y}_N \\ \boldsymbol{\Omega}\mathbf{w}_N \end{bmatrix} \right\|^2 = \left\| \begin{bmatrix} \boldsymbol{\Omega}\mathbf{N}\mathbf{x}_N \\ \boldsymbol{\Omega}\boldsymbol{\Omega}^{-1}\mathbf{N}\dot{\mathbf{x}}_N \end{bmatrix} \right\|^2$$

---

[5]Not to be confused with $\mathbf{Z}$ introduced in Exercise 6.8.

Upon expansion,

$$\|\mathbf{Z}\zeta_N\|^2 = \mathbf{x}_N^T \mathbf{N} \underbrace{\Omega^2}_{\mathbf{N}^{-1}\mathbf{K}\mathbf{N}^{-1}} \mathbf{N}\mathbf{x}_N + \dot{\mathbf{x}}_N^T \underbrace{\mathbf{N}^2}_{\mathbf{M}} \dot{\mathbf{x}}_N$$

where the definitions of Ω^2 and $\mathbf{N}$, besides their symmetries, have been recalled from Eqs. 7.3 and 7.4. Therefore,

$$\|\mathbf{Z}\zeta_N\|^2 = \mathbf{x}_N^T \mathbf{K}\mathbf{x}_N + \dot{\mathbf{x}}_N^T \mathbf{M}\dot{\mathbf{x}}_N = 2E_N$$

which is twice the total energy E_N at instant $t = t_N$. Likewise,

$$\|\mathbf{Z}\zeta_0\|^2 = \mathbf{x}_0^T \mathbf{K}\mathbf{x}_0 + \dot{\mathbf{x}}_0^T \mathbf{M}\dot{\mathbf{x}}_0 = 2E_0$$

However, in light of the orthogonality of $\mathbf{H}^N$, the magnitudes of $\mathbf{Z}\zeta_N$ and $\mathbf{Z}\zeta_0$ are identical. Therefore,

$$\|\mathbf{Z}\zeta_N\|^2 = \|\mathbf{Z}\zeta_0\|^2 \quad \Rightarrow \quad E_N = E_0$$

thereby showing that the discrete zero-input response of the undamped system under study is energy-preserving. Commercial simulation software being by its nature of a general-purpose type, entails a cumulative roundoff-error that produces a slow, but persistent growth—or decay—of the energy of the system.

7.3.2   On the Choice of the Time Step

While the ZOH and a suitable change of variable have led to a robust algorithm guaranteeing the conservation of energy in the simulation of undamped systems, the time step h must be suitably chosen; else, the simulation results may be misleading. In fact, h cannot be larger than $T_{min}/2$, where T_{min} is the minimum period associated with the natural frequencies of the system, as required by Shannon's Theorem.[6] That is, if the natural frequencies are ordered such that $\omega_1 \leq \omega_2 \leq \ldots \leq \omega_n$, then

$$h \leq \frac{\pi}{\omega_n} \tag{7.32}$$

Failure of complying with this condition will lead to the hiding of high frequencies, a phenomenon known as *aliasing*.

---

[6]The interested reader can have a glimpse of the theorem in Åström and Wittenmark [2].

Fig. 7.1 The iconic model of
the suspension system

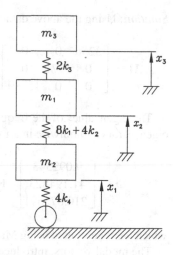

One example is included below to illustrate the performance of the algorithm.

Example 7.3.1 (Discrete-time Response of Undamped, Three-dof Suspension). The
suspension system shown in Fig. 4.8 was modeled in Example 4.5.3. The time
response of this system is simulated here under a perturbation arising by a sudden
displacement of 100 mm of the third mass in the upward direction, with the purpose
of exciting the lowest frequency of the system. Moreover, the effects of the dashpots
are neglected here.

Referring to Fig. 7.1, the undamped counterpart of Fig. 4.9, m_1, m_2, and m_3
denote the mass of: each motor-differential bridge; the chassis and one-half the mass
of the car body, respectively.

In deriving the mathematical model of the system appearing in Fig. 7.1, we define
the three-dimensional vector of generalized coordinates $\mathbf{x}$ as

$$\mathbf{x} = \begin{bmatrix} x_1 & x_2 & x_3 \end{bmatrix}^T$$

where all three components are measured from the equilibrium configuration.

The mathematical model corresponding to Fig. 7.1 takes the form of Eq. 6.1, with
matrices $\mathbf{M}$ and $\mathbf{K}$ given by

$$\mathbf{M} = \begin{bmatrix} m_2 & 0 & 0 \\ 0 & m_1 & 0 \\ 0 & 0 & m_3 \end{bmatrix}, \quad \mathbf{K} = \begin{bmatrix} 8k_1 + 4k_2 + 4k_4 & -8k_1 - 4k_2 & 0 \\ -8k_1 - 4k_2 & 8k_1 + 4k_2 + 2k_3 & -2k_3 \\ 0 & -2k_3 & 2k_3 \end{bmatrix}$$

Moreover, the manufacturer provides the numerical values given below:

$$m_1 = 1971\,\text{kg}, \quad m_2 = 3256\,\text{kg}, \quad m_3 = 15780\,\text{kg}$$

$$k_1 = 4900 \times 10^3\,\text{N/m}, k_2 = 3430 \times 10^3\,\text{N/m}, k_3 = 837 \times 10^3\,\text{N/m},$$

$$k_4 = 1783 \times 10^3\,\text{N/m}$$

Solution: Using the above data, the system matrices are readily calculated as

$$\mathbf{M} = \begin{bmatrix} 3256 & 0 & 0 \\ 0 & 1971 & 0 \\ 0 & 0 & 15780 \end{bmatrix} \text{kg}, \quad \mathbf{K} = \begin{bmatrix} 60052 & -5292 & 0 \\ -5292 & 5459 & -1674 \\ 0 & -1674 & 1674 \end{bmatrix} \text{kN/m}$$

The eigenvalues of the frequency matrix are displayed in array $\boldsymbol{\omega}$, its three unit eigenvectors columnwise in array $\mathbf{E}$:

$$\boldsymbol{\omega} = \begin{bmatrix} 9.092543 \\ 41.194823 \\ 210.875591 \end{bmatrix}, \quad \mathbf{E} = \begin{bmatrix} 0.625961 & -0.774860 & 0.088118 \\ -0.779837 & -0.621173 & 0.077450 \\ 0.005277 & 0.117198 & 0.993095 \end{bmatrix}$$

which were computed using Maple^{TM}.

The modal vectors, introduced in Eq. 6.21a, are displayed below in array[7] $\mathbf{F}$:

$$\mathbf{F} = \mathbf{N}^{-1}\mathbf{E} = \begin{bmatrix} 0.001544 & -0.013579 & 0.010970 \\ 0.001745 & -0.013992 & -0.017565 \\ 0.007906 & 0.000933 & 0.000042 \end{bmatrix}$$

It is recalled that the eigenvectors of $\boldsymbol{\Omega}$ are the result of a change of coordinates, given by Eq. 7.5, and hence, do not offer an immediate interpretation in terms of the given set of generalized coordinates, stored in array $\mathbf{x}$. To ease the interpretation of the analysis results, a transformation back into the original coordinates is thus warranted, thereby obtaining the above modal vectors. What the above natural frequencies reveal is interesting: with wheel diameters of 0.960 m, as per Example 4.5.3, and the car rolling with unavoidably small unbalanced wheels,[8] resonances are expected to occur at traveling speeds of 4.3644 m/s ($\approx$15.7 km/h), 19.7735 m/s ($\approx$71.18 km/h), and 100.8 m/s ($\approx$363.0 km/h). The first natural frequency is not apparent at a constant speed, as the subway zooms through this speed. Neither is the third, but the second natural frequency is more likely to be excited. Under the second mode, masses 1 and 2 move with approximately the same displacement in the same direction, while mass 3 moves in the opposite direction with a displacement of about one order of magnitude smaller than that of the two other masses.

Under the above perturbation, and with reference to the model of Eq. 6.1, we have

$$\mathbf{f}(t) = \mathbf{0}, \quad \mathbf{x}(0) = \begin{bmatrix} 0 & 0 & 100 \end{bmatrix}^T \text{mm}, \quad \dot{\mathbf{x}}(0) = \mathbf{0}$$

---

[7] Not to be confused with matrix $\mathbf{F}$ of Eq. 7.15b!

[8] A (statically) unbalanced wheel can be visualized as a disc with its mass concentrated at a point C offset from its center O by a distance $e \ll r$, for a radius r.

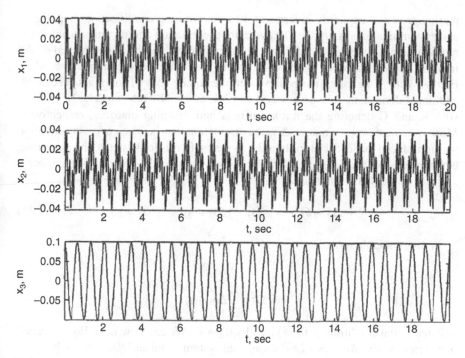

Fig. 7.2   The time response produced by the simulation algorithm

Figure 7.2 shows the time response of the system to be above excitation. Those results were generated by the algorithm introduced here. What these plots reveal is that the amplitude of the motion of the three masses remains unaltered during the first 20 s, a result of the inherent-energy-preserving feature of the algorithm.

Animations of the three modes of the model of the subway cars are included in

```
7-metro1stMode.mw
7-metro2ndMode.mw
7-metro3rdMode.mw
```

Animation of the response of the same model to a bump is included in

```
7-MetroBump.mw
```

7.4   The Discrete-Time Response of Damped Systems

The systems of interest are modeled by Eq. 4.22, which represents an initial-value problem of n second-order ODEs in the n generalized coordinates **x**. This model was cast in state-variable form in Eqs. 6.64–6.66b, which allowed the derivation of its time response explicitly in terms of the system initial conditions η_0, input $\mathbf{g}(t)$, and system matrices $\mathbf{A}$ and $\mathbf{B}$. The system representation and its time response are reproduced below for quick reference: we start by recalling the mathematical model in terms of $2n$ first-order, linear ODE in the state-variable vector $\eta = [\mathbf{y}^T,\ \mathbf{s}^T]^T$. Further, with $\mathbf{N}$ denoting the positive-definite square root of the mass matrix $\mathbf{M}$, additional definitions follow: $\mathbf{y} = \mathbf{Nx}$; $\mathbf{s} \equiv \mathbf{N\dot{x}}$; $\mathbf{\Omega}^2 \equiv \mathbf{N}^{-1}\mathbf{KN}^{-1}$; $\mathbf{\Delta} \equiv \mathbf{N}^{-1}\mathbf{CN}^{-1}$, with $\mathbf{K}$ and $\mathbf{C}$ denoting the $n \times n$ stiffness and damping matrices, respectively. Moreover, the new $n \times n$ matrix $\mathbf{\Delta}$ is the dissipation matrix. The governing equation was introduced in Eq. 5.118 for two-dof systems. For n-dof systems, the governing equation has exactly the same form, the vectors involved being now n-dimensional and the matrices of $n \times n$, namely,

$$\ddot{\mathbf{y}} + \mathbf{\Delta}\dot{\mathbf{y}} + \mathbf{\Omega}^2\mathbf{y} = \mathbf{g}(t), \quad \mathbf{y}(0) = \mathbf{y}_0, \quad \dot{\mathbf{y}}(0) = \mathbf{s}_0 \tag{7.33}$$

In state-variable form,

$$\dot{\mathbf{y}} = \mathbf{s} \tag{7.34a}$$

$$\dot{\mathbf{s}} = -\mathbf{\Omega}^2\mathbf{y} - \mathbf{\Delta s} + \mathbf{g}(t) \tag{7.34b}$$

with the initial conditions $\mathbf{y}(0) = \mathbf{y}_0$ and $\mathbf{s}(0) = \mathbf{s}_0$. We can now write Eq. 7.34a and b in *state-variable form* as a $2n$-dimensional system of linear ODEs, namely,

$$\dot{\eta} = \mathbf{A}\eta + \mathbf{Bg}(t), \quad \eta(0) = \eta_0 \tag{7.35a}$$

where

$$\mathbf{A} \equiv \begin{bmatrix} \mathbf{O} & \mathbf{1} \\ -\mathbf{\Omega}^2 & -\mathbf{\Delta} \end{bmatrix}, \quad \mathbf{B} \equiv \begin{bmatrix} \mathbf{O} \\ \mathbf{1} \end{bmatrix}, \quad \eta \equiv \begin{bmatrix} \mathbf{y} \\ \mathbf{s} \end{bmatrix} \tag{7.35b}$$

whose time response was found in Chap. 6 to be

$$\eta(t) = e^{\mathbf{A}t}\eta_0 + \int_0^t e^{\mathbf{A}(t-\tau)}\mathbf{Bg}(\tau)\,d\tau \tag{7.35c}$$

The foregoing representation of the time response is useful for gaining insight into the system response, but is not convenient for simulation, as it requires computing the exponential and the integral of a product in which the exponential appears as a factor at every instant t of interest. An alternative representation of the time response in *discrete time* is thus sought, similar to that of Eq. 3.39.

7.4.1 A Straightforward Approach

An approach similar to that of Sect. 3.4.2 is followed here: upon introduction of the ZOH for a constant time increment $t_{k+1} - t_k = h$, for $k = 1, 2, \ldots, N$, the discrete-time response of the system becomes formally similar to that derived for the second-order damped system in Eq. 3.39, namely,

$$\eta_{k+1} = \mathbf{F}\eta_k + \mathbf{G}\mathbf{g}_k \qquad (7.36a)$$

with

$$\eta_k \equiv \begin{bmatrix} \mathbf{y}_k \\ \mathbf{s}_k \end{bmatrix}, \quad \mathbf{F} \equiv e^{\mathbf{A}h}, \quad \mathbf{G} \equiv (\mathbf{F} - \mathbf{1})\mathbf{A}^{-1}\mathbf{B}, \quad \mathbf{g}_k \equiv \mathbf{g}(t_k) \qquad (7.36b)$$

Similar to the discrete-time response of second-order damped systems, Eq. 3.48, that of the system at hand becomes

$$\eta_N = \mathbf{F}^N \eta_0 + \sum_{k=0}^{N-1} \mathbf{F}^k \mathbf{G}\mathbf{g}_{N-1-k} \qquad (7.37)$$

From the experience gained in Sect. 7.3, one can realize that the form of the system matrix $\mathbf{A}$ is not appropriate for simulation, as it involves three non-zero blocks with different units: $\mathbf{1}$ is obviously dimensionless; Ω^2 has units of frequency-squared; and Δ has units of frequency. The outcome is that the powers of $\mathbf{F}$ appearing in the foregoing time response will become corrupted with roundoff-error amplification, which can appear in the form of numerical damping, or numerical excitation, for that matter. The way to cope with this shortcoming is straightforward, if the remedy found for single-dof damped systems is recalled. In this vein, a new variable $\mathbf{w}$ is defined as

$$\mathbf{w} \equiv \Omega^{-1}\mathbf{N}\dot{\mathbf{x}} = \Omega^{-1}\mathbf{s} \qquad (7.38)$$

whence the system Eqs. 7.34a, b now takes the form

$$\dot{\mathbf{y}} = \Omega\mathbf{w} \qquad (7.39a)$$

$$\dot{\mathbf{w}} = -\Omega\mathbf{y} - \Omega^{-1}\Delta\Omega\mathbf{w} + \Omega^{-1}\mathbf{g}(t) \qquad (7.39b)$$

and the system matrices become

$$\mathbf{A} = \begin{bmatrix} \mathbf{O} & \Omega \\ -\Omega & -\Omega^{-1}\Delta\Omega \end{bmatrix}, \quad \mathbf{B} = \begin{bmatrix} \mathbf{O} \\ \Omega^{-1} \end{bmatrix} \qquad (7.40)$$

where, apparently, all the entries of $\mathbf{A}$ now have units of frequency, and hence, its exponential and the integral of the latter are numerically better behaved. As the time response sought involves $\mathbf{x}$ and $\dot{\mathbf{x}}$, these state variables should be recovered from the time response of Eq. 7.34b. This is simply done by introducing the system *output* $\mathbf{z}$, defined as

$$\mathbf{z} = \begin{bmatrix} \mathbf{x} \\ \dot{\mathbf{x}} \end{bmatrix} = \Xi\zeta \qquad (7.41a)$$

with ζ and Ξ defined as

$$\zeta = \begin{bmatrix} \mathbf{y} \\ \mathbf{w} \end{bmatrix}, \quad \Xi \equiv \begin{bmatrix} \mathbf{N}^{-1} & \mathbf{O} \\ \mathbf{O} & \mathbf{N}^{-1}\Omega \end{bmatrix} \qquad (7.41b)$$

The discrete-time response of the system now takes the form

$$\zeta_{k+1} = \mathbf{H}\zeta_k + \mathbf{J}\mathbf{g}_k \qquad (7.42a)$$

with ζ_k, $\mathbf{g}_k$, $\mathbf{H}$ and $\mathbf{J}$ defined as

$$\zeta_k = \begin{bmatrix} \mathbf{y}_k \\ \mathbf{w}_k \end{bmatrix} = \begin{bmatrix} \mathbf{y}(t_k) \\ \mathbf{w}(t_k) \end{bmatrix}, \ \mathbf{g}_k = \mathbf{g}(t_k), \ \mathbf{H} = e^{\mathbf{A}h}, \ \mathbf{J} = (\mathbf{H}-\mathbf{1})\mathbf{A}^{-1}\mathbf{B} \equiv \mathbf{A}^{-1}(\mathbf{H}-\mathbf{1})\mathbf{B}$$
$$(7.42b)$$

The simulation algorithm, for $k = 1, \cdots, N$, parallels that devised for undamped systems:

Algorithm DnDOF

1. calculate the eigenvalues and eigenvectors of $\mathbf{M}$, $\{\mu_i\}_1^n$ and $\{\mathbf{m}_i\}_1^n$, respectively; $k \leftarrow 0$
2. $\overline{\mathbf{M}} \leftarrow [\mathbf{m}_1 \quad \mathbf{m}_2 \quad \cdots \quad \mathbf{m}_n]$;
3. $\mathbf{M}_d \leftarrow \mathrm{diag}(\mu_1, \mu_2, \cdots, \mu_n)$;
4. $\mathbf{N}_d^{-1} \leftarrow \mathrm{diag}(1/\sqrt{\mu_1}, 1/\sqrt{\mu_2}, \cdots, 1/\sqrt{\mu_n})$;
5. $\mathbf{M} = \overline{\mathbf{M}}\mathbf{M}_d\overline{\mathbf{M}}^T, \quad \mathbf{N}^{-1} \equiv \sqrt{\mathbf{M}^{-1}} \leftarrow \overline{\mathbf{M}}\mathbf{N}_d^{-1}\overline{\mathbf{M}}^T$;
6. $\Omega^2 \leftarrow \mathbf{N}^{-1}\mathbf{K}\mathbf{N}^{-1}$;
7. calculate the eigenvalues and eigenvectors of Ω^2, $\{\omega_i^2\}_1^n$ and $\{\mathbf{e}_i\}_1^n$, respectively;
8. $\mathbf{E} \leftarrow [\mathbf{e}_1 \quad \mathbf{e}_2 \quad \cdots \quad \mathbf{e}_n]$;
9. $\Omega_d \leftarrow \mathrm{diag}(\omega_1, \omega_2, \cdots, \omega_n)$;
10. $\Omega \leftarrow \mathbf{E}\Omega_d\mathbf{E}^T, \quad \Omega^{-1} = \mathbf{E}\Omega_d^{-1}\mathbf{E}^T$;
11. $\Delta \leftarrow \mathbf{N}^{-1}\mathbf{C}\mathbf{N}^{-1}$;
12. $\mathbf{A} \leftarrow \begin{bmatrix} \mathbf{O} & \Omega \\ \Omega & -\Omega^{-1}\Delta\Omega \end{bmatrix}$;
13. $\mathbf{H} \leftarrow e^{\mathbf{A}h}$;
14. Solve for $\mathbf{J}$ from $\mathbf{A}\mathbf{J} = (\mathbf{H}-\mathbf{1})\mathbf{B}, \ \mathbf{B} = \begin{bmatrix} \mathbf{O} \\ \Omega^{-1} \end{bmatrix}$;
15. $\Xi \leftarrow \begin{bmatrix} \mathbf{N}^{-1} & \mathbf{O} \\ \mathbf{O} & \mathbf{N}^{-1}\Omega \end{bmatrix}$;
16. $\zeta_{k+1} \leftarrow \mathbf{H}\zeta_k + \mathbf{J}\mathbf{g}_k$;
17. $\mathbf{z}_{k+1} \leftarrow \Xi\zeta_{k+1}$;
18. if $t_k < T$ then $k \leftarrow k+1$ go to 16; else stop

Example 7.4.1 (Discrete-time Response of Damped, Three-dof Suspension). We revisit here Example 7.3.1, but with shock absorbers in the primary and secondary suspensions, as per the iconic model of Fig. 4.9, with the damping coefficients found to be appropriate for coping with resonance upon exciting the second mode [3]:

$$c_1 = 39.225 \text{ kN s/m}, \quad c_2 = 123.55 \text{ kN s/m}$$

Plot the generalized coordinates $x_1(t)$, $x_2(t)$ and $x_3(t)$ for $0 \le t \le 4$ s.

Solution: The equations of motion of this system were derived in parametric form in Example 4.5.3. Substituting the numerical values of the stiffness and damping coefficients, Algorithm DnDOF algorithm is used to obtain the plots of the generalized coordinates $x_1(t)$, $x_2(t)$ and $x_3(t)$, as shown in Fig. 7.3.

7.4.2 An Approach Based on the Laplace Transform

First, some preliminary results are introduced.[9] These are related to the eigenvalue problem associated with n-dof damped systems. For completeness, a summary of the Laplace transform is given in Appendix B.

7.4.2.1 The Eigenvalue Problem of n-dof Damped Systems

For starters, the mathematical model of the system in what was dubbed normal form in Chap. 5 is recalled from Eq. 5.118, reproduced below for quick reference:

$$\ddot{\mathbf{y}} + \mathbf{\Delta}\dot{\mathbf{y}} + \mathbf{\Omega}^2\mathbf{y} = \mathbf{g}(t), \quad \mathbf{y}(0) = \mathbf{y}_0, \quad \dot{\mathbf{y}}(0) = \mathbf{s}_0 \qquad (7.43)$$

In formulating the eigenvalue problem of the system at hand, the counterpart problem of two-dof systems of Chap. 5 is recalled, but now as applied to n-dof systems. To this end, we set $\mathbf{g}(t) = \mathbf{0}$ in Eq. 7.43 and Laplace-transform the two sides of the equation thus resulting, under zero initial conditions:

$$(s^2\mathbf{1} + s\mathbf{\Delta} + \mathbf{\Omega}^2)\boldsymbol{\eta} = \mathbf{0} \qquad (7.44)$$

in which s is the complex variable of the Laplace transform, while $\boldsymbol{\eta}$ is an *eigenvector*[10] of the system (5.118). The latter will be henceforth assumed of unit norm. Thus, since $\boldsymbol{\eta}$ is, in general, complex,

$$\|\boldsymbol{\eta}\|^2 = \boldsymbol{\eta}^*\boldsymbol{\eta} = 1 \qquad (7.45)$$

---

[9]The material in this section is based largely on Angeles et al. [4].

[10]Not to be confused with the state-variable vector introduced in Eq. 7.15a.

Fig. 7.3 The time response produced by the simulation algorithm

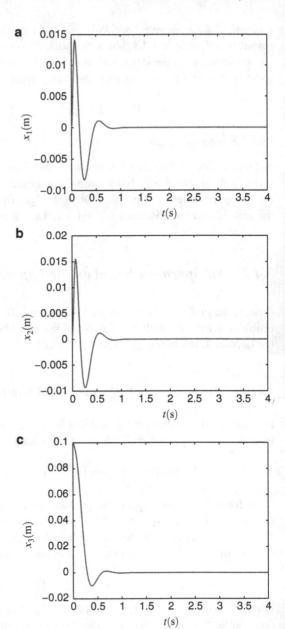

with $(\cdot)^*$ denoting *conjugate-transposition*, whether of a vector or of a matrix. Furthermore, note that the original model of the damped system is given in Eq. 7.33 Therefore, the eigenvectors of the original model, henceforth referred to as the *modal vectors* and denoted by ϕ, corresponding to η, are obtained from the change of variable of variable $\mathbf{y} = \mathbf{N}\mathbf{x}$, and denoted by ϕ. The two vectors are related by

$$\phi = \mathbf{N}^{-1}\eta, \quad \text{or} \quad \eta = \mathbf{N}\phi \qquad (7.46)$$

where it is apparent that vector ϕ is not of unit norm in the sense of Eq. 7.45. However, it is of *weighted unit norm* with respect to the mass matrix, i.e.,

$$\phi^*\mathbf{M}\phi = 1$$

By virtue of the foregoing property $\phi \neq \mathbf{0}$, and hence, the matrix in parentheses in Eq. 7.44 must be singular, which thus leads to the characteristic equation of system (7.33), namely,

$$P(s) \equiv \det(s^2\mathbf{1} + s\boldsymbol{\Delta} + \boldsymbol{\Omega}^2) = 0$$

Moreover, since matrices $\mathbf{1}$, $\boldsymbol{\Delta}$ and $\boldsymbol{\Omega}$ are all real, the coefficients of the characteristic polynomial $P(s)$ are all real as well, and hence, *the complex eigenvalues of an arbitrarily damped linear mechanical system occur in complex-conjugate pairs.*

Now, since the determinant is homogeneous of degree n in its entries, $P(s)$ is a $2n$-degree polynomial in s, its $2n$ roots being, then, the *eigenvalues* of system (7.43). The solution of the characteristic equation and the computation of the natural modes of the system is described below. The foregoing number of eigenvalues tallies with the number of eigenvalues of the $2n \times 2n$ system matrix $\mathbf{A}$ of Eq. 7.35b.

The computation of the natural frequencies of linear mechanical systems under small-amplitude vibration is nowadays a trivial issue in the case of undamped systems. In fact, in this case, the problem reduces to a standard eigenvalue problem, as shown in Sect. 5.2, for whose solution commercial software is available, e.g., Maple, MATLAB, etc.

The general case of damped systems is quite challenging in that it gives rise to a more algebraically cumbersome problem. In this case, standard software cannot be applied, at least directly, and special methods must be devised. The challenge here lies in the calculation of the complex solutions of a fairly general eigenvalue problem, associated with $n \times n$ matrices, where n can be in the hundreds. Moreover, the problem cannot be reduced to the simple calculation of the complex eigenvalues of one single matrix, but rather a complex number must be found that renders the determinant of a linear combination of $n \times n$ matrices equal to zero.

Below we introduce a method that resorts to commercially available hardware and software, and produces the desired eigenvalues as the intersections of two contours in the complex plane. Hence, these eigenvalues can be determined visually with the aid of a CAD system using the *mouse* to point at the intersections. Further refinement, if needed, can be obtained using the Newton-Raphson method to compute the roots of a system of two nonlinear equations in two unknowns.

In these cases, the rough estimates of the contour intersection points are good initial guesses, to be submitted to the code implementing the Newton-Raphson method. Explicit invariant expressions are provided for the calculation of the underlying 2×2 Jacobian matrix that is needed by this method.

Properties of the System Eigenvalues and Eigenvectors

Henceforth, given a complex number z, its conjugate is indicated by $\bar{z}$. We first recall a basic result:

Lemma 7.4.1. *Let* **A** *be a symmetric, positive-definite real $n \times n$ matrix, and* **u** *and* **v** *two n-dimensional vectors defined over the complex field. Then*

$$\overline{(\mathbf{v}^*\mathbf{A}\mathbf{u})} = \mathbf{u}^*\mathbf{A}\mathbf{v} \tag{7.47}$$

The proof of this lemma is straightforward and is, hence, omitted.

As a direct consequence of the foregoing lemma, we have, upon setting $\mathbf{v} = \mathbf{u}$ in Eq. 7.47,

$$\overline{(\mathbf{u}^*\mathbf{A}\mathbf{u})} = \mathbf{u}^*\mathbf{A}\mathbf{u} \tag{7.48}$$

Now, let us assume that $\boldsymbol{\eta}_k$ and $\boldsymbol{\eta}_l$ are two complex eigenvectors of system (7.43), of two complex eigenvalues s_k and s_l, respectively, so that each pair $(s_k, \boldsymbol{\eta}_k)$ and $(s_l, \boldsymbol{\eta}_l)$ verifies Eq. 7.44, i.e.,

$$(s_k^2 \mathbf{1} + s_k \boldsymbol{\Delta} + \boldsymbol{\Omega}^2)\boldsymbol{\eta}_k = \mathbf{0} \tag{7.49a}$$

$$(s_l^2 \mathbf{1} + s_l \boldsymbol{\Delta} + \boldsymbol{\Omega}^2)\boldsymbol{\eta}_l = \mathbf{0} \tag{7.49b}$$

Upon premultiplying Eq. 7.49a by $\boldsymbol{\eta}_l^*$ and Eq. 7.49b by $\boldsymbol{\eta}_k^*$, one obtains

$$s_k^2 \boldsymbol{\eta}_l^* \boldsymbol{\eta}_k + s_k \boldsymbol{\eta}_l^* \boldsymbol{\Delta} \boldsymbol{\eta}_k + \boldsymbol{\eta}_l^* \boldsymbol{\Omega}^2 \boldsymbol{\eta}_k = 0 \tag{7.50a}$$

$$s_l^2 \boldsymbol{\eta}_k^* \boldsymbol{\eta}_l + s_l \boldsymbol{\eta}_k^* \boldsymbol{\Delta} \boldsymbol{\eta}_l + \boldsymbol{\eta}_k^* \boldsymbol{\Omega}^2 \boldsymbol{\eta}_l = 0 \tag{7.50b}$$

The conjugate of Eq. 7.50a is

$$\bar{s}_k^2 \boldsymbol{\eta}_k^* \boldsymbol{\eta}_l + \bar{s}_k \boldsymbol{\eta}_k^* \boldsymbol{\Delta} \boldsymbol{\eta}_l + \boldsymbol{\eta}_k^* \boldsymbol{\Omega}^2 \boldsymbol{\eta}_l = 0 \tag{7.50c}$$

and hence, if we subtract sidewise Eq. 7.50b from Eq. 7.50c, we have

$$(\bar{s}_k^2 - s_l^2)\boldsymbol{\eta}_k^* \boldsymbol{\eta}_l + (\bar{s}_k - s_l)\boldsymbol{\eta}_k^* \boldsymbol{\Delta} \boldsymbol{\eta}_l = 0$$

Further, if $\bar{s}_k \neq s_l$, we can delete the common factor $\bar{s}_k - s_l$ from the left-hand side of the above equation, thus obtaining

$$\boldsymbol{\eta}_k^*[(\bar{s}_k + s_l)\mathbf{1} + \boldsymbol{\Delta}]\boldsymbol{\eta}_l = 0$$

a relation that can be interpreted as a form of *mode-orthogonality*. Notice that, contrary to the undamped case, this relation depends on the corresponding eigenvalues.

Now, when $l = k$, Eq. 7.50a becomes

$$s_k^2 \eta_k^* \eta_k + s_k \eta_k^* \Delta \eta_k + \eta_k^* \Omega^2 \eta_k = 0$$

which, by virtue of assumption (7.45), becomes

$$s_k^2 + s_k \eta_k^* \Delta \eta_k + \eta_k^* \Omega^2 \eta_k = 0$$

and, in light of Eq. 7.48, the complex-conjugate of the above equation is

$$\overline{s}_k^2 + \overline{s}_k \overline{\eta}_k^* \Delta \overline{\eta}_k + \overline{\eta}_k^* \Omega^2 \overline{\eta}_k = 0$$

whereby the term independent of s_k, by analogy with single-dof second-order systems, is labeled ω_k^2, and the coefficient of the term linear in $\overline{s}_k$ is, likewise labeled, $2\zeta_k \omega_k$, these two items pertaining to the kth mode, i.e.,

$$\omega_k = \sqrt{\eta_k^* \Omega^2 \eta_k} = \sqrt{\overline{\eta}_k^* \Omega^2 \overline{\eta}_k}, \quad \zeta_k = \frac{\eta_k^* \Delta \eta_k}{2\omega_k} = \frac{\overline{\eta}_k^* \Delta \overline{\eta}_k}{2\omega_k} \tag{7.51}$$

As a result, then, once the system eigenvalues and eigenvectors are available, computing the parameters ζ_k and ω_k of the underdamped, complex modes, is straightforward. Moreover, the above results show that the modal parameters are derived from quadratic forms associated with the frequency and the dissipation matrices.

The Solution of the System Eigenvalue Problem

Discussed here is the computation of the system eigenvalues and eigenvectors, $\{s_k, \eta_k\}_1^{2n}$. We start with a fundamental result:

Lemma 7.4.2. *Let $\{\lambda_k\}_1^n$ and $\{e_k\}_1^n$ be the set of eigenvalues of a $n \times n$ matrix A, defined over the complex field, and their corresponding eigenvectors. Then, the set of eigenvalues of A^* is $\{\overline{\lambda}_k\}_1^n$.*

The proof of this lemma is also straightforward and hence, omitted. Furthermore,

Lemma 7.4.3. *Let A be a $n \times n$ matrix over the complex field. Then,*

$$\det(A^*) = \overline{\det(A)}$$

Proof: By virtue of Lemma 7.4.2, we have

$$\det(A^*) = \prod_1^m \overline{\lambda}_k$$

But, since the product of the conjugates of various complex factors is identical to the conjugate of the product of the same factors, the above determinant can be expressed as

$$\det(\mathbf{A}^*) = \overline{\prod_1^m \lambda_k}$$

which is identical to the conjugate of the determinant of $\mathbf{A}$, i.e.,

$$\det(\mathbf{A}^*) = \overline{\det(\mathbf{A})}$$

thus completing the proof.   □

From the foregoing lemma it is now apparent that the characteristic polynomial $P(s)$ obeys

$$P(\bar{s}) \equiv \overline{P(s)}$$

Thus, the real and imaginary parts of $P(s)$ can be found using standard relations, namely,

$$\mathrm{Re}\{P(s)\} = \frac{1}{2}[P(\bar{s}) + P(s)] \quad \mathrm{Im}\{P(s)\} = \frac{1}{2}[P(\bar{s}) - P(s)]$$

Moreover, let $s = x + jy$, with $j \equiv \sqrt{-1}$. Now, in order for $P(s)$ to vanish, both its real and imaginary parts must vanish, which thus leads to

$$P_1(x,y) \equiv \frac{1}{2}[\det(\bar{s}^2\mathbf{1} + \bar{s}\boldsymbol{\Delta} + \boldsymbol{\Omega}^2) + \det(s^2\mathbf{1} + s\boldsymbol{\Delta} + \boldsymbol{\Omega}^2)] = 0 \quad (7.52a)$$

$$P_2(x,y) \equiv \frac{1}{2}[\det(\bar{s}^2\mathbf{1} + \bar{s}\boldsymbol{\Delta} + \boldsymbol{\Omega}^2) - \det(s^2\mathbf{1} + s\boldsymbol{\Delta} + \boldsymbol{\Omega}^2)] = 0 \quad (7.52b)$$

thereby deriving two polynomial bivariate equations in the two unknowns x and y. The roots of these equations are bound to occur in complex-conjugate pairs, given that the coefficients of s and s^2 in the two equations are real. We are interested only in the real solutions.

Notice that Eq. 7.52a and b define two contours, $\mathscr{C}_1$ and $\mathscr{C}_2$, in the x-y plane, whose intersections determine all the real solutions of the foregoing polynomial system. These solutions, in turn, define all real and complex eigenvalues of the problem at hand. While the intersections of the two contours can be detected by hand and eye, with the aid of a mouse, precise solutions can be obtained numerically. We will not elaborate on the numerical solution of this system, which is amply discussed in the specialized textbooks.

The Factoring of the Characteristic Polynomial

Let $\{s_k\}_1^{2m}$ be the set of complex system eigenvalues, with $s_{m+k} = \bar{s}_k$, and $\{s_k\}_{2m+1}^{2n}$ the set of real system eigenvalues. From the results of the foregoing section, it is now apparent that the characteristic polynomial $P(s)$ admits a factoring in m quadratic and $2(n - m)$ linear factors in s, containing the modal information of the system.

That is,

$$P(s) = \prod_{1}^{m}(s^2 + 2\zeta_i\omega_i s + \omega_i^2) \prod_{2m+1}^{2n}(s + s_i)$$

where ω_i and ζ_i are given in Eq. 7.51. Moreover, in terms of the modal frequencies and damping ratios, the complex eigenvalues are given as

$$s_{i,m+i} = \left(-\zeta_i \pm j\sqrt{1 - \zeta_i^2}\right)\omega_i, \quad i = 1, \ldots, m$$

Note that the only parameter of each linear factor is its time constant τ_k, defined as

$$\tau_i \equiv -\frac{1}{s_i} > 0, \quad i = 2m+1, \ldots, 2n$$

which is necessarily positive, given that both the dissipation and the frequency matrices are positive-semidefinite.

The System Transfer-function Matrix

Once the factoring of the characteristic polynomial $P(s)$ is available, finding the transfer-function matrix is straightforward. It will be made apparent that the transfer function leads to the impulse response of the system by application of the inverse Laplace transform, as well as the frequency response of the system.

To obtain the transfer function of the system at hand, we simply Laplace-transform the mathematical model of the system in normal form, Eq. 7.43, with zero initial conditions and nonzero input $\mathbf{g}(t)$, which yields

$$(s^2\mathbf{1} + s\boldsymbol{\Delta} + \boldsymbol{\Omega}^2)\mathbf{y}(s) = \mathbf{g}(s)$$

whence,

$$\mathbf{y}(s) = (s^2\mathbf{1} + s\boldsymbol{\Delta} + \boldsymbol{\Omega}^2)^{-1}\mathbf{g}(s)$$

The transfer function $\mathbf{H}(s)$—a function of s, not to be confused with constant $\mathbf{H}$ introduced in Eqs. 7.22d and 7.41b—can then be obtained as the partial derivative of the response, $\mathbf{y}(s)$, with respect to the input, $\mathbf{g}(s)$, i.e.,

$$\mathbf{H}(s) = (s^2\mathbf{1} + s\boldsymbol{\Delta} + \boldsymbol{\Omega}^2)^{-1} \tag{7.53}$$

which is, by definition, the *impulse response* of the system. From the above definition of $\mathbf{H}(s)$, and recalling the units of $\mathbf{1}$, $\boldsymbol{\Delta}$ and $\boldsymbol{\Omega}$, it is apparent that $\mathbf{H}(s)$ has units of s^2, i.e., of time-squared. As pointed out in Appendix B, moreover, the units of the Laplace transform of a function of time are those of the function times second. In this light, then $\mathbf{H}(t)$ has units of time.

Note that the *zero-state response* of the system to any arbitrary input can now be written as the convolution of $\mathbf{H}(t)$, the inverse Laplace transform of $\mathbf{H}(s)$, and $\mathbf{g}(t)$, namely,

$$\mathbf{y}(t) = \int_0^t \mathbf{H}(t-\tau)\mathbf{g}(\tau)d\tau \tag{7.54a}$$

In order to have the complete zero-state response of the system, we need $\dot{\mathbf{y}}(t)$, which is readily computed from the above expression upon application of the formula for the derivative of an integral with variable integration extremes, namely,

$$\dot{\mathbf{y}}(t) = \int_0^t \frac{\partial \mathbf{H}(t-\tau)}{\partial t}\mathbf{g}(\tau)d\tau + \mathbf{H}(t-\tau)\mathbf{g}(\tau)|_{\tau=t}$$

$$= \int_0^t \frac{\partial \mathbf{H}(t-\tau)}{\partial t}\mathbf{g}(\tau)d\tau + \mathbf{H}(0)\mathbf{g}(t) \tag{7.54b}$$

The inverse Laplace transform of $\mathbf{H}(s)$ is readily obtained via its *partial-fraction expansion*:

$$\mathbf{H}(s) = \sum_1^m (s^2 + 2\zeta_i\omega_i s + \omega_i^2)^{-1}(\mathbf{H}_i + s\mathbf{H}_{m+i}) + \sum_{2m+1}^{2n}(s+s_i)^{-1}\mathbf{H}_i \tag{7.55}$$

The constant coefficients $\{\mathbf{H}_i\}_1^{2n}$ are obtained using the standard procedure applicable to scalar transfer functions. Indeed, upon multiplying both sides of Eq. 7.55 by $(s^2 + 2\zeta_i\omega_i + \omega_i^2)$, and evaluating the two sides of the expression thus resulting at the two roots s_i and $\bar{s}_i$ of the foregoing quadratic factor, we obtain, for $i = 1, \ldots, m$,

$$\mathbf{H}_i + s_i\mathbf{H}_{m+i} = \mathbf{A}_i \tag{7.56a}$$

$$\mathbf{H}_i + \bar{s}_i\mathbf{H}_{m+i} = \bar{\mathbf{A}}_i \tag{7.56b}$$

where

$$\mathbf{A}_i \equiv (s^2 + 2\zeta_i\omega_i s + \omega_i^2)\mathbf{H}(s)\big|_{s=s_i}$$

Now, the constant coefficient matrices $\mathbf{H}_i$ and $\mathbf{H}_{m+i}$ can be computed upon subtracting and adding sidewise Eq. 7.56a and b, thus obtaining, in this order,

$$\mathbf{H}_{m+i} = \frac{\text{Im}\{\mathbf{A}_i\}}{\text{Im}\{s_i\}}, \quad \mathbf{H}_i = \text{Re}\{\mathbf{A}_i\} - \mathbf{H}_{m+i}\text{Re}\{s_i\}$$

Coefficients $\mathbf{H}_i$, for $i = 2m+1, \ldots, n$, are calculated, in turn, as

$$\mathbf{H}_i = (s+s_i)\mathbf{H}(s)\big|_{s=s_i}$$

Hence, the impulse response of the system takes the form

$$\mathbf{H}(t) = \sum_1^m \left\{ \mathbf{H}_i \mathcal{L}^{-1}\left[\frac{1}{s^2 + 2\zeta_i\omega_i s + \omega_i^2}\right] \right.$$

$$\left. + \mathbf{H}_{m+i}\frac{d}{dt}\mathcal{L}^{-1}\left[\frac{1}{s^2 + 2\zeta_i\omega_i s + \omega_i^2}\right] \right\} + \sum_{2m+1}^{2n} e^{s_i t}\mathbf{H}_i$$

The explicit form of the foregoing inverse Laplace transforms is that of under-damped systems, critically damped and overdamped modes having been included as linear terms. Moreover, if a critically-damped mode ever occurs, then its eigenvalue would be a double root of the characteristic polynomial, which would give rise to two terms containing functions e^t and te^t in the above summation. Then, the corresponding term in that summation would change accordingly. For conciseness, we do not consider any further the rather unlikely case of critical damping. Hence,

$$
\mathbf{H}(t) = \sum_1^m \frac{e^{-\zeta_i \omega_i t}}{\sqrt{1 - \zeta_i^2}} \left\{ \frac{\sin \omega_i t}{\omega_i} \mathbf{H}_i + \left[\sqrt{1 - \zeta_i^2} \cos \omega_i t - \zeta_i \sin \omega_i t \right] \mathbf{H}_{m+i} \right\}
$$
$$
+ \sum_{2m+1}^{2n} e^{s_i t} \mathbf{H}_i \tag{7.57a}
$$

its time-derivative being obtained using computer algebra, thus obtaining

$$
\dot{\mathbf{H}}(t) = \sum_1^m \frac{-\zeta_i \omega_i e^{-\zeta_i \omega_i t}}{\sqrt{1 - \zeta_i^2}} \left\{ \cos \omega_i t \mathbf{H}_i + \left[-\omega_i \sqrt{1 - \zeta_i^2} \sin \omega_i t - \zeta_i \omega_i \cos \omega_i t \right] \mathbf{H}_{m+i} \right\}
$$
$$
+ \sum_{2m+1}^{2n} s_i e^{s_i t} \mathbf{H}_i \tag{7.57b}
$$

It is noteworthy that the impulse response verifies the matrix ODE

$$
\ddot{\mathbf{H}} + \boldsymbol{\Delta}\dot{\mathbf{H}} + \boldsymbol{\Omega}^2 \mathbf{H} = \delta(t)\mathbf{1}, \quad \mathbf{H}(0) = \mathbf{O}, \quad \dot{\mathbf{H}}(0) = \mathbf{O} \tag{7.57c}
$$

where $\delta(t)$ and $\mathbf{O}$ denote, respectively, the *Dirac impulse function* and the $n \times n$ zero matrix.

Furthermore, the *zero-input response* can now be derived upon taking the Laplace transform of the two sides of Eq. 7.43 with $\mathbf{g}(t) = \mathbf{0}$ and nonzero initial conditions, thus obtaining

$$
\mathbf{y}(s) = \mathbf{H}(s)[(s\mathbf{1} + \boldsymbol{\Delta})\mathbf{y}_0 + \mathbf{s}_0],
$$

and hence, the desired response takes the form

$$
\mathbf{y}(t) = \mathbf{H}(t)(\boldsymbol{\Delta}\mathbf{y}_0 + \mathbf{s}_0) + \dot{\mathbf{H}}(t)\mathbf{y}_0 \tag{7.58a}
$$

The reader is invited to compare the similarity between the above expression and the zero-input response of a scalar second-order system.

An expression for $\dot{\mathbf{y}}(t)$, completing the zero-input response, is derived by straightforward time-differentiation of the above expression for $\mathbf{y}(t)$, namely,

$$
\dot{\mathbf{y}}(t) = \dot{\mathbf{H}}(t)(\boldsymbol{\Delta}\mathbf{y}_0 + \mathbf{s}_0) + \ddot{\mathbf{H}}(t)\mathbf{y}_0 \tag{7.58b}
$$

The total time response of the system at hand is then obtained by *superposition*, i.e., as the sum of the zero-input response given by Eq. 7.58a and b and the zero-state response of Eq. 7.54a and b, thereby obtaining

$$y(t) = \left[\mathbf{H}(t)\mathbf{\Delta} + \dot{\mathbf{H}}(t)\right]\mathbf{y}_0 + \mathbf{H}(t)\mathbf{s}_0 + \int_0^t \mathbf{H}(t-\tau)\mathbf{g}(\tau)d\tau \qquad (7.59a)$$

$$s(t) = \left[\dot{\mathbf{H}}(t)\mathbf{\Delta} + \ddot{\mathbf{H}}(t)\right]\mathbf{y}_0 + \dot{\mathbf{H}}(t)\mathbf{s}_0 + \int_0^t \frac{\partial \mathbf{H}(t-\tau)}{\partial t}\mathbf{g}(\tau)d\tau$$

$$+\mathbf{H}(0)\mathbf{g}(t) \qquad (7.59b)$$

the last term of the foregoing equation vanishing by virtue of the initial conditions (7.57c). The discrete-time response can now be derived upon calculating the response at $t = t_{k+1} = t_k + h$ and regarding the state-variable values at $t = t_k$ as the initia conditions. Moreover, $\mathbf{H}(t)$ and $\dot{\mathbf{H}}(t)$ are evaluated at $t = t_{k+1} - t_k = h$, thereby obtaining

$$\mathbf{y}_{k+1} = \left[\mathbf{H}(h)\mathbf{\Delta} + \dot{\mathbf{H}}(h)\right]\mathbf{y}_k + \mathbf{H}(h)\mathbf{s}_k + \int_{t_k}^{t_k+h} \mathbf{H}(t_k + h - \tau)\mathbf{g}_k d\tau$$

$$\mathbf{v}_{k+1} = \left[\dot{\mathbf{H}}(h)\mathbf{\Delta} + \ddot{\mathbf{H}}(h)\right]\mathbf{y}_k + \dot{\mathbf{H}}(h)\mathbf{s}_k + \int_{t_k}^{t_k+h} \frac{\partial \mathbf{H}(t_k + h - \tau)}{\partial t}\mathbf{g}_k d\tau$$

where $\mathbf{g}_k \equiv \mathbf{g}(t_k)$, matrices $\mathbf{H}(h)$ and $\dot{\mathbf{H}}(h)$ are evaluated directly from Eq. 7.57a and b, while $\ddot{\mathbf{H}}(h)$ from Eq. 7.57c, which yields

$$\ddot{\mathbf{H}}(h) = \delta(h)\mathbf{1} - \mathbf{\Delta}\dot{\mathbf{H}}(h) - \mathbf{\Omega}^2\mathbf{H}(h)$$

but $\delta(h) = 0$ from the definition of the impulse function, and hence,

$$\ddot{\mathbf{H}}(h) = -\mathbf{\Delta}\dot{\mathbf{H}}(h) - \mathbf{\Omega}^2\mathbf{H}(h) \qquad (7.60)$$

The simulation scheme for $s(t)$ thus becomes

$$\mathbf{s}_{k+1} = \left[\dot{\mathbf{H}}(h)\mathbf{\Delta} - \mathbf{\Delta}\dot{\mathbf{H}}(h) - \mathbf{\Omega}^2\mathbf{H}(h)\right]\mathbf{y}_k + \dot{\mathbf{H}}(h)\mathbf{s}_k + \left[\int_{t_k}^{t_k+h} \frac{\partial \mathbf{H}(t_k + h - \tau)}{\partial t}\right]\mathbf{g}_k d\tau$$

Again, for reasons of numerical stability, it is preferable to work with $\mathbf{w}_k = \mathbf{\Omega}^{-1}\mathbf{s}_k$, and hence, the simulation scheme becomes

$$\mathbf{y}_{k+1} = \left[\mathbf{H}(h)\mathbf{\Delta} + \dot{\mathbf{H}}(h)\right]\mathbf{y}_k + \mathbf{H}(h)\mathbf{\Omega}\mathbf{w}_k$$

$$+ \left[\int_{t_k}^{t_k+h} \mathbf{H}(t_k + h - \tau)d\tau\right]\mathbf{g}_k \qquad (7.61a)$$

$$\mathbf{w}_{k+1} = \mathbf{\Omega}^{-1} \left[\dot{\mathbf{H}}(h)\mathbf{\Delta} - \mathbf{\Delta}\dot{\mathbf{H}}(h) - \mathbf{\Omega}^2 \mathbf{H}(h) \right] \mathbf{y}_k + \mathbf{\Omega}^{-1}\dot{\mathbf{H}}(h)\mathbf{\Omega}\mathbf{w}_k \quad (7.61b)$$

$$+ \mathbf{\Omega}^{-1} \left[\int_{t_k}^{t_k+h} \frac{\partial \mathbf{H}(t_k+h-\tau)}{\partial t} \right] \mathbf{g}_k d\tau \qquad (7.61c)$$

All that remains now to complete the simulation scheme of interest is the evaluation of the integrals in the above equation. This is done using the same idea introduced in Sect. 3.4: let $\theta \equiv t_{k+1} - \tau$, which yields the integral $\mathbf{J}$ of Eq. 7.61a as

$$\mathbf{J} \equiv \int_{t_k}^{t_{k+1}} \mathbf{H}(t_{k+1} - \tau)d\tau = -\int_{h}^{0} \mathbf{H}(\theta)d\theta = \int_{0}^{h} \mathbf{H}(\theta)d\theta$$

Likewise, the integral $\mathbf{L}$ of Eq. 7.61c is obtained as

$$\mathbf{L} \equiv \int_{t_k}^{t_{k+1}} \frac{\partial \mathbf{H}(t_{k+1} - \tau)}{\partial t_{k+1}} d\tau = -\int_{h}^{0} \frac{\partial \mathbf{H}(\theta)}{\partial \theta} d\theta = \int_{0}^{h} \frac{\partial \mathbf{H}(\theta)}{\partial \theta} d\theta = \int_{0}^{h} d\mathbf{H}$$

$$= \mathbf{H}(h) - \mathbf{H}(0) = \mathbf{H}(h)$$

where the initial conditions of Eq. 7.57c have been recalled.

Expressions for the foregoing integrals are obtained with the aid of computer algebra:

$$\mathbf{J} = \sum_{1}^{m} \frac{1}{\omega_i^2 \left(1 + \zeta_i^2\right) \sqrt{1 - \zeta_i^2}} \left\{ \left[e^{-\zeta_i \omega_i h} \left(-\zeta_i \sin \omega_i h - \cos \omega_i h\right) + 1 \right] \mathbf{H}_i \right.$$

$$+ \left[\omega_i \sqrt{1 - \zeta_i^2} \left[e^{-\zeta_i \omega_i h} \left(-\zeta_i \cos \omega_i h + \sin \omega_i h\right) - \zeta_i \right] \right.$$

$$\left. + \omega_i \left[e^{-\zeta_i \omega_i h} \zeta_i \left(\zeta_i \sin \omega_i h + \cos \omega_i h\right) - \zeta_i \right] \right] \mathbf{H}_{m+i} \right\}$$

$$+ \sum_{2m+1}^{2n} \frac{1}{s_i} \left(e^{s_i h} - 1 \right) \mathbf{H}_i \qquad (7.62a)$$

and

$$\mathbf{L} = \mathbf{H}(h) = \sum_{1}^{m} \frac{1}{\sqrt{1 - \zeta_i^2}} \left\{ \frac{e^{-\zeta_i \omega_i h} \sin \omega_i h}{\omega_i} \mathbf{H}_i + \left[e^{-\zeta_i \omega_i h} \left(\sqrt{1 - \zeta_i^2} \cos \omega_i h \right. \right. \right.$$

$$\left. \left. \left. - \zeta_i \sin \omega_i h \right) - \sqrt{1 - \zeta_i^2} \right] \mathbf{H}_{m+i} \right\}$$

$$+ \sum_{2m+1}^{2n} \left(e^{s_i h} - 1 \right) \mathbf{H}_i \qquad (7.62b)$$

If we let $\mathbf{H}(h) \equiv \mathbf{H}_h$ and $\dot{\mathbf{H}}(h) \equiv \dot{\mathbf{H}}_h$, then the simulation scheme (7.61a and b) takes the form

$$\mathbf{y}_{k+1} = \left[\mathbf{H}_h\boldsymbol{\Delta} + \dot{\mathbf{H}}_h\right]\mathbf{y}_k + \mathbf{H}_h\boldsymbol{\Omega}\mathbf{w}_k + \mathbf{J}\mathbf{g}_k$$

$$\mathbf{w}_{k+1} = \boldsymbol{\Omega}^{-1}\left[\dot{\mathbf{H}}_h\boldsymbol{\Delta} - \boldsymbol{\Delta}\dot{\mathbf{H}}_h - \boldsymbol{\Omega}^2\mathbf{H}_h\right]\mathbf{y}_k + \boldsymbol{\Omega}^{-1}\dot{\mathbf{H}}_h\boldsymbol{\Omega}\mathbf{w}_k + \boldsymbol{\Omega}^{-1}\mathbf{L}\mathbf{g}_k$$

or, in terms of the state-variable vector $\boldsymbol{\zeta}$ defined in Eq. 7.41a,

$$\boldsymbol{\zeta}_{k+1} = \mathbf{F}\boldsymbol{\zeta}_k + \mathbf{G}\mathbf{g}_k \tag{7.63a}$$

where

$$\mathbf{F} = \begin{bmatrix} \mathbf{H}_h\boldsymbol{\Delta} + \dot{\mathbf{H}}_h & \mathbf{H}_h\boldsymbol{\Omega} \\ \boldsymbol{\Omega}^{-1}\left(\dot{\mathbf{H}}_h\boldsymbol{\Delta} - \boldsymbol{\Delta}\dot{\mathbf{H}}_h - \boldsymbol{\Omega}^2\mathbf{H}_h\right) & \boldsymbol{\Omega}^{-1}\dot{\mathbf{H}}_h\boldsymbol{\Omega} \end{bmatrix}, \quad \mathbf{G} = \begin{bmatrix} \mathbf{J} \\ \boldsymbol{\Omega}^{-1}\mathbf{L} \end{bmatrix} \tag{7.64a}$$

Given that $\mathbf{H}(t)$ has units of time, all the entries of $\mathbf{F}$ are dimensionless, and hence, $\mathbf{F}$ is dimensionally homogeneous. The reader should be able to prove that all the entries of $\mathbf{G}$ have units of time-squared s^2. The simulation algorithm derived in Sect. 7.4.1 is still applicable, the only steps that change are 12–14. Once vector $\boldsymbol{\zeta}_{k+1}$ is available, its counterpart $\mathbf{z}_{k+1}$ in the original state variable is obtained by means of a linear transformation:

$$\mathbf{z}_{k+1} = \mathbf{Z}\boldsymbol{\zeta}_{k+1}, \quad \mathbf{Z} \equiv \begin{bmatrix} \mathbf{N}^{-1} & \mathbf{0} \\ \mathbf{0} & \mathbf{N}^{-1}\boldsymbol{\Omega} \end{bmatrix}$$

thereby completing the simulation scheme.

Note that, once functions $\mathbf{y}(t)$ and $\dot{\mathbf{y}}(t)$ have been obtained, a transformation back to the original variables $\mathbf{x}(t)$ and $\dot{\mathbf{x}}(t)$ is needed, as per the transformations of Eq. 7.5, i.e.,

$$\mathbf{x}(t) = \mathbf{N}^{-1}\mathbf{y}(t), \quad \dot{\mathbf{x}}(t) = \mathbf{N}^{-1}\dot{\mathbf{y}}(t)$$

whence the impulse response $\mathbf{I}(t)$ of the original mathematical model, Eq. 7.33, follows. Indeed, upon taking into account the variable transformation of Eq. 7.5, we have

$$\mathbf{I}(t) = \mathbf{N}^{-1}\mathbf{H}(t)\mathbf{N}$$

With the above relations, it is now possible to derive the zero-input response of the given system in terms of the generalized-coordinate and the generalized-velocity vectors, $\mathbf{x}$ and $\dot{\mathbf{x}}$, respectively:

$$\mathbf{x}(t) = [\mathbf{I}(t)\mathbf{M}^{-1}\mathbf{C} + \dot{\mathbf{I}}(t)]\mathbf{x}_0 + \mathbf{I}(t)\mathbf{v}_0 \tag{7.65a}$$

$$\dot{\mathbf{x}}(t) = [\dot{\mathbf{I}}(t)\mathbf{M}^{-1}\mathbf{C} + \ddot{\mathbf{I}}(t)]\mathbf{x}_0 + \dot{\mathbf{I}}(t)\mathbf{v}_0 \tag{7.65b}$$

In fact, in deriving the above expressions, we have obtained the exponential of the matrix coefficient of system (7.33) when written in state-variable form, namely, as

$$\dot{\mathbf{z}} = \mathbf{A}\mathbf{x} + \mathbf{B}f(t), \quad \mathbf{z}(0) = \mathbf{z}_0 \tag{7.66a}$$

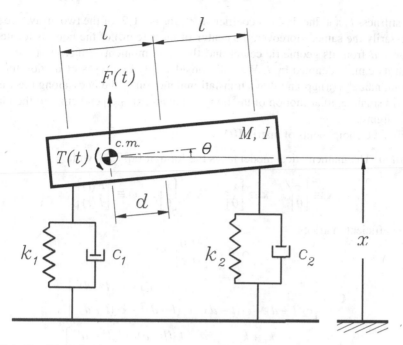

Fig. 7.4 Simplified vibrational model of an automobile suspension

with the definitions:

$$\mathbf{z} \equiv \begin{bmatrix} \mathbf{x} \\ \dot{\mathbf{x}} \end{bmatrix}, \quad \mathbf{A} \equiv \begin{bmatrix} \mathbf{O} & \mathbf{1} \\ -\mathbf{M}^{-1}\mathbf{K} & -\mathbf{M}^{-1}\mathbf{C} \end{bmatrix}, \quad \mathbf{B} \equiv \begin{bmatrix} \mathbf{O} \\ \mathbf{M}^{-1} \end{bmatrix} \tag{7.66b}$$

and $\mathbf{O}$ denoting the $n \times n$ zero matrix. Therefore,

$$e^{\mathbf{A}t} = \begin{bmatrix} \mathbf{I}(t)\mathbf{M}^{-1}\mathbf{C} + \dot{\mathbf{I}}(t) & \mathbf{I}(t) \\ \dot{\mathbf{I}}(t)\mathbf{M}^{-1}\mathbf{C} + \ddot{\mathbf{I}}(t)] & \dot{\mathbf{I}}(t) \end{bmatrix} \tag{7.66c}$$

thereby obtaining the exponential of the system matrix times t in block form, in terms of the original system matrices $\mathbf{M}$, $\mathbf{K}$ and $\mathbf{C}$. As a matter of fact, $\mathbf{K}$ does not appear explicitly in Eq. 7.66c. This matrix is implicit in terms $\mathbf{I}$ and $\dot{\mathbf{I}}$.

Example 7.4.2 (Two-dof Damped Test Pad Revisited). We revisit here the two-dof model of the test pad introduced in Example 5.7.1, and reproduced in Fig. 7.4 for quick reference. In the foregoing example, all the system parameters were given numerical values. In this example, these parameters are left in terms of one parameter, $\omega = \sqrt{k/M}$. The model consists of a body with mass M, supported by two spring-dashpot arrays and excited by an external force $F(t)$ and an external torque $T(t)$, which need not be specified, as they are not required in this example.

The stiffness k_i and the dashpot coefficient c_i, for $i = 1, 2$, of the two arrays are not necessarily the same. Moreover, the center of mass (c.m.) of the body is located a distance d from its geometric center and the mass moment of inertia of the body about its c.m. is denoted by I. We will consider only two types of motion for the system, namely, (a) up-and-down translational motion of the body along the x axis and (b) small angular motion of the body about an axis perpendicular to the plane of the figure.

Find the components of matrix $\mathbf{I}(t)$.

Solution: The mathematical model takes the form of Eq. 7.33, with

$$\mathbf{x} \equiv \begin{bmatrix} x \\ \theta \end{bmatrix}, \quad \dot{\mathbf{x}} \equiv \begin{bmatrix} \dot{x} \\ \dot{\theta} \end{bmatrix}, \quad \ddot{\mathbf{x}} \equiv \begin{bmatrix} \ddot{x} \\ \ddot{\theta} \end{bmatrix}, \quad \boldsymbol{\phi} \equiv \begin{bmatrix} F(t) \\ T(t) \end{bmatrix}$$

and coefficient matrices

$$\mathbf{M} \equiv \begin{bmatrix} M & 0 \\ 0 & I \end{bmatrix},$$

$$\mathbf{C} \equiv \begin{bmatrix} c_1 + c_2 & c_2(l+d) - c_1(l-d) \\ c_2(l+d) - c_1(l-d) & c_1(l-d)^2 + c_2(l+d)^2 \end{bmatrix},$$

$$\mathbf{K} \equiv \begin{bmatrix} k_1 + k_2 & k_2(l+d) - k_1(l-d) \\ k_2(l+d) - k_1(l-d) & k_1(l-d)^2 + k_2(l+d)^2 \end{bmatrix}$$

Moreover,

$$\mathbf{N} = \begin{bmatrix} \sqrt{M} & 0 \\ 0 & \sqrt{I} \end{bmatrix}$$

Now, using matrix $\mathbf{N}$ we can obtain the mathematical model of the system at hand in normal form (7.33), and hence, its characteristic equation. We assume the data shown below:

$$c_1 = c, \quad c_2 = 2c, \quad k_1 = 2k, \quad k_2 = k, \quad d = \frac{l}{2}, \quad I = Mr^2,$$

with r defined as the radius of gyration of the block. Using the foregoing data, the system matrices take the forms

$$\mathbf{M} = M \begin{bmatrix} 1 & 0 \\ 0 & r^2 \end{bmatrix}, \quad \mathbf{C} = c \begin{bmatrix} 3 & 5l/2 \\ 5l/2 & 19(l/2)^2 \end{bmatrix}, \quad \mathbf{K} = k \begin{bmatrix} 3 & l/2 \\ l/2 & 11(l/2)^2 \end{bmatrix},$$

whence,

$$\mathbf{N} = \sqrt{M} \begin{bmatrix} 1 & 0 \\ 0 & r \end{bmatrix}$$

Moreover, we introduce the notation

$$\lambda = \frac{l}{r}, \quad \sigma = \frac{c}{M}, \quad \omega^2 = \frac{k}{M}$$

Then, the two matrices $\mathbf{\Delta}$ and $\mathbf{\Omega}^2$ are readily computed as

$$\mathbf{\Delta} = \sigma \begin{bmatrix} 3 & 5\lambda/2 \\ 5\lambda/2 & 19(\lambda/2)^2 \end{bmatrix}, \quad \mathbf{\Omega}^2 = \omega^2 \begin{bmatrix} 3 & \lambda/2 \\ \lambda/2 & 11(\lambda/2)^2 \end{bmatrix}$$

It is now apparent that $\mathbf{\Delta}$ and $\mathbf{\Omega}^2$ do not commute under multiplication, and hence, the system at hand is not proportionally-damped.[11] The characteristic equation of the system then follows immediately:

$$\det \begin{bmatrix} s^2 + 3\sigma s + 3\omega^2 & (5/2)\lambda\sigma s + (1/2)\lambda\omega^2 \\ (5/2)\lambda\sigma s + (1/2)\lambda\omega^2 & s^2 + 19(\lambda/2)^2\sigma s + 11(\lambda/2)^2\omega^2 \end{bmatrix} = 0$$

Upon expansion, the characteristic equation becomes, with the assumed relations $\lambda = 2\sqrt{3}$ and $\sigma = \omega$,

$$P(s) \equiv s^4 + 60\omega s^3 + 132\omega^2 s^2 + 240\omega^3 s + 96\omega^4 = 0$$

The two scalar equations derived from the above characteristic equation are, then, with $s \equiv x + jy$,

$$f_1(x,y) \equiv 96\omega^4 + 240\omega^3 x + 132\omega^2(x^2 - y^2) + 60\omega x(x^2 - 3y^2)$$
$$+ x^4 + y^4 - 6x^2 y^2 = 0,$$
$$f_2(x,y) \equiv (60\omega^3 + 66\omega^2 x + 45\omega x^2 + x^3 - 15\omega y^2 - xy^2)\omega y = 0$$

It is apparent from the second equation that $y = 0$ is a possible solution. Moreover, note that the first equation is even in y, and so is the second, once the solution $y = 0$ is removed from it. This is natural, for the solutions should exhibit a symmetric array in the complex plane, about the x-axis. The superimposed contour plots of the two foregoing equations are shown in Fig. 7.5, in which $\mathscr{C}_2$ is shown with solid line. It should be noted that the x and y axes represent the real and imaginary parts of the system eigenvalues, respectively. Moreover, the two contours intersect at four points, two of which lie in the x axis, and hence, correspond to real eigenvalues. The computed eigenvalues, along with their corresponding eigenvectors, are displayed in Table 7.1.

With the system eigenvalues and eigenvectors known, we can now proceed to obtain the modal parameters. Since we have two complex and two real modes, we have one underdamped and one overdamped mode, the latter being broken down into two first-order systems with time constants τ_3 and τ_4. The natural frequency and the damping ratio of the underdamped mode are, then,

$$\omega_1 = 1.804337024\omega \text{ s}^{-1}, \quad \zeta_1 = 0.4718069257$$

---

[11] In proportionally damped systems, the damping matrix is a linear combination of the mass and the stiffness matrices, which then leads to a simple eigenvalue problem.

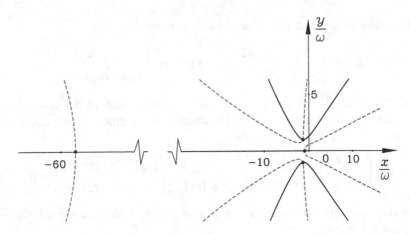

Fig. 7.5 Contours $\mathscr{C}_1$ and $\mathscr{C}_2$ of the numerical example

Table 7.1 Eigenvalues and eigenvectors

Intersection point	Real part (x)	Image part (y)	Eigenvector (η)
1	-0.851298704ω	1.59088737ω	$\begin{bmatrix} 0.190672205 - 0.967892209\,j \\ -0.0620574141 + 0.151583836\,j \end{bmatrix}$
2	-0.851298704ω	-1.59088737ω	$\begin{bmatrix} 0.190672205 + 0.967892209\,j \\ -0.0620574141 - 0.151583836\,j \end{bmatrix}$
3	-0.510275608ω	0	$\begin{bmatrix} 0.840871577 \\ 0.541234691 \end{bmatrix}$
4	-57.7871270ω	0	$\begin{bmatrix} -0.155461394 \\ -0.987841969 \end{bmatrix}$

while the time constants of the two first-order systems are

$$\tau_3 = 1.95972526436\omega \text{ s}, \quad \tau_4 = 0.0173048921492\omega \text{ s}$$

The system transfer-function matrix is readily calculated as

$$\mathbf{H}(s) = \begin{bmatrix} \dfrac{s^2 + 57s + 33}{P(s)} & -\dfrac{\sqrt{3}\,(5s+1)}{P(s)} \\ -\dfrac{\sqrt{3}\,(5s+1)}{P(s)} & \dfrac{s^2 + 3s + 3}{P(s)} \end{bmatrix}$$

Now, in order to obtain the partial-fraction expansion of $\mathbf{H}(s)$, we calculate matrices $\mathbf{A}_k$, for $k = 1, 2, 3, 4$:

$$\mathbf{A}_1 = \begin{bmatrix} (-0.1438449538 - .1311668795\,j) \times 10^7 & 276353.9107 + 167636.7853\,j \\ 276353.9107 + 167636.7853\,j & -50391.61089 - 18462.28050\,j \end{bmatrix},$$

$$\mathbf{A}_2 = \begin{bmatrix} (-0.1438449538 + .1311668795\,j) \times 10^7 & 276353.9107 - 167636.7853\,j \\ 276353.9107 - 167636.7853\,j & -50391.61089 + 18462.28050\,j \end{bmatrix},$$

$$\mathbf{A}_3 = \begin{bmatrix} 0.7282387310 & 4.627417546 \\ 4.627417546 & 29.40380979 \end{bmatrix},$$

$$\mathbf{A}_4 = \begin{bmatrix} 0.1376649089 \times 10^7 & 886092.7926 \\ 886092.7926 & 570341.7414 \end{bmatrix}$$

Hence, the $\mathbf{H}_k$ matrix coefficients take the forms

$$\mathbf{H}_1 = \begin{bmatrix} -0.1576534923 \times 10^7 & 201487.8310 \\ 201487.8310 & -22190.06370 \end{bmatrix},$$

$$\mathbf{H}_2 = \frac{1}{\omega} \begin{bmatrix} -0.2876899076 \times 10^7 & 552707.8214 \\ 552707.8214 & -100783.2218 \end{bmatrix},$$

$$\mathbf{H}_3 = \frac{1}{\omega} \begin{bmatrix} 0.7282387310 & 4.627417546 \\ 4.627417546 & 29.40380979 \end{bmatrix},$$

$$\mathbf{H}_4 = \frac{1}{\omega} \begin{bmatrix} 0.1376649089 \times 10^7 & 886092.7926 \\ 886092.7926 & 570341.7414 \end{bmatrix}$$

Thus, the components of the impulse-response matrix are

$$H_{11}(t) = \frac{1}{\omega}[1.134170188\,f_5(t)\,(483593.4317\,f_4(t) - 0.2536567356 \times 10^7\,f_3(t))$$

$$+\, 0.7282387310\,f_2(t) + 0.1376649089 \times 10^7\,f_1(t)],$$

$$H_{12}(t) = \frac{1}{\omega}[1.134170188\,f_5(t)\,(-149102.7549\,f_4(t) + 487323.5314\,f_3(t))$$

$$+\, 4.627417546\,f_2(t) + 886092.7926\,f_1(t)],$$

$$H_{21}(t) = \frac{1}{\omega}[1.134170188\,f_5(t)\,(-149102.7549\,f_4(t) + 487323.5314\,f_3(t))$$

$$+\, 4.627417546\,f_2(t) + 886092.7926\,f_1(t)],$$

$$H_{22}(t) = \frac{1}{\omega}[1.134170188\,f_5(t)\,(35252.04082\,f_4(t) - 88860.75726\,f_3(t))$$

$$+\, 29.40380979\,f_2(t) + 570341.7414\,f_1(t)]$$

where

$$f_1(t) \equiv e^{-.5102756076t}, \qquad f_2(t) \equiv e^{-57.78712698t}, \qquad f_3(t) \equiv \cos(1.804337024t),$$

$$f_4(t) \equiv \sin(1.804337024t), f_5(t) \equiv e^{-.8512987042t}$$

Further,

$$N = \begin{bmatrix} 1 & 0 \\ 0 & \frac{1}{6}\sqrt{3} \end{bmatrix} \quad \Rightarrow \quad N^{-1} = \begin{bmatrix} 1 & 0 \\ 0 & 2\sqrt{3} \end{bmatrix}$$

Thus, the components of the impulse-response matrix $I(t)$, for the original generalized coordinates, are

$$I_{11}(t) = \frac{1}{\omega}[1.134170188\, f_5(t)\,(483593.4317\, f_4(t) - .2536567356\,10^7\, f_3(t))$$
$$+ 0.7282387310\, f_2(t) + .1376649089\,10^7\, f_1(t)],$$

$$I_{12}(t) = \frac{1}{\omega}[0.3274067319\, f_5(t)\,(-149102.7549\, f_4(t) + 487323.5314\, f_3(t))$$
$$+ 1.335820384\, f_2(t) + 255792.9562\, f_1(t)],$$

$$I_{21}(t) = \frac{1}{\omega}[3.928880782\, f_5(t)\,(-149102.7549\, f_4(t) + 487323.5314\, f_3(t))$$
$$+ 16.02984460\, f_2(t) + .3069515474\,10^7\, f_1(t)],$$

$$I_{22}(t) = \frac{1}{\omega}[1.134170188\, f_5(t)\,(35252.04082\, f_4(t) - 88860.75726\, f_3(t))$$
$$+ 29.40380979\, f_2(t) + 570341.7414\, f_1(t)]$$

7.5 Exercises

For Exercises 7.2–7.10, obtain the discrete-time response and illustrate it with plots for a duration that shows the salient features of the response, when the duration is not indicated.

7.1. Compute e^{At} for the system of Example 7.4.2 using computer algebra. Then, using the foregoing expression, compute matrix H of Algorithm DnDOF. When doing this, notice that, although the latter is given by e^{Ah}, the two exponentials are different, for their arguments A are. *Hint: Notice the relation between the two different state-variable vectors, z and ζ, as given by Eq. 7.41a. Based on this relation, find the corresponding relation between the two exponentials.*

7.2. The system of Example 6.3.1 when the wheels encounter a bump identical to that of Example 3.4.1. Plot the time response for 20 s, assuming that the subway train is traveling at constant speeds of 50, 71 and 80 km/h.

7.3. The vehicle modeled in Example 6.2.2, to study its behavior under roll motion. To this end, assume that $v_0 = 1$ m/s and $\omega = 1$ Hz. Compare your result with the response obtained in Example 6.3.1.

7.4. The system described in Exercise 5.11, for a response defined in terms of dimensionless generalized coordinates $\xi_i \equiv x_1/l$, for $i = 1, 2$, assuming $\mathbf{x} = \mathbf{0}$ and $\dot{\mathbf{x}} = \mathbf{0}$. Plot ξ_i, for $i = 1, 2$, for the time intervals $0 \leq t \leq 5/\omega$ and $1000 \leq t \leq 1005/\omega$. Compare your results with those given by the total response given in Sect. 5.6.

7.5. Repeat Exercise 7.4, but now taking into account damping, which can be modeled by replacing the springs of Fig. 5.24 with a parallel array of a spring and a dashpot of coefficient c, so that $c/m = 0.002\omega$.

7.6. The system described in Exercise 5.12, for a response defined in terms of dimensionless generalized coordinates $\xi_i \equiv x_1/l$, for $i = 1, 2$, assuming $\mathbf{x} = \mathbf{0}$ and $\dot{\mathbf{x}} = \mathbf{0}$. Plot ξ_i, for $i = 1, 2$, for the time intervals $0 \leq t \leq 5/\omega$ and $1000 \leq t \leq 1005/\omega$. Compare your results with those of the total response given in Sect. 5.6.

7.7. Repeat Exercise 7.6, but now taking into account damping, which can be modeled by replacing the springs of Fig. 5.24 with a parallel array of a spring and a dashpot of coefficient c, so that $c/m = 0.002\omega$.

7.8. The system described in Exercise 5.15, with $v \equiv N_1/N_2$ and $\sqrt{k/(J_1 + J_2 v2)} = 10\,\text{Hz}$ under a sudden disturbance $\theta_1(0) = \theta_2(0) = 0$, $\dot{\theta}_1 = 1\,\text{rad/s}$, $\dot{\theta}_2 = 0$.

7.9. The system of Example 5.5.2, with $\omega_f = 100\,\text{Hz}$, under the assumption that $f_0 = 100 m_1 g$. With the purpose of keeping the overall system as heavy as the original one, m_2 should be made 1% of m_1.

7.10. The system of Example 4.2.1, if with some modifications, is used to model the human arm under "normal" walking conditions. To this end, the forearm is assumed to have a length $\sqrt{2}/2a$, with a denoting the length of the arm.[12] Moreover, the shoulder joint O_1 is assumed to undergo a harmonic vertical displacement of amplitude 50 mm and frequency given by twice the length of one step, i.e., 900 mm, but no horizontal displacement—laboratory conditions assumed here. Moreover, the subject is assumed to walk at a speed of 1 step/s. Use anthropometric data for the mass of the arm and the forearm to produce an undamped two-dof system, with reasonable values for the masses and the moments of inertia of the arm and the forearm.

7.11. The system of Example 7.4.2 is revisited here. Simulate the time response of this system under the assumption that, with the system at rest at $t = 0$, impulsive loads $F(t) = F_0\delta(t)$ and $T(t) = T_0\delta(t)$ are applied. Consider three cases: (1) $F_0 = Mg/\omega$, $T_0 = 0$; (2) $F_0 = 0$, $T_0 = Mgr/\omega$; (3) $F_0 = Mg/\omega$, $T_0 = Mgr/\omega$. In all three cases, assume the numerical values given in that example and $\omega = 1\,\text{Hz}$.

---

[12]Leonardo da Vinci concluded that the ratio of the forearm length to that of the arm is 71.4%, not too far from the assumption adopted here.

References

1. Al-Widyan K, Angeles J, Ostrovskaya S (2003) A Robust simulation algorithm for conservative linear mechanical systems. Int J Multiscale Comput Eng 1(2):289–309
2. Åström K, Wittenmark B (1997) Computer-controlled systems: theory and design. Prentice-Hall Inc, Upper Saddle River
3. Angeles J, Espinosa I (1981) Suspension-system synthesis for mass transport vehicles with prescribed dynamic behavior. ASME Paper 81-DET-44, In: Proceeding 1981 ASME design engineering technical conference, Hartford, 20–23 September 1981
4. Angeles J, Etemadi Zanganeh K, Ostrovskaya S (1999) The analysis of arbitrarily-damped linear mechanical systems. Arch Appl Mech 69(8):529–541

Chapter 8
Vibration Analysis of Continuous Systems

And the spiral in every thing
disperses its vibration as it turns:
motion knows no rest.

Paz, O., 1960, *Libertad Bajo Palabra,*
Lecturas Mexicanas 4, Mexico City.[1] Translated by the author.

8.1 Introduction

The *continuum* is an abstraction that finds extensive applications in mechanics, for it serves to model many mechanical systems, such as fluids and structural elements of the most complex shapes. In fact, all mechanical systems encountered by the engineer are most accurately modeled by continua, but, in some instances, the type of motion most likely to occur is describable by a finite set of independent generalized variables. This is why we can interpret mechanical systems with *n*-dof as *approximations* of continuous systems; the latter are characterized by the presence of infinitely many dof. For example, the membrane in a loudspeaker is most accurately modeled as a continuum, but, because of the symmetries of its shape and those imposed by its constraints, we can, in many instances, model it as a simple, single-dof mass-spring-dashpot system. If non-symmetries are present, or if we are interested in modeling the imperfections of the design and the assembly, then a continuous model becomes mandatory.

We will not be concerned in this chapter with arbitrary continuous systems, but rather with the simplest ones, i.e., those with one single spatial variable. In spite of this limited scope, we will be able to model many practical mechanical systems, such as prismatic bars under axial and torsional motion; strings under lateral motion; and beams under flexural motion. In all these cases, the geometry of the system,

---

[1] From Octavio Paz's 1944 poem *Condición de Nube*.

J. Angeles, *Dynamic Response of Linear Mechanical Systems: Modeling, Analysis and Simulation*, Mechanical Engineering Series, DOI 10.1007/978-1-4419-1027-1_8,
© Springer Science+Business Media, LLC 2011

which will be referred to as *the element* indistinctly, is assumed to be simple, namely, of a constant cross section, whose detailed shape, in most cases, is not needed. As will become apparent in the sections below, all that matters for the purposes of our analysis is the *global* properties of the cross section, such as its area, its polar moment of inertia, and other similar area properties. Therefore, the configuration of the overall element is completely known when the location of a landmark point on its cross section is known as a function of a variable, say x, defined along a line perpendicular to the cross section, and time t. We will thus define the displacement, whether translational, in which case it is denoted by $u(x, t)$ or $y(x, t)$, or angular, in which case it is denoted by $\theta(x, t)$, as a function of two variables, x and t. Here, x takes any values between a *left* value, say a and a *right* value, say b. That is,

$$u = u(x, t), \quad \theta = \theta(x, t), \quad a \leq x \leq b, \quad t \geq 0 \tag{8.1}$$

Therefore, we can regard the continuum as the limiting case, in which $n \to \infty$, of a *finite* set of n generalized coordinates $\{x_k\}_1^n$, where

$$a \equiv x_1 < x_2 < \cdots < x_n \equiv b \tag{8.2}$$

We shall see that all properties studied for n-dof systems, e.g., the existence of natural frequencies and natural modes, orthogonality of the latter, etc., find their counterpart in the case of continuous systems. A major difference will be made apparent in the case of continuous systems: the number of natural frequencies and natural modes becomes now infinite.

8.2   Mathematical Modeling

We derive below the mathematical models of the systems that we will consider in this chapter: (1) bars under *axial vibration*; (2) bars under *torsional vibration*; (3) strings under *transverse vibration*; and (4) beams under *flexural vibration*. In all cases below, we assume that the motions under study produce *small enough* strains on the element at hand so as to allow us to assume safely that the constitutive material operates in the *linearly elastic* regime. This leads to finite axial or angular displacements in bars, and small transverse displacements in strings and beams, assumptions that will be invoked when deriving integrals for the mathematical models to be obtained in the subsections below.

8.2.1   *Bars Under Axial Vibration*

We first consider the prismatic bar shown in Fig. 8.1a. This bar has cross-section area A_s, constant Young modulus of elasticity E and mass μ per unit length. Moreover,

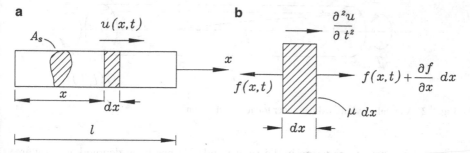

Fig. 8.1 (**a**) A prismatic bar under axial vibration; (**b**) the FBD of a differential element of length dx

for the sake of conciseness, we will assume that μ is also a constant. From the FBD of the differential element shown in Fig. 8.1b, it is apparent that

$$\frac{\partial f(x,t)}{\partial x}dx = \mu dx \frac{\partial^2 u(x,t)}{\partial t^2} \tag{8.3}$$

Furthermore, let us denote with $\sigma(x,t)$ the stress at the cross section located at distance x from the left end, at time t. This stress is assumed to be *uniformly distributed* throughout the cross section, i.e.,

$$\sigma(x,t) \equiv \frac{f(x,t)}{A_s} \tag{8.4}$$

Likewise, we let $\varepsilon(x,t)$ be the strain at the same cross section, assumed to be *uniformly distributed* throughout the cross section, i.e.,

$$\varepsilon(x,t) \equiv \frac{\partial u(x,t)}{\partial x} \tag{8.5}$$

Next, we invoke *Hooke's Law*:

$$\sigma(x,t) = E\varepsilon(x,t) \tag{8.6}$$

Upon substitution of Eqs. 8.4–8.6 into Eq. 8.3, and dropping of the common factor dx from both sides of the same equation, we obtain

$$EA_s \frac{\partial^2 u(x,t)}{\partial x^2} = \mu \frac{\partial^2 u(x,t)}{\partial t^2} \tag{8.7}$$

which can be recast in the form

$$\frac{\partial^2 u(x,t)}{\partial x^2} = \beta^2 \frac{\partial^2 u(x,t)}{\partial t^2}, \quad 0 < x < l, \quad t \geq 0 \tag{8.8a}$$

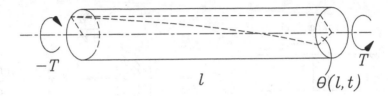

Fig. 8.2 A circular cylindrical bar under torsional vibration

where the constant appearing in the right-hand side has been denoted as a square because it is the quotient of two positive quantities, and is thus defined as

$$\beta \equiv \sqrt{\frac{\mu}{EA_s}} \tag{8.8b}$$

which, as the reader is invited to verify, has units of speed-inverse. Note that the mathematical model derived above consists of a partial differential equation (PDE) in u. The integral of this equation yields a function $u(x,t)$ determining the motion of the bar under study both in the one-dimensional x-domain and in the time-domain. Therefore, in order to find a particular integral of this equation, we need both *initial* and *boundary* conditions. Moreover, since the PDE of interest is of the second order in both x and t, we need two initial conditions and two boundary conditions. Details on these conditions will be discussed in Sect. 8.3.

8.2.2   Bars Under Torsional Vibration

Let us now consider the *circular cylindrical* bar of Fig. 1.6, reproduced here as Fig. 8.2, undergoing torsional vibration.

The FBD of a differential element of length dx and differential mass moment of inertia about the axis dI, of the bar under study, is included in Fig. 8.3. In this figure, the torque acting at the cross section located a distance x from the left end is defined as $\tau(x,t)$. We further define $\gamma(x,t)$ as the *angular deflection per unit length*, also known as the *shear deformation*; $\gamma(x,t)$ is measured in units of rad/m, and is given by

$$\gamma(x,t) = \frac{\partial\theta}{\partial x} \tag{8.9}$$

Let us now consider an external fiber of the bar that prior to deformation is parallel to the axis of the bar, and let ϕ denote the angle made by that fiber with its initial configuration, as depicted in Fig. 8.3. Then, for "small" values of ϕ, and noticing that the radius of the cross section is r, we have, from the same figure,

$$\phi \approx \tan\phi = \frac{rd\theta}{dx}$$

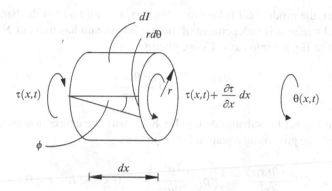

Fig. 8.3 FBD of a differential element of a circular cylindrical bar under torsion

and, if we recall that θ is a function of two variables, x and t, then

$$\phi = r\frac{\partial \theta}{\partial x} \equiv r\gamma \qquad (8.10)$$

Further, application of Euler's Law, Eq. 1.4b, to the element of Fig. 8.3 yields

$$\frac{\partial \tau(x,t)}{\partial x}dx = (dI)\frac{\partial^2 \theta(x,t)}{\partial t^2} \qquad (8.11)$$

where

$$dI = \frac{1}{2}r^2 dm, \quad \text{with} \quad dm = \rho\pi r^2 dx$$

and ρ defined as the constant mass density of the bar, which has units of kg/m^3. Therefore,

$$dI = \frac{1}{2}\rho\pi r^4 dx \qquad (8.12)$$

Furthermore, we invoke Hooke's Law for torsion, which takes the form

$$\tau(x,t) = GJ\gamma(x,t) \qquad (8.13)$$

and rightfully has units of Nm.

In the above equation, G is a *constant* known as the *shear modulus*, and has units of stress, i.e., of N/m^2, while J is another constant denoting, in the case of a circular cross section, the *area moment of inertia* of the cross section about the axis of the bar, i.e.,

$$J = \frac{1}{2}\pi r^4$$

Moreover, the product GJ is known as the *torsional stiffness* of the bar. Note that the torsional stiffness is independent of the bar length and has units of $\mathrm{N\,m^2}$. Upon substitution of Eq. 8.9 into Eq. 8.13, we obtain

$$\frac{\partial \tau(x,t)}{\partial x} = GJ\frac{\partial^2 \theta(x,t)}{\partial x^2} \tag{8.14}$$

Moreover, if Eq. 8.14 is substituted into Eq. 8.11, with the expression (8.12) derived above for dI, the governing equation is obtained as

$$GJ\frac{\partial^2 \theta(x,t)}{\partial x^2} = \frac{1}{2}\pi\rho r^4\frac{\partial^2 \theta(x,t)}{\partial t^2}, \quad 0<x<l, \quad t\geq 0 \tag{8.15}$$

or

$$\frac{\partial^2 \theta(x,t)}{\partial x^2} = \beta^2\frac{\partial^2 \theta(x,t)}{\partial t^2}, \quad 0<x<l, \quad t\geq 0 \tag{8.16a}$$

which is formally identical to Eq. 8.8a, with constant β now defined as

$$\beta \equiv \sqrt{\frac{\pi\rho r^4}{2GJ}} = \sqrt{\frac{\rho}{G}} \tag{8.16b}$$

and has, as in Sect. 8.2.1, units of speed-inverse.

Again, the mathematical model derived above is a PDE in $\theta(x,t)$ of the second order in x and t. All that was said for the mathematical model of the prismatic bar under axial vibration, then, is applicable to the mathematical model of the circular cylindrical bar under torsional vibration.

8.2.3   Strings Under Transverse Vibration

Shown in Fig. 8.4a is a string of uniform mass per unit length μ undergoing transverse motion. We are interested in motions entailing *small* absolute values of the slope $y'(x)$, and *negligible* displacements in the horizontal direction. Moreover, we assume henceforth that the effects of gravity are negligible, and hence, gravity will not be considered. This assumption implies that the *prevailing* values of the inertia forces are much greater than those due to gravity. As a consequence of the foregoing assumptions, the particles of the string can be assumed to be in *static equilibrium* in the horizontal direction, although they are in *dynamic equilibrium* in the vertical direction. Furthermore, by virtue of the small-slope assumption, the length ds of a differential element of the string can be safely approximated as its horizontal projection, dx. Now, let us look at the FBD of a differential element of the string, of length ds, that is subjected to a tension acting on the left end, τ_L, and one acting on the right end, τ_R, as shown in Fig. 8.4b. By virtue of the small-slope assumption, the horizontal projection of τ_L can be safely approximated by τ_L itself,

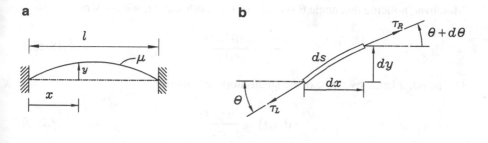

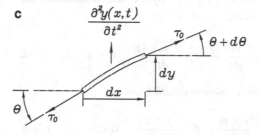

Fig. 8.4 A string under transverse vibration: (**a**) general layout; (**b**) a differential element under horizontal static equilibrium; (**c**) the same element under vertical dynamic equilibrium

the same approximation holding for the horizontal projection of τ_R. Moreover, since we are assuming a static equilibrium of the differential element under study, in the horizontal direction, the two tensions are bound to balance each other, i.e.,

$$\tau_L = \tau_R \qquad (8.17)$$

As a consequence, then, *if no external forces act on the string*, the tension along the whole string is uniform. Henceforth, we denote this uniform tension by τ_0. Note that this assumption is not valid in the case of a string hanging from the ceiling, for, in that case, the weight of the element induces a difference, in fact a gradient, in the tension acting upward and the tension acting downward on the element.

Next, let us write the Newton equation for the element shown in Fig. 8.4c in the vertical direction:

$$\tau_0 \sin(\theta + d\theta) - \tau_0 \sin\theta = \mu ds \frac{\partial^2 y(x,t)}{\partial t^2} \qquad (8.18)$$

Furthermore, under the small-slope assumption, $\sin\theta$ and $\sin(\theta + d\theta)$ can be safely approximated by θ and $\theta + d\theta$, respectively, and, as mentioned above, the length ds can be approximated by its horizontal projection dx. Thus, Eq. 8.18 reduces to

$$\tau_0 d\theta = \mu dx \frac{\partial^2 y(x,t)}{\partial t^2} \qquad (8.19)$$

Moreover, noticing that angle θ is a function of both x and t, we can write

$$\frac{d\theta}{dx} \rightarrow \frac{\partial\theta(x,t)}{\partial x} \tag{8.20a}$$

On the other hand, by virtue of the small-slope assumption,

$$\theta(x,t) = \frac{\partial y(x,t)}{\partial x} \tag{8.20b}$$

Upon dividing both sides of Eq. 8.19 by dx, and substituting Eqs. 8.20a and b into the equation thus resulting, we obtain

$$\tau_0 \frac{\partial^2 y(x,t)}{\partial x^2} = \mu \frac{\partial^2 y(x,t)}{\partial t^2}, \quad 0 < x < l, \quad t \geq 0 \tag{8.21}$$

which can be recast in the now standard form

$$\frac{\partial^2 y(x,t)}{\partial x^2} = \beta^2 \frac{\partial^2 y(x,t)}{\partial t^2}, \quad 0 < x < l, \quad t \geq 0 \tag{8.22a}$$

with constant β now defined as

$$\beta \equiv \sqrt{\frac{\mu}{\tau_0}} \tag{8.22b}$$

Once more, the mathematical model derived here is a PDE in $y(x,t)$ of the second order in x and t. Again, β has units of speed-inverse.

8.2.4   Beams Under Flexural Vibration

Shown in Fig. 8.5a is a linearly elastic beam undergoing flexural motion. What this means is that all particles of the beam, similar to those of the string, are in horizontal static equilibrium, but in vertical dynamic equilibrium. Here, we neglect the mass moment of inertia of the element about an axis perpendicular to the plane of the figure, which, together with our assumption of small deformations and a suitable geometry of the beam—maximum height of the beam section, in the case of variable cross sections, smaller than one tenth the length of the beam—leads to what is known as an Euler-Bernoulli beam. We will further assume that the beam has a uniform cross section throughout its length, and that the mass distribution μ per unit length is constant. Note, moreover, that by virtue of the theory of beams, the normal stress acting on a beam section is not uniform, but varies linearly from top to bottom. In any instance, we are not interested in the normal stress at a point, but rather in the resultant normal force $f(x,t)$ and the resultant bending moment $M(x,t)$ acting on the whole cross section located a distance x from the left. Note that both

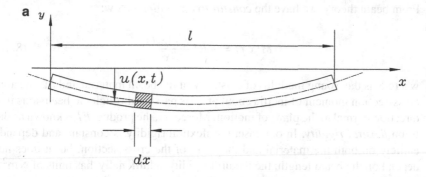

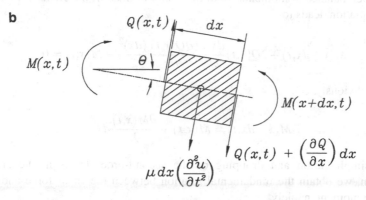

Fig. 8.5 An elastic beam under flexural vibration: (a) general layout; (b) FBD of a differential element

the shear force and the bending moment are the resultant of the distributed shear and normal stresses acting throughout the cross section. By the same token, we are not interested in the tangential stress acting at each point of the cross section, but rather in the shear force $Q(x,t)$ acting on the whole section. One more fundamental assumption in our study, under which the strains in the beam are *small*, implies that the slope of the *neutral axis* of the beam, a.k.a. the *elastica*, is "small." This is known as the *small-slope* assumption.

Now, the dynamic equilibrium of the differential element of Fig. 8.5b leads to

$$Q(x,t) + \frac{\partial Q(x,t)}{\partial x}dx - Q(x,t) = \mu \frac{\partial^2 u(x,t)}{\partial t^2} \tag{8.23}$$

We now recall the relations below:

- From the small-slope assumption,

$$\theta(x,t) \approx \frac{\partial u(x,t)}{\partial x} \tag{8.24}$$

- From beam theory, we have the *constitutive equation* below:

$$M(x,t) = -EI\frac{\partial^2 u(x,t)}{\partial x^2} \tag{8.25}$$

where E is the Young modulus of elasticity of the beam material and I is the area cross-section moment of inertia about an axis passing through the beam axis in a direction normal to the plane of motion. Moreover, the product EI is known as the beam *flexural rigidity*. In our case, the flexural rigidity is constant and depends entirely on both the material and the form of the cross section, but it does not depend on the beam length; the flexural rigidity, additionally, has units of $\mathrm{N\,m^2}$.

Further, balance of moments about the center of mass of the element, i.e., the Euler equation, leads to

$$M(x+dx,t) - 2Q(x,t)\frac{dx}{2} - \frac{\partial Q(x,t)}{\partial x}\frac{(dx)^2}{2} - M(x,t) = 0$$

where, obviously,

$$M(x+dx,t) = M(x,t) + \frac{\partial M(x,t)}{\partial x}dx$$

Upon simplification and dropping of the second-order terms in the above equation, we obtain the fundamental relation between the shear force and the bending moment, namely,

$$Q(x,t) = \frac{\partial M(x,t)}{\partial x} \tag{8.26}$$

Equations 8.25 and 8.26 thus yield one more relation, namely,

$$\frac{\partial Q(x,t)}{\partial x} = -EI\frac{\partial^2 u_{xx}(x,t)}{\partial x^2} = -EI\frac{\partial^4 u(x,t)}{\partial x^4} \tag{8.27}$$

- The Newton equation in the vertical direction leads to

$$\mu dx\frac{\partial^2 u(x,t)}{\partial t^2} = \frac{\partial Q(x,t)}{\partial x}dx$$

- The governing equation now takes the form

$$EI\frac{\partial^4 u(x,t)}{\partial x^4} + \mu\frac{\partial^2 u(x,t)}{\partial t^2} = 0 \tag{8.28}$$

which can be cast in the simpler form

$$\frac{\partial^4 u(x,t)}{\partial x^4} + \beta^4\frac{\partial^2 u(x,t)}{\partial t^2} = 0 \tag{8.29a}$$

with constant β defined as

$$\beta \equiv \left(\frac{\mu}{EI}\right)^{1/4} \tag{8.29b}$$

Note that, contrary to the three previous cases, the governing equation of the beam under flexural vibration is now a PDE of the *fourth order* in x, but it is still of the second order in t. Hence, in order to fully determine one particular integral of the foregoing equation, we need now *four* boundary conditions. This issue will be discussed further Sect. 8.3. As the reader is invited to verify, β has now units of $\sqrt{s}/m$.

8.3 Natural Frequencies and Natural Modes

We study now the mathematical models derived in Sect. 8.2. To do this, we distinguish two cases, models comprising second-order partial derivatives and those comprising fourth-order partial derivatives.

8.3.1 Systems Governed by Second-Order PDE

In deriving the mathematical models of the systems of this class, we denoted with $u(x,t)$ and $\theta(x,t)$ the axial and the angular displacements of a bar, respectively; we used $y(x,t)$ to denote the vertical displacement of a string. In the discussion below, we denote by $w(x,t)$ any of the three foregoing functions of x and t, namely, u, θ or y. Moreover, all the systems of the class under study are governed by a PDE of the same form, namely,

$$\frac{\partial^2 w(x,t)}{\partial x^2} = \beta^2 \frac{\partial^2 w(x,t)}{\partial t^2}, \quad 0 < x < l, \quad t \geq 0 \tag{8.30}$$

where β was already defined in Eqs. 8.8b, 8.16b and 8.22b.

In trying to find integrals for Eq. 8.30, we resort to the *method of separation of variables*, which consists in expressing the integral sought as the product of two functions, each of a single variable, i.e., as

$$w(x,t) = W(x)T(t) \tag{8.31}$$

Upon substitution of Eq. 8.31 into Eq. 8.30, we otain

$$W''(x)T(t) = \beta^2 W(x)\ddot{T}(t)$$

which can be rearranged as

$$\frac{1}{\beta^2}\frac{W''(x)}{W(x)} = \frac{\ddot{T}(t)}{T(t)}$$

Now, in the above equation, we have a ratio of two functions of x equated with a ratio of two functions of t. Since x and t are independent variables, the only possibility we have to verify the above equation is that both be constant, i.e.,

$$\frac{1}{\beta^2}\frac{W''(x)}{W(x)} = \frac{\ddot{T}(t)}{T(t)} = C = \text{const}$$

which thus yields two ODEs, namely,

$$\ddot{T}(t) - CT(t) = 0$$
$$W''(x) - C\beta^2 W(x) = 0$$

So far, we have left C arbitrary. Further, we recall our assumptions regarding the variables involved, i.e., the axial displacement u or the angular displacement θ of the cross section of a bar, or the transverse displacement y of a string. In any instance, we have assumed that these displacements are either small or finite. As a consequence, the above equations should yield *bounded*, i.e., stable integrals, which requires that the constant coefficient multiplying the function itself in the second term of the left-hand side in each of the two foregoing equations be positive. This means that C **must be negative**.

Moreover, we notice that the above-mentioned constant has units of frequency-squared, and hence, we let

$$C = -\omega^2 \tag{8.32}$$

which thus leads to the two equations below:

$$\ddot{T}(t) + \omega^2 T(t) = 0 \tag{8.33a}$$
$$W''(x) + \lambda^2 W(x) = 0 \tag{8.33b}$$

with constant λ defined as

$$\lambda \equiv \beta\omega \tag{8.33c}$$

and ω being a frequency as yet to be determined.

Therefore, Eq. 8.33a defines an *initial-value problem* for $T(t)$, whose solution we already derived in Chap. 2. That is, if we are given initial conditions $T(0) = T_0$ and $\dot{T}(0) = V_0$, then the integral of the said equation is

$$T(t) = T_0 \cos \omega t + \frac{V_0}{\omega} \sin \omega t \tag{8.34}$$

On the other hand, Eq. 8.33b defines a *boundary-value problem* for $W(x)$. Here, we need *two* conditions on either $W(x)$ or $W'(x)$ in either of the extremes of the spatial

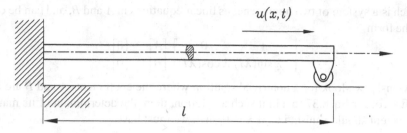

Fig. 8.6 A fixed-free bar under axial vibration

domain of interest, i.e., at $x = 0$ or at $x = l$. In any instance, the integral of the equation at hand takes the form

$$W(x) = A\cos(\lambda x) + B\sin(\lambda x) \qquad (8.35)$$

with constants A and B to be determined from the boundary conditions (BCs). The type of boundary conditions for a specific case depends entirely on the loading and the support of the system at hand, as we describe below.

8.3.1.1   Bars Under Axial Vibration

In the most typical case, we have a bar with, say, its left end fixed and its right end free, as shown in Fig. 8.6, the boundary conditions thus being

$$u(0,t) = 0, \quad \sigma(l,t) = 0$$

Note that, by virtue of Eqs. 8.4, 8.5, and 8.6, the second of the above BCs leads to

$$\frac{\partial u(x,t)}{\partial x}\Big|_{x=l} = 0$$

As a consequence, then, the two boundary conditions on $W(x)$ are

$$W(0) = 0, \quad W'(l) = 0 \qquad (8.36)$$

Upon imposition of the foregoing BCs on the expression for $W(x)$ given in Eq. 8.35, we have

$$A\cos(0) + B\sin(0) = 0$$
$$-A\lambda\sin(\lambda l) + B\lambda\cos(\lambda l) = 0$$

which is a system of two homogeneous linear equations in A and B, and can be cast in the form

$$\begin{bmatrix} 1 & 0 \\ -\lambda \sin(\lambda l) & \lambda \cos(\lambda l) \end{bmatrix} \begin{bmatrix} A \\ B \end{bmatrix} = \begin{bmatrix} 0 \\ 0 \end{bmatrix} \tag{8.37}$$

Obviously, we look for a nontrivial solution where the coefficients A and B are not both zero; for Eq. 8.37 to admit such a solution, then, the determinant of its matrix coefficient should vanish, i.e., if $\lambda \neq 0$, then we must have

$$\cos(\lambda l) = 0 \tag{8.38a}$$

which is the *characteristic equation* of the given problem. Its roots are the *characteristic values* or *eigenvalues*, of the boundary-value problem (BVP) defined by Eq. 8.33b and the BCs of Eq. 8.36. The characteristic equation holds if and only if

$$\lambda l = \frac{\pi}{2}, \frac{3\pi}{2}, \dots, \quad \text{or} \quad \lambda_k = \frac{(2k-1)\pi}{2l}, \quad k = 1, 2, \dots \tag{8.38b}$$

We have, therefore, an *infinity* of characteristic values λ_k, which thus lead to *infinitely-many solutions* of the BVP at hand; moreover, it is apparent from Eq. 8.37 that $A = 0$, and hence, $W_k(x)$ takes the form

$$W_k(x) = B_k \sin\left(\frac{(2k-1)\pi}{2l}x\right), \quad k = 1, 2, \dots \tag{8.39}$$

The above set of characteristic solutions are known as the *eigenfunctions* of the problem. Moreover, because these functions define the shape of the motion of the bar at hand, the same functions are also known as the *natural modes* of the bar. Note that the natural modes depend not only on the nature of the bar under study and its physical parameters, but also on the BCs involved. Now, from Eq. 8.33c, ω is found to admit infinitely many solutions as well, which are represented as ω_k. These are the *natural frequencies* of the system at hand, which are defined as

$$\omega_k = \frac{\lambda_k}{\beta} = \frac{(2k-1)\pi}{2l\beta} = \frac{2k-1}{2}\pi\sqrt{\frac{EA_s}{\mu l^2}}, \quad k = 1, 2, \dots \tag{8.40}$$

Also note that ω_k, as given above, is a quantity with units identical to those of the square root therein. The corresponding radical, in turn, can be regarded as the quotient of the quantity EA_s/l, which is nothing but the stiffness of the rod when regarded as a lumped spring, as displayed in Eq. 1.9, divided by μl, which is the mass of the bar. Thus, ω_k has consistently units of frequency, each of its (infinitely many) values being termed a *natural frequency* of the bar under study. We thus have an infinity of natural frequencies, a natural mode being associated to each of them. Moreover, by virtue of Eq. 8.34, each natural frequency defines a function of time $T_k(t)$ of the form

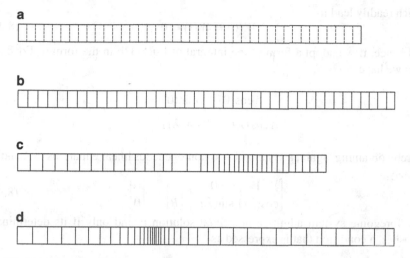

Fig. 8.7 Natural modes of a bar under axial vibration: (**a**) undeformed configuration; (**b**) first mode; (**c**) second mode; (**d**) third mode

Fig. 8.8 A circular cylindrical bar, with its two ends fixed, under torsional vibration

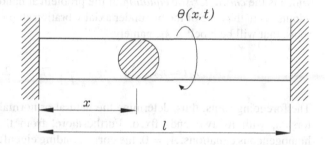

$$T_k(t) = T_{0k} \cos \omega_k t + \frac{V_{0k}}{\omega_k} \sin \omega_k t \qquad (8.41)$$

Depicted in Fig. 8.7 are the undeformed configuration and the first three natural modes of the bar of Fig. 8.6. In that figure, we show a uniform pattern painted on the undeformed bar in Fig. 8.7a; this pattern is deformed as the bar takes the shapes of its first three modes.

8.3.1.2 Bars Under Torsional Vibration

We now consider the BCs associated with the bar under torsional vibration depicted in Fig. 8.8 whose two ends are fixed to walls constituting an inertial frame.

Here, the boundary conditions are

$$\theta(0,t) = \theta(l,t) = 0$$

which readily lead to

$$W(0) = W(l) = 0 \tag{8.42}$$

and hence, if we adopt a form of the integral of Eq. 8.33b in the form of Eq. 8.35, then we have

$$A\cos(0) + B\sin(0) = 0$$

$$A\cos(\lambda l) + B\sin(\lambda l) = 0$$

thereby obtaining, again, a system of two homogeneous linear equations in A and B, namely,

$$\begin{bmatrix} 1 & 0 \\ \cos(\lambda l) & \sin(\lambda l) \end{bmatrix} \begin{bmatrix} A \\ B \end{bmatrix} = \begin{bmatrix} 0 \\ 0 \end{bmatrix} \tag{8.43}$$

The foregoing system admits a nontrivial solution if and only if its determinant vanishes, a condition that is expressed as

$$\sin(\lambda l) = 0$$

which is the *characteristic equation* of the problem at hand. Note that this equation admits, as in the case of the bar under axial vibration studied above, infinitely many roots, that will be labelled λ_k, namely,

$$\lambda_k l = \pi k, \quad \text{or} \quad \lambda_k = \frac{\pi k}{l}, \quad k = 1, 2, \ldots \tag{8.44}$$

The foregoing roots, thus, determine the natural or normal modes of the bar under torsion, with its two ends fixed. Furthermore, from the first of the foregoing homogeneous equations, $A_k = 0$, the corresponding eigenfunctions then being

$$W_k(x) = B_k \sin\left(\frac{\pi k x}{l}\right), \quad k = 1, 2, \ldots \tag{8.45}$$

Shown in Fig. 8.9 are the first three modes of the system under study.

Again, substitution of Eq. 8.45 into Eq. 8.33c, with λ and ω now subscripted, leads to the natural frequencies associated with the above modes:

$$\omega_k = \frac{\pi k}{\beta l} = \pi k \sqrt{\frac{G\pi r^2}{\mu l^2}}, \quad k = 1, 2, \ldots \tag{8.46}$$

It is noteworthy that the above square root has in its radical the quotient of the torsional stiffness of a circular cylindrical bar, as introduced in Sect. 1.5.1, when regarded as a lumped torsional spring, namely, $(1/2)\pi G(r^4/l)$, divided by the mass moment of inertia of the bar, i.e., $(1/2)\mu l r^2$. Hence, the above square root has units of frequency. Again, each natural frequency ω_k leads to a function of time $T_k(t)$ of the form of Eq. 8.41.

Fig. 8.9 Natural modes of a
bar under torsional vibration:
(**a**) undeformed
configuration; (**b**) first mode;
(**c**) second mode; (**d**) third
mode

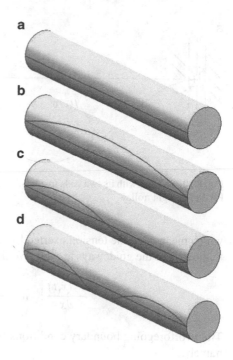

8.3.1.3   Strings Under Transverse Vibration

We now derive the natural modes and natural frequencies of vibration of the
string sketched in Fig. 8.10a. In the analysis below we recall that gravity has been
neglected at the outset. Moreover, each end of the string is fixed to a massless trolley
that can slide freely in a direction parallel to that of the motion of the string, by
means of the wheels rolling on a guideway. Further, we assume that the string moves
in a vertical plane; by virtue of our hypotheses, then, gravity does not appear in the
model that we will formulate below.[2]

The BCs are now derived by resorting to the FBD of the trolleys. We focus on
the FBD of the right trolley, depicted in Fig. 8.10b, which makes it apparent that the
force exerted by the guideway on the trolley wheels, and hence, on the trolley itself,
is normal to the axis of the guideways, for we have assumed that the inertia of the
trolley is negligible. Since the trolley is massless, it must be in static equilibrium,

---

[2]Even in the presence of gravity, we can always decompose the motion of the string into a rigid-
body motion downwards, due to gravity, and a vibratory motion, the latter being the subject of this
chapter.

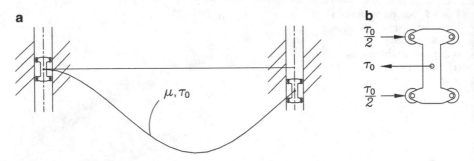

Fig. 8.10 A string with its two ends free to slide along parallel guideways: (**a**) general layout; (**b**) FBD of its right trolley

which means that the tension exerted by the string onto the trolley must be normal to the axis of the guideways as well. We therefore have the BCs below:

$$\frac{\partial y(x,t)}{\partial x}\bigg|_{x=0} = \frac{\partial y(x,t)}{\partial x}\bigg|_{x=l} = 0 \tag{8.47}$$

The two foregoing boundary conditions thus lead to corresponding BCs on $W(x)$, namely,

$$W'(0) = 0, \quad W'(l) = 0 \tag{8.48}$$

Upon imposing the above BCs onto function $W(x)$ as given by Eq. 8.35, we obtain the conditions below:

$$-A\sin(0) + B\cos(0) = 0 \tag{8.49}$$

$$-A\sin(\lambda l) + B\cos(\lambda l) = 0 \tag{8.50}$$

The two above equations are similar to those derived above for other cases, i.e., linear homogeneous, and can thus be cast in the form

$$\begin{bmatrix} 0 & 1 \\ -\sin(\lambda l) & \cos(\lambda l) \end{bmatrix} \begin{bmatrix} A \\ B \end{bmatrix} = \begin{bmatrix} 0 \\ 0 \end{bmatrix} \tag{8.51}$$

which admits nontrivial solutions if and only if its determinant vanishes, i.e., if and only if

$$\sin(\lambda l) = 0 \tag{8.52}$$

thereby obtaining the characteristic equation of the problem at hand. Its roots yield the infinitely many eigenvalues displayed below:

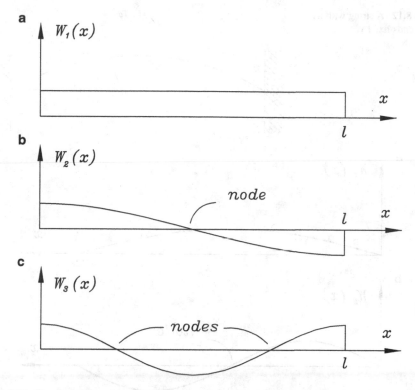

Fig. 8.11 The first three natural modes of vibration of the string supported on free-sliding trolleys

$$\lambda l = 0, \pi, \ldots \quad \text{or} \quad \lambda_k = \frac{\pi}{l}(k-1), \quad k = 1, 2, \ldots \qquad (8.53)$$

the eigenfunctions associated with the foregoing eigenvalues being derived upon noticing that, from Eq. 8.49, $B_k = 0$, and hence,

$$W_k(x) = A_k \cos\left(\frac{\pi(k-1)}{l}x\right), \quad k = 1, 2, \ldots \qquad (8.54)$$

the first three of which are displayed in Fig. 8.11. The presence of a *rigid mode* in this example is noteworthy. Again, note that each natural frequency ω_k leads to a function of time $T_k(t)$ as in the previous cases. The natural frequencies are derived from the general relation, Eq. 8.33c, i.e., as

$$\omega_k = \frac{\lambda_k}{\beta} = \sqrt{\frac{\tau_0}{\mu}}\lambda_k \qquad (8.55)$$

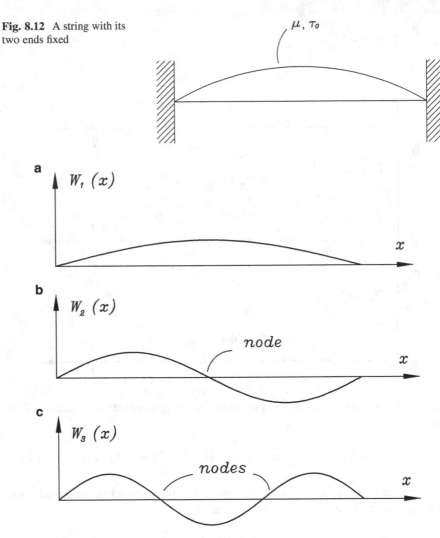

Fig. 8.12 A string with its two ends fixed

Fig. 8.13 The natural modes of a string with its two ends fixed: (**a**) first mode; (**b**) second mode; (**c**) third mode

As a second example, we consider the string with its two ends fixed, as shown in Fig. 8.12. It is left as an exercise to the reader to show that the eigenfunctions of the string of Fig. 8.12 are identical to those of the bar under torsional vibration with its two ends fixed. The three natural modes of this string are displayed in Fig. 8.13.

Now let us study the BCs for the case in which the trolley of the right-hand side of Fig. 8.10 has a non-negligible mass M. The FBD of the trolley under these conditions is shown in Fig. 8.14.

In Fig. 8.14, $\ddot{y}_l \equiv \partial y / \partial t^2 |_{x=l}$. Thus, applying Newton's second law to the trolley in the direction of motion,

$$M\ddot{y}_l = -\tau_0 \sin \theta_l$$

Fig. 8.14 The FBD of the
heavy right-hand side trolley

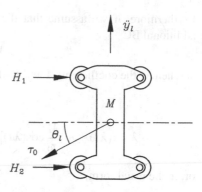

Under the assumption that θ_l is "small", then

$$\sin \theta_l \approx \theta_l \approx \tan \theta_l = y'_l \equiv \left.\frac{\partial y}{\partial x}\right|_{(l,t)}$$

We thus have

$$M\left.\frac{\partial^2 y}{\partial t^2}\right|_{(l,t)} = -\tau_0 \left.\frac{\partial y}{\partial x}\right|_{(l,t)} \tag{8.56}$$

or, if we assume, as usual, that

$$y(x,t) = W(x)T(t) \tag{8.57}$$

then, Eq. 8.56 leads to the desired BC, namely,

$$MW(l)\ddot{T}(t) = -\tau_0 W'(l)T(t)$$

But, since $T(t)$ obeys Eq. 8.33a, we have

$$\ddot{T}(t) = -\omega^2 T(t)$$

and hence, the above BC becomes

$$\omega^2 MW(l)T(t) = \tau_0 W'(l)T(t)$$

However, $T(t)$ is not identically zero—it may vanish instantaneously, though—and
hence, we can delete this function from the above equation to obtain the BC

$$W'(l) - \omega^2 \frac{M}{\tau_0}W(l) = 0 \tag{8.58a}$$

Furthermore, if we assume that the left-end of the string is fixed, we have the additional BC

$$W(0) = 0 \qquad (8.58b)$$

and hence, the coefficients A and B of the general solution (8.35) obey the relations

$$A\cos(0) + B\sin(0) = 0$$

$$-[\lambda \sin(\lambda l) + \omega^2 \frac{M}{\tau_0} \cos(\lambda l)]A + [\lambda \cos(\lambda l) - \omega^2 \frac{M}{\tau_0} \sin(\lambda l)]B = 0$$

or, in the usual form,

$$\begin{bmatrix} 1 & 0 \\ -\lambda \sin(\lambda l) - \omega^2 \frac{M}{\tau_0} \cos(\lambda l) & \lambda \cos(\lambda l) - \omega^2 \frac{M}{\tau_0} \sin(\lambda l) \end{bmatrix} \begin{bmatrix} A \\ B \end{bmatrix} = \begin{bmatrix} 0 \\ 0 \end{bmatrix}$$

For nontrivial solutions, then, the determinant of the above matrix should vanish, which thus yields the characteristic equation

$$\lambda \cos(\lambda l) - \omega^2 \frac{M}{\tau_0} \sin(\lambda l) = 0$$

But, from Eq. 8.33c, $\omega^2 = \lambda^2/\beta^2 = (\tau_0 \lambda^2)/\mu$, and hence, the above equation leads to

$$\lambda \cos(\lambda l) - \frac{\lambda^2 M}{\mu} \sin(\lambda l) = 0$$

or, for $\lambda \neq 0$,

$$\cos(\lambda l) - \frac{\lambda M}{\mu} \sin(\lambda l) = 0$$

Moreover, the numerator and denominator of the function appearing in the second term of the foregoing equation are now multiplied by l, which thus leads to

$$\tan(\lambda l) = \frac{m}{M} \frac{1}{\lambda l} \qquad (8.59)$$

where m is the total mass of the string. The above equation is transcendental, its infinitely-many roots being most easily found by graphical means upon superimposing the plots of the two functions $f(\lambda l) = \tan(\lambda l)$ and $g(\lambda l) = (m/M)/(\lambda l)$; the eigenvalues λ_1, λ_2, etc. can then be obtained from the abscissae of the intersections of the two plots. Finding these values is left to the reader as an exercise.

One more case worth analyzing is the same system of Fig. 8.10 when the *massless* right-hand trolley is suspended from a lumped spring of stiffness k_s, as shown in Fig. 8.15.

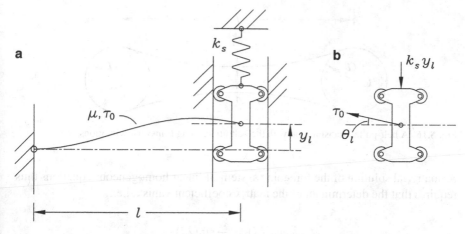

Fig. 8.15 String with left-end fixed and right-end constrained by a guideway and suspended from a lumped spring: (**a**) general layout; (**b**) FBD of trolley

From the FBD of Fig. 8.15b, with $y_l \equiv y(l,t)$ and $\theta_l \approx \tan\theta_l = \partial y/\partial x|_{x=l} \equiv y'_l$,

$$-k_s y_l + \tau_0 y'_l = 0$$

or, if we recall Eq. 8.57,

$$-k_s W(l)T(t) + \tau_0 W'(l)T(t) = 0$$

and, since $T(t)$ is not identically zero, we can delete it from the above equation to obtain the BC

$$W'(l) = \frac{k_s}{\tau_0}W(l) \tag{8.60}$$

Furthermore, the BC at the left-end is, again, $W(0) = 0$, the two BCs thus becoming

$$A\cos(0) + B\sin(0) = 0$$

$$-[\lambda\sin(\lambda l) + \frac{k_s}{\tau_0}\cos(\lambda l)]A + [\lambda\cos(\lambda l) - \frac{k_s}{\tau_0}\sin(\lambda l)]B = 0$$

or, in the usual form,

$$\begin{bmatrix} 1 & 0 \\ -\lambda\sin(\lambda l) - \frac{k_s}{\tau_0}\cos(\lambda l) & \lambda\cos(\lambda l) - \frac{k_s}{\tau_0}\sin(\lambda l) \end{bmatrix} \begin{bmatrix} A \\ B \end{bmatrix} = \begin{bmatrix} 0 \\ 0 \end{bmatrix}$$

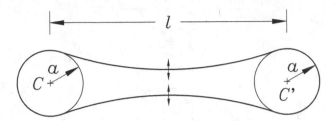

Fig. 8.16 A belt-pulley transmission with the belt undergoing transverse vibrations

A nontrivial solution of the foregoing system of linear homogeneous equations thus requires that the determinant of the matrix coefficient vanish, i.e.,

$$\lambda \cos(\lambda l) = \frac{k_s}{\tau_0} \sin(\lambda l)$$

or

$$\tan(\lambda l) = \frac{\tau_0}{k_s l}(\lambda l) \tag{8.61}$$

thereby ending up, again, with a transcendental equation whose roots yield the eigenvalues of the problem under study. These roots can be found, again, by super-position of the plots of the functions $f(\lambda l) = \tan(\lambda l)$ and $g(\lambda l) = -[\tau_0/(k_s l)]\lambda l$. The infinitely many intersections of the two foregoing plots, then, yield the eigenvalues λ_1, λ_2, etc.

Another system of engineering relevance is the belt-pulley transmission shown in Fig. 8.16 whose transverse vibrations can be treated as those of a string.[3]

In order to establish the boundary conditions of the belt, we look at the geometry around the bottom separation points S and S' of the pulleys, as sketched in Fig. 8.17a,b, respectively, where

$$y_S \equiv y(x_S, t), \quad y_{S'} \equiv y(x_{S'}, t)$$

It is now apparent from Fig. 8.17a that the slope of the belt at S is given by

$$\left.\frac{\partial y}{\partial x}\right|_{(x_S, t)} = \tan \theta = \frac{x_S}{a - y_S}$$

and hence,

$$(a - y_S)\left.\frac{\partial y}{\partial x}\right|_{(x_S, t)} = x\Big|_{x_S}$$

---

[3]Special assumptions on the belt geometry and construction must be introduced here for the string model to be valid. For example, a "belt" made of a rubber O-ring would be a candidate for this analysis.

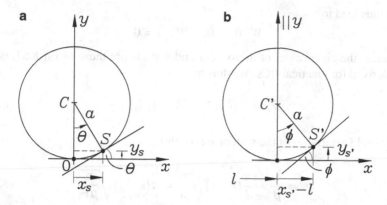

Fig. 8.17 Geometric relations at the separation point of the belt from the pulley: (**a**) bottom left-end; (**b**) bottom right-end

Upon expansion,

$$a\frac{\partial y}{\partial x}\bigg|_{(x_S,t)} - y_S\frac{\partial y}{\partial x}\bigg|_{(x_S,t)} = x_S$$

But, at S, both $y \approx 0$ and $\partial y/\partial x \approx 0$, if the belt is to remain taut. Likewise, $x_S \approx 0$, and hence, the foregoing relation leads to

$$\frac{\partial y}{\partial x}\bigg|_{(0,t)} = 0 \tag{8.62}$$

Now, if we look at the right pulley, in Fig. 8.17b, the slope at the separation point S' is given by

$$\frac{\partial y}{\partial x}\bigg|_{(x_{S'},t)} = \tan\phi = \frac{x_{S'} - l}{a - y_{S'}}$$

i.e.,

$$\left[(a-y)\frac{\partial y}{\partial x}\right]\bigg|_{(x_{S'},t)} = x\big|_{x_{S'}} - l$$

Again, if the belt is to remain taut, $y \approx 0$ and $\partial y/\partial x \approx 0$ at S', while $x_{S'} \approx l$, the foregoing relation thus leading to

$$\frac{\partial y}{\partial x}\bigg|_{(l,t)} = 0$$

with identical relations for the upper section of the belt. Now it is apparent that the BCs for the pulley are

$$\frac{\partial y}{\partial x}\bigg|_{(0,t)} = 0, \quad \frac{\partial y}{\partial x}\bigg|_{(l,t)} = 0$$

which thus lead to

$$W'(0) = 0, \quad W'(l) = 0$$

and hence, the eigenvalues of the system under study are those of Eq. 8.53, which were derived for identical BCs. We thus have

$$\lambda_k = \frac{\pi}{l}(k-1), \quad k = 1, 2, \ldots \tag{8.63a}$$

The natural frequencies of the system are, in turn,

$$\omega_k = \frac{\lambda_k}{\beta} = \frac{\pi}{l}(k-1)\sqrt{\frac{\tau_0}{\mu}} = \pi(k-1)\sqrt{\frac{\tau_0}{\mu l^2}} \tag{8.63b}$$

Again, it is noteworthy that the radical is the quotient of the stiffness, τ_0/l, divided by a mass, μl. In summary, the eigenfunctions of the system under study are

$$W_k = A_k \cos\left(\frac{(k-1)\pi}{l}x\right), \quad k = 1, 2, \ldots \tag{8.63c}$$

Notice the presence of a rigid mode for this system.

8.3.2   Systems Governed by Fourth-Order PDEs: Beams Under Flexural Vibration

The mathematical model for beams under flexural vibration is displayed in Eq. 8.28. If we apply again the technique of variable separation to this equation, we obtain

$$W^{iv}(x)T(t) + \beta^4 W(x)\ddot{T}(t) = 0$$

which can be readily rewritten in the alternative form

$$\frac{1}{\beta^4}\frac{W^{iv}(x)}{W(x)} = -\frac{\ddot{T}(t)}{T(t)} = \omega^2$$

where we have used our experience gained in the study of systems governed by second-order PDEs to produce the last equation. We obtain, again, two ODEs, one for $W(x)$ and one for $T(t)$, namely,

$$\ddot{T}(t) + \omega^2 T(t) = 0 \tag{8.64a}$$

$$W^{iv}(x) - \lambda^4 W(x) = 0 \tag{8.64b}$$

with λ defined as

$$\lambda \equiv \beta\sqrt{\omega} \tag{8.64c}$$

Fig. 8.18 A clamped-free beam with uniform cross section

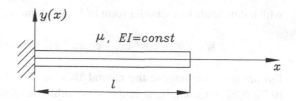

Note that the pair of Eqs. 8.64a and b is similar to the pair of Eqs. 8.33a and b, except that the second of those equations is of fourth order. The integration of the equation in $T(t)$ needs no further discussion, that in $W(x)$ is now addressed. The general form of the integral of Eq. 8.64b is known to be [1]

$$W(x) = A\cos(\lambda x) + B\sin(\lambda x) + C\cosh(\lambda x) + D\sinh(\lambda x) \qquad (8.65)$$

whose four coefficients A, B, C, and D are to be determined from the BCs of the given problem.

Let us consider the clamped-free beam of Fig. 8.18, of constant cross section of area A_s, flexural rigidity EI and uniform mass per unit length μ.

The boundary conditions of the beam under study are readily derived: At the clamped end, both the displacement $u(x,t)$ and the slope $\partial u(x,t)/\partial x$ vanish, while, at the free end, both the bending moment $M(x,t)$ and the shear force $Q(x,t)$ vanish. We thus have, at the clamped end,

$$u(0,t) = 0, \qquad \left.\frac{\partial u(x,t)}{\partial x}\right|_{x=0} = 0$$

In setting up the BCs for the free end, we realize that we have such conditions in terms of bending moment and shear force, not in terms of displacement and its derivatives. We see here an essential difference in the two sets of BCs; this difference is stressed by terming the first set, i.e., those associated with displacement and slope *geometric* BCs. Those associated with bending moment and shear force are termed *natural* BCs. Furthermore, if we recall the constitutive equations (8.25), we can set up the natural BCs as

$$\left.\frac{\partial^2 u(x,t)}{\partial x^2}\right|_{x=l} = 0, \qquad \left.\frac{\partial^3 u(x,t)}{\partial x^3}\right|_{x=l} = 0$$

Now, the four BCs derived above lead to corresponding BCs on $W(x)$, namely,

$$W(0) = 0, \quad W'(0) = 0, \quad W''(l) = 0, \quad W'''(l) = 0 \qquad (8.66)$$

Upon imposing the geometric BCs derived above onto $W(x)$, we obtain, at $x = 0$,

$$A + C = 0, \quad B + D = 0$$

which thus leads to a simpler form of $W(x)$, namely,

$$W(x) = A\left[\cos(\lambda x) - \cosh(\lambda x)\right] + B\left[\sin(\lambda x) - \sinh(\lambda x)\right] \qquad (8.67)$$

Furthermore, we impose the natural BCs, at $x = l$, onto the form of $W(x)$ given in Eq. 8.67, thus obtaining two more conditions on the two remaining coefficients, A and B, namely,

$$\begin{bmatrix} \cos(\lambda l) + \cosh(\lambda l) & \sin(\lambda l) + \sinh(\lambda l) \\ -\sin(\lambda l) + \sinh(\lambda l) & \cos(\lambda l) + \cosh(\lambda l) \end{bmatrix} \begin{bmatrix} A \\ B \end{bmatrix} = \begin{bmatrix} 0 \\ 0 \end{bmatrix} \qquad (8.68)$$

What Eq. 8.68 represents is a system of two linear *homogeneous* equations in A and B. Since the trivial solution $A = B = 0$ yields the equilibrium configuration $w(x,t) = 0$ of the beam, which we already know, we are after values of these coefficients that do not vanish simultaneously. This means that the matrix coefficient of the above equation must be singular, and hence,

$$\det \begin{bmatrix} \cos(\lambda l) + \cosh(\lambda l) & \sin(\lambda l) + \sinh(\lambda l) \\ -\sin(\lambda l) + \sinh(\lambda l) & \cos(\lambda l) + \cosh(\lambda l) \end{bmatrix} = 0$$

which is an equation in λ only, free of A and B. The above equation is, then, the characteristic equation of the problem at hand. Upon expansion of the determinant, this equation leads to

$$\cos(\lambda l)\cosh(\lambda l) + 1 = 0$$

or

$$\cos(\lambda l) = -\frac{1}{\cosh(\lambda l)} \qquad (8.69)$$

Solving for λ from the above equation can be done, as in similar cases, resorting to a graphical approach. To this end, we regard the left- and the right-hand sides of the above characteristic equation as independent functions of (λl), and plot them vs. (λl). Upon superimposing the two plots, the solutions sought are given by the abscissae of their intersections, as shown in Fig. 8.19. It is apparent that we obtain, again, infinitely many eigenvalues. Moreover, the large eigenvalues, from the third on, can be safely approximated by the roots of the cosine function.

The first three modes of the beam under study are shown in Fig. 8.20. The natural frequencies of the beam at hand are now obtained from Eq. 8.64c, namely,

$$\omega = \frac{\lambda^2}{\beta^2} \quad \text{or} \quad \omega_k = \frac{\lambda_k^2}{\beta^2}, \quad k = 1, 2, \ldots$$

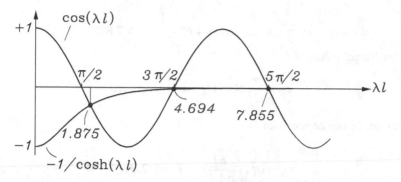

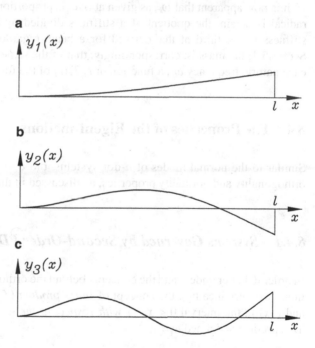

Fig. 8.19 The infinitely-many eigenvalues of the vibrating clamped-free beam

Fig. 8.20 The first three
natural modes of vibration of
a clamped-free beam: (a) first
mode; (b) second mode;
(c) third mode

Now we recall the expression derived for β, and displayed in Eq. 8.16b, which
thus yields

$$\omega_k = \lambda_k^2 \sqrt{\frac{EI}{\rho A_s}}, \quad k = 1, 2, \ldots$$

Furthermore, λ_k can be expressed as

$$\lambda_k = \frac{\alpha_k}{l}, \quad k = 1, 2, \ldots$$

with

$$\alpha_1 \approx 1.875, \quad \alpha_2 \approx 4.694, \quad \alpha_3 \approx 7.855$$

and, for 'large' values of k,

$$\alpha_k \approx \frac{\pi}{2}(2k-1), \quad k = 4,5,6,\ldots$$

Therefore, ω_k can be rewritten as

$$\omega_k = \alpha_k^2 \sqrt{\frac{EI}{\rho A_s l^4}}, \quad k = 1,2,3,\ldots$$

It is now apparent that ω_k, as given above, is proportional to a square root whose radical is, again, the quotient of a stiffness divided by a mass. In this case, the stiffness is one third of that derived for a beam regarded as a lumped spring, in Sect. 1.5.1; the mass is, correspondingly, that of the same beam. We have, thus, for each natural frequency ω_k, a function of t, $T_k(t)$ of the form of Eq. 8.41.

8.4   The Properties of the Eigenfunctions

Similar to the normal modes of n-dof systems, those of continuous systems obey orthogonality and normality properties, as discussed in this section.

8.4.1   Systems Governed by Second-Order PDEs

In order to better understand the concepts behind the orthogonality of the eigenfunctions, we introduce first the concept of *inner product* (f, g) of two functions $f(x)$ and $g(x)$ in the interval $0 < x < l$ *with respect to the weighing function* $w(x) > 0$. This product is defined as

$$(f, g) \equiv \int_0^l w(x) f(x) g(x) dx \tag{8.70}$$

If the foregoing integral turns out to vanish, the two functions $f(x)$ and $g(x)$ are said to be *orthogonal with respect to* $w(x)$ in $0 < x < l$. Moreover, if we set $g(x) = f(x)$ in Eq. 8.70, then we obtain the square of the *weighted Euclidean norm* of $f(x)$, i.e.,

$$\|f(x)\|_w^2 \equiv \int_0^l w(x) f^2(x) dx \tag{8.71}$$

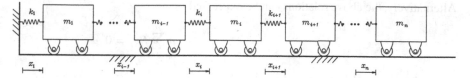

Fig. 8.21 A n-dof mass-spring system

Note that, by virtue of the hypothesis $w(x) > 0$ in $0 < x < l$, it is apparent that the right-hand side of the above equation is positive; it vanishes if and only if $f(x) = 0$ in $0 < x < l$. Moreover, for $w(x) = 1$, the foregoing norm yields the root-mean-square value of $f(x)$ in $0 < x < l$.

We now go back to the n-dof case and consider the system of Fig. 8.21. The mass and stiffness matrices of this system are readily derived as

$$\mathbf{M} = \begin{bmatrix} m_1 & 0 & 0 & \cdots & 0 \\ 0 & m_2 & 0 & \cdots & 0 \\ \vdots & \vdots & \vdots & \ddots & 0 \\ 0 & 0 & 0 & \cdots & m_n \end{bmatrix}, \quad \mathbf{K} = \begin{bmatrix} k_1 + k_2 & -k_2 & 0 & \cdots & 0 & 0 \\ -k_2 & k_2 + k_3 & -k_3 & \cdots & 0 & 0 \\ \vdots & \vdots & & \ddots & \cdots & \vdots & \vdots \\ 0 & 0 & 0 & \cdots & -k_n & k_n \end{bmatrix}$$

More specifically, let us assume that all masses and springs are identical, i.e.,

$$m_1 = m_2 = \cdots = m_n = m, \quad k_1 = k_2 = \cdots = k_n = k$$

Let us further assume that a modal analysis of the system under study has been conducted, which yielded the natural frequencies $\{\omega_i\}_1^n$ and the modal vectors $\{\mathbf{f}_i\}_1^n$. For simplicity of notation, let us define, for $i \neq j$,

$$\mathbf{f} \equiv \mathbf{f}_i \equiv \begin{bmatrix} f_1 \\ f_2 \\ \vdots \\ f_n \end{bmatrix}, \quad \mathbf{g} \equiv \mathbf{f}_j \equiv \begin{bmatrix} g_1 \\ g_2 \\ \vdots \\ g_n \end{bmatrix}$$

Since these vectors are orthogonal with respect to the mass and stiffness matrices, we have

$$\mathbf{f}^T \mathbf{M} \mathbf{g} = 0, \quad \mathbf{f}^T \mathbf{K} \mathbf{g} = 0$$

or, in component form,

$$m(f_1 g_1 + f_2 g_2 + \cdots + f_n g_n) = 0$$

$$k[f_1(2g_1 - g_2) + f_2(-g_1 + 2g_2 - g_3) + \cdots$$

$$+ f_{n-1}(-g_{n-2} + 2g_{n-1} - g_n) + f_n(-g_{n-1} + g_n)] = 0$$

Alternatively, the above relations can be expressed as

$$m\sum_{1}^{n} f_i g_i = 0$$

$$k\left[f_1(2g_1 - g_2) + \sum_{2}^{n-1} f_i(-g_{i-1} + 2g_i - g_{i+1}) \right.$$

$$\left. + f_n(-g_{n-1} + g_n) \right] = 0$$

Let us further assume that the system of Fig. 8.21 is an approximation of the system of Fig. 8.6, so that

$$m = \mu \Delta x, \quad k = \kappa \Delta x, \quad \kappa \equiv \frac{EA_s}{l^2}$$

where μ and κ are the constant mass and stiffness distributions per unit length. We have, therefore,

$$\mu \sum_{1}^{n} f_i g_i \Delta x = 0 \qquad (8.72a)$$

$$\frac{EA_s}{l^2}\left[f_1(2g_1 - g_2) + \sum_{2}^{n-1} f_i(-g_{i-1} + 2g_i - g_{i+1}) \right.$$

$$\left. + f_n(-g_{n-1} + g_n) \right] \Delta x = 0 \qquad (8.72b)$$

On the other hand, let us approximate the integral of Eq. 8.70 in the form of the sum of the areas of n rectangles, the ith of which has a height $z_i \equiv w(x_i)f(x_i)g(x_i)$ and all have the same base Δx, which thus yields

$$(f, g) \approx \sum_{1}^{n} w_i f_i g_i \Delta x \qquad (8.73a)$$

where
$$w_i \equiv w(x_i), \quad f_i \equiv f(x_i), \quad g_i \equiv g(x_i), \quad \Delta x \equiv \frac{l}{n} \qquad (8.73b)$$

Upon comparing Eqs. 8.72a and 8.73a, it is apparent that, when $\Delta x \to 0$, the summation of the left-hand side of the former tends to the integral of Eq. 8.70, with a proper identification of $w(x)$, $f(x)$, and $g(x)$, i.e., we have that, if $f(x)$ and $g(x)$ denote the ith and the jth normal modes of the system of Fig. 8.6, and $w(x) = \mu$, then, by letting $f(x) \equiv W_i(x)$ and $g(x) \equiv W_j(x)$, where $W_i(x)$ and $W_j(x)$ denote the ith and the jth modes of the system,

$$(W_i, W_j) \equiv \int_0^l \mu W_i(x) W_j(x) dx = 0, \quad i \neq j \qquad (8.74a)$$

which means that, just as in the n-dof case, the natural modes of the system of Fig. 8.6 are orthogonal with respect to the mass distribution μ. Likewise, for $j = i$,

$$(W_i, W_i) \equiv \int_0^l \mu W_i^2(x)\,dx = 1, \quad i = 1, 2, \ldots, \text{etc.} \tag{8.74b}$$

whence it becomes apparent that the eigenfunctions have units of $\text{kg}^{-1/2}$. Now, in order to better interpret Eq. 8.72b in the continuous case, let us recall the formula for the *central-difference approximation* of the second derivative of a function [2], where a sample of equally spaced abscissae $\{x_i\}_1^n$ has been defined, with $x_0 = 0$ and $x_n = 1$, the interval length being denoted by Δx, i.e.,

$$g_i'' \equiv g''(x)\big|_{x_i} \approx \frac{g_{i-1} - 2g_i + g_{i+1}}{(\Delta x)^2}, \quad i = 2, \ldots, n-1 \tag{8.75a}$$

and

$$g_1'' \approx \frac{g_0 - 2g_1 + g_2}{(\Delta x)^2}, \quad g_n'' \approx \frac{g_{n-1} - 2g_n + g_{n+1}}{(\Delta x)^2} \tag{8.75b}$$

where $g_0 = 0$, but we do not have a value for g_{n+1}. In order to determine this value, we resort to the nature of the continuous counterpart, Fig. 8.6, of the n-dof system under study. In this light, we can readily realize that the slope at the free end should vanish, just as dictated by the BCs of that system, Eq. 8.36. This condition leads to $g_{n+1} = g_{n-1}$, the expressions for g_1'' and g_n'' thus becoming

$$g_1'' \approx \frac{-2g_1 + g_2}{(\Delta x)^2}, \quad g_n'' \approx \frac{2g_{n-1} - 2g_n}{(\Delta x)^2} \tag{8.75c}$$

On the other hand, upon multiplying both sides of Eq. 8.72b by n^2, we have

$$EA_s \frac{n^2}{l^2} \left[f_1(2g_1 - g_2) + \sum_2^{n-1} f_i(-g_{i-1} + 2g_i - g_{i+1}) \right.$$

$$\left. + f_n(-g_{n-1} + g_n) \right] \Delta x = 0$$

Now, if we subtract $EA_s(n^2/l^2)f_n(g_{n-1} - g_n)\Delta x$ from the two sides of the foregoing equation, we obtain

$$EA_s \frac{n^2}{l^2} \left[f_1(2g_1 - g_2) + \sum_2^{n-1} f_i(-g_{i-1} + 2g_i - g_{i+1}) \right.$$

$$\left. + f_n(-2g_{n-1} + 2g_n) \right] \Delta x = -EA_s \frac{n^2}{l^2} f_n(g_{n-1} - g_n)\Delta x$$

where we readily recognize n^2/l^2 as $1/(\Delta x)^2$, and hence, by virtue of expressions (8.75a and c), the above relation becomes

$$-EA_s \sum_1^n f_i g_i'' \Delta x = -EA_s f_n \frac{g_{n-1} - g_n}{(\Delta x)^2} \Delta x$$

which, in light of Eq. 8.75c, becomes further

$$EA_s \sum_1^n f_i g_i'' \Delta x = \frac{1}{2} EA_s f_n g_n'' \Delta x \qquad (8.76)$$

If we now let $\Delta x \to 0$, the foregoing summation becomes an integral similar to that of Eq. 8.70, but with $g''(x)$ instead of $g(x)$, i.e.,

$$\int_0^l EA_s f(x) g''(x) dx = 0$$

If, moreover, we let $f(x)$ and $g(x)$ denote the ith and the jth eigenfunctions of the system of Fig. 8.6, and notice that EA_s now plays the role of the weighing function $w(x)$, then Eq. 8.70 takes the form

$$\int_0^l EA_s W_i(x) W_j''(x) dx = 0 \qquad (8.77)$$

Upon integration of the left-hand side of Eq. 8.77 by parts, that equation takes the form

$$EA_s W_i(x) W_j'(x) \big|_0^l - \int_0^l EA_s W_i'(x) W_j'(x) dx = 0$$

Furthermore, in view of the BCs of the problem at hand, it is a simple matter to verify that

$$EA_s W_i(x) W_j'(x) \big|_0^l \equiv EA_s [W_i(l) W_j'(l) - W_i(0) W_j'(0)] = 0 \qquad (8.78)$$

equation (8.77) thus becoming

$$\int_0^l EA_s W_i'(x) W_j'(x) dx = 0 \qquad (8.79a)$$

In direct analogy with Eq. 6.24a, we should have

$$\int_0^l EA_s [W_i'(x)]^2 dx = \omega_i^2 \qquad (8.79b)$$

In summary, then, Eqs. 8.74a and 8.79a state the orthogonality of the natural modes of the bar of Fig. 8.6 under axial vibration. Moreover, if the natural modes verify the normality conditions of Eqs. 8.74b and 8.79b, then these modes, i.e., the eigenfunctions of the system at hand, are termed the *normal modes* of this system.

We just introduced the concept of mode-orthogonality for a specific example. Mode-orthogonality, however, is a general concept and can be proven to be valid in all continuous systems leading to linear PDEs. The concept thus applies to all cases studied in this chapter. A fundamental relation that we shall invoke in this regard is Eq. 8.78. The reader is invited to verify that this relation holds for all systems studied in Sect. 8.4.1 leading to a second-order boundary-value problem (BVP), with coefficient EA_s suitably replaced by its corresponding counterpart, which depends on the nature of the system at hand, as explained below.

The orthogonality of modes of bars under torsional vibration and of strings under transverse vibration, with respect to the mass distribution, take essentially the same form as Eqs. 8.74a and b. However, notice that for bars under torsional vibration, the variable of interest, $\theta(x,t)$, is nondimensional, and hence, properly speaking, the normal modes $W_i(x)$ have units of $kg^{-1/2}m^{-1}$. Thus, for Eq. 8.74b to hold, μ must be replaced by a distribution of moment of inertia which, in this case, is $(1/2)\mu r^2$.

As to orthogonality with respect to stiffness distribution, note that the weighing function in integrals (8.79a and b) takes the form κl^2, the corresponding coefficient for bars under torsional vibration and for strings under transverse vibration being determined as explained below.

We first look at bars under torsional vibration, for which it is apparent from Eq. 8.46 and the ensuing discussion, that κ is now a torsional vibration per unit length, namely,

$$\kappa = \frac{1}{2}\pi G \frac{r^4}{l^2}$$

and hence, the orthogonality conditions (8.79a and b) take the forms

$$\int_0^l \frac{1}{2}\pi G r^4 W_i'(x)W_j'(x)dx = 0 \tag{8.80a}$$

$$\int_0^l \frac{1}{2}\pi G r^4 [W_i'(x)]^2 dx = \omega_i^2 \tag{8.80b}$$

Likewise, for strings under transverse vibration it is apparent, from Eq. 8.63b and the ensuing discussion, that κ is, in this case,

$$\kappa = \frac{\tau_0}{l^2} \tag{8.81}$$

and hence, the orthogonality of modes with respect to the stiffness distribution takes the forms

$$\int_0^l \tau_0 W_i'(x)W_j'(x)dx = 0 \tag{8.82a}$$

$$\int_0^l \tau_0 [W_i'(x)]^2 dx = \omega_i^2 \tag{8.82b}$$

Strictly speaking, μ and κ need not be constant and, in fact, they are not constant in many instances. For example, in the case of the string of Fig. 8.10 with a heavy trolley at one end of mass M, the mass of the string can be regarded as a function of x, with a Dirac function, in the x-domain, to account for the lumped mass M, i.e.,

$$\mu(x) = \bar{\mu} + M\delta(x - l)$$

where we have relabelled the uniform mass distribution as $\bar{\mu}$ in order to avoid confusion with function $\mu(x)$. Note that, similar to an impulse defined in the time domain, which has units of frequency, an impulse in the x-domain has units of length-inverse, and hence, the two terms of the right-hand side of the above equation are dimensionally homogeneous. The presence of the "impulsive" mass of magnitude M in $\mu(x)$ has an important consequence in the orthogonality of the eigenfunctions of this system. Indeed, orthogonality now takes the form

$$\int_0^l \mu(x)W_i(x)W_j(x)dx \equiv \int_0^l [\bar{\mu} + M\delta(x - l)]W_i(x)W_j(x)dx$$

and hence, the above integral becomes

$$\int_0^l \mu(x)W_i(x)W_j(x)dx = \bar{\mu}\int_0^l W_i(x)W_j(x)dx + MW_i(l)W_j(l)$$

Therefore, the orthogonality of the eigenfunctions now becomes

$$\bar{\mu}\int_0^l W_i(x)W_j(x)dx + MW_i(l)W_j(l) = \begin{cases} 1, & \text{for } i = j; \\ 0, & \text{otherwise.} \end{cases} \tag{8.83}$$

With the aid of the above orthogonality condition, we determine now the eigenfunctions of the corresponding system, that of Fig. 8.10 with its left-end fixed and with its right-end pinned to a heavy trolley. We thus have, with $A = 0$,

$$W_k(x) = B_k \sin(\lambda_k x), \quad k = 1, 2, \ldots \tag{8.84}$$

with λ_k determined as the kth root of Eq. 8.59. Thus, for $k = j = i$, Eq. 8.83 leads to

$$\bar{\mu}\int_0^l W_i^2(x)dx + MW_i^2(l) = 1$$

or, after substitution of Eq. 8.84 into the above orthogonality condition, we obtain

$$\bar{\mu}\int_0^l B_i^2 \sin^2(\lambda_i x)dx + MB_i^2 \sin^2(\lambda_i l) = 1$$

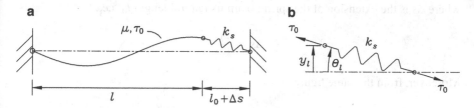

Fig. 8.22 A string under transverse vibration, with one end pinned and the other end coupled to a lumped spring: (**a**) layout of the system; (**b**) FBD of the spring

Upon expansion of the above integral and solving for B_i^2 from the equation thus resulting, we have

$$B_i^2 = \frac{2\lambda_i}{\lambda_i l \overline{\mu} + [2\lambda_i M \sin(\lambda_i l) - \overline{\mu} \cos(\lambda_i l)] \sin(\lambda_i l)}$$

Now, from the associated characteristic equation,

$$\sin(\lambda_i l) = \frac{\overline{\mu}}{\lambda_i M} \cos(\lambda_i l)$$

Substitution of the above expression into the expression for B_i^2 derived above yields, after simplifications,

$$B_i^2 = \frac{2\lambda_i^2 M}{\lambda_i^2 l \overline{\mu} M + \overline{\mu}^2 \cos^2(\lambda_i l)}$$

and hence,

$$W_i(x) = \sqrt{\frac{2\lambda_i^2 M}{\lambda_i^2 l \overline{\mu} M + \overline{\mu}^2 \cos^2(\lambda_i l)}} \sin(\lambda_i l x) \qquad (8.85)$$

Further, let us study the orthogonality condition with respect to the stiffness distribution. To this end, let us consider the string of Fig. 8.22, which is pinned at the left-end and coupled to a lumped spring of stiffness k_s at its right-end, the spring being prestressed by an amount Δs and having a natural length l_0.

The boundary condition (BC) at the left end was already derived when the motion of the string of Fig. 8.12 was studied, and is reproduced below for quick reference:

$$W(0) = 0$$

Now, the BC at the right-end is established from the FBD of the spring, as shown in Fig. 8.22b. From this figure, it is apparent that

$$\tau_0 = k_s \Delta s$$

where Δs is the extension of the spring from its natural length l_0, i.e.,

$$\Delta s = \frac{\tau_0}{k_s}$$

Moreover, from the same figure,

$$y_l = (l_0 + \Delta s)\tan\theta_l = -\left(l_0 + \frac{\tau_0}{k_s}\right)y_l' \tag{8.86}$$

The two BCs derived above, then, lead to

$$A\cos(0) + B\sin(0) = 0$$

$$\left[\cos(\lambda l) - \left(l_0 + \frac{\tau_0}{k_s}\right)\lambda\sin(\lambda l)\right]A + \left[\sin(\lambda l) + \left(l_0 + \frac{\tau_0}{k_s}\right)\lambda\cos(\lambda l)\right]B = 0$$

Upon setting up the two BCs in the usual form, we have

$$\begin{bmatrix} 1 & 0 \\ \cos(\lambda l) - \left(l_0 + \frac{\tau_0}{k_s}\right)\lambda\sin(\lambda l) & \sin(\lambda l) + \left(l_0 + \frac{\tau_0}{k_s}\right)\lambda\cos(\lambda l) \end{bmatrix}\begin{bmatrix} A \\ B \end{bmatrix} = \begin{bmatrix} 0 \\ 0 \end{bmatrix}$$

and hence, the characteristic equation of this system is

$$\sin(\lambda l) + \left(l_0 + \frac{\tau_0}{k_s}\right)\lambda\cos(\lambda l) = 0$$

whence,

$$\tan(\lambda_k l) = -\left(l_0 + \frac{\tau_0}{k_s}\right)\lambda_k, \quad k = 1, 2, \ldots \tag{8.87}$$

Now, the orthogonality condition with respect to the stiffness distribution is obtained. The latter is given in our case by Eq. 8.81 for the case in which the distribution is constant. However, in the case at hand there is a lumped spring at the right end, and hence, κ is no longer constant, but a function of x, given by

$$\kappa(x) = \frac{\tau_0}{l} + k_s\delta(x - l)$$

Therefore,

$$\kappa(x)l^2 = \tau_0 + l^2 k_s\delta(x - l)$$

the orthogonality condition sought thus being expressed as

$$\int_0^l [\tau_0 + l^2 k_s\delta(x - l)]W_i'(x)W_j'(x)dx = \begin{cases} \omega_i^2, & \text{for } i = j; \\ 0, & \text{otherwise.} \end{cases}$$

Upon expansion, the above integral, I, becomes

$$I \equiv \int_0^l \tau_0 W_i'(x) W_j'(x) dx + k_s l^2 W_i'(l) W_j'(l)$$

the orthogonality condition under study thus becoming

$$\int_0^l \tau_0 W_i'(x) W_j'(x) dx + k_s l^2 W_i'(l) W_j'(l) = \begin{cases} \omega_i^2, & \text{for } i = j; \\ 0, & \text{otherwise.} \end{cases} \tag{8.88}$$

Finally, the eigenfunctions are found with the aid of the orthogonality condition with respect to the mass distribution. From the BC leading to $A_k = 0$, we obtain

$$W_k(x) = B_k \sin(\lambda_k x)$$

with λ_k obtained as the kth root of the characteristic equation (8.87). Now, we find B_k from the orthogonality condition with respect to the mass distribution:

$$\int_0^l \mu B_k^2 \sin^2(\lambda_k x) dx = 1$$

which yields

$$\frac{1}{2} \mu B_k^2 \left[l - \frac{1}{2\lambda_k} \sin(2\lambda_k l) \right] = 1$$

and hence,

$$B_k = \sqrt{\frac{4\lambda_k}{\mu[2\lambda_k l - \sin(\lambda_k l)]}}$$

Therefore, the eigenfunctions take the form

$$W_k(x) = \sqrt{\frac{4\lambda_k}{\mu[2\lambda_k l - \sin(\lambda_k l)]}} \sin(\lambda_k x) \tag{8.89a}$$

their derivative with respect to x then becoming

$$W_k'(x) = \sqrt{\frac{4\lambda_k}{\mu[2\lambda_k l - \sin(\lambda_k l)]}} \lambda_k \cos(\lambda_k x) \tag{8.89b}$$

8.5   Exercises

The exercises given below may require a graphical solution found as the intersections of two functions, as illustrated with one example in Sect. 8.3.2.

8.1. Shown in Fig. 8.23 is an elastic rod of uniform cross section with constant mass μ per unit length, Young modulus E, cross-section area A_s, length l, and a concentrated mass M attached to its right end. Moreover, we assume that $\mu l / M = 0.1$. Under axial vibrations, a section of the rod, located at point x, experiences a displacement $u(x,t)$. For purposes of analysis, we factor $u(x,t)$ in the form

$$u(x,t) = U(x)F(t), \quad 0 < x < l$$

(a) Find ordinary differential equations for $U(x)$ and $F(t)$.
(b) Establish boundary conditions for $U(x)$.
(c) For the numerical values $\beta = 1.9812 \times 10^{-4}$ s/m and $l = 10.0$ m, obtain "good" estimates of the first ten natural frequencies and plot the first three modes.
(d) Set up the orthogonality conditions of the eigenfunctions and test them by numerical integration.

8.2. Consider a uniform beam of flexural rigidity EI, mass per unit length ρA_s, and length l, as shown in Fig. 8.24. This beam is clamped at one end and supported by a spring of stiffness k at the other end. State the boundary conditions and derive the associated characteristic equation, under the assumption that the spring is unloaded at the configuration displayed in the figure.

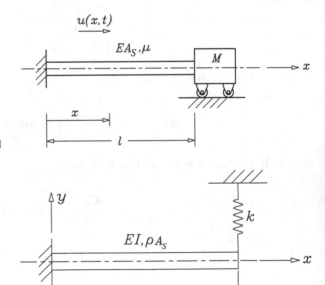

Fig. 8.23 Elastic rod with concentrated mass at one end

Fig. 8.24 A uniform beam supported at one end by a concentrated spring

Fig. 8.25 A string supported by two massless, rigid links

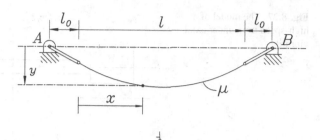

Fig. 8.26 The iconic model of a long-haul truck

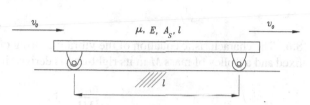

8.3. Shown in Fig. 8.25 is a string of uniform mass per unit length μ and length l, attached to two rigid, massless links of length l_0 that can rotate freely and without friction about their pinned edges at A and B. Under *small-amplitude* transverse vibrations, the projection of each link onto the x-axis is essentially equal to the link length l_0. Moreover, a point of the string, of abscissa x, experiences a displacement $y(x,t)$. For purposes of analysis, we factor $y(x,t)$ in the form

$$y(x,t) = W(x)F(t), \quad 0 < x < l$$

(a) Find ordinary differential equations for $W(x)$ and $F(t)$ that will allow you to find $y(x,t)$.
(b) State the boundary conditions for $W(x)$, and derive the associated characteristic equation.
(c) For values of $l_0/l = 0.01, 0.1, 0.2, 0.5$, find the first three natural frequencies of the system, and plot the corresponding eigenfunctions, for $\beta = 0.04\,\mathrm{m^{-1}s}$.
(d) Does the string show a rigid mode? If so, describe it.

8.4. Shown in Fig. 8.26 is the iconic model of an unloaded long-haul truck. This model consists of an elastic rod of uniform cross section A_s with constant mass μ per unit length, Young modulus E, and length l, that travels in such a way that its ends are kept at a uniform speed v_0. Establish the eigenvalue problem associated with the axial vibrations and find the corresponding eigenfrequencies and natural modes.

8.5. Assume that the B end of the shaft of the two-rotor turbine shown in Fig. 5.15 is fixed, and that the A end is mounted on roller bearings that allow the A rotor to turn freely and without friction. Moreover, the shaft has a uniform, circular cross section; shear modulus G; and radius r. Derive the characteristic equation of the system for the torsional vibration of the shaft; find values for the ratios ω_i/ω_1, for $i = 2, 3$, with ω_i denoting the ith natural frequency of the system, for $i = 1, 2, 3$, and plot the first three modes.

Fig. 8.27 The model of a
high-speed train under axial
vibration

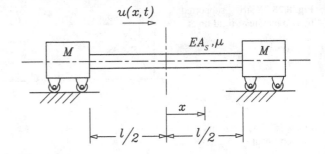

8.6. The characteristic equation of the vibrating string of Fig. 8.10 with its left-end fixed and a trolley of mass M in its right-end is derived in Eq. 8.52. For a value

$$\frac{\tau_0}{\omega^2 Ml} = 1$$

plot the first three natural modes of the string.

8.7. Derive the characteristic equation of the vibrating string of Fig. 8.15 under the assumption that the trolley has a mass M.

8.8. Derive the characteristic equation of the cantilever beam of Fig. 8.18 when its right-end carries a mass M.

8.9. A high-speed train is usually composed of two identical locomotives at its ends and a series of a dozen passenger cars in-between. A simple iconic model of such a system, for purposes of analysis of its axial vibration, is shown in Fig. 8.27, consisting of a linearly elastic bar of uniform cross-section area A_s, Young modulus of elasticity E, and mass per unit length μ. Derive the characteristic equation of the system at hand, considering that the boundary conditions of the model thus described take the form

$$Mu_{tt}(-l/2,t) = EA_s u_x(-l/2,t), \quad Mu_{tt}(l/2,t) = -EA_s u_x(l/2,t)$$

where u_v indicates the partial derivative of u with respect to variable v.

References

1. Boyce WE, DiPrima RC (1991) Elementary differential equations and boundary-value problems, 5th edn. Wiley, New York
2. Kahaner D, Moler C, Nash S (1989) Numerical methods and software. Prentice Hall, Englewood Cliffs, NJ

Appendix A
Matrix Functions

Noli turbare circulos meos
Attributed to Archimedes.[1]

A.1 Introduction

Here we introduce the concept of *analytic function* of a square matrix and methods to compute it. We illustrate the concept with a number of examples pertaining to 2×2 matrices that can be handled with longhand calculations. For symmetric matrices, we introduce the Mohr circle to compute not only their eigenvalues and eigenvectors, but also their analytic functions. Moreover, we include shortcuts applicable to specific types of matrices, e.g., matrices with simple structures, with, e.g., a limited number of non-zero entries.

A.2 Preliminary Concepts

Given a $n \times n$ matrix $\mathbf{A}$, it is often necessary to compute an *analytic function* $\mathbf{F}(\mathbf{A})$ of $\mathbf{A}$. Probably the best known analytic function of a square matrix is the *matrix exponential*, whose numerical calculation has been the subject of intensive research. In fact, this calculation can be accomplished in many different ways, nineteen of which were identified by Moler and Van Loan [1], but there are more [2]. The simplest and most straightforward method of computing analytic functions of 2×2 matrices is based on the *Cayley-Hamilton Theorem*. Before recalling this fundamental result of linear algebra, some preliminary definitions and concepts are introduced below.

---

[1]*Do not disturb my circles*. Claimed to be Archimedes' last words before he was murdered.

J. Angeles, *Dynamic Response of Linear Mechanical Systems: Modeling, Analysis and Simulation*, Mechanical Engineering Series, DOI 10.1007/978-1-4419-1027-1_9, © Springer Science+Business Media, LLC 2011

Let $\mathbf{x}$ and $\mathbf{y}$ be n-dimensional vectors related by

$$\mathbf{y} = \mathbf{A}\mathbf{x} \tag{A.1}$$

In general, $\mathbf{y}$ has an orientation and a magnitude different from that of $\mathbf{x}$. However, it may happen that some vectors $\mathbf{x} \neq \mathbf{0}$ are transformed by $\mathbf{A}$ into multiples of $\mathbf{x}$, in which case we can write

$$\mathbf{A}\mathbf{x} = \lambda\mathbf{x} \tag{A.2}$$

where λ is a scalar. Equation A.2 can be rewritten as

$$(\lambda\mathbf{1} - \mathbf{A})\mathbf{x} = 0 \tag{A.3}$$

where $\mathbf{1}$ is the $n \times n$ identity matrix. Equation A.3 represents, thus, a *homogeneous* system of n linear algebraic equations in n unknowns, the latter being all grouped within vector $\mathbf{x}$. Since we are interested in nontrivial solutions, the matrix coefficient should be singular, i.e.,

$$\det(\lambda\mathbf{1} - \mathbf{A}) = 0 \tag{A.4a}$$

Upon expansion of the foregoing determinant, one obtains

$$\det(\lambda\mathbf{1} - \mathbf{A}) \equiv \lambda^n + c_{n-1}\lambda^{n-1} + \cdots + c_1\lambda + c_0 = 0 \tag{A.4b}$$

i.e., the determinant is an nth-degree polynomial in λ, which is called the *characteristic polynomial* of $\mathbf{A}$. The vanishing condition imposed on this polynomial, appearing in Eq. A.4b, is called the *characteristic equation* of $\mathbf{A}$, its n roots, whether real or complex, distinct or repeated, being called the *eigenvalues* of $\mathbf{A}$. Let these roots be labelled λ_i, for $i = 1, \ldots, n$. For each λ_i, there exists *at least* one non-zero vector $\mathbf{x}_i$ for which Eq. A.2 holds, i.e.,

$$\mathbf{A}\mathbf{x}_i = \lambda_i\mathbf{x}_i \tag{A.5}$$

Each $\mathbf{x}_i$ is termed an *eigenvector* of $\mathbf{A}$. Now we can state the main result:

   Theorem (Cayley-Hamilton). *Let Eq. A.4b be the characteristic equation of* $\mathbf{A}$. *Matrix* $\mathbf{A}$ *verifies its own characteristic equation, i.e.,*

$$\mathbf{A}^n + c_{n-1}\mathbf{A}^{n-1} + \cdots + c_1\mathbf{A} + c_0\mathbf{1} = \mathbf{O} \tag{A.6}$$

where $\mathbf{O}$ *is the* $n \times n$ *zero matrix.*
   The proof of the Cayley-Hamilton Theorem is straightforward for $n \times n$ matrices with n linearly independent eigenvectors, especially for symmetric matrices. For matrices with an *incomplete* set of eigenvectors, i.e., with less than n linearly independent eigenvectors, the proof is more elaborate and falls beyond the scope of the book. The diligent reader is referred to the pertinent literature [3] for an outline of the proof.

What the foregoing theorem states is very important, namely, the nth power of $\mathbf{A}$ is a *linear combination* of its first n powers, since, from Eq. A.6 we can solve for $\mathbf{A}^n$ as

$$\mathbf{A}^n = -c_0\mathbf{A}^0 - c_1\mathbf{A}^1 - c_2\mathbf{A}^2 - \cdots - c_{n-1}\mathbf{A}^{n-1} \tag{A.7}$$

with $\mathbf{A}^0 = \mathbf{1}$ being identified as the *first* power of $\mathbf{A}$.

As a consequence of Eq. A.4b, λ^p, for any integer $p \geq n$, can be expressed as a linear combination of the first n powers of λ, including the first one, $\lambda^0 = 1$, i.e.,

$$\lambda^p = k_0 + k_1\lambda + \cdots + k_{n-1}\lambda^{n-1} \tag{A.8a}$$

Furthermore, as a consequence of Eqs. A.7 and A.8a, the pth power of $\mathbf{A}$, for $p \geq n$, can be correspondingly written as a linear combination of the first n powers of $\mathbf{A}$, i.e.,

$$\mathbf{A}^p = k_0\mathbf{1} + k_1\mathbf{A} + \cdots + k_{n-1}\mathbf{A}^{n-1} \tag{A.8b}$$

where the k_i coefficients are the same in Eqs. A.8a and b, but, in general, $k_i \neq c_i$, with c_i as given in Eqs. A.4b, A.6 and A.7. Of course, if $p = n$, then $k_i = c_i$.

Further results that will be found useful in computing analytic functions of square matrices are included below:

Fact 1: *Let $\mathbf{A}$ be an arbitrary $n \times n$ matrix, its eigenvalues and eigenvectors being $\{\lambda_i\}_1^n$ and $\{\mathbf{e}_i\}_1^n$, respectively. Moreover, the latter set is assumed to be normalized, i.e., with each $\mathbf{e}_i$ of unit magnitude. The eigenvalues $\{\mu_i\}_1^n$ of $\mathbf{A}^k$ are simply $\mu_i = \lambda_i^k$, for $i = 1, \ldots, n$, while $\mathbf{A}$ and $\mathbf{A}^k$, for integer k share the same eigenvectors.*

Fact 2: *Let s be a scalar and $\mathbf{A}$ a $n \times n$ matrix. Then, $\mathbf{A}$ and $s\mathbf{A}$ share the same eigenvectors, the eigenvalues of the latter being s times the eigenvalues of the former.*

The proofs of the two foregoing facts are straightforward and can be skipped.

A.3 Calculation of Analytic Matrix Functions of a Matrix Argument

As a direct application of the Cayley-Hamilton Theorem, let us consider any *analytic* function $f(\lambda)$ of λ. If this function is analytic, then it can be expanded in series, namely,

$$f(\lambda) = f(0) + f'(0)\lambda + \frac{1}{2!}f''(0)\lambda^2 + \cdots + \frac{1}{k!}f^{(k)}(0)\lambda^k + \cdots \tag{A.9}$$

From relation (A.8a) it is clear that the foregoing *series* reduces to the *sum*

$$f(\lambda) = f_0 + f_1\lambda + f_2\lambda^2 + \cdots + f_{n-1}\lambda^{n-1} \qquad \text{(A.10a)}$$

with $f_0, f_1, \ldots, f_{n-1}$ being constant coefficients, as yet to be determined. Similarly, let $\mathbf{F}(\mathbf{A})$ be the corresponding matrix function of $\mathbf{A}$, i.e., if, for example, $f(\lambda) = e^\lambda$, then $\mathbf{F}(\mathbf{A}) = e^{\mathbf{A}}$. By virtue of the Cayley-Hamilton Theorem and Eq. A.10a, we have

$$\mathbf{F}(\mathbf{A}) = f_0\mathbf{1} + f_1\mathbf{A} + f_2\mathbf{A}^2 + \cdots + f_{n-1}\mathbf{A}^{n-1} \qquad \text{(A.10b)}$$

whose coefficients are the same as those appearing in Eq. A.10a.

In vibration analysis, the matrices whose analytic functions are to be computed represent physical quantities like frequency. For example, when studying the vibration of multi-dof systems, we come across the *frequency matrix*, whose entries have all units of s^{-1}, and hence, its kth power has entries with units of s^{-k}. Obviously, the kth and the $(k+1)$st powers of this matrix cannot be added, for they have different units. Hence, in performing calculations, we must make sure that all matrices whose powers are to be added are dimensionally homogeneous. The simplest way of rendering them so is by scaling all of them so that their entries will be dimensionless.

We have thus a method to compute an analytic matrix function $\mathbf{F}(\mathbf{A})$ of matrix $\mathbf{A}$: Compute the coefficients $f_0, f_1, \ldots, f_{n-1}$ of the associated function $f(\lambda)$ and express $\mathbf{F}(\mathbf{A})$ as a linear combination of the first n powers of $\mathbf{A}$ with the coefficients mentioned above. To do this, we first evaluate $f(\lambda)$ for each eigenvalue λ_i, thereby deriving a system of *linear equations* that can then be solved for these coefficients. There are two cases, namely,

1. The n eigenvalues are distinct. Here, we can derive n linearly independent equations in the given coefficients, i.e.,

$$f_0 + \lambda_1 f_1 + \cdots + \lambda_1^{n-1} f_{n-1} = f(\lambda_1)$$

$$f_0 + \lambda_2 f_1 + \cdots + \lambda_2^{n-1} f_{n-1} = f(\lambda_2)$$

$$\vdots$$

$$f_0 + \lambda_n f_1 + \cdots + \lambda_n^{n-1} f_{n-1} = f(\lambda_n) \qquad \text{(A.11a)}$$

which can be rewritten in vector form as

$$\mathbf{\Lambda f} = \phi \qquad \text{(A.11b)}$$

where Λ is an $n \times n$ matrix, whereas $\mathbf{f}$ and ϕ are n-dimensional vectors, all of which are defined below.[2]

$$
\Lambda = \begin{bmatrix} 1 & \lambda_1 & \cdots & \lambda_1^{n-1} \\ 1 & \lambda_2 & \cdots & \lambda_2^{n-1} \\ \vdots & \vdots & \ddots & \vdots \\ 1 & \lambda_n & \cdots & \lambda_n^{n-1} \end{bmatrix}, \quad \mathbf{f} = \begin{bmatrix} f_0 \\ f_1 \\ \vdots \\ f_{n-1} \end{bmatrix}, \quad \phi = \begin{bmatrix} f(\lambda_1) \\ f(\lambda_2) \\ \vdots \\ f(\lambda_n) \end{bmatrix} \tag{A.11c}
$$

Since Eq. A.11a are linearly independent, a solution $\mathbf{f}$ to the foregoing system exists, which is, additionally, unique.

2. Some eigenvalues are repeated. Assume that $\lambda_1 = \lambda_2 = \cdots = \lambda_r (r < n)$, but the remaining ones, $\lambda_{r+1}, \lambda_{r+2}, \ldots, \lambda_n$, are distinct. Then, we compute the $r - 1$ derivatives of $f(\lambda)$ with respect to λ and evaluate them, together with the function itself, at λ_1. To these equations, we add equations corresponding to Eq. A.11c for $\lambda_{r+1}, \ldots, \lambda_n$, to obtain a system of n linearly independent equations in the n unknown coefficients, namely,

$$
f_0 + \lambda_1 f_1 + \lambda_1^2 f_2 + \cdots + \lambda_1^{r-1} f_{r-1} + \cdots + \lambda_1^{n-1} f_{n-1} = f(\lambda_1)
$$

$$
f_1 + 2\lambda_1 f_2 + \cdots + (r-1)\lambda_1^{r-2} f_{r-1} + \cdots + (n-1)\lambda_1^{n-2} f_{n-1} = f'(\lambda_1)
$$

$$
2f_2 + \cdots + (r-2)(r-1)\lambda_1^{r-3} f_{r-1} + \cdots + (n-2)(n-1)\lambda_1^{n-3} f_{n-1} = f''(\lambda_1)
$$

$$
\vdots
$$

$$
2 \cdots (r-2)(r-1) f_{r-1} + \cdots + (n-r+1)(n-r+2)\cdots(n-1)\lambda_1^{n-r} f_{n-1} = f^{(r-1)}(\lambda_1)
$$

$$
f_0 + \lambda_{r+1} f_1 + \lambda_{r+1}^2 f_2 + \cdots + \lambda_{r+1}^{n-1} f_{n-1} = f(\lambda_{r+1})
$$

$$
\vdots
$$

$$
f_0 + \lambda_n f_1 + \lambda_n^2 f_2 + \cdots + \lambda_n^{n-1} f_{n-1} = f(\lambda_n)
$$

---

[2]The structure of Λ is so frequent in system theory that it bears a name, *Vandermonde matrix*. Scientific software provides means to create a Vandermonde matrix by entering, in general, only the name (computer algebra) or the value of the argument λ and the dimension n of the square matrix.

The foregoing system is similar to system (A.11b), except that now $\boldsymbol{\Lambda}$ and ϕ are defined as

$$\boldsymbol{\Lambda} = \begin{bmatrix} 1 & \lambda_1 & \lambda_1^2 & \cdots & \lambda_1^{r-1} & \cdots & \lambda_1^{n-1} \\ 0 & 1 & 2\lambda_1 & \cdots & (r-1)\lambda_1^{r-2} & \cdots & (n-1)\lambda_1^{n-2} \\ 0 & 0 & 2 & \cdots & (r-2)(r-1)\lambda_1^{r-3} & \cdots & (n-2)(n-1)\lambda_1^{n-3} \\ \vdots & \vdots & \vdots & \ddots & \vdots & \ddots & \vdots \\ 0 & 0 & 0 & \cdots & (r-1)! & \cdots & (n-1)!\lambda_1^{n-r}/(n-r)! \\ 1 & \lambda_{r+1} & \lambda_{r+1}^2 & \cdots & \lambda_{r+1}^{r-1} & \cdots & \lambda_{r+1}^{n-1} \\ \vdots & \vdots & \vdots & \ddots & \vdots & \ddots & \vdots \\ 1 & \lambda_n & \lambda_n^2 & \cdots & \lambda_n^{r-1} & \cdots & \lambda_n^{n-1} \end{bmatrix}$$

$$\phi = \begin{bmatrix} f(\lambda_1) & f'(\lambda_1) & f''(\lambda_1) & \cdots & f^{(r-1)}(\lambda_1) & f(\lambda_{r+1}) & \cdots & f(\lambda_n) \end{bmatrix}^T x \qquad (A.12)$$

As a result of Facts 1 and 2, and Eq. A.10b, we now have

Fact 3: *Let $\mathbf{F}(\mathbf{A})$ be the analytic matrix function of $\mathbf{A}$ derived from the analytic scalar function $f = f(\lambda)$. Then, if we let $\{\phi_i\}_1^n$ be the set of eigenvalues of $\mathbf{F}(\mathbf{A})$ and $\{\lambda_i\}_1^n$ those of $\mathbf{A}$, we have*

$$\phi_i = f(\lambda_i), \quad i = 1, \ldots, n \qquad (A.13)$$

Moreover, $\mathbf{F}(\mathbf{A})$ and $\mathbf{A}$ share the same eigenvectors.

A.3.1   Special Case: 2×2 Matrices

The special case of 2×2 matrices deserves attention because: (1) 2×2 non-singular matrices can be inverted in closed form; (2) 2×2 matrices entail the essential properties of general $n \times n$ matrices, while being much easier to manipulate; and (3) when these are, additionally, symmetric, their analytic functions can be computed *graphically*, namely, with the aid of the Mohr circle.

So, let us assume that we have a 2×2 arbitrary matrix $\mathbf{A}$, given by

$$\mathbf{A} \equiv \begin{bmatrix} a_{11} & a_{12} \\ a_{21} & a_{22} \end{bmatrix} \equiv \begin{bmatrix} \mathbf{c}_1 & \mathbf{c}_2 \end{bmatrix} \equiv \begin{bmatrix} \mathbf{r}_1^T \\ \mathbf{r}_2^T \end{bmatrix} \qquad (A.14)$$

where $\mathbf{c}_k$ and $\mathbf{r}_k^T$ represent its kth two-dimensional column and row vectors, respectively. Note that here we have indicated vectors as column arrays, a practice that is followed throughout the book.

The *trace* of $\mathbf{A}$ is a very important *invariant* [3] quantity that is defined as the sum of the diagonal entries of the matrix, i.e.,

$$\text{tr}(\mathbf{A}) \equiv a_{11} + a_{22} \tag{A.15}$$

The determinant of the foregoing matrix, also a matrix invariant, is given in turn by

$$\det(\mathbf{A}) \equiv a_{11}a_{22} - a_{12}a_{21} \equiv -\mathbf{r}_1^T \mathbf{E} \mathbf{r}_2 \tag{A.16}$$

with $\mathbf{E}$ defined in Sect. 1.6 as the 2×2 orthogonal matrix that rotates two-dimensional vectors in their plane through an angle of $90°$ counterclockwise, which is reproduced below for quick reference:

$$\mathbf{E} \equiv \begin{bmatrix} 0 & -1 \\ 1 & 0 \end{bmatrix} \tag{A.17}$$

Matrix $\mathbf{E}$, as shown in the Section recalled above, has interesting properties, namely,

$$\mathbf{E}^T = -\mathbf{E}, \quad \mathbf{E}^{-1} = \mathbf{E}^T = -\mathbf{E} \tag{A.18}$$

Now, the inverse of $\mathbf{A}$ is calculated just by (1) interchanging its diagonal entries, (2) reversing the signs of its off-diagonal entries, and (3) dividing the resultant matrix by $\Delta \equiv \det(\mathbf{A})$, i.e.,

$$\mathbf{A}^{-1} = \frac{1}{\Delta} \begin{bmatrix} a_{22} & -a_{12} \\ -a_{21} & a_{11} \end{bmatrix} \tag{A.19a}$$

which can be represented in terms of its columns and rows, alternatively, as

$$\mathbf{A}^{-1} = \frac{1}{\Delta} \mathbf{E} \begin{bmatrix} -\mathbf{r}_2 & \mathbf{r}_1 \end{bmatrix} \equiv \frac{1}{\Delta} \begin{bmatrix} \mathbf{c}_2^T \\ -\mathbf{c}_1^T \end{bmatrix} \mathbf{E} \tag{A.19b}$$

The characteristic equation of $\mathbf{A}$ takes the form

$$P(\lambda) \equiv \det \begin{bmatrix} \lambda - a_{11} & -a_{12} \\ -a_{21} & \lambda - a_{22} \end{bmatrix} \tag{A.20}$$

or, upon expansion, as

$$P(\lambda) = \lambda^2 - \text{tr}(\mathbf{A})\lambda + \Delta \tag{A.21}$$

---

[3] Invariance means that, under a change of vector basis, the trace does not change. More precisely, a quantity is invariant when it follows certain rules under a change of frame. A scalar is invariant when it does not change under a change of frame.

and hence, the two eigenvalues of this matrix are

$$\lambda_{1,2} = \frac{\text{tr}(\mathbf{A})}{2} \pm \sqrt{\left(\frac{\text{tr}(\mathbf{A})}{2}\right)^2 - \Delta} \qquad (A.22)$$

If we want to calculate the coefficients f_0 and f_1 of the matrix function $\mathbf{F}(\mathbf{A})$ of Eq. A.10b, we end up with a 2×2 matrix $\mathbf{\Lambda}$ and a two-dimensional vector ϕ, that were defined in Eq. A.11c for the general $n \times n$ case, namely,

$$\mathbf{\Lambda} = \begin{bmatrix} 1 & \lambda_1 \\ 1 & \lambda_2 \end{bmatrix}, \quad \mathbf{f} = \begin{bmatrix} f_0 \\ f_1 \end{bmatrix}, \quad \phi = \begin{bmatrix} f(\lambda_1) \\ f(\lambda_2) \end{bmatrix} \qquad (A.23a)$$

and hence,

$$\begin{bmatrix} f_0 \\ f_1 \end{bmatrix} = \frac{1}{\lambda_2 - \lambda_1} \begin{bmatrix} \lambda_2 & -\lambda_1 \\ -1 & 1 \end{bmatrix} \begin{bmatrix} f(\lambda_1) \\ f(\lambda_2) \end{bmatrix} \qquad (A.23b)$$

or

$$f_0 = \frac{\lambda_2 f(\lambda_1) - \lambda_1 f(\lambda_2)}{\lambda_2 - \lambda_1} \qquad (A.24a)$$

$$f_1 = \frac{f(\lambda_2) - f(\lambda_1)}{\lambda_2 - \lambda_1} \qquad (A.24b)$$

If matrix $\mathbf{A}$ has only one linearly independent eigenvector, the equations from which the coefficients f_0 and f_1 are computed take the form

$$\begin{bmatrix} 1 & \lambda_1 \\ 0 & 1 \end{bmatrix} \begin{bmatrix} f_0 \\ f_1 \end{bmatrix} = \begin{bmatrix} f(\lambda_1) \\ f'(\lambda_1) \end{bmatrix} \qquad (A.25)$$

and hence, its solution readily follows, namely,

$$f_1 = f'(\lambda_1) \qquad (A.26a)$$

$$f_0 = f(\lambda_1) - \lambda_1 f_1 \qquad (A.26b)$$

The foregoing expressions are given here for reference, as a means to verify numerical results. In the examples that follow, we do not use these formulas, but rather work on a case-by-case basis. The reader is invited to verify the results given in Sect. A.4 with the aid of these expressions.

A.3.2 Examples

In the examples below, matrix $\mathbf{A}$ is given, and a matrix function $\mathbf{F}(\mathbf{A})$ is sought.

A.3.2.1 Example A.3.1

$$\mathbf{A} = \begin{bmatrix} 3 & 4 \\ 4 & -3 \end{bmatrix}, \quad \mathbf{F}(\mathbf{A}) = \mathbf{A}^{100}$$

1. Characteristic equation:

$$\det(\lambda \mathbf{1} - \mathbf{A}) \equiv \lambda^2 - 25 = 0$$

Hence,

$$\lambda_1 = 5, \quad \lambda_2 = -5$$

2. Write $f(\lambda) = \lambda^{100}$, for λ_1 and λ_2, in the form (A.11a):

$$\begin{bmatrix} 1 & 5 \\ 1 & -5 \end{bmatrix} \begin{bmatrix} f_0 \\ f_1 \end{bmatrix} = \begin{bmatrix} 5^{100} \\ (-5)^{100} \end{bmatrix}$$

i.e.,

$$\begin{bmatrix} f_0 \\ f_1 \end{bmatrix} = 5^{100} \begin{bmatrix} 1 \\ 0 \end{bmatrix}$$

and hence,

$$f_0 = 5^{100}, \quad f_1 = 0$$

3. Write $\mathbf{A}^{100}$ as a linear combination of $\mathbf{1}$ and $\mathbf{A}$, namely, as

$$\mathbf{A}^{100} = f_0 \mathbf{1} + f_1 \mathbf{A} = 5^{100} \mathbf{1} + 0 \mathbf{A} = \begin{bmatrix} 5^{100} & 0 \\ 0 & 5^{100} \end{bmatrix}$$

Thus,

$$\mathbf{A}^{100} = \begin{bmatrix} 7.9 \times 10^{69} & 0 \\ 0 & 7.9 \times 10^{69} \end{bmatrix}$$

A.3.2.2   Example A.3.2

A is given as in Example A.3.1, but $F(A)$ is now defined as e^A. Thus, Λ remains, but ϕ changes as indicated below:

$$\phi = \begin{bmatrix} e^5 & e^{-5} \end{bmatrix}^T$$

Hence,

$$f_0 = \frac{e^5 + e^{-5}}{2} \equiv \cosh 5, \quad f_1 = \frac{e^5 - e^{-5}}{10} \equiv \frac{\sinh 5}{5}$$

and so,

$$e^A = f_0 \mathbf{1} + f_1 \mathbf{A} = \begin{bmatrix} \cosh 5 + (3/5)\sinh 5 & (4/5)\sinh 5 \\ (4/5)\sinh 5 & \cosh 5 - (3/5)\sinh 5 \end{bmatrix}$$

A.3.2.3   Example A.3.3

$$\mathbf{A} = \begin{bmatrix} \sigma & -\sigma \\ \sigma & \sigma \end{bmatrix}, \quad \sigma = \frac{\sqrt{2}}{2}, \quad F(\mathbf{A}) = \mathbf{A}^{1000}$$

$$\det(\mathbf{A} - \lambda \mathbf{1}) = \det \begin{bmatrix} \sigma - \lambda & -\sigma \\ \sigma & \sigma - \lambda \end{bmatrix} = 0$$

Thus,

$$\lambda^2 - 2\sigma\lambda + 2\sigma^2 = 0$$

whence,

$$\lambda_{1,2} = \sigma \pm \sqrt{\sigma^2 - 2\sigma^2}$$

or

$$\lambda_{1,2} = \sigma \pm j\sigma = \frac{\sqrt{2}}{2}(1 \pm j1) = e^{\pm j\pi/4}$$

Thus, coefficients f_0 and f_1 are computed from the equations below:

$$f_0 + \lambda_i f_1 = \lambda_i^{1000}, \quad i = 1, 2$$

i.e.,

$$f_0 + e^{j\pi/4} f_1 = e^{j250\pi} = 1 \tag{A.27}$$

$$f_0 + e^{-j\pi/4} f_1 = e^{-j250\pi} = 1 \tag{A.28}$$

Now, if Eq. A.28 is subtracted from Eq. A.27, we obtain

$$2\left(\sin\frac{\pi}{4}\right)f_1 = 0, \qquad f_1 = 0$$

Substituting $f_1 = 0$ in Eq. A.27 then gives

$$f_0 = 1$$

which thus produces,

$$\mathbf{A}^{1000} = \mathbf{1}$$

where $\mathbf{1}$ is the 2×2 identity matrix. Can you explain this result?

A.3.2.4   Example A.3.4: The Zero-input Response of a Second-order Damped System

Consider the system:

$$\dot{x} = v$$

$$\dot{v} = -\omega_n^2 x - 2\zeta\omega_n v$$

$$r(0) = x_0, \quad v(0) = v_0$$

This system can be written in vector form as

$$\dot{\mathbf{x}} = \mathbf{A}\mathbf{x}, \quad \mathbf{x}(0) = \mathbf{x}_0$$

with $\mathbf{A}$ and $\mathbf{x}$ defined below:

$$\mathbf{A} = \begin{bmatrix} 0 & 1 \\ -\omega_n^2 & -2\zeta\omega_n \end{bmatrix}, \quad \mathbf{x} = \begin{bmatrix} x(t) \\ v(t) \end{bmatrix}$$

Furthermore, the response of the system can be written as

$$\mathbf{x}(t) = e^{\mathbf{A}t}\mathbf{x}_0$$

Thus, in order to evaluate the response of the system, all that we need is the exponential of matrix $\mathbf{A}t$, which is computed in the usual manner. The characteristic equation of $\mathbf{A}$ is first derived:

$$\det(\lambda\mathbf{1} - \mathbf{A}) = \lambda^2 + 2\zeta\omega_n\lambda + \omega_n^2 = 0$$

from which,

$$\lambda_1 = (-\zeta + \sqrt{\zeta^2 - 1})\omega_n, \quad \lambda_2 = (-\zeta - \sqrt{\zeta^2 - 1})\omega_n$$

and, recalling Fact 2, the eigenvalues of $\mathbf{A}t$ are $\{\lambda_k t\}_1^2$. Now, Eq. A.10a are written for $e^{\lambda_k t}$ and $k = 1, 2$ as

$$\begin{bmatrix} 1 & \lambda_1 t \\ 1 & \lambda_2 t \end{bmatrix} \begin{bmatrix} f_0 \\ f_1 \end{bmatrix} = \begin{bmatrix} e^{\lambda_1 t} \\ e^{\lambda_2 t} \end{bmatrix}$$

whence,

$$f_0 = \frac{\lambda_2 e^{\lambda_1 t} - \lambda_1 e^{\lambda_2 t}}{(\lambda_2 - \lambda_1)}, \quad f_1 = \frac{-e^{\lambda_1 t} + e^{\lambda_2 t}}{(\lambda_2 - \lambda_1)t}$$

Moreover,

$$\lambda_2 - \lambda_1 = -2\sqrt{\zeta^2 - 1}\,\omega_n = -2\rho\,\omega_n; \quad \rho \equiv \sqrt{\zeta^2 - 1}$$

and hence,

$$f_0 = \frac{e^{-\zeta \omega_n t}}{\rho}\left(\zeta\frac{e^{\rho \omega_n t} - e^{-\rho \omega_n t}}{2} + \rho\frac{e^{\rho \omega_n t} + e^{-\rho \omega_n t}}{2}\right) \qquad \text{(A.29a)}$$

$$f_1 = \frac{e^{-\zeta \omega_n t}}{\rho \omega_n t}\frac{e^{\rho \omega_n t} - e^{-\rho \omega_n t}}{2} \qquad \text{(A.29b)}$$

Thus, $e^{\mathbf{A}t}$ takes on the form

$$e^{\mathbf{A}t} = f_0\mathbf{1} + f_1\mathbf{A}t$$

Below we consider three cases, namely,

1. Underdamped system: $\zeta < 1$. In this case, ρ is imaginary and can be written as

$$\rho = jr, \quad r \equiv \sqrt{1 - \zeta^2}\,(\text{real}), \quad j \equiv \sqrt{-1}$$

and now f_0 and f_1 take the forms:

$$f_0 = \frac{e^{-\zeta \omega_n t}}{jr}(j\zeta \sin r\omega_n t + jr \cos r\omega_n t)$$

$$= \frac{e^{-\zeta \omega_n t}}{\sqrt{1 - \zeta^2}}\left(\zeta \sin \omega_n \sqrt{1 - \zeta^2}t + \sqrt{1 - \zeta^2} \cos \omega_n \sqrt{1 - \zeta^2}t\right)$$

$$f_1 = \frac{e^{-\zeta \omega_n t}}{jr\omega_n}j \sin r\omega_n t = \frac{e^{-\zeta \omega_n t}}{\omega_n\sqrt{1 - \zeta^2}} \sin \omega_n \sqrt{1 - \zeta^2}t$$

The matrix exponential, then, takes the form

$$
e^{\mathbf{A}t} = \frac{e^{-\zeta\omega_n t}}{R} \begin{bmatrix} \zeta\sin\omega_d t + R\cos\omega_d t & (1/\omega_n)\sin\omega_d t \\ -\omega_n\sin\omega_d t & -\zeta\sin\omega_d t + R\cos\omega_d t \end{bmatrix}, \quad \text{(A.30a)}
$$

$$
R \equiv \sqrt{1-\zeta^2} \qquad\qquad\qquad\qquad\qquad \text{(A.30b)}
$$

ω_d being known as the *damped frequency* and is defined as

$$
\omega_d \equiv \sqrt{1-\zeta^2}\,\omega_n \qquad\qquad\qquad \text{(A.30c)}
$$

Thus, $\mathbf{x}(t)$ reduces to

$$
\mathbf{x}(t) = (f_0\mathbf{1} + f_1\mathbf{A}t)\mathbf{x}_0
$$

i.e.,

$$
x(t) = \frac{e^{-\zeta\omega_n t}}{\sqrt{1-\zeta^2}}\left(\zeta\sin\omega_d t + \sqrt{1-\zeta^2}\cos\omega_d t\right)x_0 + \frac{e^{-\zeta\omega_n t}}{\omega_d}(\sin\omega_d t)v_0
$$

$$
v(t) = \frac{-\omega_n e^{-\zeta\omega_n t}}{\sqrt{1-\zeta^2}}(\sin\omega_d t)x_0 + \frac{e^{-\zeta\omega_n t}}{\sqrt{1-\zeta^2}}\left(-\zeta\sin\omega_d t + \sqrt{1-\zeta^2}\cos\omega_d t\right)v_0
$$

Exercise; Prove that $v(t) = dx/dt$ from the above expressions.

2. Critically damped system: $\zeta = 1$. Here, $\lambda_1 = \lambda_2 = -\omega_n$, i.e., we end up with a repeated eigenvalue and, hence, $\rho = 0$. Furthermore, note that, from Eqs. A.29a and b, it is not possible to compute the coefficients f_0 and f_1 when $\zeta = 1$, for those equations yield indeterminacies in this case. Hence, we have to proceed as indicated above for the case of repeated eigenvalues. We thus need the derivative of both sides of Eq. A.10a with respect to λ. Below we write that equation and its derivative for the case at hand:

$$
f_0 + \lambda f_1 = e^{\lambda}
$$

$$
f_1 = e^{\lambda}
$$

whence,

$$
f_1 = e^{\lambda}, \quad f_0 = (1-\lambda)e^{\lambda}
$$

We further replace λ by λt and recall that $\lambda = -\omega_n$, which leads to

$$
e^{\mathbf{A}t} = (1+\omega_n t)e^{-\omega_n t}\mathbf{1} + e^{-\omega_n t}\mathbf{A}t
$$

i.e.,

$$e^{\mathbf{A}t} = \begin{bmatrix} (1+\omega_n t)e^{-\omega_n t} & te^{-\omega_n t} \\ -\omega_n^2 te^{-\omega_n t} & (1-\omega_n t)e^{-\omega_n t} \end{bmatrix} \tag{A.31}$$

Thus, the time response of the system at hand becomes

$$x(t) = (1+\omega_n t)e^{-\omega_n t}x_0 + te^{-\omega_n t}v_0$$
$$v(t) = -\omega_n^2 te^{-\omega_n t}x_0 + (1-\omega_n t)e^{-\omega_n t}v_0$$

3. Overdamped system: $\zeta > 1$. Since ρ is real in this case, f_0 and f_1, as given by Eqs. A.29a and b, can be written as

$$f_0 = \frac{e^{-\zeta\omega_n t}}{\rho}(\rho\cosh\rho\omega_n t + \zeta\sinh\rho\omega_n t)$$

$$f_1 = \frac{e^{-\zeta\omega_n t}}{\rho\omega_n}\sinh\rho\omega_n t$$

and hence,

$$e^{\mathbf{A}t} = \frac{e^{-\zeta\omega_n t}}{\rho}\left[(\rho\cosh\rho\omega_n t + \zeta\sinh\rho\omega_n t)\mathbf{1} + \frac{1}{\omega_n}\sinh\rho\omega_n t\mathbf{A}\right]$$

or

$$e^{\mathbf{A}t} = \frac{e^{-\zeta\omega_n t}}{\rho}\begin{bmatrix} \rho\cosh\rho\omega_n t + \zeta\sinh\rho\omega_n t & \frac{1}{\omega_n}\sinh\rho\omega_n t \\ -\omega_n\sinh\rho\omega_n t & \rho\cosh\rho\omega_n t - \zeta\sinh\rho\omega_n t \end{bmatrix} \tag{A.32}$$

the time response of interest thus being

$$x(t) = \frac{e^{-\zeta\omega_n t}}{\rho}(\rho\cosh\rho\omega_n t + \zeta\sinh\rho\omega_n t)x_0 + \frac{e^{-\zeta\omega_n t}}{\rho\omega_n}\sinh\rho\omega_n t v_0$$

$$v(t) = -\frac{\omega_n e^{-\zeta\omega_n t}}{\rho}(\sinh\rho\omega_n t)x_0 + \frac{e^{-\zeta\omega_n t}}{\rho}(\rho\cosh\rho\omega_n t - \zeta\sinh\rho\omega_n t)v_0$$

A.3.2.5   Example A.3.5

$$\mathbf{A} = \begin{bmatrix} 2 & 1 \\ 1 & 2 \end{bmatrix}, \quad \mathbf{F}(\mathbf{A}) = \ln(\mathbf{A})$$

The characteristic equation of $\mathbf{A}$ is first derived:

$$\det(\lambda\mathbf{1} - \mathbf{A}) = \det\begin{bmatrix} \lambda - 2 & -1 \\ -1 & \lambda - 2 \end{bmatrix} = \lambda^2 - 4\lambda + 3 = 0$$

Hence, the eigenvalues of $\mathbf{A}$ are:

$$\lambda_{1,2} = 2 \pm \sqrt{4-3} = 2 \pm 1 = 1,3$$

Now, $\mathbf{F}(\mathbf{A})$ is written as a linear combination of $\mathbf{1}$ and $\mathbf{A}$, with coefficients f_0 and f_1 that are computed from:

$$f_0 + \lambda_i f_1 = \ln(\lambda_i), \quad i = 1,2$$

i.e.,

$$f_0 + f_1 = 0 \tag{A.33a}$$

$$f_0 + 3f_1 = \ln(3) \tag{A.33b}$$

From Eq. A.33a,

$$f_0 = -f_1 \tag{A.34}$$

Upon substitution of Eq. A.34 into Eq. A.33b, we determine the two coefficients as

$$f_1 = \frac{1}{2}\ln(3)$$

$$f_0 = -\frac{1}{2}\ln(3)$$

Thus,

$$\ln\mathbf{A} = \frac{1}{2}\ln(3)\left(\begin{bmatrix} -1 & 0 \\ 0 & -1 \end{bmatrix} + \begin{bmatrix} 2 & 1 \\ 1 & 2 \end{bmatrix}\right) = \frac{1}{2}\ln(3)\begin{bmatrix} 1 & 1 \\ 1 & 1 \end{bmatrix} x \equiv \mathbf{B}$$

and hence, the following relation holds:

$$e^{\mathbf{B}} = \mathbf{A}$$

As a matter of verification, let $\lambda_{1,2}$ be eigenvalues of $\mathbf{B}$ and $\mu_{1,2}$ those of $2\mathbf{B}/\ln(3)$. One can readily verify that

$$\lambda_i = \frac{1}{2}\ln(3)\mu_i, \quad i = 1,2$$

As a further verification, given $\mathbf{B}$, let us compute $\exp(\mathbf{B})$. To this end, first the eigenvalues of $\mathbf{B}$ are determined. The characteristic equation of $\mathbf{B}$ is readily derived:

$$\det \begin{bmatrix} \mu - 1 & -1 \\ -1 & \mu - 1 \end{bmatrix} = \mu^2 - 2\mu = 0$$

from which,

$$\mu_1 = 0, \qquad \mu_2 = 2$$

and hence, one can verify immediately that

$$\lambda_1 = 0, \quad \lambda_2 = \ln(3)$$

Moreover, $\exp(\mathbf{B})$ is written as a linear combination of $\mathbf{1}$ and $\mathbf{B}$, with coefficients g_0 and g_1 that are computed from the equations below:

$$g_0 + \lambda_i g_1 = e^{\lambda_i}, \quad i = 1, 2$$

i.e.,

$$g_0 + 0 = 1$$
$$g_0 + \ln(3) g_1 = e^{\ln(3)} (= 3)$$

Upon solving the foregoing equations for the coefficients sought, one obtains

$$g_0 = 1, \quad g_1 = \frac{2}{\ln(3)}$$

and hence,

$$e^{\mathbf{B}} = \begin{bmatrix} 1 & 0 \\ 0 & 1 \end{bmatrix} + \begin{bmatrix} 1 & 1 \\ 1 & 1 \end{bmatrix} = \begin{bmatrix} 2 & 1 \\ 1 & 2 \end{bmatrix} = \mathbf{A}$$

thereby verifying the computed results.

A.3.2.6   Example A.3.6

Let

$$\mathbf{A} = \begin{bmatrix} -3 & 4 \\ 4 & -9 \end{bmatrix}, \quad \mathbf{F}(\mathbf{A}) = \ln(\mathbf{A})$$

First, the characteristic equation of $\mathbf{A}$ is

$$\det(\lambda \mathbf{1} - \mathbf{A}) = 0$$

Upon expansion, the foregoing equation leads to

$$\lambda^2 + 12\lambda + 11 = 0$$

from which one can readily compute the eigenvalues of $\mathbf{A}$:

$$\lambda_1 = -1, \quad \lambda_2 = -11$$

Now, $\ln(\mathbf{A})$ is written as

$$\ln(\mathbf{A}) = f_0\mathbf{1} + f_1\mathbf{A}$$

where f_0 and f_1 are computed from the equations below:

$$f_0 + f_1\lambda_1 = \ln(\lambda_1)$$
$$f_0 + f_1\lambda_2 = \ln(\lambda_2)$$

i.e., with $j \equiv \sqrt{-1}$,

$$f_0 - f_1 = j\pi$$
$$f_0 - 11f_1 = \ln(11) + j\pi$$

from which the values sought are obtained as

$$f_0 = j\pi - \alpha, \quad f_1 = -\alpha, \quad \alpha \equiv \frac{\ln(11)}{10}$$

and hence,

$$\ln(\mathbf{A}) = \begin{bmatrix} j\pi - \alpha & 0 \\ 0 & j\pi - \alpha \end{bmatrix} - \alpha \begin{bmatrix} -3 & 4 \\ 4 & -9 \end{bmatrix} = \begin{bmatrix} 2\alpha + j\pi & -4\alpha \\ -4\alpha & 8\alpha + j\pi \end{bmatrix}$$

As a matter of verification again, let μ_1 and μ_2 be the eigenvalues of $\ln(\mathbf{A})$. Clearly,

$$\mu_1 = \ln(\lambda_1), \quad \mu_2 = \ln(\lambda_2)$$

as expected.

A.3.2.7 Example A.3.7

$$\mathbf{A} = \begin{bmatrix} \sigma & \omega \\ \omega & -\sigma \end{bmatrix}, \quad \mathbf{F}(\mathbf{A}) = e^{\mathbf{A}}$$

As reported in Kailath's book, the exponential of $\mathbf{A}$, computed in symbolic form with *Macsyma*, is given as

$$e^{\mathbf{A}} = \begin{bmatrix} \cosh\tau + (\sigma\sinh\tau/)\tau & (\omega\sinh\tau/\tau) \\ (\omega\sinh\tau)/\tau & \cosh\tau - (\sigma\sinh\tau)/\tau \end{bmatrix}, \quad \text{with} \quad \tau \equiv \sqrt{\omega^2 + \sigma^2}$$

The same result is now obtained using the Cayley-Hamilton Theorem. To this end, the eigenvalues of $\mathbf{A}$ are first determined. This is done from the characteristic equation of $\mathbf{A}$, namely,

$$\det(\lambda\mathbf{1} - \mathbf{A}) = 0$$

i.e.,

$$\lambda^2 - \omega^2 - \sigma^2 = 0$$

from which one readily determines the two eigenvalues of $\mathbf{A}$, namely,

$$\lambda_1 = \tau, \quad \lambda_2 = -\tau$$

with τ as previously defined. Next, $e^{\mathbf{A}}$ is written as

$$e^{\mathbf{A}} = f_0\mathbf{1} + f_1\mathbf{A}$$

with f_0 and f_1 determined as usual, i.e., from

$$f_0 + f_1\lambda_1 = e^{\lambda_1}$$
$$f_0 + f_1\lambda_2 = e^{\lambda_2}$$

or, in vector form,

$$\begin{bmatrix} 1 & \lambda_1 \\ 1 & \lambda_2 \end{bmatrix} \begin{bmatrix} f_0 \\ f_1 \end{bmatrix} = \begin{bmatrix} e^{\lambda_1} \\ e^{\lambda_2} \end{bmatrix}$$

from which one readily obtains

$$\begin{bmatrix} f_0 \\ f_1 \end{bmatrix} = \frac{1}{\lambda_2 - \lambda_1} \begin{bmatrix} \lambda_2 & -\lambda_1 \\ -1 & 1 \end{bmatrix} \begin{bmatrix} e^{\lambda_1} \\ e^{\lambda_2} \end{bmatrix} = \frac{1}{\lambda_2 - \lambda_1} \begin{bmatrix} \lambda_2 e^{\lambda_1} - \lambda_1 e^{\lambda_2} \\ -e^{\lambda_1} + e^{\lambda_2} \end{bmatrix}$$

and hence,

$$f_0 = \frac{-\tau e^{\tau} - \tau e^{-\tau}}{-2\tau} = \frac{e^{\tau} + e^{-\tau}}{2} = \cosh\tau$$

$$f_1 = \frac{-e^{\tau} + e^{-\tau}}{-2\tau} = \frac{e^{\tau} - e^{-\tau}}{2\tau} = \frac{\sinh\tau}{\tau}$$

Hence, $e^{\mathbf{A}}$ can be readily written as

$$e^{\mathbf{A}} = \cosh \tau \mathbf{1} + \frac{\sinh \tau}{\sqrt{\omega^2 + \sigma^2}} \mathbf{A}$$

which leads exactly to the result reported by Kailath.

Interestingly, if $\mathbf{A}$ is of 2×2 and *symmetric*, then the foregoing calculations can be done geometrically, using Mohr's circle, as described in Sect. A.4.

A.3.2.8 Example A.3.8: A Matrix with a Repeated Eigenvalue

$$\mathbf{A} = \begin{bmatrix} 0 & 1 \\ 0 & 0 \end{bmatrix}$$

Now we want to evaluate $\mathbf{F}(\mathbf{A})$, which is defined as

$$\mathbf{F}(\mathbf{A}) = \int_0^h e^{\mathbf{A}t}\, dt$$

where h is a constant.

Here, we have a matrix with the double eigenvalue $\lambda = 0$ and an analytic scalar function $f(\lambda)$ given by

$$f(\lambda) = \int_0^h e^{\lambda t}\, dt = \frac{e^{\lambda h} - 1}{\lambda} \equiv \frac{N(\lambda)}{D(\lambda)}$$

We now calculate $\mathbf{F}$ as

$$\mathbf{F} = f_0 \mathbf{1} + f_1 \mathbf{A}$$

with coefficients f_0 and f_1 obtained upon evaluating $f(\lambda)$ and $f'(\lambda)$ at $\lambda = 0$, namely,

$$f_0 + \lambda f_1 = f(0)$$
$$f_1 = f'(0)$$

However, note that, in calculating $f(0)$, we end up with an indeterminacy, because

$$N(0) = 0, \quad D(0) = 0$$

This indeterminacy can be resolved by resorting to L'Hospital's rule:

$$f(0) = \frac{N'(\lambda)}{D'(\lambda)}\bigg|_{\lambda=0} = \frac{he^{\lambda h}}{1}\bigg|_{\lambda=0} = h$$

On the other hand,

$$f'(\lambda) = \frac{1}{D(\lambda)}[N'(\lambda) - f(\lambda)D'(\lambda)]$$

$$= \frac{1}{\lambda}\left[he^{\lambda h} - \frac{e^{\lambda h}-1}{\lambda}(1)\right] = \frac{h\lambda e^{\lambda h} - e^{\lambda h} + 1}{\lambda^2} \equiv \frac{P(\lambda)}{Q(\lambda)}$$

Again, the evaluation of the above expression leads to an indeterminacy, because

$$P(0) = 0, \quad Q(0) = 0$$

To resolve the foregoing indeterminacy, we resort again to L'Hospital's rule:

$$f'(0) = \left.\frac{P'(\lambda)}{Q'(\lambda)}\right|_{\lambda=0} = \left.\frac{he^{\lambda h} + h^2\lambda e^{\lambda h} - he^{\lambda h}}{2\lambda}\right|_{\lambda=0} = \left.\frac{h^2 e^{\lambda h}}{2}\right|_{\lambda=0} = \frac{h^2}{2}$$

Therefore,

$$f_0 = h, \quad f_1 = \frac{h^2}{2}$$

the final result being

$$\int_0^h e^{\mathbf{A}t}\,dt = h\begin{bmatrix}1 & 0 \\ 0 & 1\end{bmatrix} + \frac{h^2}{2}\begin{bmatrix}0 & 1 \\ 0 & 0\end{bmatrix} = \begin{bmatrix}h & h^2/2 \\ 0 & h\end{bmatrix}$$

A.4 Use of Mohr's Circle to Compute Analytic Matrix Functions

Prior to discussing the calculation of matrix functions using Mohr's circle, we need to recall a few facts concerning the construction of this circle for the calculation of the eigenvalues and eigenvectors of a *symmetric*, but not necessarily positive-semidefinite 2×2 matrix. Let, for example, $\mathbf{A}$ be given as

$$\mathbf{A} = \begin{bmatrix}a_{11} & a_{12} \\ a_{12} & a_{22}\end{bmatrix} \tag{A.35}$$

The Mohr circle of the above matrix is constructed by following the steps described below:

1. Define the orthogonal coordinate axes X and Y
2. Locate the points A, B, C and D at $(a_{11},0)$, $(a_{22},0)$, (a_{11},a_{12}) and $(a_{22},-a_{12})$, respectively

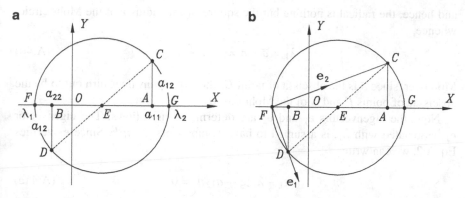

Fig. A.1 Mohr circle of a 2×2 symmetric matrix

3. Determine point E on the X axis as the intersection of the line defined by points C and D with the X axis
4. The Mohr circle of the given matrix is the circle centered at E of radius $r = \overline{DE}$

The foregoing construction is indicated in Fig. A.1a. From this figure we can determine the eigenvalues and the eigenvectors of $\mathbf{A}$. In fact, the eigenvalues of $\mathbf{A}$ are given at the intersections of the Mohr circle with the X axis, namely, as points F and G, whose abscissae are λ_1 and λ_2, respectively. Moreover, the eigenvector associated with λ_1 is the unit vector obtained from the vector directed from F to D, and indicated as $\mathbf{e}_1$ in Fig. A.1b. The eigenvector associated with λ_2 is simply vector $\mathbf{e}_1$ rotated through an angle of 90° counterclockwise and appearing as $\mathbf{e}_2$ in Fig. A.1b. Below we prove these results.

The characteristic equation of matrix $\mathbf{A}$ can be written as given in Eq. A.21 for more general 2×2 matrices, not necessarily symmetric, if by changing a_{21} by a_{12}, thereby obtaining

$$P(\lambda) = \lambda^2 - 2\overline{a}\lambda + \Delta \tag{A.36}$$

with $\overline{a}$ and Δ defined as the mean value of the diagonal entries of $\mathbf{A}$ and the determinant of $\mathbf{A}$, respectively, i.e.,

$$\overline{a} \equiv \frac{a_{11} + a_{22}}{2} \equiv \frac{\mathrm{tr}(\mathbf{A})}{2}, \quad \Delta \equiv a_{11}a_{22} - a_{12}^2 \equiv \det(\mathbf{A}) \tag{A.37}$$

Thus, the two roots of the foregoing characteristic polynomial are

$$\lambda_1 = \overline{a} - \sqrt{\overline{a}^2 - \Delta}, \quad \lambda_2 = \overline{a} + \sqrt{\overline{a}^2 - \Delta} \tag{A.38}$$

Now, the radical R above turns out to be

$$R \equiv \overline{a}^2 - \Delta = \left(\frac{a_{11} - a_{22}}{2}\right)^2 + a_{12}^2 \equiv r^2 \tag{A.39}$$

and hence, the radical is nothing but the square of the radius r of the Mohr circle, whence,

$$\lambda_1 = \bar{a} - r, \; \lambda_2 = \bar{a} + r. \tag{A.40}$$

Moreover, since $\bar{a}$ is the abscissa of point E, the two eigenvalues turn out to be the abscissae of points F and G of the Mohr circle.

Now, the eigenvectors e_1 and e_2 are determined as follows: The eigenvector e_1, associated with λ_1, is assumed to have components $[\xi, \; \eta]^T$. Since e_1 verifies Eq. A.2, we can write

$$(a_{11} - \lambda_1)\xi + a_{12}\eta = 0 \tag{A.41a}$$

$$a_{12}\xi + (a_{22} - \lambda_1)\eta = 0 \tag{A.41b}$$

Now, the two foregoing equations are linearly dependent, and hence, only one can be used to determine ξ and η. However, one single equation of these is not enough, for we have two unknowns. The two unknowns are then determined by recalling that we have assumed at the outset that the eigenvectors are normalized, i.e., of unit magnitude, and hence,

$$\xi^2 + \eta^2 = 1$$

Moreover, we use Eq. A.41a to determine e_1 by noticing that this equation can be written as a dot product of vector $\mathbf{a} = \overrightarrow{FC}$ by vector e_1, where $\mathbf{a}$ has the components shown below:

$$\mathbf{a} \equiv \begin{bmatrix} a_{11} - \lambda_1 \\ a_{12} \end{bmatrix} \tag{A.42}$$

and hence, Eq. A.41a can be rewritten in the form

$$\mathbf{a}^T e_1 = 0 \tag{A.43}$$

which means that vector e_1 is perpendicular to vector $\mathbf{a}$. In order to ease matters, we recall now the matrix $\mathbf{E}$ reproduced in Eq. A.17 that rotates vectors $90°$ counterclockwise without changing their magnitudes.

What Eq. A.43 states is that vector e_1 is nothing but vector $\mathbf{a}$ rotated through $90°$ either clockwise or counterclockwise and divided by the magnitude of $\mathbf{a}$, $\|\mathbf{a}\|$, in order to render it of unit magnitude. Now, from the Mohr circle, $\|\mathbf{a}\| = \overline{FC}$ and hence, we can express e_1 as vector $-\mathbf{E}\mathbf{a}$ divided by $\overline{FC}$, namely,

$$e_1 = -\frac{\mathbf{E}\mathbf{a}}{\overline{FC}} \tag{A.44}$$

Thus, geometrically we can determine e_1 as the unit vector directed from F to D in Fig. A.1b. Furthermore, inorder to determine e_2 all we have to do is just remember

Fig. A.2 Mohr cicle of $\mathbf{F}(\mathbf{A})$ for $a_{11} > a_{22}$ and $a_{12} > 0$

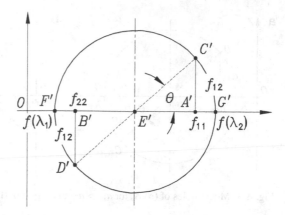

that the two eigenvectors of symmetric $\mathbf{A}$ are mutually orthogonal. Thus, $\mathbf{e}_2$ turns out to be simply the unit vector directed from F to C, i.e.,

$$\mathbf{e}_2 = \frac{\mathbf{a}}{\|\mathbf{a}\|} \tag{A.45}$$

and hence, $\mathbf{e}_2$ is as indicated in Fig. A.1b, thereby completing the intended proof.

From the Cayley-Hamilton Theorem and Fact 1, the eigenvalues of $\mathbf{F}(\mathbf{A})$ are simply $f(\lambda_i)$ and its eigenvectors are those of $\mathbf{A}$. Now, since the two matrices $\mathbf{A}$ and $\mathbf{F}(\mathbf{A})$ share the same eigenvectors, the points A', B', ..., G' in the Mohr circle of $\mathbf{F}(\mathbf{A})$ must be located at the vertices of a polygon *similar* to that determined by points A, B, ..., G of the Mohr circle of $\mathbf{A}$. Hence, the Mohr circle of $\mathbf{F}(\mathbf{A})$ can be readily derived once that of $\mathbf{A}$ is available. Moreover, the intersections of the former with the horizontal axis are at $\phi_i \equiv f(\lambda_i)$. The Mohr circle of $\mathbf{F}(\mathbf{A})$ is indicated in Fig. A.2.

Furthermore, from Figs. A.1 and A.2, the entries of $\mathbf{F}(\mathbf{A})$ can be readily obtained. Indeed, since $\mathbf{A}$ is symmetric, $\mathbf{F}(\mathbf{A})$ is symmetric as well, its entries being denoted by f_{11}, f_{12}, and f_{22}. Now, let r_F denote the radius of the Mohr circle of $\mathbf{F}$. From Figs. A.1 and A.2, then, the relations given below are readily derived:

$$f_{11} = \frac{\phi_1 + \phi_2}{2} - \frac{r_F}{r}\sqrt{r^2 - a_{12}^2} \tag{A.46a}$$

$$f_{22} = \frac{\phi_1 + \phi_2}{2} + \frac{r_F}{r}\sqrt{r^2 - a_{12}^2} \tag{A.46b}$$

$$f_{12} = \frac{r_F}{r}a_{12} \tag{A.46c}$$

where, if $a_{12} < 0$, then so is f_{12} and the two Mohr circles take the form of Figs. A.3a and b. The eigenvector $\mathbf{e}_1$ of both $\mathbf{A}$ and $\mathbf{F}$, in this case, is the unit vector directed from F to C or from F' to C' of Figs. A.3a and b, respectively. Moreover, $\mathbf{e}_2$ is just $\mathbf{e}_1$ rotated 90° counterclockwise, i.e., the unit vector directed from F to D of Fig. A.3a or, correspondingly, from F' to D' of Fig. A.3b.

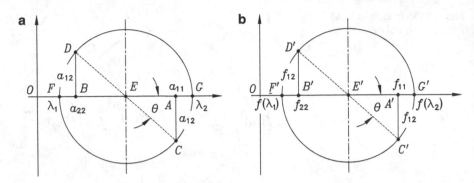

Fig. A.3 Mohr circles of (**a**) **A**, for $a_{11} > a_{22}$ and $a_{12} < 0$; (**b**) **F(A)**

Code for the production of the Mohr circle of a 2×2 symmetric matrix is availale in `ApA-MohrCircle.mw`.

Furthermore, we have one more result:

Fact 4: *Any square matrix* **A** *commutes with any of its analytic matrix functions* **F(A)**, i.e.,

$$\mathbf{AF(A)} = \mathbf{F(A)A} \tag{A.47}$$

This result can be proven by noticing that: (1) any analytic function of **A** can be expressed as a sum of the first n powers of **A**, and (2) the obvious fact that any square matrix commutes with its powers, i.e.,

$$\mathbf{AA}^k = \mathbf{A}^k\mathbf{A}$$

As a direct consequence of this fact, we have

Fact 5: *Let* $\mathbf{\Phi} \equiv \mathbf{F(A)}$ *be any analytic function of the* $n \times n$ *matrix* **A**, *and* **B** *a similarity transformation*[4] *of* **A**, *given by the* $n \times n$ *invertible matrix* **L**:

$$\mathbf{B} = \mathbf{LAL}^{-1} \tag{A.48}$$

Then,

$$\mathbf{\Psi} \equiv \mathbf{F(B)} = \mathbf{L\Phi L}^{-1} \tag{A.49}$$

Proof. Since $\mathbf{\Phi}$ is an analytic function of **A**, there exist coefficients $\{\phi_k\}_0^n$ such that

$$\mathbf{\Phi} = \sum_0^n \phi_k \mathbf{A}^k \tag{A.50}$$

---

[4] A similarity transformation occurs wherever a change of variable $\mathbf{y} = \mathbf{Lx}$ is introduced. Similarity transformations are studied in basic linear-algebra courses.

Likewise, $\boldsymbol{\Psi}$ can be expressed as

$$\boldsymbol{\Psi} = \sum_0^n \psi_k \mathbf{B}^k \tag{A.51}$$

However,

$$\mathbf{B}^k = \left(\mathbf{L}\mathbf{A}\mathbf{L}^{-1}\right)^k = \overbrace{\mathbf{L}\mathbf{A}\underbrace{\mathbf{L}^{-1}\mathbf{L}}_{1}\mathbf{A}\underbrace{\mathbf{L}^{-1}\mathbf{L}}_{1}\dots\underbrace{\mathbf{L}^{-1}\mathbf{L}}_{1}\mathbf{A}\mathbf{L}^{-1}}^{k\,\text{factors}} \tag{A.52}$$

whence,

$$\mathbf{B}^k = \mathbf{L}\mathbf{A}^k\mathbf{L}^{-1} \tag{A.53}$$

Therefore,

$$\boldsymbol{\Psi} = \sum_0^n \psi_k \mathbf{L}\mathbf{A}^k\mathbf{L}^{-1} = \mathbf{L}\left(\sum_0^n \psi_k \mathbf{A}^k\right)\mathbf{L}^{-1} \tag{A.54}$$

In order to produce relation (A.49), all that remains is to prove that $\psi_k = \phi_k$, for $k = 0, 1, \dots, n$. By virtue of relation (A.49), and the definition of the *characteristic equation* of $\mathbf{A}$, Eq. A.4a, this equation holds if $\mathbf{A}$ is substituted by $\mathbf{B}$. The consequence is that $\mathbf{A}$ and $\mathbf{B}$ share the same eigenvalues. Moreover, in light of the results of Sect. A.3, $\psi_k = \phi_k$, for $k = 0, 1, \dots, n$, thereby proving this fact.

A.4.1   Examples

In order to illustrate the foregoing concepts, we compute here some analytic functions of a few symmetric 2×2 matrices.

A.4.1.1   Example A.4.1

Let $\mathbf{A}$ be given as

$$\mathbf{A} = \begin{bmatrix} 25 & 48 \\ 48 & 97 \end{bmatrix}$$

We want to compute the positive-definite square root of matrix $\mathbf{A}$, whose Mohr circle is illustrated in Fig. A.4a. From this figure, the eigenvalues of $\mathbf{A}$ are 1 and 121, its eigenvectors being $[0.8944, -0.4472]^T$ and $[0.4472, 0.8944]^T$. The Mohr circle of $\sqrt{\mathbf{A}}$ is readily constructed in Fig. A.4b as indicated below: the intersections of this circle with the horizontal axis are $\sqrt{1}$ and $\sqrt{121}$, i.e., 1 and 11, the eigenvectors $\mathbf{e}_1$

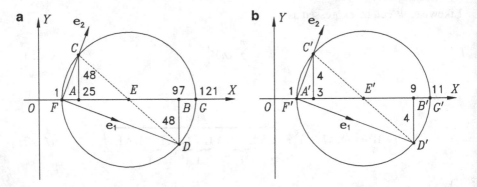

Fig. A.4 Mohr circles of matrices (**a**) **A** and (**b**) **F**(**A**) of Example A.4.1

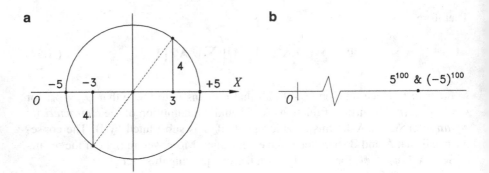

Fig. A.5 Mohr circles of matrices **A** and **F**(**A**) of Example A.3.1

and $\mathbf{e}_2$ being determined as discussed above. From this information, the entries of $\sqrt{\mathbf{A}}$ are readily computed, namely,

$$\sqrt{\mathbf{A}} = \begin{bmatrix} 3 & 4 \\ 4 & 9 \end{bmatrix}$$

Note that, according with the Mohr-circle construction, if the signs of the two off-diagonal entries of **A**, as given above, are reversed, so are those of $\sqrt{\mathbf{A}}$.

A.4.1.2 Example A.4.2

We repeat here Example A.3.1, but resorting to the Mohr circle. The Mohr circle of matrix **A**, as given in that example, is displayed in Fig. A.5a. The eigenvalues of **A** are thus readily found as $\lambda_1 = -5$ and $\lambda_2 = 5$. The eigenvalues of $\mathbf{A}^{100}$ are then simply 5^{100} and $(-5)^{100}$, i.e., they are identical and equal to 7.9×10^{69}. Thus, the

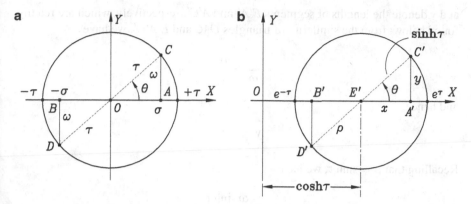

Fig. A.6 Mohr circles of matrices $\mathbf{A}$ and $F(\mathbf{A})$ of Example A.3.7

Mohr circle of $\mathbf{A}^{100}$ reduces to a point, as shown in Fig. A.5b, which means that the associated matrix is proportional to the identity matrix, i.e.,

$$\mathbf{A}^{100} = 7.9 \times 10^{69}\mathbf{1} \tag{A.55}$$

A.4.1.3   Example A.4.3

As a further illustration of the power of the Mohr-circle construction, we repeat here Example A.3.7 using this technique. Thus, the Mohr circle of matrix $\mathbf{A}$ is displayed in Fig. A.6a, that of $e^{\mathbf{A}}$ in Fig. A.6b. In Fig. A.6a, the eigenvalues of $\mathbf{A}$ are simply $\lambda_{1,2} = \pm\tau$, where τ is the radius of the Mohr circle of Fig. A.6a, i.e.,

$$\tau \equiv \sqrt{\sigma^2 + \omega^2}$$

the eigenvalues of $e^{\mathbf{A}}$ simply being $\mu_{1,2} = e^{\pm\tau}$, as shown in Fig. A.6b. Hence, the radius ρ of the Mohr circle of the foregoing exponential is simply

$$\rho \equiv \frac{e^{\tau} - e^{-\tau}}{2} \equiv \sinh\tau$$

Moreover, the centre of the same circle is located a distance d to the right of the origin, that is given by

$$d \equiv \frac{e^{\tau} + e^{-\tau}}{2} \equiv \cosh\tau$$

The entries of $e^{\mathbf{A}}$ are now found from the coordinates of the points A', B' and C' of Fig. A.6b, which are the counterparts of points A, B and C of Fig. A.6a, i.e., the two segments DC and $D'C'$ make the same angle with the horizontal axis. Let x

and y denote the lengths of segments $E'A'$ and $A'C'$, respectively, which are readily found below: from the similarity of triangles OAC and $E'A'C'$, we have

$$\frac{y}{\omega} = \frac{\rho}{\tau}$$

and hence,

$$y = \frac{\rho}{\tau}\omega$$

Recalling that $\rho = \sinh \tau$, we have

$$y = \frac{\omega \sinh \tau}{\tau}$$

Furthermore, from the similarity of the same foregoing triangles, we have

$$\frac{x}{\sigma} = \frac{\sinh \tau}{\tau}$$

and hence,

$$x = \frac{\sigma \sinh \tau}{\tau}$$

the abscissae f_{11} and f_{22} of points A' and B' thus being

$$f_{11} = \cosh \tau + \frac{\sigma \sinh \tau}{\tau}$$

$$f_{22} = \cosh \tau - \frac{\sigma \sinh \tau}{\tau}$$

which are the diagonal entries of the exponential sought. Likewise, the off-diagonal entry of the said exponential, f_{12} is simply y, i.e.,

$$f_{12} = y = \frac{\omega \sinh \tau}{\tau}$$

Thus, the matrix sought takes the form

$$e^{\mathbf{A}} = \begin{bmatrix} \cosh \tau + (\sigma \sinh \tau)/\tau & (\omega \sinh \tau)/\tau \\ (\omega \sinh \tau)/\tau & \cosh \tau - (\sigma \sinh \tau)/\tau \end{bmatrix}$$

which is exactly that obtained with *Macsyma*—see [4].

A.4.1.4 Example A.4.4: The Harmonic Functions of a Positive-definite Matrix

Here we are given a symmetric positive-definite matrix that we encounter quite often in the vibration analysis of two-degree-of-freedom systems, namely, the *frequency matrix* shown below:

$$\Omega = \begin{bmatrix} \omega_{11} & \omega_{12} \\ \omega_{12} & \omega_{22} \end{bmatrix} \tag{A.56}$$

where, for positive-definiteness, we must have

$$\text{tr}(\Omega) = \omega_{11} + \omega_{22} > 0, \quad \det(\Omega) = \omega_{11}\omega_{22} - \omega_{12}^2 > 0$$

and so, we can expect that the two eigenvalues of Ω, ω_1 and ω_2, will be positive. The Mohr circle of Ω is shown in Fig. A.7a. Shown in Figs. A.7b and c are the Mohr circles of $\cos\Omega t$ and $\sin\Omega t$.

We start with some definitions: the center of the Mohr circle of Ω is located at a distance equal to the mean value of the diagonal entries of Ω. It turns out that this distance is also the mean value of the two eigenvalues of Ω. In the study of vibrations of systems with two degrees of freedom, the eigenvalues of their frequency matrix are nothing but the two *natural frequencies*, or *eigenfrequencies* of the system at hand. Thus, we can call this distance the *mean frequency* of the system and represent it by $\overline{\omega}$, the diameter of the Mohr circle of the same matrix being equal to the difference of the two cigenfrequencies. We thus call the half of this difference the *frequency radius* and represent it by ρ. We have formally defined these quantities as

$$\overline{\omega} \equiv \frac{\omega_1 + \omega_2}{2} \equiv \frac{\omega_{11} + \omega_{22}}{2} \tag{A.57a}$$

$$\rho \equiv \frac{\omega_2 - \omega_1}{2} \equiv \sqrt{\left(\frac{\omega_{11} - \omega_{22}}{2}\right)^2 + \omega_{12}^2} \tag{A.57b}$$

From the Mohr circle of the $\cos\Omega t$ and $\sin\Omega t$ matrices we can then infer their entries, namely,

$$\cos\Omega t = \begin{bmatrix} a + u_C & v_C \\ v_C & a - u_C \end{bmatrix}, \quad \sin\Omega t = \begin{bmatrix} b + u_S & v_S \\ v_S & b - u_S \end{bmatrix}$$

and all we are left with is the calculation of the various parameters involved, namely, $a, b, \ldots, u_S$ and v_S. We do these calculations geometrically, as described below:

First, we label r_C and r_S the radii of the Mohr circle of the $\cos\Omega t$ and $\sin\Omega t$ matrices, which are calculated as

Fig. A.7 The Mohr circles of (a) a positive-definite matrix and of its (b) cosine and (c) sine functions

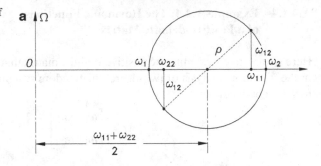

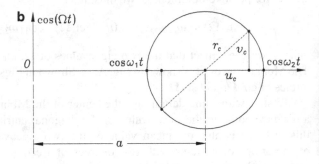

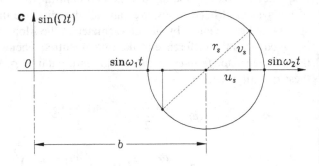

$$r_C = \frac{\cos(\omega_2 t) - \cos(\omega_1 t)}{2}, \quad r_S = \frac{\sin(\omega_2 t) - \sin(\omega_1 t)}{2}$$

Moreover, the distance of the centers of the Mohr circles of the $\cos \Omega t$ and $\sin \Omega t$ matrices, a and b, are given by

$$a = \frac{\cos(\omega_2 t) + \cos(\omega_1 t)}{2}, \quad b = \frac{\sin(\omega_2 t) + \sin(\omega_1 t)}{2}$$

Furthermore, the Mohr circle of the frequency matrix makes apparent the relations below:

$$\omega_1 = \overline{\omega} - \rho, \quad \omega_2 = \overline{\omega} + \rho$$

Upon substitution of the latter relations into the expressions for a, b, r_C and r_S, we obtain

$$a = \cos(\overline{\omega}t)\cos(\rho t)$$
$$b = \sin(\overline{\omega}t)\cos(\rho t)$$
$$r_C = -\sin(\overline{\omega}t)\sin(\rho t)$$
$$r_S = \cos(\overline{\omega}t)\sin(\rho t)$$

Now, from the geometry of Figs. A.7b and c, we readily derive expressions for u_C, v_C, u_S and v_S, namely,

$$u_C = -\frac{\omega_{11} - \omega_{22}}{2\rho}\sin(\overline{\omega}t)\sin(\rho t)$$

$$v_C = -\frac{\omega_{12}}{\rho}\sin(\overline{\omega}t)\sin(\rho t)$$

$$u_S = \frac{\omega_{11} - \omega_{22}}{2\rho}\cos(\overline{\omega}t)\sin(\rho t)$$

$$v_S = \frac{\omega_{12}}{\rho}\cos(\overline{\omega}t)\sin(\rho t)$$

and hence, the desired matrices take the forms

$$\cos(\Omega t) = \begin{bmatrix} c(\overline{\omega}t)c(\rho t) - \overline{\omega}s(\overline{\omega}t)s(\rho t) & -\dfrac{\omega_{12}}{\rho}s(\overline{\omega}t)s(\rho t) \\ -\dfrac{\omega_{12}}{\rho}s(\overline{\omega}t)s(\rho t) & c(\overline{\omega}t)c(\rho t) + \overline{\omega}s(\overline{\omega}t)s(\rho t) \end{bmatrix}$$

$$\sin(\Omega t) = \begin{bmatrix} s(\overline{\omega}t)c(\rho t) + \overline{\omega}c(\overline{\omega}t)s(\rho t) & \dfrac{\omega_{12}}{\rho}c(\overline{\omega}t)s(\rho t) \\ \dfrac{\omega_{12}}{\rho}c(\overline{\omega}t)s(\rho t) & s(\overline{\omega}t)c(\rho t) - \overline{\omega}c(\overline{\omega}t)s(\rho t) \end{bmatrix}$$

where c and s stand for cos and sin, respectively, while $\overline{\omega}$—pronounced *varpi*—stands for

$$\overline{\omega} \equiv \frac{\omega_{11} - \omega_{22}}{2\rho}$$

thereby completing the intended calculations.

A.5 Shortcuts for Special Matrices

Besides symmetric matrices, other matrices lend themselves to fast techniques to compute their analytic functions, particularly their exponentials and trigonometric functions. For example, skew-symmetric matrices, to which the Mohr circle cannot be applied, are easy to handle because their diagonal entries are zero and their off-diagonal entries have identical absolute values, but are of opposite signs. Other matrices with several zeros lend themselves to similar shortcuts, as shown with the examples below.

A.5.1 Example A.5.1

Given

$$A = \begin{bmatrix} 0 & -1 \\ 1 & 0 \end{bmatrix} \tag{A.58}$$

find the exponential of At. We do this by straightforward series expansions, namely, by computing the exponential as

$$e^{At} = 1 + At + \frac{1}{2!}A^2t^2 + \cdots + \frac{1}{k!}A^kt^k + \cdots \tag{A.59}$$

It is a simple matter to show that

$$A^2 = -1$$
$$A^3 = -A$$
$$A^4 = 1$$
$$\vdots$$

etc.

and hence,

$$e^{At} = 1 + At - \frac{1}{2!}1t^2 - \frac{1}{3!}At^3 + \frac{1}{4!}1t^4 + \frac{1}{5!}At^5 - \cdots \tag{A.60}$$

$$= \left(1 - \frac{1}{2!}t^2 + \frac{1}{4!}t^4 - \cdots\right)1 + \left(t - \frac{1}{3!}t^3 + \frac{1}{5!}t^5 - \cdots\right)A \tag{A.61}$$

The series inside the parentheses above are readily recognized to be $\cos t$ and $\sin t$, and hence,

$$e^{At} = \cos t\,1 + \sin t\,A = \begin{bmatrix} \cos t & -\sin t \\ \sin t & \cos t \end{bmatrix}$$

which is a matrix representing a rotation in the plane through a ccw angle of t.

A.5.2   Example A.5.2

Let

$$\mathbf{A} = \begin{bmatrix} 0 & 1 \\ 0 & 0 \end{bmatrix} \tag{A.62}$$

Find the exponential of $\mathbf{A}t$. We do it by straightforward calculation of the series expansion, as in the previous example. Here, we note that $\mathbf{A}^2 = \mathbf{O}$, i.e., $\mathbf{A}^2$ vanishes, and so do all higher powers of $\mathbf{A}$, the exponential sought thus reducing to

$$e^{\mathbf{A}t} = \mathbf{1} + \mathbf{A}t = \begin{bmatrix} 1 & t \\ 0 & 1 \end{bmatrix} \tag{A.63}$$

A.5.3   Example A.5.3

For $\mathbf{A}$ given as

$$\mathbf{A} = \begin{bmatrix} 1 & 1 \\ 0 & 1 \end{bmatrix} \tag{A.64}$$

find the exponential of $\mathbf{A}t$.

In this case it is a simple matter to verify that

$$\mathbf{A}^k = \begin{bmatrix} 1 & k \\ 0 & 1 \end{bmatrix}$$

and hence, the diagonal entries of the series expansion of the exponential of $\mathbf{A}t$ yield simply e^t, while the $(2,1)$ entry of the same expansion vanishes and the $(1,2)$ entry, e_{12} turns out to be

$$e_{12} = t + \frac{2}{2!}t^2 + \frac{3}{3!}t^3 + \cdots + \frac{k}{k!}t^k + \cdots$$

which can be readily shown to yield

$$e_{12} = te^t \tag{A.65}$$

the exponential sought thus becoming

$$e^{\mathbf{A}t} = \begin{bmatrix} e^t & te^t \\ 0 & e^t \end{bmatrix} = \begin{bmatrix} 1 & t \\ 0 & 1 \end{bmatrix} e^t$$

A.5.4   Example A.5.4

For

$$A = \begin{bmatrix} 1 & 0 \\ 0 & 0 \end{bmatrix}$$

find $\cos At$ and $\sin At$.

This is very simple to do because the matrix is symmetric and is already in diagonal form. Thus,

$$\cos At = \begin{bmatrix} \cos t & 0 \\ 0 & \cos(0) \end{bmatrix} = \begin{bmatrix} \cos t & 0 \\ 0 & 1 \end{bmatrix}$$

$$\sin At = \begin{bmatrix} \sin t & 0 \\ 0 & \sin(0) \end{bmatrix} = \begin{bmatrix} \sin t & 0 \\ 0 & 0 \end{bmatrix}$$

References

1. Moler CB, Van Loan C (1978) Nineteen dubious ways to compute the exponential of a matrix. SIAM Reviews 20(4):801–836
2. Zung E, Angeles J (1988) Simulation of finite-dimensional linear dynamical systems using zero-order holds and numerical stabilization methods. Comput Math Appl 16(4):307–320
3. Halmos PR (1974) Finite-dimensional vector spaces. Springer, New York
4. Kailath T (1980) Linear systems. Prentice-Hall, Englewood-Cliffs

Appendix B
The Laplace Transform

I did it my way.

Paul Anka, 1969.

B.1   Introduction

The Laplace transform is a *linear transformation* that maps functions in the domain of the real variable t, denoting *time*, into functions in the domain of the *complex variable s*, defined as

$$s = \sigma + j\omega \tag{B.1}$$

in which both σ and ω are real variables attaining values from $-\infty$ to $+\infty$.

Properly speaking, we should distinguish between the *one-sided Laplace transform* and the *double-sided Laplace transform*, represented as $\mathscr{L}_-(s)$ and $\mathscr{L}(s)$, respectively. However, we will use only the one-sided Laplace transform, and so, we need not specify the type of transform that we are talking about; neither will we need any subscript to represent the one-sided transform.

Formally, then, the Laplace transform—that is, the one-sided Laplace transform—of a real function $f(t)$ is denoted as $F(s)$, and defined as

$$F(s) \equiv \mathscr{L}[f(t)] \equiv \int_{t=0^-}^{\infty} f(t)e^{-st}\,dt \tag{B.2}$$

whence it is apparent that the physical units of the Laplace transform of a function $f(t)$ are those of this function times t. For example, if $f(t)$ represents a force, then its Laplace transform has units of impulse.

In the above definition, the exponent of e should be dimensionless; otherwise, the series expansion of that exponential would be meaningless—it would be the infinite sum of terms with different units!—and hence, the units of s are apparently those of frequency, i.e., s^{-1}, which is one reason why we denote the complex part of s as ω. One more item that should be apparent from that definition, Eq. B.2, is

J. Angeles, *Dynamic Response of Linear Mechanical Systems: Modeling, Analysis and Simulation*, Mechanical Engineering Series, DOI 10.1007/978-1-4419-1027-1_10, © Springer Science+Business Media, LLC 2011

that not every real function $f(t)$ need have a Laplace transform. For that transform to exist, the integral should be finite, i.e., the integral should exist.

Moreover, if we are given a function of s, say $F(s)$, we can determine the real function $f(t)$ whose Laplace transform is $F(s)$. To find $f(t)$ for a given $F(s)$, we resort to the *inverse Laplace transform*, denoted $\mathscr{L}^{-1}[F(s)]$, namely,

$$f(t) = \mathscr{L}^{-1}[F(s)] = \frac{1}{2\pi j} \int_{s=\sigma_0 - j\infty}^{s=\sigma_0 + j\infty} F(s)e^{st}\,ds \tag{B.3}$$

where σ_0 is any real value of σ, that is chosen large enough so as to make the above integral converge to a function $f(t)$.

As we will show presently, however, the practical way of finding the inverse Laplace transform is by means of tables of Laplace-transform pairs, the foregoing expression for that transform being rather of theoretical interest. Table X of Cannon's book [1] includes a basic list of *Laplace-transform pairs*, i.e., a list of pairs of the most frequently encountered functions of time and their respective Laplace transforms. From this list it is possible to obtain the inverse Laplace transform of a given function of s, without resorting to expression (B.3).

Example B.1.1 (Laplace Transform of the Decaying Exponential). For the function

$$f(t) = e^{-at}$$

where $a \equiv \alpha + j\beta$ is an arbitrary complex number, with $\alpha > 0$, state the condition(s) under which its Laplace transform (B.2) exists, and find this transform.

Solution: By a straightforward application of the definition, we find

$$F(s) = \mathscr{L}[e^{-at}] = \int_{t=0^-}^{t=\infty} e^{-at} e^{-st}\,dt = \int_{t=0^-}^{t=\infty} e^{-[\alpha+\sigma+j(\beta+\omega)]t}\,dt$$

Upon integration,

$$F(s) = -\left[\frac{e^{-[\alpha+\sigma+j(\beta+\omega)]t}}{\alpha+\sigma+j(\beta+\omega)}\right]_{t=0^-}^{\infty} = -\frac{\lim_{t\to\infty}[e^{-(\alpha+\sigma)t}e^{-j(\beta+\omega)t}] - 1}{\alpha+\sigma+j(\beta+\omega)}$$

Below we calculate the above limit:

$$\lim_{t\to\infty}[e^{-(\alpha+\sigma)t}e^{-j(\beta+\omega t)}] = [\lim_{t\to\infty} e^{-(\alpha+\sigma)t}][\lim_{t\to\infty} e^{-j(\beta+\omega)t}]$$

$$= [\lim_{t\to\infty} e^{-(\alpha+\sigma)t}] \lim_{t\to\infty}[\cos(\beta+\omega)t - j\sin(\beta+\omega)t]$$

Now, the second limit of the rightmost-hand side is a complex number whose real and imaginary parts oscillate between -1 and $+1$, regardless of the value of t. This limit, then, does not exist, but this does not bother us because the sine and cosine functions remain finite at any time. On the other hand, the first limit is either unbounded when $\alpha + \sigma < 0$ or zero when $\alpha + \sigma > 0$. Therefore, the above limit exists, and hence, the Laplace transform of the given function exists if and only if

$$\sigma > -\alpha$$

What the above condition states is that any value of σ above $-\alpha$ will allow us to recover the given function of time upon computing the inverse Laplace transform as given in Eq. B.3. Now, under the assumption that the above existence condition holds,

$$\mathscr{L}[e^{-at}] = \frac{1}{s+a} \tag{B.4a}$$

Note that, since we are calculating the one-sided Laplace transform, the same result would have been obtained if we had calculated the Laplace transform of $e^{-at}u(t)$, with $u(t)$ defined as the unit step function, i.e.,

$$\mathscr{L}[e^{-at}u(t)] = \frac{1}{s+a} \tag{B.4b}$$

The reader is invited to derive the above result. However, note that the double-sided Laplace transform of the above two functions are different.

Now it is a simple matter to obtain the Laplace transform of the unit step function. To this end, let us make $a = 0$ in Eq. B.4b, thereby obtaining

$$\mathscr{L}[u(t)] = \frac{1}{s} \tag{B.5}$$

Note that, while the unit step function is dimensionless, its Laplace transform has units of time.

B.1.1 Properties of the Laplace Transform

We first recall that the integration operator, like the differentiation operator, is *linear homogeneous* in that (a) the integral (derivative) of a sum of two functions is equal to the sum of the integrals (derivatives) of these functions, and (b) the integral (derivative) of a function scaled by a constant factor α, real or complex, is equal to the integral (derivative) of that function scaled by the same factor α. Property (a) is known as *additivity*; property (b) as *linear-homogeneity*. We thus have that if $F_i(s)$ represents the Laplace transform of $f_i(t)$, for $i = 1, 2$, and α is a constant, then

$$\mathscr{L}[f_1(t) + f_2(t)] = F_1(s) + F_2(s) \quad \text{additivity} \tag{B.6a}$$
$$\mathscr{L}[\alpha f(t)] = \alpha F(s) \quad \text{linear homogeneity} \tag{B.6b}$$

The above two properties amount to *superposition*.

We can now obtain the Laplace transform of $\sin \omega t$ from the Laplace transform of the exponential by a straightforward application of the above properties. To this end, we recall the exponential representation of the sine function:

$$\sin \omega t \equiv \frac{1}{2j}(e^{j\omega t} - e^{-j\omega t}) \tag{B.7}$$

whence

$$\mathscr{L}[\sin \omega t] = \frac{1}{2j}(\mathscr{L}[e^{j\omega t}] - \mathscr{L}[e^{-j\omega t}])$$

Now, the first Laplace transform of the right-hand side is derived by setting $a = -j\omega$ in Eq. B.4b; the second Laplace transform is obtained likewise, upon setting $a = j\omega$ in the same equation, which thus yields

$$\mathscr{L}[\sin \omega t] = \frac{1}{2j}\left(\frac{1}{s-j\omega} - \frac{1}{s+j\omega}\right)$$

and can be readily simplified to

$$\mathscr{L}[\sin \omega t] = \frac{\omega}{s^2 + \omega^2} \tag{B.8}$$

Additional relations pertaining to the Laplace transform are displayed below, with $F(s)$ denoting the Laplace transform of $f(t)$.

$$\mathscr{L}\left[\frac{df(t)}{dt}\right] = sF(s) - f(0^-) \tag{B.9}$$

$$\mathscr{L}\left[\frac{d^2 f(t)}{dt^2}\right] = s^2 F(s) - sf(0^-) - \dot{f}(0^-) \tag{B.10}$$

$$\vdots$$

$$\mathscr{L}\left[\frac{d^k f(t)}{dt^k}\right] = s^k F(s) - s^{k-1} f(0^-) - s^{k-2}\dot{f}(0^-) - \cdots$$

$$-\left[\frac{d^{k-1} f(t)}{dt^{k-1}}\right]_{t=0^-} \tag{B.11}$$

$$\mathscr{L}\left[\int_t f(\theta)d\theta\right] = \frac{1}{s}F(s) + \int_t f(\theta)d\theta\Big|_{t=0^-} \tag{B.12}$$

$$\mathscr{L}[f(t-T)] = e^{-Ts}F(s) \tag{B.13}$$

the last item of the above list being the Laplace transform of the *delay*.

A relation that turns out to be of the utmost importance is the Laplace transform of the convolution of two functions $f(t)$ and $g(t)$, represented by $f(t) * g(t)$.

For quick reference, we recall below the definition of the convolution, already introduced in Chap. 2:

$$f(t) * g(t) \equiv \int_0^\infty f(\tau)g(t - \tau)d\tau \equiv \int_0^\infty f(\tau - t)g(\theta)d\tau \qquad (B.14)$$

If we let $F(s)$ and $G(s)$ represent the Laplace transforms of $f(t)$ and $g(t)$, respectively, then we have

$$\mathscr{L}[f(t) * g(t)] = F(s)G(s) \qquad (B.15)$$

The reader is invited to verify the above relations.

By application of Property (B.9), we can find the Laplace transform of the cosine function. To this end, we recall that

$$\cos \omega t = \frac{1}{\omega} \frac{d(\sin \omega t)}{dt}$$

whence

$$\mathscr{L}[\cos \omega t] = \frac{1}{\omega}[s\mathscr{L}[\sin \omega t] - \sin \omega(0^-)] = \frac{s}{s^2 + \omega^2} \qquad (B.16)$$

Likewise, the Laplace transform of the impulse and the ramp functions can be readily derived as those of the derivative and the integral, respectively, of the unit step function, which yields

$$\mathscr{L}[\delta(t)] = 1 - u(0^-) = 1 \qquad (B.17)$$

$$\mathscr{L}[r(t)] = \frac{1}{s^2} + \int_{0^-}^{0^-} u(\theta)d\theta = \frac{1}{s^2} \qquad (B.18)$$

In the foregoing relations, each pair of a function of time and its corresponding Laplace transform is called a *Laplace-transform pair*. Additional Laplace-transform pairs are available in many places, e.g., in Table X and Appendix J of Cannon's book—see [1]—and in computer algebra software (Macsyma, Maple, or Mathematica).

B.2   Time Response via the Laplace Transform

The use of the Laplace transform to obtain the time response of a linear, time-invariant dynamical system is best illustrated with the aid of examples, as we do in the balance of this appendix.

Example B.2.1 (A Fluid-clutch System). The system of Example 1.6.11, consisting of a motor driving a rotary load via a fluid clutch, as shown in Fig. 1.24 and

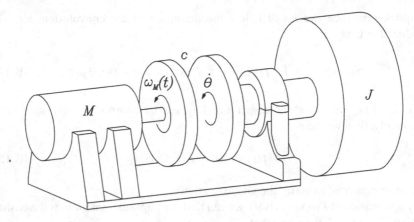

Fig. B.1 Fluid clutch with attached rotor (repeated from Fig. 1.24)

reproduced below as Fig. B.1 for quick reference, is revisited here. The clutch is to be tested by driving it with a velocity of the motor that varies harmonically with a frequency ω and an amplitude A. Thus, the motor delivers an input angular velocity to the clutch disk of the form $A\cos\omega t$. Moreover, the clutch disk is assumed to be turning at a constant 1500 rpm when it is engaged by the motor turning with an angular velocity of amplitude 300 rpm at a frequency of 1 Hz. The time constant of the system has been estimated to be 2 s.

Solution: The mathematical model of the system under study was derived in Example 1.6.11, as reproduced below:

$$\dot{p} + \frac{1}{\tau}p = \frac{1}{\tau}A\cos\omega t, \quad p(0^-) = p_0, \quad t \ge 0$$

where $p = \dot{\theta}$.

Upon Laplace-transforming both sides of the above equation, we obtain

$$sP(s) - p_0 + \frac{1}{\tau}P(s) = \frac{A}{\tau}\frac{s}{s^2 + \omega^2}$$

which can be rearranged as

$$\left(s + \frac{1}{\tau}\right)P(s) = p_0 + \frac{A}{\tau}\frac{s}{s^2 + \omega^2}$$

and hence, upon solving for $P(s)$,

$$P(s) = \frac{p_0}{s + 1/\tau} + \frac{A}{\tau}\frac{1}{s + 1/\tau}\frac{s}{s^2 + \omega^2}$$

Now the response $p(t)$ is derived by inversion of the above Laplace transform, which is eased by the properties of this transform. Thus, from the superposition property, Eqs. B.6a and b,

$$p(t) = p_0 \mathscr{L}^{-1} \left[\frac{1}{s + 1/\tau} \right] + \frac{A}{\tau} \mathscr{L}^{-1} \left[\frac{1}{s + 1/\tau} \frac{s}{s^2 + \omega^2} \right]$$

But the first inverse Laplace transform is readily recognized to be the exponential of $-t/\tau$, i.e.,

$$\mathscr{L}^{-1} \left[\frac{1}{s + 1/\tau} \right] = e^{-t/\tau}$$

while the second inverse transform is that of the product of two Laplace transforms, and hence, is a convolution, namely,

$$\mathscr{L}^{-1} \left[\frac{1}{s + 1/\tau} \frac{s}{s^2 + \omega^2} \right] = \mathscr{L}^{-1} \left[\frac{1}{s + 1/\tau} \right] * \mathscr{L}^{-1} \left[\frac{s}{s^2 + \omega^2} \right] = e^{-t/\tau} * \cos \omega t$$

That is

$$\mathscr{L}^{-1} \left[\frac{1}{s + 1/\tau} \frac{s}{s^2 + \omega^2} \right] = \int_0^\infty e^{-\theta/\tau} \cos \omega(t - \theta) d\theta$$

Now, the evaluation of the foregoing integral can be done in various forms. We can, for example, recall the exponential representation of the cosine function:

$$\cos \omega t = \frac{1}{2} (e^{j\omega t} + e^{-j\omega t})$$

and substitute it into the above integral, thereby obtaining the integral of a complex exponential form. This integral is readily derived upon recalling

$$\int_0^t e^{a\theta} d\theta = \frac{e^{at} - 1}{a}$$

where a can be any complex number and θ is a dummy variable of integration. After some algebraic manipulations, the above convolution integral is obtained as

$$\int_0^\infty e^{-\theta/\tau} \cos \omega(t - \theta) d\theta = \frac{1}{(1/\tau)^2 + \omega^2} \left(-\frac{1}{\tau} e^{-t/\tau} + \frac{1}{\tau} \cos \omega t + \omega \sin \omega t \right)$$

the time response sought thus becoming

$$p(t) = p_0 e^{-t/\tau} + \frac{A}{\tau} \frac{1}{(1/\tau)^2 + \omega^2} \left(-\frac{1}{\tau} e^{-t/\tau} + \frac{1}{\tau} \cos \omega t + \omega \sin \omega t \right)$$

Upon substitution of the given numerical values, we have

$$p(t) = \frac{1500(2\pi)}{60}e^{-0.5t} + \frac{300(2\pi)}{60 \times 22}\frac{1}{(0.5)^2+(2\pi)^2}[-0.5e^{-0.5t}+0.5\cos(2\pi t)$$
$$+2\pi\sin(2\pi t)]$$

or

$$p(t) = 157.08e^{-0.5t}+0.39538[-0.5e^{-0.5t}+0.5\cos(6.2832t)+6.2832\sin(6.2832t)]$$

which is the time response sought.

B.2.1   The Inverse Laplace Transform via Partial-Fraction Expansion

While Example B.2.1 allowed for a straightforward derivation of the time response of the fluid-clutch system therein, most applications involve a rather complicated Laplace transform not readily identifiable in a table of Laplace-transform pairs. In these cases, a *partial-fraction expansion* (PFE) of the corresponding expression, along with application of the superposition properties of the Laplace transform ease the determination of the inverse Laplace transform of interest. We recall below the partial-fraction expansion of a *rational function*, i.e., a function consisting of the quotient of two polynomials of different degrees, in general. From a quick look at a table of Laplace-transform pairs, it will become apparent that, except for the Laplace transform of the delay, Eq. B.13, all Laplace transforms take the form of rational functions in the variable s. Moreover, the Laplace transforms of the derivatives of a function aside, the rational functions occurring in our study are of the *proper* type, i.e., rational functions of s, for which the degree of the numerator is at most equal to the degree of the denominator. Furthermore, the very nature of dynamical systems leading to differential equations in the time domain, which is a consequence of their nonanticipative behavior, invariably leads to systems whose impulse response has a Laplace transform that is a proper rational function of the variable s. As a consequence, the transfer functions of dynamical systems, to be formally introduced in the section below, are proper rational functions of s.

Thus, we will be confronted with the calculation of the inverse Laplace transform of rational functions of s, of the form

$$R(s) = \frac{P(s)}{Q(s)} \tag{B.19a}$$

where

$$\deg[P(s)] = m \le \deg[Q(s)] = n \tag{B.19b}$$

Furthermore, we assume henceforth, without loss of generality, that (a) $Q(s)$ is *monic*, i.e., that the coefficient of its highest-degree term, s^n, is unity, and (b) *we know the roots* of $Q(s)$, $\{q_k\}_1^n$, which allows us to express $Q(s)$ in factored form as

$$Q(s) = (s - q_1)(s - q_2)\cdots(s - q_n) \tag{B.20}$$

Note that, if the denominator of $R(s)$ is not a monic polynomial, i.e., if its leading coefficient is different from unity, then we can divide both $P(s)$ and $Q(s)$ by that coefficient and render $Q(s)$ in monic form.

The partial-fraction expansion of $R(s)$ is now derived. To this end we consider two distinct cases: either $Q(s)$ has distinct roots or it has at least one repeated root.

- Case I: $Q(s)$ has distinct roots. In this case, the partial-fraction expansion of $R(s)$ takes the form

$$R(s) = \frac{R_1}{s - q_1} + \frac{R_2}{s - q_2} + \cdots + \frac{R_n}{s - q_n}. \tag{B.21}$$

the problem now being to compute the coefficients $\{R_k\}_1^n$. This is a simple task, for all we do to compute, say R_k, is multiply both sides of Eq. B.21 by the corresponding denominator, $s - q_k$, thereby obtaining

$$(s - q_k)R(s) = (s - q_k)\frac{R_1}{s - q_1} + (s - q_k)\frac{R_2}{s - q_2} + \cdots + R_k + \cdots + (s - q_k)\frac{R_n}{s - q_n} \tag{B.22}$$

It is now apparent that R_k can be readily obtained if we evaluate both sides of Eq. B.22 at $s = q_k$, which thus yields

$$R_k = (s - q_k)R(s)\Big|_{s=q_k} \tag{B.23}$$

Once all R_k coefficients of the PFE (B.21) are available, the inverse transform $r(t)$ of $R(s)$ is derived as

$$r(t) = R_1 e^{q_1 t} + R_2 e^{q_2 t} + \cdots + R_n e^{q_n t} \tag{B.24}$$

- Case II: $Q(s)$ has a repeated root. For the sake of conciseness, we illustrate the PFE in this case under the assumption that only one root of $Q(s)$ is repeated, the more general case of various repeated roots being an obvious extension of this case. Moreover, we start by assuming that the repeated root of $Q(s)$ appears only twice, and that this root is q_1, i.e., we assume that $Q(s)$ has been factored as

$$Q(s) = (s - q_1)^2(s - q_3)\cdots(s - q_n) \tag{B.25}$$

where q_2 does not appear explicitly because $q_2 = q_1$.

The PFE of $R(s)$ now takes the form

$$R(s) = \frac{R_{11}}{s-q_1} + \frac{R_{12}}{(s-q_1)^2} + \frac{R_3}{s-q_3} + \cdots + \frac{R_n}{s-q_n} \qquad (B.26)$$

To compute R_{12}, we proceed as before, but now we multiply both sides of Eq. B.26 by $(s-q_1)^2$, which thus yields

$$(s-q_1)^2 R(s) = (s-q_1)R_{11} + R_{12} + (s-q_1)^2 \frac{R_3}{s-q_3} + \cdots + (s-q_1)^2 \frac{R_n}{s-q_n}$$

$$(B.27)$$

Apparently, R_{12} can be found from the foregoing equation upon setting $s = q_1$ in its two sides, thereby obtaining

$$R_{12} = (s-q_1)^2 R(s) \Big|_{s=q_1} \qquad (B.28)$$

Now, in order to find R_{11}, all we do is differentiate both sides of Eq. B.27 with respect to s, which yields

$$\frac{d}{ds}\left[(s-q_1)^2 R(s)\right] = R_{11} + 2(s-q_1)\frac{R_2}{s-q_2} + (s-q_1)^2 \frac{d}{ds}\left(\frac{R_2}{s-q_2}\right) + \cdots$$

$$+ 2(s-q_1)\frac{R_n}{s-q_n} + (s-q_1)^2 \frac{d}{ds}\left(\frac{R_n}{s-q_n}\right)$$

$$= R_{11} + 2(s-q_1)\left(\frac{R_2}{s-q_2} + \cdots + \frac{R_n}{s-q_n} +\right)$$

$$+ (s-q_1)^2 \frac{d}{ds}\left(\frac{R_2}{s-q_2} + \cdots + \frac{R_n}{s-q_n}\right)$$

It is now apparent that

$$R_{11} = \frac{d}{ds}\left[(s-q_1)^2 R(s)\right]\Big|_{s=q_1} \qquad (B.29)$$

the remaining coefficients, $\{R_k\}_2^n$, being computed exactly as in Case I. Likewise, the inverse transform $r(t)$ of $R(s)$ is now readily derived as

$$r(t) = R_{11}e^{q_1 t} + R_{12}\mathcal{L}^{-1}\left[\frac{1}{(s-q_1)^2}\right] + R_3 e^{q_3 t} + \cdots + R_n e^{q_n t}$$

the remaining question now being what the inverse transform of the second term of the right-hand side of the above expression is. To answer this question, let us express the above term as the product of two identical factors, and recall that the Laplace transform of a product of two functions of s, each being in turn the

Laplace transform of a given function of time, is the Laplace transform of the convolution of the two functions of time, and hence,

$$\mathscr{L}^{-1}\left[\frac{1}{(s-q_1)^2}\right] = \mathscr{L}^{-1}\left[\frac{1}{s-q_1}\right] * \mathscr{L}^{-1}\left[\frac{1}{s-q_1}\right]$$

But

$$\mathscr{L}^{-1}\left[\frac{1}{s-q_1}\right] = e^{q_1 t}$$

and so,

$$\mathscr{L}^{-1}\left[\frac{1}{(s-q_1)^2}\right] = e^{q_1 t} * e^{q_1 t}$$

or

$$\mathscr{L}^{-1}\left[\frac{1}{(s-q_1)^2}\right] = \int_{0^-}^{t} e^{q_1(t-\tau)}e^{q_1\tau}d\tau = \int_{0^-}^{t} e^{q_1(t-\tau+\tau)}d\tau$$

which, upon integration, leads to

$$\mathscr{L}^{-1}\left[\frac{1}{(s-q_1)^2}\right] = te^{q_1 t} \tag{B.30}$$

The reader is invited to prove that the inverse Laplace transform of a rational expression similar to the one at hand, but with a multiple root q_k of multiplicity m_k is given by

$$\mathscr{L}^{-1}\left[\frac{1}{(s-q_k)^{m_k}}\right] = \frac{1}{m_k-1}t^{m_k-1}e^{q_k t} \tag{B.31}$$

It is now apparent that the inverse transform of the function $R(s)$ of Eq. B.26 is given by

$$r(t) = R_{11}e^{q_1 t} + R_{12}te^{q_1 t} + R_3 e^{q_3 t} + \cdots + R_n e^{q_n t} \tag{B.32}$$

Should a given root q_1 of $Q(s)$ appear more than twice, say, m_1 times, then the PFE of $Q(s)$ would be

$$R(s) = \frac{R_{11}}{s-q_1} + \frac{R_{12}}{(s-q_1)^2} + \cdots + \frac{R_{1m_1}}{(s-q_1)^{m_1}} + \frac{R_{m_1+1}}{s-q_{m_1+1}} + \cdots + \frac{R_n}{s-q_n} \tag{B.33}$$

Coefficient R_{1m_1} is now computed as

$$R_{1m_1} = (s-q_1)^{m_1}R(s)\Big|_{s=q_1} \tag{B.34}$$

the remaining coefficients being found upon successively differentiating the product $(s-q_1)^{m_1}R(s)$ with respect to s and evaluating this derivative at $s = q_1$.

The coefficients associated with nonrepeated roots, i.e., all R_k coefficients with a single subscript in Eq. B.33, are computed, again, exactly as indicated in Case I. Therefore, the inverse transform $r(t)$ of the expression $R(s)$ of Eq. B.33 is given by

$$r(t) = R_{11}e^{q_1 t} + R_{12}te^{q_1 t} + \cdots + R_{1m_1}\frac{1}{m_1 - 1}t^{m_1 - 1}e^{q_1 t} + R_{m_1+1}e^{q_{m_1}+1 t}$$

$$+ \cdots + R_n e^{q_n t} \tag{B.35}$$

Example B.2.2 (The Fluid-clutch System Revisited). Find the time response of the fluid-clutch system of Example B.2.1 by application of PFE to the Laplace transform found in that example.

Solution: For quick reference, we reproduce below the Laplace transform derived in Example B.2.1:

$$P(s) = \frac{p_0}{s + 1/\tau} + \frac{A}{\tau}\frac{1}{s + 1/\tau}\frac{s}{s^2 + \omega^2}$$

Apparently, PFE is needed only to compute the inverse Laplace transform of the second term of the right-hand side of the above equation. Let that term be denoted by $R(s)$, i.e.,

$$R(s) = \frac{s}{(s + 1/\tau)(s^2 + \omega^2)}$$

which can be rewritten in factored form as

$$R(s) = \frac{s}{(s + 1/\tau)(s + j\omega)(s - j\omega)}$$

and hence, the desired PFE takes the form

$$R(s) = \frac{R_1}{s + 1/\tau} + \frac{R_2}{s + j\omega} + \frac{R_3}{s - j\omega}$$

its coefficients being determined below:

$$R_1 = (s + 1/\tau)R(s)\Big|_{s=-1/\tau} = \frac{-1/\tau}{(1/\tau)^2 + \omega^2}$$

$$R_2 = (s + j\omega)R(s)\Big|_{s=-j\omega} = \frac{-j\omega}{(1/\tau - j\omega)(-j2\omega)} = \frac{1}{2(1/\tau - j\omega)} \equiv \frac{1/\tau + j\omega}{2[(1/\tau)^2 + \omega^2]}$$

$$R_3 = (s - j\omega)R(s)\Big|_{s=j\omega} = \frac{j\omega}{(1/\tau + j\omega)j2\omega} = \frac{1}{2(1/\tau + j\omega)} \equiv \frac{1/\tau - j\omega}{2[(1/\tau)^2 + \omega^2]}$$

Note that R_2 and R_3 are complex coefficients, but they are conjugate, i.e.,

$$R_3 = \bar{R}_2$$

We thus have

$$R(s) = \frac{-1/\tau}{[(1/\tau)^2 + \omega^2](s + 1/\tau)} + \frac{R_2}{s + j\omega} + \frac{\overline{R}_2}{s - j\omega}$$

$$= \frac{-1/\tau}{[(1/\tau)^2 + \omega^2](s + 1/\tau)} + \frac{R_2(s - j\omega)}{s^2 + \omega^2} + \frac{\overline{R}_2(s + j\omega)}{s^2 + \omega^2}$$

$$= \frac{-1/\tau}{[(1/\tau)^2 + \omega^2](s + 1/\tau)} + \frac{2\text{Re}\{R_2(s - j\omega)\}}{s^2 + \omega^2}$$

$$= \frac{-1/\tau}{[(1/\tau)^2 + \omega^2](s + 1/\tau)} + \frac{1}{s^2 + \omega^2}\text{Re}\left\{\frac{s/\tau + \omega^2 + j\omega(s - 1/\tau)}{(1/\tau)^2 + \omega^2}\right\}$$

which reduces to

$$R(s) = \frac{-1/\tau}{[(1/\tau)^2 + \omega^2](s + 1/\tau)} + \frac{\omega^2 + s/\tau}{[(1/\tau)^2 + \omega^2](s^2 + \omega^2)}$$

and so,

$$P(s) = \frac{p_0}{s + 1/\tau} - \frac{A/\tau^2}{[(1/\tau)^2 + \omega^2](s + 1/\tau)} + \frac{(A/\tau)(\omega^2 + s/\tau)}{[(1/\tau)^2 + \omega^2](s^2 + \omega^2)}$$

or

$$P(s) = \left(p_0 - \frac{A/\tau^2}{(1/\tau)^2 + \omega^2}\right)\frac{1}{s + 1/\tau} + \frac{A}{\tau}\frac{\omega^2 + s/\tau}{[(1/\tau)^2 + \omega^2](s^2 + \omega^2)}$$

Hence,

$$p(t) = \left(p_0 - \frac{A/\tau^2}{(1/\tau)^2 + \omega^2}\right)\mathcal{L}^{-1}\left[\frac{1}{s + 1/\tau}\right] + \frac{A}{\tau}\mathcal{L}^{-1}\left[\frac{\omega^2 + s/\tau}{[(1/\tau)^2 + \omega^2](s^2 + \omega^2)}\right]$$

$$= \left(p_0 - \frac{A/\tau^2}{(1/\tau)^2 + \omega^2}\right)\mathcal{L}^{-1}\left[\frac{1}{s + 1/\tau}\right] + \frac{A\omega}{\tau[(1/\tau)^2 + \omega^2]}\mathcal{L}^{-1}\left[\frac{\omega}{s^2 + \omega^2}\right]$$

$$+ \frac{A}{\tau^2[(1/\tau)^2 + \omega^2]}\mathcal{L}^{-1}\left[\frac{s}{s^2 + \omega^2}\right]$$

Now it is a simple matter to find the individual inverse Laplace transforms:

$$p(t) = \left(p_0 - \frac{A/\tau^2}{(1/\tau)^2 + \omega^2}\right)e^{-t/\tau} + \frac{A\omega}{\tau[(1/\tau)^2 + \omega^2]}\sin\omega t + \frac{A}{\tau^2[(1/\tau)^2 + \omega^2]}\cos\omega t$$

which coincides with the expression found previously for the same time response.

Example B.2.3 (Second-order Underdamped Systems). Find the inverse Laplace transform of

$$R(s) = \frac{1}{s^2 + 2\zeta\omega_n s + \omega_n^2} \qquad (B.36)$$

for the case in which $\zeta < 1$.

Solution: We first find the roots q_1 and q_2 of the denominator:

$$q_{1,2} = -\zeta\omega_n \pm \sqrt{\zeta^2\omega_n^2 - \omega_n^2} = -\zeta\omega_n \pm j\omega_n\sqrt{1 - \zeta^2}$$

or

$$q_1 = (-\zeta + j\sqrt{1 - \zeta^2})\omega_n, \quad q_2 = (-\zeta - j\sqrt{1 - \zeta^2})\omega_n$$

Thus,

$$R(s) = \frac{1}{[s + (\zeta - j\sqrt{1 - \zeta^2})\omega_n][s + (\zeta + j\sqrt{1 - \zeta^2})\omega_n]}$$

Hence, the PFE of $R(s)$ is

$$R(s) = \frac{R_1}{s + (\zeta - j\sqrt{1 - \zeta^2})\omega_n} + \frac{R_2}{s + (\zeta + j\sqrt{1 - \zeta^2})\omega_n}$$

Therefore,

$$R_1 = [s + (\zeta - j\sqrt{1 - \zeta^2})\omega_n]R(s)\Big|_{s=(-\zeta+j\sqrt{1-\zeta^2})\omega_n}$$

$$= \frac{1}{s + (\zeta + j\sqrt{1 - \zeta^2})\omega_n}\Big|_{s=(-\zeta+j\sqrt{1-\zeta^2})\omega_n}$$

$$= \frac{1/\omega_n}{j2\sqrt{1 - \zeta^2}} = -\frac{j}{2\sqrt{1 - \zeta^2}\omega_n}$$

$$R_2 = [s + (\zeta + j\sqrt{1 - \zeta^2})\omega_n]R(s)\Big|_{s=(-\zeta-j\sqrt{1-\zeta^2})\omega_n}$$

$$= \frac{1}{s + (\zeta - j\sqrt{1 - \zeta^2})\omega_n}\Big|_{s=-(\zeta-j\sqrt{1-\zeta^2})\omega_n}$$

$$= \frac{1/\omega_n}{-j2\sqrt{1 - \zeta^2}} = \frac{j}{2\sqrt{1 - \zeta^2}\omega_n}$$

We thus have

$$\frac{1}{s^2 + 2\zeta\omega_n s + \omega_n^2} = -\frac{j/(2\sqrt{1 - \zeta^2}\omega_n)}{s + (\zeta - j\sqrt{1 - \zeta^2})\omega_n} + \frac{j/(2\sqrt{1 - \zeta^2}\omega_n)}{s + (\zeta + j\sqrt{1 + \zeta^2})\omega_n}$$

Now it is a simple matter to derive the inverse Laplace transform of the given function of s:

$$\mathscr{L}\left[\frac{1}{s^2+2\zeta\omega_n s+\omega_n^2}\right] = -\frac{j}{2\sqrt{1-\zeta^2}\,\omega_n}e^{-(\zeta-j\sqrt{1-\zeta^2})\omega_n t}$$

$$+\frac{j}{2\sqrt{1-\zeta^2}\,\omega_n}e^{-(\zeta+j\sqrt{1-\zeta^2})\omega_n t}$$

$$=\frac{je^{-\zeta\omega_n t}}{2\sqrt{1-\zeta^2}\,\omega_n}\left(e^{-j\sqrt{1-\zeta^2}\omega_n t}-e^{j\sqrt{1-\zeta^2}\omega_n t}\right)$$

$$=\frac{je^{-\zeta\omega_n t}}{2\sqrt{1-\zeta^2}\,\omega_n}\left(-j2\sin\sqrt{1-\zeta^2}\omega_n t\right)$$

$$=\frac{e^{-\zeta\omega_n t}}{\sqrt{1-\zeta^2}\,\omega_n}\sin\sqrt{1-\zeta^2}\omega_n t$$

which is the desired inverse.

The foregoing examples illustrate one important fact in connection with the PFE of a real rational function containing a quadratic factor in its denominator: although the coefficients of the individual terms of its PFE are complex when the roots of the quadratic factor are complex, these coefficients are complex-conjugate. As a consequence, the sum of the corresponding terms of the PFE becomes real, as it should. This means that quadratic factors of $Q(s)$ deserve a special treatment that saves us the time to compute complex coefficients, as explained below.

B.2.1.1 The Case of Quadratic Factors

We consider the general case of a quadratic factor in the denominator of $R(s)$:

$$R(s) = \frac{P(s)}{(s^2+2\zeta\omega_n s+\omega_n^2)(s-q_3)\cdots(s-q_n)} \tag{B.37}$$

where we have, on purpose, written the quadratic denominator exactly as the characteristic equation of the second-order system, as derived in Example A.3.4, because this factor will appear in systems containing such a system as a subsystem. From our experience with Example B.2.2, it is apparent that a quadratic factor will admit a PFE with a term that has a numerator linear in s and that quadratic factor in the denominator, whence the PFE of the above expression takes the form

$$R(s) = \frac{R_1+R_2 s}{s^2+2\zeta\omega_n s+\omega_n^2}+\frac{R_3}{s-q_3}+\cdots+\frac{R_n}{s-q_n} \tag{B.38}$$

Now, in order to compute the coefficients of the term with the quadratic denominator, we first multiply both sides of Eq. B.38 by the denominator of the first term of that equation, thus obtaining

$$(s^2 + 2\zeta\omega_n s + \omega_n^2)R(s) = R_1 + R_2 s + (s^2 + 2\zeta\omega_n s + \omega_n^2)\frac{R_3}{s - q_3} + \cdots$$

$$+ (s^2 + 2\zeta\omega_n s + \omega_n^2)\frac{R_n}{s - q_n}$$

Further, let q_1 and q_2 be the two roots of the quadratic term, which can be real or complex, but, if real, they are assumed to be distinct, the case of a repeated root having been previously discussed. We thus evaluate the above expression for the two roots of that term, which yields

$$R_1 + R_2 q_1 = (s^2 + 2\zeta\omega_n s + \omega_n^2)R(s)\Big|_{s=q_1} \tag{B.39a}$$

$$R_1 + R_2 q_2 = (s^2 + 2\zeta\omega_n s + \omega_n^2)R(s)\Big|_{s=q_2} \tag{B.39b}$$

thereby obtaining two equations in two unknowns, R_1 and R_2, of the form

$$\begin{bmatrix} 1 & q_1 \\ 1 & q_2 \end{bmatrix}\begin{bmatrix} R_1 \\ R_2 \end{bmatrix} = \begin{bmatrix} A \\ B \end{bmatrix} \tag{B.40a}$$

with A and B denoting the right-hand sides of Eqs. B.39a and b, which are thus known. The determinant Δ of the matrix coefficient of the above system is readily computed as

$$\Delta = q_2 - q_1 \tag{B.40b}$$

which, by virtue of the assumption that the two roots are distinct, is different from zero, and hence the matrix is nonsingular; the system thus admits a unique solution for the two unknowns. We thus have, upon solving the foregoing system for R_1 and R_2,

$$\begin{bmatrix} R_1 \\ R_2 \end{bmatrix} = \frac{1}{q_2 - q_1}\begin{bmatrix} q_2 & -q_1 \\ -1 & 1 \end{bmatrix}\begin{bmatrix} A \\ B \end{bmatrix} \tag{B.41a}$$

or

$$R_1 = \frac{Aq_2 - Bq_1}{q_2 - q_1} \tag{B.41b}$$

$$R_2 = \frac{-A + B}{q_2 - q_1} \tag{B.41c}$$

Example B.2.4. Derive the PFE of the function $R(s)$ of Example B.2.2 using the quadratic-factor approach.

Solution: With this approach, we write $R(s)$ in the form

$$R(s) = \frac{R_1}{s + 1/\tau} + \frac{R_2 + R_3 s}{s^2 + \omega^2}$$

in which R_1 was already found using the standard procedure for nonrepeated roots, namely,

$$R_1 = \frac{-1/\tau}{(1/\tau)^2 + \omega^2}$$

the two remaining coefficients being calculated as outlined above. To this end, we first realize that the two roots of the square factor $s^2 + \omega^2$ are $q_1 = j\omega$ and $q_2 = -j\omega$, and set up the corresponding equations:

$$R_2 + R_3 q_1 = (s^2 + \omega^2)R(s)\Big|_{s=q_1}$$

$$R_2 + R_3 q_2 = (s^2 + \omega^2)R(s)\Big|_{s=q_2}$$

where

$$(s^2 + \omega^2)R(s)\Big|_{s=q_1} = \frac{s}{s^2 + \omega^2}\Big|_{s=j\omega} = \frac{\omega^2 + j\omega/\tau}{(1/\tau)^2 + \omega^2}$$

$$(s^2 + \omega^2)R(s)\Big|_{s=q_2} = \frac{s}{s^2 + \omega^2}\Big|_{s=-j\omega} = \frac{\omega^2 - j\omega/\tau}{(1/\tau)^2 + \omega^2}$$

and hence, the system of equations in the unknown coefficients is

$$\begin{bmatrix} 1 & j\omega \\ 1 & -j\omega \end{bmatrix} \begin{bmatrix} R_2 \\ R_3 \end{bmatrix} = \frac{1}{(1/\tau)^2 + \omega^2} \begin{bmatrix} \omega^2 + j\omega/\tau \\ \omega^2 - j\omega/\tau \end{bmatrix}$$

The determinant Δ of the matrix coefficient is readily found as

$$\Delta = -j2\omega$$

and hence,

$$\begin{bmatrix} R_2 \\ R_3 \end{bmatrix} = \frac{1}{-j2\omega} \frac{1}{(1/\tau)^2 + \omega^2} \begin{bmatrix} -j\omega & -j\omega \\ -1 & 1 \end{bmatrix} \begin{bmatrix} \omega^2 + j\omega/\tau \\ \omega^2 - j\omega/\tau \end{bmatrix}$$

which, after some simplifications, leads to

$$\begin{bmatrix} R_2 \\ R_3 \end{bmatrix} = \frac{1}{(1/\tau)^2 + \omega^2} \begin{bmatrix} \omega^2 \\ 1/\tau \end{bmatrix}$$

or

$$R_2 = \frac{\omega^2}{(1/\tau)^2 + \omega^2}, \quad R_3 = \frac{1/\tau}{(1/\tau)^2 + \omega^2}$$

the PFE sought thus being

$$R(s) = -\frac{1/\tau}{(1/\tau)^2 + \omega^2} + \frac{\omega^2 + s/\tau}{(1/\tau)^2 + \omega^2} \frac{1}{s^2 + \omega^2}$$

which is identical to the one found in Example B.2.2.

B.2.2 The Final- and the Initial-Value Theorems

These theorems, like several other results recalled in the book, are given without a proof. The interested reader can find these proofs, e.g., in Cannon's book, cited in [1]. An interesting application of the Laplace transform lies in its ability to tell values of the time response of a system at $t = 0$ and at $t \to \infty$. This is done by application of two basic theorems in connection with the Laplace transform, namely, the *Final-value Theorem* (FVT) and the *Initial-value Theorem* (IVT), as recalled below.

Under the assumption that *all the roots of the denominator $Q(s)$ of the Laplace transform $X(s)$ of a time response $x(t)$ lie either in the left-hand side of the complex plane or at the origin*, the FVT states that the value of $x(t)$ after a *long time* has elapsed, can be found without resorting to the inverse Laplace transform, but rather using the Laplace transform itself, namely, as

$$\lim_{t \to \infty} x(t) = \lim_{s \to 0}[sX(s)] \tag{B.42}$$

Likewise, the IVT states that the initial value of the response, $x(0^+)$, can be determined directly from the Laplace transform $X(s)$, without resorting to the inverse Laplace transform, namely,

$$x(0^+) = \lim_{t \to 0^+} x(t) = \lim_{s \to \infty}[sX(s)] \tag{B.43}$$

Example B.2.5 (The Fluid-clutch System... Again!). We recall the Laplace transform $P(s)$ of $p(t)$ as found in Example B.2.1, namely,

$$P(s) = \frac{p_0}{s + 1/\tau} + \frac{A}{\tau} \frac{1}{s + 1/\tau} \frac{s}{s^2 + \omega^2}$$

Now,

(a) Find the angular velocity $p(t)$ of the rotor both just after the clutch was engaged, and when a long time has elapsed; and
(b) Find the angular acceleration $\dot{p}(t)$ just after the clutch was engaged.

Solution:

(a) To find $p(0^+)$, we apply the IVT:

$$p(0^+) = \lim_{t \to 0^+} p(t) = \lim_{s \to \infty} [sP(s)]$$

i.e.,

$$p(0^+) = \lim_{s \to \infty} \left(\frac{p_0 s}{s + 1/\tau} + \frac{A}{\tau} \frac{1}{s + 1/\tau} \frac{s^2}{s^2 + \omega^2} \right)$$

or

$$p(0^+) = \lim_{s \to \infty} \left[\frac{p_0}{1 + 1/(s\tau)} + \frac{A}{\tau} \frac{1/s}{1 + 1/(s\tau)} \frac{1}{1 + (\omega/s)^2} \right]$$

which readily leads to

$$p(0^+) = p_0$$

thereby finding that the angular velocity of the rotor is continuous at time $t = 0$, for $p(0^+) = p(0^-)$. Now, in order to find $\lim_{t \to \infty} p(t)$, we need the FVT. Before applying it, we must verify that the roots of the denominator of the Laplace transform $P(s)$ comply with the hypothesis of the FVT—displayed as emphasized text in the second paragraph of this subsection. To this end, we need $P(s)$ expressed as a rational function with a denominator $Q(s)$. It is apparent that the denominator, in this case, is

$$Q(s) = (s + 1/\tau)(s^2 + \omega^2)$$

whose roots are $q_1 = -1/\tau$, $q_2 = j\omega$, and $q_3 = -j\omega$. While q_1 lies in the left-hand side of the complex plane, neither q_2 nor q_3 does, and hence, the theorem is not applicable. The reason here is that $\lim_{t \to \infty} p(t)$ does not exist. Indeed, if we look at the time response $p(t)$ derived in Example B.2.1, it is apparent that, after a long-enough time has elapsed, the exponential terms have decayed and can be neglected, the said response thus becoming

$$p(t) = 0.39538[0.5\cos(31.416t) + 6.2832\sin(31.416t)], \quad \text{for large } t$$

which, with the aid of relations (2.104b), can be expressed as

$$p(t) = 2.4921 \cos(6.2832t - 1.4914), \quad \text{for large } t$$

and hence, for large t, $p(t)$ does not attain a definite value, for it oscillates harmonically between -2.4921 and $+2.4921$.

(b) Now, to calculate $\dot{p}(0^+)$, we apply again the IVT:

$$\dot{p}(0^+) = \lim_{s \to \infty}\{s\mathcal{L}^{-1}[\dot{p}(t)]\} = \lim_{s \to \infty}[s^2 P(s) - sp(0^+)]$$

$$= \lim_{s \to \infty}\left[\frac{p_0 s}{1 + 1/(s\tau)} + \frac{A}{\tau}\frac{1}{1 + 1/(s\tau)}\frac{1}{1 + (\omega/s)^2} - sp_0\right]$$

$$= \left(-\frac{p_0/\tau}{1 + 1/(s\tau)} + \frac{A}{\tau}\right) = \frac{A - p_0}{\tau}$$

The reader is asked to compare this result with that obtained by substitution of $t = 0^+$ in the mathematical model of this system, as given in Example B.2.1.

Reference

1. Cannon RH (1967) Dynamics of physical systems. McGraw-Hill, New York

Index

Symbols

2×2 identity matrix, 22
$\cos \Omega t$, 525
$\ln(\mathbf{A})$, 512
$\mathbf{K}$, 295
π, 193
$\sin \Omega t$, 525
$e^{\mathbf{A}}$, 506
ith modal vector, 311
n-dof mechanical systems, 268
n-dof system, 145, 419
n-dof undamped system, 419
q, 28
$\dot{q}$, 290
$\mathbf{q}$, 267, 290
$\dot{\mathbf{q}}$, 290
$\ddot{\mathbf{q}}$, 289
Ω^2, 316
Ω^2, 322
$\dot{q}$, 267
$\mathbf{p}$, 289, 290
q, 289
sgn, 54, 110
'small' perturbations of the equilibrium states, 283

A

A/D, *see* analog to digital
absolute acceleration, 11
abstraction, 7
accelerometer design, 172
active force, 29
additivity, 86, 128
algorithm
   DnDOF, 434
   UDnDOF, 426

Algorithm Damped-1dof, 247
aliasing, 428
analog system, 234
analog-to-digital (A/D) converter, 234
analysis, 6
analytic function $f(\lambda)$ of λ, 499
analytic function $\mathbf{F}(\mathbf{A})$ of $\mathbf{A}$, 497
analytic matrix function
   method to compute, 500
angular deflection, 458
angular velocity, 9, 10
area moment of inertia, 459
asymptotic stability, 68
asymptotically stable, 284
asymptotically stable system, 68
asymptotically unstable system, 66
autonomous system, 31, 284

B

bandwidth, 171
bars
   under axial vibration, 456
   under torsional vibration, 456
BC, *see* boundary condition
beams
   under flexural vibration, 456
beat phenomenon, 344
Belleville spring, 14
belt-pulley transmission, 328, 350
black-box, 127
black-box representation, 87
Bode plots, 147
   of first- and second-order systems, 161
bogie-half-car, 8
boundary condition, 4, 491
boundary-value problem, 466

J. Angeles, *Dynamic Response of Linear Mechanical Systems: Modeling, Analysis and Simulation*, Mechanical Engineering Series, DOI 10.1007/978-1-4419-1027-1, © Springer Science+Business Media, LLC 2011

bump, 302
    function, 243
BVP, *see* boundary-value problem

C

c.o.m, *see* center of mass
cable, 12
Cannon's book, *see* Cannon, R.H.
Cannon, R.H., 304
cantilever beam, 13
Carl Sagan, 193
causal, 236
causality, 236
Cayley-Hamilton Theorem, 98, 241, 497
center of mass, 28
characteristic equation, 468
characteristic equation of $\mathbf{A}$, 498
characteristic polynomial
    of the dynamic matrix, 313
    of $\mathbf{A}$, 498
characteristic solutions, 468
characteristic values, 468
Cholesky factoring, 308
classical modal method, 347
    applied to the total response, 364
commutativity, 129
compressible and incompressible fluid, 4
computation of the Fourier coefficients, 189
computer algebra, 156, 160
conditions
    boundary, 458
    initial, 458
configuration
    -dependent damping coefficient, 32
    of a system, 25
conservative forces, 31
constant-coefficient mechanical systems, 6
constitutive equation, 7, 464
constitutive equations
    of mechanical elements, 11
contact, 193
continua, 4, 455
continuous
    -time system, 234, 421
    train of impulses, 128
continuous model, 455
continuous systems, 455, 456
continuous-time convolution, 237
continuum, 455
continuum mechanics, 4
controlled
    forces or torques, 26
    motion, 28

rates, 266
    variables, 266
convolution, 127, 129, 442
    Duhamel integral, 115, 355
    for critically damped systems, 130
    for overdamped systems, 130
    of underdamped systems, 130
Coriolis and centrifugal forces, 35
cosine law, 37
Coulomb
    damping, 54
    dry-friction damping, 30
    friction, 6
    friction cum geometric nonlinearity, 56
critically damped system, 91, 509
cross product, 22
CT system, *see* continuous-time system
cycloidal slope, 215

D

damped n-dof system
    discrete-time response, 431
    eigenvalue problem, 435
    simulation via an extension of single-dof
        systems, 432
    simulation via the Laplace transform, 435
damped natural frequency, 99
damped suspension
    discrete time-response, 435
damped systems, 97, 245, 307
damped two-dof systems, 366
damping
    constant, 18
    matrix, 285
damping ratio, 69
dashpots, 5
db, *see* decibel
dec, *see* decade
decade, 164
decibel, 164
definite system, 358
degree of freedom, 7, 25, 263
delayed
    impulse response, 128
    input, 128
delta function, *see* Dirac function
derivative of the impulse response, 137
design of foundations, 167
design of pneumatic hammers, 167
deterministic, 26
difference equation, 238
differential-algebraic systems, 5
digital system, 234

Dirac function, 112
discrete Fourier transform, 192
discrete time, 234, 432
discrete-time
    linear dynamical system, 234
    system, 236, 248, 421
discrete-time response, 237
    of n-dof undamped systems, 421, 423
    of a damped suspension, 250
    of first-order systems, 237
    of undamped n-dof systems, 422
    of undamped second-order systems, 241
discretization methods, 5
dissipation function, 29, 30
dissipative force, 29, 30
distributed normal stress, 463
distributed shear stress, 463
distributed-parameter models, 4, 5
distributed-parameter systems, 5
distributivity, 129
dof
    see degree of freedom, 145
dot product, see scalar product
doublet function, 112
doublet response, 137
    of second-order damped systems, 136
    of second-order systems, 134
    of second-order undamped systems, 132
drill for deep-boring, 303
driving force, 29
Duhammel integral, 129
dynamic equilibrium, 460
dynamic matrix, 312
dynamical system, 4

E

eigenfrequencies, 525
eigenfunctions, 468
    properties, 484
eigenmatrix, 322
eigenvalue problem, 312, 313
eigenvalues, 468
eigenvalues of Ω, 525
eigenvalues of $\mathbf{A}$, 498
elastic potential energy, 29
elastica, 463
energy functions, 26
engineering approximation, 8
epicyclic gear train, 33
equilibrium configuration, 292
equilibrium analysis
    of the eccentric plate, 62
    of the overhead crane, 60

equilibrium configuration, 67
equilibrium configurations
    of the gantry robot, 279
    of the two-link robot, 278
equilibrium states, 278
    of mechanical systems, 59
    of the actuator mechanism, 61
equivalent dashpot coefficient, 20
equivalent stiffness of parallel array, 20
Euclidean norm, 242
Euler's Law, 459
Euler-Bernoulli beam, 462
even function, 182
excitations, 3
existence and unicity of the solution, 358
exponential of $\mathbf{A}t$, 98, 528

F

fast Fourier transform, 193
FBD, see free-body diagram
FFT, see fast Fourier transform
FFT analysis, 193
final-value theorem, 548
first law of thermodynamics, 194
first-order
    LTI dynamical systems, 203
    ODEs, 6
    systems, 116, 235
flexural rigidity, 464
floating-point arithmetic, 241
floor function, 193
flow-induced drag, 30
fluid-clutch system, 542
force
    -controlled source, 29
    -driven overhead crane, 42
    -driven system, 42
    sources, 29
forced response, 86
Fourier algorithm, 195
Fourier analysis, 182, 183
    of a monotonic function, 187
    of a square wave, 186
    of a train of impulses, 184
    of the pyr(x) function, 187
Fourier coefficients, 192
Fourier expansion, 183
Fourier series, 145
Fourier transform, 145
free response, 86
free-body diagram, 18
frequency matrix, 310, 500
frequency radius, 315, 525

frequency ratio, 153
frequency response, 145, 147
functional, 236
FVT, *see* final-value theorem

G
gear transmission, 304
generalized, *see* generalized velocity
    active force, 30
    coordinate, 6, 25, 264
    Coriolis and centrifugal forces, 267
    damping matrix, 263
    dissipative force, 265
    driving force, 265
    forces, 30, 289
    mass, 28
    mass matrix, 263
    momentum, 266
    velocity, 30
generalized coordinate, 28
Geneva mechanism, 131
Geneva wheel, 134, 303
    discrete-time response, 254
geometric BCs, 481
governing equation, 38, 271, 464
gyroscopic effects, 290
gyroscopic forces, 284

H
harmonic
    excitation, 146
    functions of a positive-definite matrix, 525
    motions of single-dof systems, 313
    oscillator, 92
harmonic response, 145
    applications, 164
    of first-order systems, 149
    of second-order systems, 155
    of undamped systems, 152, 154, 155
Heaviside function, 113
Hessian matrices, 289
high-pass filter, 174
Hooke's Law, 457
hydraulic clutch, 51
hysteretic damping, 30, 53

I
iconic model, 8
idealization, 7
identification of damping from the time
        response, 100

improper orthogonal matrix, 241
impulse, 128
impulse response, 115, 120, 127
    of n-dof damped systems, 441
    of first- and second-order LTI systems, 115
IMSL, 372
independent generalized coordinates, 25
independent generalized speeds, 25
indexing mechanism of a production machine,
        302
inertia matrix, 9
inertial measurement units (IMU), 6
initial-value problem, 466
initial-value theorem, 548
input, 3
instability, 158
invariant quantity, 503
inverse Laplace transform
        via partial-fraction expansion, 538
inviscid and viscous fluid, 4
IVT, *see* initial-value theorem

J
Joseph Fourier (1768–1830), 145
journal bearings, 24

K
König's Theorem, 27
König's theorem, 44
kinematic analysis, 28
kinetic energy, 26, 265
    of the system, 266

L
L'Hospital's rule, 158, 178, 515
Lagrange equations, 25, 266
    of linear mechanical systems, 288
    vector form, 263
Lagrangian, 26, 267
Laplace transform, 531
    additivity, 533
    basic properties, 534
    denominator with repeated roots, 539
    double-sided, 531
    inverse, 532, 538
    linear homogeneity, 533
    numerator with distinct roots, 539
    of $\cos \omega t$, 535
    of $\sin \omega t$, 534
    of second-order underdamped systems, 544
    of the convolution, 534

of the decaying exponential, 532
of the delay, 534
of the unit step function, 533
one-sided, 531
pairs, 532
properties, 533
quadratic factors, 545
superposition, 533
the impulse function, 535
the ramp function, 535
LHS, 119
linear dashpots, 30
linear homogeneity, 129
linear mechanical systems, 5
linear springs, 29, 30
linear time-invariant system, 85
linear transformation, 531
linear, soft, and hard springs, 15
linearity, 115, 127
linearity and time-invariance, 176, 179
linearization
    about equilibrium states, 66
    of the governing equations, 283
linearized
    equation of a one-dof system, 68
    model of a two-link robot, 285
linearized model
    of a gantry robot, 286
linearly elastic, 456
linearly homogeneous, 86
linearly viscous damping, 30
load, 29
locomotive wheel array, 35
logarithmic decrement, 101
low-pass filters, 164
LTIS, see linear time-invariant system
lumped-parameter models, 5

M

Macsyma, 372
magnification factor, 162
    of the transmitted force, 166
    of the transmitted motion, 169
magnitude, 146
Maple, 372
marginal stability, 68
marginally stable, 284
    equilibrium state, 66
mass density, 459
mass matrix, 266, 285
mass subjected to a time-varying force, 253
mass-spring-dashpot system in a gravity field,
        74

mass-transit system
    undamped discrete-time response, 429
Mathematica, 372
mathematical model, 35, 233
mathematical modeling, 8
Matlab, 193, 372
matrix
    exponential, 497
    function $\mathbf{F}(\mathbf{A})$, 500
    functions, 497
    notation, 21
    with a repeated eigenvalue, 515
mean frequency, 315, 525
mechanical modeling, 7
mechanical system, 4
mechanical system configuration, 264
mechanical transmissions, 4
memory, 4
memoryless systems, 4
modal analysis, 318
    of a damped test pad, 372
    of a two-dof gantry robot, 318
modal coordinates, 364
modal equation, 314
modal matrix, 322
modal vectors, 311, 313, 316, 430
mode, 314
mode-orthogonality, 489
model, 233
model for the vertical vibration of mass-transit
        cars, 295
modeling, 6
modeling process, 7
Mohr circle, 307, 313, 497, 516, 518
    of $\cos \Omega t$, 525
    of $\sin \Omega t$, 525
    of Ω, 525
momentum-preserving system, 355
    revisited, 365
motion, 25
motion sources, 29
motion-controlled source, 29
motion-driven system, 42
motor-cam transmission, 58
multibody systems, 7, 307

N

n-tuplet function, 113
natural BCs, 482
natural frequencies, 310, 525
natural frequencies of two-dof undamped
        systems, 308
natural frequency, 69

natural modes of two-dof undamped systems,
    308
negative-definite matrix, 284
negative-semidefinite matrix, 284
neutral axis, 463
Newton equation, 464
Newton-Euler equations, 8, 10
nominal behavior, 6
non-zero initial conditions, 141
non-zero input, 141
nonlinear spring, 15
normal form
    of the governing equations, 310
    of the mathematical model, 420
normal stress, 462
numerical quadrature, 189

O

oct, *see* octave
octave, 164
odd function, 182
ODE, *see* ordinary differential equations
ordinary differential equations, 5, 6
orthogonal matrix, 22
orthogonality and normality, 484
orthogonality condition, 491
orthogonality of the eigenvectors, 314
oscillating follower, 229
output, 3, 87
overdamped system, 91, 510
overhead crane, 39

P

Parseval's identity, *see* Parseval's Theorem
Parseval's Theorem, 193
partial derivatives
    fourth-order, 465
    second-order, 465
partial differential equation, 4, 458
partial-fraction expansion, 538
    of a rational function, 538
particular solution of the second-order damped
        system, 155
PDE, *see* partial differential equations
PDE of the fourth order, 465
perfectly elastic shock, 338
periodic
    function, 182
    input, 202
    response, 181
    response of an air compressor, 205

response of first- and second-order LTIS,
    202
persistent time-varying input, 145
PFE, *see* partial-fraction expansion
phase, 146
planar motion, 21, 265
pneumatic hammer, 164
Poisson ratio, 15
positive-definite
    matrix, 284
    square root of **M**, 309
positive-definite matrix
    harmonic functions, 525
positive-semidefinite
    frequency matrix, 354
    matrix, 284
potential, 31
    energy: elastic and gravitational, 28
potentiometers, 4
power
    developed by a moment, 29
    dissipated by a damped second-order
        system, 161
    supplied to a system, 26, 29
Principle of Conservation of Energy, 194
proper orthogonal matrix, 241
proper rational function, 538
proportional damping, 368
pulse, 140
pulse-like input, 174
purely flexible mode, 329

Q

quadratic expressions, 266
quadratic form, 288
quadruplet function, 113
quick-return cam mechanism, 229

R

ramp response, 139
    of first-order systems, 132
    of an overdamped second-order system,
        139
    of second-order undamped systems, 135
reflection, 241
resonance, 155
response, 3
    to abrupt and impulsive inputs, 130
    of a damped, second-order system to a unit
        doublet, 137
    of a second-order underdamped system to a
        ramp, 139

of a second-order, critically damped system
    to a ramp, 139
of first-order systems, 129
of second-order damped systems, 130
of second-order undamped systems, 130
to constant and linear inputs, 159
to the unilateral cosine function, 152
to the unilateral sine function, 156
Riemann integral, 129
rigid and deformable solid, 4
rigid mode, 263, 295, 323, 473
rigid ring suspended from a pin, 23
rms, *see* root-mean-square
rod, 12
root-mean-square, 194
rotation, 241, 420
rotational damping coefficient, 18
rotational dashpot, 18
roundoff errors, 241

S

sampled signal, 234
sampling interval, 234, 243
saturation function, 55
scalar moment of inertia, 10
scalar product, 23
Scotch yoke, 223
second-order
    damped systems, 120
    ODEs, 6
    systems, 152, 239
    undamped systems, 119, 204
second-order damped systems
    discrete-time response, 249
seismograph design, 173
semidefinite stiffness matrix, 295
semidefinite systems, 354
semigraphical approach, 372
separation of variables, 465
series
    equivalent of two springs, 19
    expansion of $\cos \omega_n t$, 92
    expansion of $\sin \omega_n t$, 92
series and parallel array, 20
    of linear dashpots, 20
    of linear springs, 18
seven steps, 34, 35, 268
seven-step procedure, 35
shaft, 13
Shannon's Theorem, 428
shear deformation, 458
shear force, 463
shear modulus, 459

shortcuts for special matrices, 527
sign-indefinite matrix, 284
signed angles, 25
signed distances, 25
simulation, 233
    of n-dof systems, 419
    of single-dof systems, 233
    time-step choice, 428
simulation scheme
    for undamped second-order systems, 240
single-degree-of-freedom system, 6, 25, 307
skew-symmetric matrix, 22, 527
small-perturbation analysis, 6
small-slope assumption, 460, 463
spectral analysis, 145, 183
    of the displacement of an air compressor,
        195
spring stiffness, 12
springs, 5, 11
square matrix
    analytic function, 497
square root of **M**, 309
square-root factoring, 308
stability, 66
stability analysis
    of the actuator mechanism, 70
    of the eccentric plates, 72
    of the overhead crane, 69
stable system, 66
staircase approximation, 235, 246
state, 4, 25
    -variable form, 375, 431
    -variable vector, 26, 264
    variable, 25, 264
static equilibrium, 460
steady-state
    part, 146
    response, 203
    response of an undamped system to a
        sinusoidal input, 158
Steiner's Theorem, 9, 10
step response
    of critically damped systems, 139
    of first-order system, 131
    of overdamped systems, 139
    of second-order undamped systems, 135
    of the second-order underdamped system,
        138
    of underdamped systems, 138
stiffness matrix, 285
strain, 457
stress, 457
strings
    under transverse vibration, 456

superposition, 6, 86, 115, 140, 203, 326, 332
system, 3
    impulse response in state-variable form,
        144
    matrix, 419
    output, 433
    with a time-varying equilibrium state, 65
    with positive-definite frequency matrix,
        333
system with a time-varying equilibrium state,
    64

T
tangential stress, 463
test pad, 122
    revisited, 447
    two-dof, 338
theory of beams, 462
time constant, 90
time delay, 87
time invariance, 87
time response, 85
    of critically damped systems, 99
    of first- and second-order systems, 85
    of overdamped systems, 100
    of underdamped second-order system to a
        constant input, 160
    via the Laplace transform, 535
time-invariance, 115, 127, 246
time-invariant, 6
time-series, 192
time-varying equilibrium state, 280
time-varying inputs, 145
torque, 30
torque source, 29
torsional spring, 14
torsional stiffness, 14, 460
torsional vibrations of aircraft wings, 294
total generalized force, 31
total kinetic energy, 26, 28
total potential energy, 26
total response
    of two-dof system, 363
    of damped two-dof systems, 375
    of the system, 144
total time response
    of dynamical systems, 141
    of first-order systems, 141
    of second-order damped systems, 143
    of second-order undamped systems, 142
trace of $\mathbf{A}$, 503
transducers, 4
transient part, 146

translational
    damping coefficient, 18
    dashpot, 18
    spring, 13
    stiffness, 13
transmitted force, 165
transverse motion, 460
trapezoidal rule, 189
triplet function, 113
two-dimensional form, 22
two-dof gantry robot, 273
two-dof model of a terrestrial vehicle, 302
two-dof systems, 307
two-dof test pad, 348
two-dof undamped linear mechanical system,
    308
two-link robotic arm, 269

U
undamped
    linear mechanical systems, 307
    second-order system, 120
    suspension discrete-time response, 242
    systems, 91, 239
    terrestrial vehicle, 175
undamped two-dof systems, x
underdamped
    impulse response second-order system, 121
    system, 91, 508
unilateral harmonic functions, 147
unilateral sine input, 156
unit doublet, 112
unit eigenvectors, 310
unit impulse, 111, 112, 126
unit ramp function, 114
unit step function, 113
unstable system, 66

V
Vandermonde matrix, 501
vector
    notation, 21
    of generalized coordinates, 263
    of generalized forces, 267
    of generalized speeds, 263
    of modal coordinates, 347
velocity meter design, 169
vibration absorber, 360
vibration analysis of two-dof systems, 307
viscous damping, 30

W

weighted magnitude, 312
whirling of shafts, 290, 376
Willis' formula, 23

Y

Young modulus of elasticty, 456

Z

zero-input response, 86, 307
    of damped systems in state-variable form,
      143

    of first-order LTIS, 88
    of second-order damped systems, 507
    of second-order LTIS, 91
    of two-dof systems, 323
zero-order hold, 234, 419, 422, 432
zero-state response, 86, 127
    in state-variable form, 143
    of LTIS, 111
    of two-dof systems, 353
    to arbitrary input, 144
ZOH, *see* zero-order hold
ZSR, *see* zero-state response